Sustainability Learning for Action and Community Engagement

Sarah Mead Wyman · Joshua Korenblat
Editors

Sustainability Learning for Action and Community Engagement

palgrave
macmillan

Editors
Sarah Mead Wyman
State University of New York
New Paltz, NY, USA

Joshua Korenblat
State University of New York
New Paltz, NY, USA

ISBN 978-3-032-11697-0 ISBN 978-3-032-11698-7 (eBook)
https://doi.org/10.1007/978-3-032-11698-7

This work was supported by Primary Source and State University of New York.

Cover illustration: Artwork by Joshua Korenblat. Background artwork adapted from Piet Mondrian's Flowering Apple Tree, 1912. Mondrian's original work is in the public domain. All adapted and new elements are to be considered as a single image.

This Palgrave Macmillan imprint is published by the registered company Springer Nature Switzerland AG
The registered company address is: Gewerbestrasse 11, 6330 Cham, Switzerland

To Asha and the children who will see Earthrise beyond our sightlines.

Foreword

Environmental Literacy, Indigenous Wisdom, and Educating for Survival: A View from Alaska

Very little in my culture taught me that my existence depends on the existence of earth.

Susan Griffin

Our educational mission is to produce human beings who are at home in their place, their environment, their world.

Oscar Angayuqaq Kawagley, Yup'ik Elder, first Alaska Native Ph.D.

As a White woman and lifelong Alaskan born a month before the territory of Alaska became a US state (1959), I had the great good fortune of missing out on many elements of modern North American culture for most of my formative years. Alaska is roughly 1/5 the size of the combined "Lower 48" (continental) states and, at that time, had fewer than 250,000 people living in its vast and wild lands. Immigrants from the Lower 48 were then scrambling to reconstruct American cultural and economic institutions in their new northern home (on top of sacred and unceded Native lands), but these institutions were often unsubstantial, as was adult oversight. We children were free-range, often left on our own to explore the sprawling lands and waters that impressively overshadowed human occupation. This sometimes put us in great peril, but it also allowed us to retain our primary relationship to the living world.

Even more fortunately, I was exposed to and schooled in the educational systems, philosophies, and ways of knowing and living of Alaska's seven major Indigenous nations who still inhabit their ancestral lands (Alaska almost entirely escaped the reservation system). I saw human communities that had survived and thrived "since time immemorial," as Native Elders say, in some of the harshest environments on the planet through honoring and teaching *right relationship* with all of life, both human and more-than-human. And functioning at a scale that allowed for the maintenance of a healthy balance between these human and more-than-human communities. The psychological, social, and ecological genius of these cultures can hardly be overstated, and I have spent much of my career attempting to elevate these voices, this wisdom and knowledge, including traditional Indigenous ways of teaching and learning. There is a saying that "when an Elder dies, a library burns." In my opinion, we would be wise to humbly request to learn every scrap of knowledge held in those libraries before it is too late.

As many Indigenous authors, scholars, Elders, and other thinkers have articulated, modern Western paradigms for organizing human life dangerously reverse the laws for living. Alaska Native Elders call our current society the "reverse or inside out" society because it has its priorities upside down and backwards. First and foremost, these Elders teach, we must recognize the centrality of our dependent relationship on the living world—the plants, animals, waters, lands, and climate which make human life possible. As more and more of us are coming to realize, continuous consumption, population growth, relentless extraction of minerals/oil/gas, and the endless production of waste and greenhouse gas emissions are leading to catastrophe. Humans will only be able to survive and thrive on this planet to the extent that we regain some balance with those living systems and teach each new generation the attitudes and behaviors necessary to do so: humility, reciprocity, gratitude, cooperativeness, restraint, continuous consideration of the common good, a respect for the intelligence and inherent value of other creatures and humans, and more. Missing this most basic understanding of the fundamental necessities for human existence results in the kinds of distorted behaviors and leadership from which we now suffer, and this loss threatens life as we know it. Teaching students almost exclusively about Western thought through book learning and intellectual, individualistic approaches perpetuates the beliefs and practices that sustain our current dangerous path. Teaching in a way that empowers them to maintain their connection to the living Earth and their human communities, think for themselves, and take powerful action in the world gives them a chance at a hopeful future.

Combined with my own direct relationship with the living Earth, Alaska's Native cultures offered me a reference point and counterpoint to the cultural tidal wave that shocked and overwhelmed my community with the building of the trans-Alaska pipeline in 1973. The stunning upsurge in job- and profit-seekers, with the attendant upheaval in development, gave me a disturbingly rapid education in the fundamentals of American-style capitalism and how it can impact ecosystems and communities. In some very real way, we were living out the last gasp of American western expansionism that had built the country and sought to erase Indigenous nations and sensibilities. Alaskan license plates then and now read "North to the Future" and "the Last Frontier." My hometown of Anchorage exploded overnight, bringing with it Taco Bells and traffic and condominium complexes and cocaine. As a teenager living through this dislocation, I was shocked to discover that fast food outlets mattered more than wetlands and that the glimmer of oil money caused many people to sell their souls and communities to global oil corporations in exchange for fancy houses, important job titles, and fat paychecks.

Growing up in Alaska prior to the presence of global corporate oil companies and fully formed American institutions and in the presence of 10,000 year-old Indigenous cultures, I retained the clear-eyed vision of a kindergartener. Water IS life. Wetlands DO matter more than fast food chains. Staying in balance with our living Earth systems is not only logical but a fundamental necessity of self-preservation as a species. The human *self* does not end at the borders of our skin; we are literally dependent on a stable climate, clean water, healthy soil, and the contributions of our fellow species to our shared life support systems. If the bees go, we go with them.

Knowing this on a visceral level, I have spent most of my life engaged in free-range activism on behalf of our living systems and Indigenous knowledge and wisdom. Over the decades, as the tragedy at the core of our "endless growth"-based economic system has become increasingly obvious, more and more people are doing this kindergarten-level math at the core of climate change and other environmental impacts and have felt increasingly compelled to speak up and act. This includes educators who are realizing that, in the words of US environmental professor David Orr, we must prioritize ecological literacy, leadership, and behavior change "so that all the other struggles may go on." In higher education, one of our cherished values is a respect for science and the scientific process. Our own scientists are shouting at us that, if we do not educate our young people for a way of living and an economy that honors our dependency on Earth's ecosystems and take quick and intelligent action to move into right relationship with its lands, waters, and creatures, our species may not survive the century. Or those members of our species who do survive will be living lives we might not recognize or wish to experience.

Spurred by everything from love for our children to a sense of morality to sheer self-centeredness, I can see no reason why any reasonably connected human being would argue against educating our young people for their own survival. However, my experience in teaching and higher education has been that most of us—unsurprisingly—are as disconnected from these fundamental realities as the rest of the world, having ourselves been uneducated in ecological literacy; we are teaching for a world long gone by. The pace of both technological and climate change is accelerating so fast that much of what we are teaching—as incredibly valuable as it is—is no longer relevant to the survival of younger generations. If citizens (of any age) do not know where their water comes from and how to protect it; where their food comes from and how to grow it; what their soil needs and how to restore it; how to work collaboratively and respectfully with other human beings; how to reduce energy consumption and generate it from renewable sources—then much of what we are teaching them now will not matter.

In books like this, we get a chance to ask ourselves: what is the purpose of education in the twenty-first century? There are lots of answers to that question. For most folks, the purpose of a Western education is to help each new generation of young people acquire the skills and knowledge necessary to be able to support themselves and their families, participate in some meaningful way in their communities, solve problems, assess and vote on the policies and political representation that determine our shared future, and, if they are lucky, enjoy the internal and external satisfactions of literature, history, art, music, and more. At its most basic, education in the dominant culture is supposed to help young people get and keep decent-paying jobs. And, since relatively few people are able to attend college, higher education has largely been seen as a way to help young people secure higher paying jobs and become the professional—if not the elite—class within a society.

But education—particularly higher education—has also been viewed as the place where leaders are developed. It has aspired to the loftier goal of (in the words of John Samuel, former CEO of the Nelson Mandela Foundation in South Africa) "inculcating in each new generation the habits of democracy:" critical thinking; public speaking; the ability to honor more than one viewpoint; information, scientific and historical literacy; systems analysis, etc. It is common knowledge that a requirement for democracies to survive is an educated citizenry; in other words, an informed and critically thinking public is requisite to prevent despotism. We are learning about that anew as we speak.

For obvious reasons, educational systems (and cultures) necessarily tend to be matched to the economies they serve. We educate our children to be productive and to survive in the specific contexts into which they are born; as a result, education in modern times has been geared toward helping young people survive and thrive within late stage, post-industrial capitalism. That economic paradigm has established itself over the relatively recent past through the flourishing of global extractive industries based on harvesting minerals, oil, trees, fish, plants, water, crops, and more out of the living earth on large scales with little or no perceived need to restore, replenish, or protect the ecosystems from which it takes these elements. It has also relied upon the falsehood of a hierarchy of value among human beings and between human beings and other species. We have been educating our students to succeed and lead in this system. As some Native Elders have said, we have, in many ways, been "dumbing them down"—teaching them to thrive in a system that is threatening our very survival and the survival of life as we know it on this planet rather than in the ways that sustain human—and all—life.

This book, and others like it, seeks to change this and to help us move rapidly toward an educational system that reconnects our students with ecological realities and human communities in ways that support our common survival. There is no more important work in education these days. When coupled with our efforts to address equity and justice, both in our institutions and our world, these attempts are what really matters. I hope you read this, use this, spread this, add to this, and act upon what is shared here as quickly as possible. All that we care about depends upon our doing that.

Quyana. Thank you.

Libby Roderick
Director, Difficult Dialogues
Initiative
University of Alaska
Anchorage, US

Preface: Earthrise

When the Apollo 8 astronauts photographed Earth from space in 1968, watching our planet rise across an unforeseen horizon, they gave us a new perspective on our home. Despite the distance, they could see seas, mountains, and deserts that pattern the surface. From that vantage point, however, they could not see the green leaves that indicate the health of our ecosystems, the seasonal uttering out of Nature's expression. They could not detect the leverage points or places we as humans might intervene for maximum impact, like a crowbar between boulders or fallen buildings, to guide the change or correct damage to people, creatures, and Earth. These scientists might not have been looking for such opportunities, as their photographs predate today's urgent recognition of planetary crisis. Yet they were undoubtedly searching for understanding and knowledge to bring back, borne of this new perspective.

Today, the Earth and its creatures face increased peril due to human-induced global warming and our failure to care adequately for nature and each other. The perceived separation between humans and nature that has guided many of our industrialized world cultures has fostered mindsets based in individualist self-interest, neglect, and extractive practices that have cost us dearly. Poet Amanda Gorman has captured the urgency of our polycrisis in her poem (2021), "Earthrise." In *Orbital* (2023), Samantha Harvey allegorically describes our shameless flight from responsibility: "A full-throttle scarper in [a] billionaire's rocket, out and away,

away from the junk, away from the breaking burning storming scintillating earth like fleeing the scene of a crime" (199). The question of *how* to intervene in a calamity greater than our minds can hold, a complexity beyond our imaginings below and beyond the Earth's crust, informs this book.

We know what to do. Humans have the technology, tools, and ingenuity to eradicate most of the challenges to climate stability, biodiversity survival, and many of the alarming responses that have leveled cities, burned forests, flooded communities, invaded ecosystems, extinguished species, and forced migrations of people in search of water, safety, or other social determinants of health. Human conflict and apathy are harder to manage, as evidenced by battles over depleted resources, human rights violations, extreme wealth inequity, economic precarity, and ensuing wars. On an individual level, our youth demonstrate the effects of ecoanxiety, despair, isolation, and additional threats to mental health and well-being. On a planetary level, Bruno Latour (2018) has named a new force, the *terrestrial*, as a way to render visible the earthly energy that shifts definitions, attracts attention, rises up, and speaks to us, at the moment our common world threatens to slip out of reach.

So, we turn to education and community engagement for ways to comprehend complex systems, share information, generate knowledge, and address changes in mindset and behavior necessary to address these crises. These requisite shifts must not be prescriptive but instead constructed by learners (including ourselves) who come to their own conclusions, like astronauts peering through portholes at the only home we have.

Earthrise is like sunrise, without distance or separation. We might remember what Tiokasin Ghosthorse (Cheyenne River Lakota Nation) reminds us: we must not see Earth as a metaphor for present circumstances. We must *become* Earth. For that is what we are.

We will not all agree on approaches or interpretations. Nevertheless, as a species, we have the opportunity and energy to move forward with intention and care in ways that will increase the chances for future generations to witness Earthrise as a scene of flourishing, solidarity, and ongoing potential.

Earthrise

Caves of emerald ice
meet the morning light
and whether you met
the grizzlies there or elsewhere
I'll never know.
They dance in a frozen field,
the floes caught in an Idaho freeze
high in mountains
where astronauts practice
moon landings.
They lumber and twirl
on a landscape
gray and cratered.
Anders shot a photograph
so we now see from a distance
the space where travelers watched
the Earth rise
over a lunar horizon
but could not count
bears crouched in wildflower meadows.

Sarah Mead Wyman
SUNY New Paltz
New York, US

Acknowledgments

This book would not have been possible without our colleagues and collaborators. This includes our department colleagues at SUNY New Paltz: Vicki Tromanhauser, Wyatt Krause, Anne Galperin, Amy Papaelias, Woojin Lee, Chad Bridgewater, Aaron Nelson, and Alanna Dukes. Special thanks to Tara Dates in Budgets and Administration and to Cori Wilhelm, Kim Scalzo, and Tony DeFranco of the SUNY IITG OER Impact Grant team, whose generous support made this work possible. We also thank our editors, Milana Vernikova, Jayalakshmi Raju, and Chandralekha Mahamel Raja, and Cynthia Kroonen.

We are grateful to our Earthrise Commons team: Kristoff Lalicki, Genevieve Flynn, Olivia Wilson, and Stephan Macaluso (earthrisecommons.org), and to our dedicated book authors.

Finally, we thank our families for their support, love, and care.

Primary Source: Educating Global Citizens: grant for open access publication https://primarysource.org/.

State University of New York: 2024 Innovative Instruction Technology Grant (IITG) Open Educational Resources (OER) Impact Grant.

State University of New York: PRODiG grant for research by Jessica Kate Crowell and Kate Bellody.

Stop Talking: Indigenous Ways of Teaching and Learning and Difficult Dialogues in Higher Education (U Alaska 2013) for permission to reprint Ilarion Merculieff's "The Way of the Real Human Being" http://hdl.handle.net/11122/14354.

Nature Quilt, Ashokan Center, (2023) Suzanne Stokes

Contents

Notes on Contributors

Naif Alghamdi is an assistant professor of Architecture in the Department of Architecture and Building Science at the College of Architecture and Planning, King Saud University (KSU), Riyadh, Saudi Arabia. He currently leads the Riyadh Architectural Documentation Project and serves on the Editorial Advisory Board of the International Journal of Sustainability in Higher Education (Emerald, UK). He is also a regular reviewer for several international peer-reviewed journals. Alongside his academic role, Dr. Alghamdi works as an active architectural consultant, collaborating with professional practices across Saudi Arabia. He holds postgraduate degrees in project management and architecture from The Bartlett, University College London (UCL), and a Ph.D. from the Department of Management in the Built Environment, Delft University of Technology, The Netherlands. His research interests focus on sustainability, university campuses, architectural documentation, and universal accessibility. He has presented his work in numerous international workshops, seminars, and conferences, and has published journal articles, book chapters, and scholarly reviews.

Arnau Amat is a researcher and professor of science and environmental education in the department of Didàctica de les Arts i les Ciències at Universitat de Vic-Universitat Central de Catalunya. He coordinates the elementary teaching degree. He earned his Ph.D. in science education at the Universitat Autònoma de Barcelona. His research focuses on sociocultural approaches in science education for sustainability, science teaching education, and the role of emotions in science and environmental learning. Currently, he leads a project to promote and investigate biodiversity in schoolyards. Prior to that, he worked as an environmental educator and high school science teacher.

Michael Asbill specializes in public art, often produced in collaboration with others that forges connections between the natural environment and everyday social spaces. He is currently working on two projects. *Processing Collapse* makes extinction palpable by presenting sculptural objects made from North American Ash trees within the library system of a region infested with Emerald Ash Borer beetles. *Loo with a View*, a partnership with the Soil Factory in Ithaca, NY, is reimagining sanitation practices. Asbill designs elevated composting toilets toward mainstreaming bionutrient capture and circularity. Michael Asbill is an assistant professor in the sculpture program at SUNY New Paltz, New York.

Anne C. Balant Ph.D. CCC-A is an associate professor of Communication Disorders at SUNY New Paltz, New York. She earned her bachelor's degree in physics from Bryn Mawr College and her Ph.D. in speech and hearing sciences at CUNY Graduate Center. She worked in laser physics and noise control engineering at IBM and became certified as a clinical audiologist before joining the SUNY New Paltz faculty. Her research interests include noise control and psychoacoustics with a focus on increasing public acceptance and reducing the human impacts of wind farms. She leads her town's NYS DEC Climate Smart task force.

Kara Loeb Belinsky, Ph.D., is an associate professor of biology at SUNY New Paltz. Her research focus is songbird behavior and conservation including past work on the evolution of singing behavior in warblers and thrushes and current work on how human development affects wildlife biodiversity using songbirds as indicator species. She leads an active lab of undergraduate researchers, and currently runs a long-term bird banding site (MAPS), a nest box network, and grant-funded projects sampling bird and bee diversity at a municipal solar array and the effects of increased hiking on mountain-top breeding birds in the Catskill Mountains.

Kate Bellody is the student success librarian at the Sojourner Truth Library at SUNY New Paltz. Kate coordinates student-centered programs and services within the library that foster engagement and academic success. She previously served as the library's Reference Program Coordinator. In 2017, she received a Rising Star Award from the New Jersey

Library Association for her community outreach and programming initiatives at Hudson County Community College. Kate's work centers on librarianship as a human service, where she facilitates opportunities for exploration and connection.

Ripley Butterfield is a socially engaged artist and writer working with public libraries. With collaborator Michael Asbill, she is making sculptures from the remnants of declining tree species for their region-wide public art project, *Processing Collapse*. Ripley is currently a senior undergraduate student at the State University of New York at New Paltz. processingcollapse.com

April B. Coughlin, Ph.D. is an associate professor at the State University of New York at New Paltz in the School of Education. She has been a "wheeler" since the age of six, is a disability rights advocate, and worked as a high school English teacher in NYC public schools. Her research, teaching, and advocacy focus on access and equity for students with disabilities/disabled students in schools; healthcare for women with disabilities; and transit accessibility.

Jessica Kate Crowell is an assistant professor in the Department of Digital Media and Journalism at SUNY New Paltz and holds a Ph.D. in media studies from Rutgers University. Her qualitative research and teaching address the intersection of media, social justice, and sustainability, including the impact of media on work and the role of sustainability education in furthering diversity, equity, and inclusion. Jessica's research has appeared in books and academic journals. Jessica believes in engaged scholarship and as such has authored policy papers on information access in low-income communities for state and federal governments in addition to NGOs.

Anthony S. Dandridge is a Lecturer in SUNY New Paltz's Department of Black Studies, and an Afrocentric-Africana philosopher who emphasizes culture in his teaching. He cofounded and directs the Black Lives Matter at School at New Paltz Collective. As a Sustainability Faculty Fellow, he integrates environmental justice and interdisciplinary sustainability as a progressive pedagogy rooted in the interdependence of our reality. He also serves as a COIL instructor, UUP delegate, and member of the Statewide DEI Committee. In addition, he co-directs Mindful Campus, fostering contemplative practices among faculty, and contributes to campus and community efforts advancing D.E.I.J.B.

Megan A. Ferguson is professor of chemistry at SUNY New Paltz, where she also teaches environmental science and engineering. She has been involved with New Paltz's Sustainability Living Learning Community for several years and strives to incorporate sustainability into all her classes. She earned M.S. and Ph.D. degrees in Environmental Science and Engineering from the California Institute of Technology, and B.S. degrees in Chemistry and Geography & Environmental Systems from the University of Maryland Baltimore County.

Christiana Fortune-Reader is a multi-faceted musician, passionate about the intersection of teaching, performing, community building, and arts advocacy. She teaches at SUNY New Paltz, focusing on violin, viola, chamber music, and serving as orchestra director. Her research interests examine music as a tool for social change and community building, in early childhood, among college-aged students, and within larger communities of professional ensembles. She holds degrees from the Eastman School of Music, the University of Southern California, and the Cleveland Institute of Music, having studied with Donald McInnes, Jeffrey Irvine, Lynne Ramsey, Karen Ritscher, and Carol Rodland. https://christianafortunereader.com/.

Andrea Frank is an artist and professor at SUNY New Paltz, where she is also a co-coordinator of the Sustainability Faculty Learning Community and core-collaborator of Eddy at New Paltz. In her artistic practice, she explores complexity, ecology, and collective resonances through a range of media and collaborative embodied processes. She holds workshops and exhibits her work internationally. Recent publications include: LOSSY: Toward a Sensory Ecology (Aracne Editrice, Italy, 2020) and Beloved Child: The War Book (Visual Studies Workshop, NY, 2020). www.andreafrank.net.

Henry Gage, Jr., is a strategic leader and community advocate with a strong background in sustainable development, organizational growth, and collaborative problem-solving. As a key partner in **Build Green Now** and **Voices for Progress**, Henry combines his expertise in business operations, public engagement, and environmental policy to advance projects that deliver tangible benefits to both communities and the planet. At **Build Green Now**, he supports large-scale demonstrations of climate-friendly building practices, including the integration of industrial hemp and lime as a safe, efficient, and cost-effective insulation solution. His work involves coordinating with legislators, private sector innovators, and grassroots leaders to accelerate adoption of sustainable construction methods and create replicable models for energy-efficient housing.

At **Voices for Progress**, Henry plays a pivotal role in expanding the reach and impact of the AI-powered project management platform. He focuses on building partnerships with civic organizations, non-profits, and educational institutions, ensuring the platform serves as a bridge between local needs and expert-driven solutions. His efforts center on empowering volunteers, fostering workforce skill development, and creating opportunities for civic participation in projects with measurable public benefit.

- Build Green Now—www.BuildGreenNow.net
- Voices for Progress—www.VoicesforProgress.net

Tiokasin Ghosthorse is a member of the Itazipcola and Mnicoujou Lakota of the Sakowin Oyate and who draws upon the language, history, culture, philosophy, and traditions of the Indigenous peoples of the Western Hemisphere. He spoke as a 15-year-old at the United Nations in Lake Geneva, Switzerland. Tiokasin was a 2016 nominee for a Nobel Peace Prize from the International Institute of Peace Studies and Global Philosophy. He is Founder, Host, and Executive Producer of First Voices Radio. Tiokasin is also a co-author of the forthcoming *Earth Mind* and published *Butterfly Against the Wind*. https://firstvoicesindigenousradio.org/production/tiokasin-ghosthorse

Jack L. Harris is Assistant Teaching Professor and program director in sustainability in the Department of Human Ecology at Rutgers University. He previously worked at SUNY New Paltz as an assistant professor and sustainability faculty fellow. Jack's research and teaching focus on the intersection of the environment, economy, and society with a particular emphasis on community networks and hyperlocal organizations. Current research centers on problems of Scottish sustainability in historical and cultural context. He is the author of *Hyperlocal Organizing: Collaborating for Recovery over Time* (Lexington 2022), which explores how local knowledge and community networks impact post-disaster recovery over time.

Demond M. Hill Ph.D. is an assistant professor of Health Equity in the Eliot-Pearson Department of Child Study and Human Development at Tufts University, Medford, MA, USA. As an interdisciplinary scholar, his research focuses on emotion science, education, mental health, and emotional well-being amid racial inequality. Utilizing critical theories and humanizing research methods, his primary goal is to (1) design public spaces and places of human flourishing with a focus on creating culturally relevant therapeutic environments that promote healing and holistic wellness and (2) develop holistic, humanizing, and culturally liberating tools to promote human flourishing among Black students in schools. He is unapologetically committed to collectively creating a liberatory world.

Paul Hirsch is Associate Professor and graduate program director in the Department of Environmental Studies at SUNY Environmental Science and Forestry (ESF). He designs and facilitates strategies to address complexity in environmental problem-solving. His current work is on opportunities for trust-building across diverse people with shared connections to natural places and environmental systems. At ESF, Dr. Hirsch oversees the development of online graduate programs in Environmental Leadership, Justice, and Communication. He holds a Ph.D. in public policy from Georgia Tech, an MS from the Odum School of Ecology at the University of Georgia, and a BS from Cornell University.

Lisa C. Hix, DSc, MBA, is Professor and Co-Chair, Sustainable Product Design/Architecture Department, Keene State College. Her background in industrial design (Rhode Island School of Design), a career in new product design and engineering (marine and lighting industries) and in design and manufacturing consulting led to teaching in higher education and to her DSc in Work Environment: Cleaner Production/Pollution Prevention. Her research led to creating a Bachelor/Science program in Sustainable Product Design and Innovation and developing learning tools and experiences to build student awareness, knowledge, and skills in applying sustainability principles into front-end design decision-making. https://www.keene.edu/site/directories/profile/facstaff/235/.

Will Hong is an associate professor in the department of digital media and journalism at SUNY New Paltz. He holds graduate degrees from Princeton University and Tisch School of the Arts/NYU and spent two decades in the film and television industry in New York City. Hong received a professional certificate in Campus Sustainability Leadership from the University of Vermont in 2015 and has led webinars and workshops on sustainability pedagogy innovation at AASHE, ICSD (Earth Institute/Columbia University), the Sustainability Curriculum Consortium Faculty Conference, and Duke University. He lives in upstate New York with his spouse and their sensitive beagle.

Emilie Houssart is an artist and curator based in upstate New York. She teaches printmaking and sculpture as a lecturer at SUNY New Paltz, where she is a returning Sustainability Faculty Fellow and member of experimental collective *Eddy*. As a Dutch-British immigrant in the Hudson Valley, her land-based performance and living installation works disrupt psychic and physical legacies of colonizer culture, generating dialogue about alternative ways of being. Houssart's initiative *The DIRT Project* hosts playful, collaborative experiences for all ages to nurture relationships with subterranean life. emiliehoussart.com @_thedirtproject_

Andrea Kantrowitz is Director of the Art Education Program at SUNY New Paltz, and lectures internationally on art and cognition. Author of *Drawing Thought: How Drawing Helps us Observe, Discover and Invent* (MIT Press 2022), she was a teaching artist in NYC public schools for many years, involved in research on how art education could be designed to have a significant positive impact for students growing up in poverty. She holds a doctorate in art education and cognitive studies from Columbia University and an MFA from Yale. Her artwork is represented by The Painting Center in NYC and Kenise Barnes Fine Art. andreakantrowitz.com

Eric Keeling teaches ecology and related courses at SUNY New Paltz. He holds a B.S. in Conservation & Resource Studies from UC Berkeley, a M.M. in Music Performance from San Diego State University, and a Ph.D. in Organismal Biology & Ecology from the University of Montana. His main research interests are in forest ecology and conservation biology. Eric grew up in a family of musicians, hikers, and academics. His father, Charles and brother, Ralph are associated with the *Mauna Loa CO2 Record*, also known as the *Keeling Curve*.

Joshua Korenblat is an associate professor of graphic design at SUNY New Paltz, co-founder of Earthrise Learning Commons (earthrisecommons.org), Eddy at New Paltz member and specializes in the vivid storytelling of complex information. A former reviews editor for *Design and*

Culture, he also researches how designers shape culture. He has co-taught a Sustainability Living Learning Community course and authored book chapters on design and sustainability. He earned an MFA from the Maryland Institute College of Art, an MAT from Brown in education, and an MA from Johns Hopkins in writing. He has worked at *National Geographic Magazine* and *Science News* and cofounded and served as art director at Graphicacy, a data design firm.

Heather L. Lai is an associate professor at the State University of New York at New Paltz in Mechanical Engineering. She teaches undergraduate engineering students kinematics, system dynamics, computer simulation, measurement, and wind energy acoustics with the goal of helping them to understand the value of tackling dynamic problems by thinking in terms of system behavior. Teaching in collaboration with other faculty, Heather strives to encourage her students to think holistically about the global challenges they will face as engineers. Currently, Heather is engaged in research related to wind farm acoustics, specifically in measuring and characterizing wind farm noise impacting local residents living near wind farms. Additionally, Heather is a leader in her local ASME chapter and is active in citizen science initiatives.

Edward Lawson, Jr., JD/MBA, Department of Black Studies and Scholars Mentorship Program at SUNY New Paltz, founded and directs FACES: Family and Community Engagement Services in Newburgh, New York. A nationally recognized trainer, Lawson has conducted over 250 workshops, keynotes, and community engagement events across 20 states and two countries including implicit bias training, his *Diversity, Equity and Inclusion Framework*, and webinars on Domestic Violence, Special Needs Children, Child Support, and Childhood Development. Lawson serves as Executive Director for the Vin Baker Bouncing Back Foundation. At the National Responsible Fatherhood Clearinghouse (NRFC), he led President Obama's Fatherhood Buzz Barbershop Initiative and contributed to the *Responsible Fatherhood Toolkit; Resources from the Field*.

Nosipho Faith Makhakhe, Ph.D., is a senior lecturer at the Durban University of Technology Centre for General Education in South Africa. She holds a Ph.D. from the University of KwaZulu Natal in health promotion and a Master of Social Science (Cum Laude). Her primary research interest is in understanding the lifestyles of vulnerable and key populations and the social factors that expose them to sexually transmitted infections particularly HIV in South Africa. She is currently venturing into the field of climate change and the impact that climate change has on the mental health of young people in the Global South.

Ilarion Merculieff was born on the Pribilof Island of St. Paul in Alaska, raised in a traditional Unangan (Aleut) way, and given the traditional name Kuuyux by the last Elder who held that name among his people. The name means a carrier of ancient knowledge into modern times, a messenger. He graduated from the University of Washington, having established the first Indigenous student education program there. Ilarion has served Unangan and other Indigenous peoples locally, nationally, and internationally over a fifty-year career devoted to the environment, human rights, community wellness, economic development, and cultural enhancement.

Lisa Mitten, M.A., is Campus Sustainability Coordinator at the State University of New York at New Paltz and founder of the Sustainability Faculty Fellows Learning Community. She holds an M.A. in Education from Hofstra University and an M.A. in Public Policy from the University of Maryland. She coordinates the student Sustainability Ambassadors program and consults with institutional management on facilities and campus design. She serves as a board member at the Ulster County Resource Recovery Agency and as Treasurer for Seed Song Farm in Kingston, NY. She is the winner of the SUNY Chancellor's award for Excellence in Professional Service (2021).

Jared Nelson, Ph.D., is an Assistant Professor of Sustainable Product Design and Architecture at Keene State College and a SUNY New Paltz Sustainability Faculty Fellow. He uses his multidisciplinary background to specialize in materials and solid mechanics with research focused on advances in bio-industrial materials toward realizing sustainability. He has spent nearly two decades contributing to the better understanding of material responses of fiber-reinforced plastics toward design and development of improved engineered products. Most recently, he and his students have been studying variation in natural plant fibers to be used as sustainable replacements for synthetic fibers.

Matthew Newcomb, Ph.D., teaches rhetoric, writing, and literary theory courses with a focus on argument, environmental writing, and science writing at SUNY New Paltz. He is a Sustainability Faculty Fellow and has worked as a writing program administrator. Publications on sustainability topics include work on sustainability as a principle for writing studies, evangelical rhetoric and environmentalism, and implications of writing on climate-related disasters. He is also the author of *Religion, Narrative, and the Environmental Humanities: Bridging the Rhetoric Gap* from Routledge (2023).

Andrea M. Noel, Ph.D., is Professor Emerita in the Department of Childhood and Early Childhood Education at the State University of New York at New Paltz. She has expertise in Child Learning and Development and has much international experience including Australia, Germany, as a Fulbright Scholar in Hungary, and most recently as a grant recipient from the Polish National Agency for Academic Exchange. Dr. Noel was a Sustainability Fellow at SUNY New Paltz, and infused sustainability content into her classes. She has published widely in the fields of early childhood education and learning.

Brian Obach is a professor of sociology at the State University of New York at New Paltz. He specializes in the study of social movements, environmental sociology, the sociology of food and agriculture, and political economy. He is the author of *Organic Struggle* (MIT Press 2015), an examination of the sustainable agriculture movement, and *Labor and the Environmental Movement* (MIT Press 2004), an analysis of cooperation and conflict between labor unions and environmentalists. He has published articles in *Sociological Perspectives*, *Organization and Environment*, the *Journal of Civil Society*, *Social Science Quarterly* and other scholarly journals and popular publications. Dr. Obach has published several articles on pedagogy in the American Sociological Association's teaching journal, *Teaching Sociology*. He is the former director of the Environmental Studies Program and Chair of the Department of Sociology at SUNY New Paltz. He earned his Ph.D. at the University of Wisconsin, Madison. He is a member of the all-professor rock band, Questionable Authorities, and he is working toward an MFA in drawing and painting at SUNY New Paltz.

Joel Oppenheimer is a Licensed Clinical Social Worker and Assistant Director at the Psychological Counseling Center at SUNY New Paltz. His career spans from homeless shelters in San Francisco, California, to residential drug treatment programs in New York. Through his extensive experience working with men, trauma, and groups, he sees connection and relationships as vital for a sustainable earth. He created the Positive Masculinity Project (2018), a campus intervention to decrease sexual assault by changing culture. PMP is based on Transformative Justice, Relational Culture, Anti-Violence, and De-colonization to hospice toxic masculinity and to heal and birth a new way of being in relationship.

Ş. İlgü Özler is the founding director of the SUNY Global Engagement Program in NYC and a professor of Political Science and International Relations at SUNY New Paltz. Her research focuses on civic engagement. She has done research on urban slums and squatter areas in Turkey, Mexico, and Chile. Her recent work focuses on Spanish and Turkish civil society organizations and political participation as well as human rights and global governance. Her publications appeared in the following journals: *Sociological Perspectives*, *Ethics and International Affairs*, *Representation*, *Journal of Civil Society*, *Democratization*, *Latin American Perspectives*, *Global Environmental Politics*, *Mexican Studies* and *Turkish Studies*. She was a Fulbright scholar at Universidad Autónoma de Madrid in 2022. She was the recipient of a Charles E. Scheidt Faculty Fellowship in Atrocity Prevention 2022–2023. For 20 years, she taught a course on the United Nations and organized and participated in over 600 briefings for her students at the United Nations. She engages her students in applied work in global affairs through the Global Engagement Program in affiliation with over 60 different international organizations. She is the founder of Mid-Hudson Valley local chapters of Amnesty International USA (served as coordinator 2011–2017) and United Nations Association (served as president 2013–2017). She served on the Board of Directors of Amnesty USA (2018–21 and 2023–26). Özler received her Ph.D. in Political Science from University of California, Los Angeles (2003).

Samrat Pathania, a former software engineer, currently teaches mathematics, physics, environmental science, and computer science at Wallkill High School in Wallkill, New York. He is a New York State Master Teacher Emeritus and co-founder of the non-profit Environmental Justice Forward. Samrat is a passionate advocate for inquiry-based and climate solutions-centered STEM education, inside and outside the classroom and serves on the board of The Ashokan Center, New York's oldest outdoor environmental education center https://ashokancenter.org/. He is the co-author of *A Full Axiomatic Development of High School Geometry*

(Springer 2023). Samrat is the Program Manager for Science and Sustainability at Mid-Hudson Teacher Center at SUNY New Paltz. He is also a clean energy coach with New Yorkers for Clean Power.

Emily Puthoff is a sculpture professor and co-director of the Sculpture Program at SUNY New Paltz where she mentors students in sculpting sustainable lives and careers. As co-founder of Hudson Valley Bee Habitat (HVBH), she creatively inspires the conservation of pollinators by collaborating with communities to create public pollinator gardens and sculptures. Passionate about fostering a more just, wild, and regenerative future, Emily is committed to nurturing beauty, sustenance, and care in both the local community and the broader ecosystem through her pedagogy and creative conservation efforts at Hudson Valley Bee Habitat.

Melissa Yang Rock, CM, Ph.D., is an associate professor of geography and environmental studies at SUNY New Paltz as well as an affiliate faculty with the Asian Studies program and the Women's, Gender, and Sexuality Studies Department and a Certified Midwife. Rock's teaching, research, and activist commitments center on collaborative and interdisciplinary movements toward social and environmental health, equity, and justice.

Libby Roderick, is Director Emerita of the Difficult Dialogues Initiative and co-founder of the faculty development center at the University of Alaska Anchorage and of the Difficult Dialogues National Resource Center, which builds capacities for university communities to engage on the critical issues of our times. She is Associate Editor of *Start Talking: A Handbook for Engaging Difficult Dialogues in Higher Education*, co-author of *Stop Talking: Indigenous Ways of Teaching and Learning and Difficult Dialogues in Higher Education;* and editor of *Alaska Native Cultures and Issues* and *Toxic Friday: Resources for Addressing Faculty Bullying in Higher Education.* Libby is an internationally recognized workshop leader, singer/songwriter and activist with degrees from Yale and Alaska Pacific University. https://libbyroderick.com/.

Kimberly Sanford is a graduate of the School of Education at SUNY New Paltz. She teaches high school English in New York City and instructs undergraduates in disability studies at New Paltz. Kimberly is an emerging disability studies scholar and apprentice to Dr. April B. Coughlin. Together they earned the Academic Year Undergraduate

Research Experience grant in 2021 and presented at the 27th Multicultural Education Conference at SUNY New Paltz. Additionally, they published "Expanding Access to Disability Studies: Reflections from Both Sides of the Classroom" in the *Journal of Teaching Disability Studies* at CUNY in 2022.

Anat Shiftan, an Israeli artist who works in clay and print, explores the subjects of ambivalence in floral and zoological imagery; the textures of sexuality, life and death; and power and subversion, as reflections of complex political and social attitudes. Shiftan received her BA in English Literature and Philosophy from the Hebrew University, Jerusalem and her MFA in Ceramics from the Cranbrook Academy of Art and Design. She taught at Bezalel Academy for Art and Design, University of Michigan, and SUNY New Paltz. Shiftan received the Michigan Grant for Individual Artists and is represented by Hostler Burrows gallery NY/LA. Website: https://anatshiftan.info/.

Nicole A. Simon is a professor in the STEM disciplines at SUNY Nassau Community College, where she received the SUNY Chancellor's Award for Creative and Scholarship. She holds a doctorate in Educational Technology Management and Instructional Design. Nicole is a certified OSCQR Trainer and is Creative Commons certified. Nicole has been COILing for over 10 years and is a COIL Trainer. She teaches a Sustainability course focused on the UN SDGs using COIL projects. Her research focuses on the uses of technology and virtual experimentation. LinkedIn:https://www.linkedin.com/in/nicole-simon-2683b316.

Laia Solé Coromina is a visual artist, researcher, and professor of art education in the department of Didàctica de les Arts i les Ciències at Universitat de Vic-Universitat Central de Catalunya. She has investigated contemporary art practices as sites of learning, creative barriers in higher education students, and is currently involved in research projects on art and sustainability. Her work has been exhibited at the Drawing Center in New York City, Cuchifritos Gallery in New York City, Fundació Chirivella-Soriano in València, Arts Santa Mònica in Barcelona, among others. She earned a Fine Arts degree from Universitat de Barcelona and a doctorate degree from Teachers College Columbia University.

Suzanne Stokes, an artist, professor and foundation coordinator at SUNY New Paltz, is an advocate for community-based arts and dedicated to sustainability education. She is the founder/artistic director of the shadow-based theatre group Cave Dogs and creates monotypes focusing on issues about water. Performances, exhibitions, and workshops include Sibikwa Arts Centre, The Centre for the Less Good Idea, SA; Reykjavik Fringe Festival, Iceland; Academy for Untamed Creativity, Denmark; Skissernaus Museum, Sweden; The Brooklyn Children's Museum; Women's Studio Workshop; and Fringe Festivals across the US. She received grants from The Jim Henson Foundation, NYC, Franklin Furnace, NYC and the NLT Foundation, MA. www.cavedogs.org.

Roland Tormey is a senior scientist at EPFL (Ecole polytechnique fédérale de Lausanne), in Switzerland where he leads the Teaching Support Centre and researches engineering education. He previously worked for fifteen years in teacher education. His research has focused on diversity and equality issues in learning, as well as on the role of emotion in teaching and learning. Recent work includes co-authoring *Facilitating Experiential Learning in Higher Education: Teaching and Supervising in Labs, Fieldwork, Studios, and Projects* (2021), and co-editing *The Routledge International Handbook of Engineering Ethics Education* (2025). https://people.epfl.ch/roland.tormey?lang=en.

Andrea Varga is an associate professor of theatre design, a member of the Honors Faculty, and a Sustainability Faculty Fellow at SUNY New Paltz. She holds an MFA in Theatre Design from Florida State University and, for over 25 years, worked as a costume designer and assistant designer on Broadway, Off-Broadway, in television and academia. As an educator/artist she works to incorporate sustainability practice into her craft, courses, and artwork to connect past with present through tactile visual storytelling.

Christine Wamsler is Professor of Sustainability Science at Lund University Centre for Sustainability Studies (LUCSUS), Founder and Director of the Contemplative Sustainable Futures Program, and former Co-Director of the Societal Resilience Centre. She is an internationally-renowned expert in sustainable development and associated (material and cognitive) transformation processes, with 25 years of experience, both in theory and practice. Christine has been ranked as the world's 6th most influential scholar in the field of environmental science. She is leading the emergent field of inner transformation, mindfulness, and sustainability and has co-developed related theories, approaches, leadership, and educational programs for climate action and sustainability. Her work has shaped international debates and increased knowledge on personal, collective, institutional, and policy transformations in a context of global environmental change. She has led many international projects and published more than 200 academic papers, book chapters, and books on these issues. Her publications are regularly cited and used in theory and policy development, including by the Intergovernmental Panel on Climate Change (IPCC) and the Intergovernmental Science-Policy Platform on Biodiversity and Ecosystem Services (IPBES). Christine serves as a senior scientific advisor for the UNDP Conscious Food Systems Alliance, the Global Compassion Coalition, the Inner Green Deal, the Inner Development Goals (IDGs) Initiative; and the Science Centre (Science Village). Her workplace, LUCSUS, is a pioneer in transdisciplinary research and collaboration that combines critical perspectives with solutions-based approaches to address sustainability challenges.

For an overview of her work: https://www.lucsus.lu.se/christine-wamsler and https://portal.research.lu.se/en/persons/christine-wamsler.

On integrated inner-outer transformation: https://www.contemplative-sustainable-futures.com/.

Dr. Sally Warren is vice president of Build Green Now and Voices for Progress, two initiatives dedicated to advancing sustainable development, community collaboration, and climate-friendly innovation. Her work bridges environmental science, green construction, and policy advocacy, with a strong focus on practical solutions that benefit both people and the planet.

At Build Green Now, Dr. Warren leads research and pilot projects using industrial hemp and lime for external insulation to improve energy efficiency, lower heating and cooling costs, and increase fire resistance in buildings. Along with Henry Gage, Jr., and her team, she collaborates with legislators, community organizations, and industry partners to bring eco-conscious construction into the mainstream and create scalable models for green housing.

Through Voices for Progress, Dr. Warren helps develop and oversee an AI-powered project management platform that enables communities, non-profits, and businesses to plan, organize, and scale impactful initiatives. This includes creating tools for volunteer engagement, skill-building, and connecting experts with projects that benefit from their guidance.

Her leadership combines scientific insight with practical implementation, ensuring that sustainable solutions move from concept to real-world application. Dr. Warren's efforts emphasize public benefit, workforce development, and environmental stewardship, inspiring collaborative action to build a cleaner, more resilient future.

- Build Green Now—www.BuildGreenNow.net
- Voices for Progress—www.VoicesforProgress.net

Sarah Mead Wyman teaches Comparative Literature with a focus on poetry, literary theory, human ecology, and the visual arts at SUNY New Paltz, US. She is a co-founder of Earthrise Learning Commons (earthrisecommons.org), an Eddy at New Paltz core collaborator and co-coordinates the Sustainability Faculty Learning Community and leads the Faculty/Staff Adventure Group. Recent sustainability publications include: "Collaborative Online International Learning (COIL) for Sustainability Education: Promoting Intercultural Competency in English Literary Studies," *Intercultural Education* 2025, 1-19 (https://doi.org/10.1080/14675986.2025.2514383), "Introducing Sustainability Topics with Ursula Le Guin's 'The Ones Who Walk Away from Omelas' and Richard Powers' 'The Seventh Event," in *Literature as a Lens for Climate Change: Using Narratives to Prepare the Next Generation*, Ed. R. Young, Lexington, 2022, and "Reframing Nature Within the Garden Walls: Feminist Ecological Citizenship in the Work of Louise Glück, Jeanne Larsen, and Anat Shiftan." *Feminist Formations*, 2020 (32)2. Additional critical work in literature and verbal/visual art intersections has appeared in *The Comparatist, African American Review, ANQ, Theatre History Studies, Modern Drama, Journal of Learning Through the Arts, Word & Image*, *CLCWeb*, and other venues. Her poetry volumes are *Sighted Stones* (FLP 2018) and *Fried Goldfinch* (Codhill 2021). https://hawksites.newpaltz.edu/sarahwyman/.

Carrie Young, MPA, Ph.D., is a teaching assistant professor in the Graduate School's Master of Applied Professional Studies Program at the University of North Carolina at Chapel Hill. She has served in the Peace Corps in Mali, West Africa, and participated in interdisciplinary and transdisciplinary sustainability research for Cornell University in Kenya, Zambia, and Malawi, aimed at strengthening communication and education efforts around global sustainability. Her research explores collaborative methods for increasing equity and engagement, especially among underrepresented communities. At Cornell, she led the design of the innovative Cornell Atkinson—Environmental Defense Fund Sustainability Leadership Program to increase the impact of emerging researchers. She has over a decade of university-level teaching experience; her professional experience includes *National Geographic*, *SC Wildlife* magazine, and Tapestry Theatre Company.

List of Figures

List of Tables

CHAPTER 1

Green Leaves and Leverage Points

Sarah Mead Wyman

This book demonstrates how educators and community activists can step up and bridge our ecological, social, and spiritual divides: an intentional response to the United Nations' call for action in higher education and a commitment to address our students' demands. If we approach one definition of ***sustainability*** as "the ability of a system to produce what we need and want far into the future" it becomes clear that the unanticipated consequences of ***normalized*** behavior in modern industrialized societies will not sustain human and more-than-human life (Ehrenfeld 2024; Abram 1997). How can we foster critical thinking to build awareness of a socio-environmental system so complex that it defies human understanding? As members of the extended higher education community, we co-create transdisciplinary and site-specific ***education and action for sustainability*** to address academic divisions and enable generative conversations across campus, in our neighborhoods, and worldwide. This dialogue has been happening in earnest since the early 1990s, and definitions and frameworks for teaching and learning continue to

S. M. Wyman (✉)
State University of New York, New Paltz, NY, USA
e-mail: wymans@newpaltz.edu

S. M. Wyman and J. Korenblat (eds.), *Sustainability Learning for Action and Community Engagement*,
https://doi.org/10.1007/978-3-032-11698-7_1

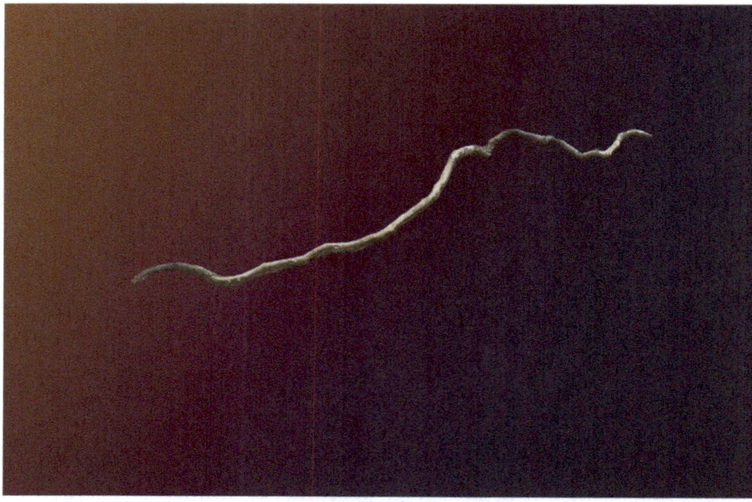

Fig. 1.1 *Untitled (Curve)* (2016) Andrea Frank

evolve. *Environment* means more than green leaves in our twenty-first-century context. Each person, each creature, embedded within a shared physical space, impacts and is impacted by ecological, material, and socio-environmental forces in play.

Our approach, Education and *Action* for Sustainability (EAfS) (Rowe 2019), constitutes a dimension of critical sustainability practice distinct from Education for Sustainability (EfS), Climate Education, Environmental Studies, Education for Sustainable Development (ESD), and Indigenous Environmental Sustainability Education (ESE). Yet, it draws on aspects of each of these theories of learning.[3] EAfS emphasizes (1) systems thinking, (2) shifts in behavior and mindset away from the dominant culture of extraction, entitlement, exploitation, and consumption, and (3) constructivist/project-based learning to foster twenty-first-century skills of agency, creative innovation, and resilience. Rather than presenting prescriptive recipes for illusionary solutions or dictating values and behaviors, or definitions of *sustainability* itself, we invite engagement from those who may be skeptical or wary about politically charged agendas.

As our transnational authors demonstrate, we do the work of critical sustainability education and outreach by developing relationships of trust in the spirit of listening, exchanging information, and creating the new knowledge necessary to shift ways of thinking and bring lasting change (Cachelin et al. 2014). Our contributors expand beyond *thinking as knowledge production* to more holistic, non-hierarchical, and embodied approaches that interpret *environments* in terms of integrated ecology and culture. As our networks increase, we include international perspectives from around the globe and from many sectors of society to present sustainability education as a realm of invention and a means to live *with* Earth, as articulated by Tiokasin Ghosthorse (Cheyenne River Lakota Nation).

In this book, professors, students, counselors, and other changemakers report on opportunities and impacts they facilitate through teaching, advising, mentoring, and institutional reforms that are learner-centered and transformative. Their work is informed by the care ethics, sense of justice, and experiential commitments of such great educators as Arne Naess, Nel Noddings, bell hooks, Maxine Greene, Paulo Freire, Urie Bronfenbrenner, and John Dewey, among others. As collaborating authors, we emphasize individual encounters with concepts and materials in the context of past experience and present knowledge generation. We learn as students ourselves, working with colleagues to define the emerging EAfS discipline. We follow the teachings of peace leaders His Holiness the 14th Dalai Lama and Archbishop Desmond Tutu (Dalai Lama & Tutu 2016) when they invite us to joy and responsibility:

> No dark fate determines our future. We do. Each day and each moment, we are able to create and re-create our lives and the very quality of human life on our planet. This is the power we wield. (ix)

Sustainability approaches depend upon *place* as a relatively bountiful, precarious, politicized, stable, threatened, enriched, impoverished physical spot on the planet. As our authors attest, thinking about sustainability practices, mindsets, and actions will look and feel different in Durban, South Africa, Recife, Brazil, Geneva, Switzerland, or Kashmir, India. Many of our contributors are involved in naming and (re)defining *sustainability* in terms of interdisciplinary action and engagement in the Hudson Mahicantuck Valley, New York, US, itself a region of extremes in relative privilege, access, and well-being, however sheltered from the most severe effects of climate instability. As we stand or sit on Lenni, Lenape, and Delaware soil of the unceded Esopus Munsee territory, we attempt to progress beyond land acknowledgments (Native Governance Center) to commit to the work of investigating wisdom traditions around the planet that hold keys to a future we can welcome and share.

All of us are indigenous to some spot on the planet, as Robin Wall Kimmerer (Potawatomi Nation) indicates in *Braiding Sweetgrass* (Kimmerer 2013). Still, not all peoples have experienced forced migration, extreme weather events, or war. As we examine complex environmental, social, and economic systems in play, including increased global warming, the spread of toxins, species extinction, and people in flight from conflict who carry their knowledge with them, our authors foresee no simple solutions. We insist, nevertheless, upon opportunities to leverage awareness, understanding, resources, human ingenuity, attitudes of care and responsibility to shift the current, unsustainable status quo.[9] Education and action for sustainability align with and support Working Class Studies, Disability Studies, Human Ecology, Biosemiotics, Migrant Studies, Eco-poetics and other emerging fields. Ultimately, education and action for sustainability are vital aspects of Peace Studies and indispensable for planetary survival and thriving.

The discipline of Sustainability Studies grows and deepens through innovations in education and community action. By confronting the climate crisis and its *threat multiplier* impacts on social cohesion, well-being, and economic security, educators join learners in the most urgent work of our time. We tap into student interest and demonstrate ways to equip people of all ages with the information and skills necessary to lead the way forward to a more equitable society where people live in right relation with a fragile environment and with each other. This movement has roots as deep as the history of human populations who have lived with earth, engaged in conflict, and cared for each other across continents. As in Walt Whitman's *Leaves of Grass* (1855–1892), our shared sustainability narrative changes over time, with a consistent emphasis on inclusive relationships, confrontation with the facts, and an expansive sense of human potential. As we leave unsustainable ways and practices behind, we create *new stories* as generative energy that may, ultimately, level the Keeling Curve or slow the progression of global warming with its ecological and social consequences.

As Grace Lee Boggs asks, "What time is it on the clock of the world?" (n.d). This book and its answers emerge from the COVID-19 pandemic-era's wake-up call to acknowledge the human body's precarity in the face of environmental instability and to center the social injustices that have been amplified. Our shared experience of contagion has demonstrated that we are truly interconnected and interdependent in ways too complex to fully understand. International movements for social justice, sparked by racial violence in the US during the summer of 2020, as well as clear disparities in COVID infection and mortality rates have called attention to demographic inequities and their often-lethal consequences. *Frontline communities*, those most vulnerable to environmental contagions, land dispossession, and other forms of disenfranchisement and oppression, must determine their own ways forward and have the environmental justice literacy to do so (Gunn-Wright 2020; Penniman 2023).[6]

We marshal, therefore, our wealth of resources, our own extensive training as academics and community organizers, drop into our bodies alongside students, youth, and other practitioners, and seek creative ways to shift the culture, to plot a path forward. This work demands an orientation away from many traditional institutions' hierarchical and often profit-driven structures. Instead, we operate through collaborative partnerships with regional community members as we reconsider outdated notions of teaching and learning.

We reimagine the distribution of agency, power, and authority embedded in conventional academic structures, governance organizations, and regional cultures that disenfranchise members of their own constituencies. We center our innovative work on identifying and promoting social determinants of health that lead to greater human, creature, and environmental thriving. Educators and students with the means and freedom to participate are not our only audience; these projects can flow with the mainstream curriculum (Kindergarten-12th grade and beyond) so that all learners, especially those most affected, can be involved.

This book provides a unique contribution to sustainability studies by:

- Integrating approaches to mindsets and practices—ways of being and doing that empower educators and community leaders interested in promoting environmental, social, and/or economic learning and engagement.

- Sharing a pedagogy of holistic, student-centered learning that emphasizes problem-solving, active inquiry, intercultural understanding, and creative expression as valuable responses and areas for generative thought.
- Exploring a transnational perspective on education and action for sustainability that amplifies diverse voices often excluded from academic discussion, including students, artists, marginalized individuals, and practitioners in the field.
- Demonstrating the generative power of working between and across traditional divisions in institutional structures and ways of knowing.
- Spanning geographies: Our authors represent at least 15 nations and many ethnicities, occupations, and social positions.
- Committing to intentional, collaborative approaches in the classroom, community, and this book's creation.
- Providing a toolbox for educators, community leaders, administrators, and workforce developers who wish to develop sustainability education and action in their institutions or communities. This includes materials on our interactive Earthrise Commons website (earthrisecommons.org):
    - A set of sustainability frameworks that present ways to conceptualize intersections between environmental, social, and economic dimensions of sustainability
    - A schema of sustainability competencies designed by Brundiers (2021), Wiek (2016), et al. to establish coherence in our learning experiences and objectives
    - A list of recommended learning outcomes to revise existing curriculum in any discipline to address the concerns and intentions of Education and Action for Sustainability (EAfS)
    - A collection of learning models, lesson plans, teaching templates, resources, and community engagement strategies to be adapted and adopted in contexts around the Earth.
- Unflinching attention to the climate crisis and our students' demands for a sustainability-informed curriculum that addresses present environmental and social challenges as it hinges content in any discipline to real-world circumstances and events.

Approach/Methodology

We combine approaches to experiential learning in the context of Sustainability Studies and practical tools to help make transformational experiences happen in classrooms, applied learning venues, and other community settings. Our research-to-action project demonstrates that ideas only matter when we witness them activated by minds and bodies that will take them further into the world and the future than any of us can do alone (Lorde 1984).

To this end, we leverage transdisciplinary, embodied learning, creativity, generative inquiry, and student-centered pedagogies. Our theoretical orientation exists within the framework of *critical sustainability*, as defined by Jeff Rose and Adrienne Cachelin (Rose and Cachelin 2018):

> Critical sustainability is developed as a series of nature-society relations that highlight social and environmental justice concepts and practices. While not entirely new, critical sustainabilities are distinct from existing literatures in that they underscore the necessity of engaging not only with socio-environmental relations, but also with the dominant political economies that so powerfully shape these relations. (518)

We motivate learners by combining interdisciplinary knowledge, transformative pedagogies, and ancient wisdom traditions as curricular methods and models to be tried out and modified by educators worldwide.

Being a Sustainability Educator: Pursuing Questions and Shaping Values

In part 1, "Being a Sustainability Educator: Pursuing Questions and Shaping Values," we present teaching genius in the form of reflective essays on learning, storytelling, loving, and making: perspectives on what it means to live in right relationship with Earth and with each other. Authors amplify voices from around the world, including those of children, many of whom have not had a seat at the table. The sooner we pool our knowledge and bring each other into the conversation, the sooner we compel our learners, colleagues, and politicians to listen, observe, take action, and do what is necessary to reverse damage and promote planetary and creature thriving.

Sustainability as Peace Studies

Any of the chapters throughout the book may be read as a response to Eric Keeling's question: *what are we hoping to sustain?* (Chap. 2). These approaches present Sustainability Studies as an aspect of Peace Studies, grounded in a culture of care and living in kinship with Earth and Earth's creatures. Joshua Korenblat uses care ethics and a capabilities approach to address our relationship with and responsibility to the *more-than-human* world (Chap. 10). These essays reflect effervescent joy, familial bonds (however defined), and the courage to connect with others through intentional dialogue and creative expression. We acknowledge the difficulties of discourse: we will not always choose the best words; our attempts to relate will falter, but we commit our many languages as tensile instruments to redefine our understanding, reach each other, and address the seemingly impossible predicament of the present polycrisis.

Parts Two–Six: Doing Pedagogy

By encouraging questions and critical thinking in our classrooms and in the field, we build learning spaces that are co-creative and inclusive. We follow bell hooks' *Teaching to Transgress: Education as the Practice of Freedom* (hooks Bell 1994) when we ask each other as learners to bring ourselves fully to re-envisioning the world. We return to the critical pedagogy of Paulo Friere's *Pedagogy of the Oppressed* (1968) to activate relationships between student, educator, and social context. As experienced sustainability educators and activists, we provide methods and models to deliver teaching and programming templates that transform institutional, public, "wild," and virtual learning spaces. We describe explorations and activities that have proven helpful and frequently transformative in the higher education learning environment or in applied fields around the globe. This work has impassioned students to assert themselves as change agents and this energy returns to us. Even in a seated class with heavy course content, small-focus shifts and opportunities for active learning can yield impactful results as Megan Ferguson demonstrates in her introductory Chemistry class (Chap. 17). Many of our projects involve engagement outside the traditional academic setting. Most would be appropriate for community organizing, workforce development, apprenticeship training, and creative expression in other contexts, including adaptations in kindergarten—12th grade classrooms.

The Learning Models

In part two, "Creating Transformational Learning Experiences," educators provide curricular springboards for teaching sustainability-focused material within their existing courses. Paul Hirsch and Andrea Varga present models for building courses centered on sustainability topics in online and seated learning modalities (Chaps. 11 & 12). Additional teaching examples in Physics, Child Development, Chemistry, Communications, Creative Writing, and Social Business Entrepreneurship use the UN Sustainable Development Goals (or *Global Goals*) and other frameworks. Learners collaborate to consider how our world's intersecting social, economic, and environmental forces can lead to greater human and creature well-being.

In part three, "Enacting Principles of Human and Environmental Flourishing in Teaching," learners lean into their change agency as they use systems thinking to understand complex social structures and opportunities for transformation. Classrooms become spaces to explore how entrenched, dominant worldviews that promote individualism, extraction, and exploitation (or a *dig-burn-dump* mentality, in Just Transition terminology), might be displaced in favor of the wisdom traditions of Indigenous cultures. This includes the Afrocentric World View, defined by Anthony S. Dandridge (Chap. 20). Learners explore concepts of climate justice, food justice, and the legacies of ecological damage, patriarchal oppression, and racism through modalities of music, philosophy, campus activism, and graphic design.

Jessica Kate Crowell and Kate Bellody offer focus-group survey results that present new information on US students' perspectives on sustainability education (Chap. 27). In part four, "Building a Sustainability Culture in Institutions," authors demonstrate the impact of an organized, intentional, and collaborative effort to prioritize sustainability in disciplines, programs, and university structures. Lisa Mitten, Campus Sustainability Coordinator, describes how she established a vibrant faculty learning community at SUNY New Paltz by investing funds and energy in professional training, using a cohort model to build interdisciplinary collaboration (Chap. 28). Additional authors show how infusing large, introductory programs with sustainability content helps students make transdisciplinary connections and develop a common language.

The campus comes alive in part five, "Engaging Communities: Campus and Region as Living Laboratories," where site-specific sustainability work defines applied and experiential learning opportunities outside the typical classroom. In new spaces for experimentation and discovery, participants can rethink teaching and learning as they observe their bodies in relationship to objects, the natural world, and other humans. Learners become active participants in problem-solving when they use information to create new knowledge. The authors discuss ways an academic institution can successfully undertake community-led partnerships where all parties engage in defining and understanding the concepts of *sustainability, health,* and *justice*. This work demands intentional efforts to de-center academic, hegemonic, or other dominant perspectives as we operate across perceived divisions.

In part six, "Reimagining and Remixing Materials for Circular Cycles: Sourcing- Design- Disposal," authors consider the sourcing, toxicity, energy load, (re)use, and disposal of materials used in the art studio, in the technical field, and in our global consumer economy. Faculty in various Art, STEM, Geography, and Environmental Studies disciplines discuss and demonstrate how students and others become more aware and conscientious about materials use and the relative cost to the planet and creatures involved. Cultural and corporate shifts in product design and disposal can make an enormous difference now that green practices and materials, including alternative substances, are available and affordable. The question of materials literacy, which includes understanding complex capitalist systems, constitutes one aspect of *environmental justice literacy* that empowers learners.

Community Engagement

The community context may be the space of the active-learning classroom, the urban garden where a Habitat for Artists is under construction, or a Pribilof seascape inhabited by seagulls, sea lions, halibut, and eider ducks. These projects break down the imaginary walls that surround our universities. They burst the *educational pipeline*, as Andrea Frank describes it, designed to generate revenue and graduates (Chap. 3). And they reframe educational systems from industries that produce a workforce—to places of learning for curious, creative, and caring citizens. Informed by a STREAM methodology, we are concerned with integrating traditional learning approaches with ongoing knowledge production in the regions where they are located. We add R (for relationship, relevance, rigor, recreation, regeneration, and resilience) to STEAM's insertion of arts education in STEM. This allows us to meet each other on terms that make our collaborations inclusive, listening- and heart-centered.

An entire discipline of Community-Engaged Critical Research attests that this work is both highly productive and ethically nuanced. Our authors acknowledge the primary requirement that all engagement—including programming, research, grant writing, and project development—center the hopes, dreams, and genius of the people who benefit from it: *nothing about us without us*. The consequences of not including all voices, of limiting access to decision-making, and of ignoring, neglecting, or destroying reservoirs of cultural knowledge (including those housed in forbidden thought or language), have brought us to this crisis point.

Existing Research-to-Action

Education and Action for Sustainability (EAfS) provides motivation and opportunity for growth and resilience on many levels. Caroline Hickman, Panu Pihkala et al. have carried out a global study of 10,000 children and youth that determined 84% were at least "moderately worried" about climate change (Hickman et al. 2021). The Harvard University 2022 US national survey on mental health highlights a lack of meaning and purpose for many Gen Z youth, with 58% reporting they experienced "little or no purpose or meaning in life" in the previous month (Weissbourd et al. 2024, 14). About half of the young adults stated that it is "pretty true" or "very true" that a "lot of the adults in my life talk about social problems but do little or nothing about them" (24–25). One young adult survey respondent commented:

> *I would really like to have some sort of meaning from working towards a goal or cause that betters humanity as a whole. I see the world falling apart and no direction for humanity, and I'd like to do something about it.* (26)

The survey creators' primary recommendation for addressing mental health challenges is "Cultivating meaning and purpose in young people, including by engaging them in caring for others and service" (Weissbourd et al. 2024, 19). This strategy coincides with the goals of our book.

Leaders in critical sustainability theory have likewise called on universities to engage in the sustainability crisis directly and have informed our work (Johnson 2024; Leal Filho 2022; Hawken 2021; Armon et al.

2020; Johnson and Wilkinson 2020; Cachelin et al. 2019; Rowe 2019; Clarke 2017; Barth et al. 2016; Agyeman 2003; Bullard, 1990, Latour, 2018, and others). We follow in the path of transformational thinkers, known and once known, who have contributed to discourses, strategies, and positive change.

Conclusion

Our authors attend to which voices are heard, which community champions are at the table when students start talking or regional partnerships evolve. When do youth lead the way, and what might that look like? How are our most wise and experienced community members integrated into the conversation? Which additional demographic tends to be left out due to a lack of understanding, a desire to control the discourse, or bias?

As in David Orr's *Dangerous Years* (Orr 2016), we balance a strong cry of emergency with a fierce determination to drive change and an acknowledgment that we will not be able to reverse the damage done; our descendants will accustom themselves to a future world we cannot easily imagine. Nevertheless, nothing could be more important than our determination to mitigate climate instability and work for positive social change toward equity, security, and collective partnership. Through our EAfS curriculum and conversation, we create a new, intimate relationship with the ecosystems in which we are enmeshed and with each other, including those we will never encounter in person.

This book thus provides mindsets, methods, and models for students, educators, and administrators in higher education institutions and beyond to use existing resources to create new ideas and experiences that address the urgent need to slow and eradicate climate instability, call attention to unnecessary suffering, and improve the potential for individual, community, and planetary recovery and prosperity. The global challenges of recent years have brought attention to the cost of failure to partner authentically and justly at the regional and global levels. This challenging era has left academics, students, and the public depleted, disturbed, and often uninspired. Our physical bodies were threatened or eradicated; voices swelled and were silenced; our forests burned, our pastures dried, and our oceans still float miles of plastic waste. As we emerge into an unfamiliar world, youth-led partnerships in classrooms, construction zones, or leafy city parks will prompt us to abandon unsustainable practices and chart a course toward greater human, creature, and planetary flourishing.

References

Abram, David. 1997. *The Spell of the Sensuous*. Vintage.

Agyeman, Julian. 2003. *Just Sustainabilities: Development in an unequal world*. Cambridge: MIT Press.

Armon, Joan, Stephen Scoffham, and Chara Armon. 2020. *Prioritizing sustainability education: A comprehensive approach*. Routledge.

Barth, Mattias, Gerd Michelsen, Marco Rieckmann, and Ian Thomas. 2016. *Routledge handbook of higher education for sustainable development*. Routledge. https://www.routledge.com/Routledge-Handbook-of-Higher-Education-for-Sustainable-Development/Barth-Michelsen-Rieckmann-Thomas/p/book/9781138597549

Boggs, Grace Lee. "What time is it on the clock of the world?" https://whattimeisitarchive.com/

Bullard, Robert. 1990. *Dumping in Dixie: Race, class and environmental quality*. Boulder, CO: Westview Press.

Cachelin, A., A. Darling, B. Clark, and K. Hackford-Peer. 2019. Weaving equity and sustainability into the fabric of higher education: The University of Utah experience. *Sustainability* 12 (6): 304–309.

Cachelin, A., J. Rose, and K. Paisley. 2014. Disrupting neoliberal discourse in critical sustainability education: A qualitative analysis of intentional language framing. *Environmental Education Research* 21 (8): 1127–1142.

Clarke, David. 2017. Educating beyond the cultural and the natural: (re) framing the limits of the possible in environmental education. In *Reimagining sustainability in precarious times*, ed. K. Malone, S. Truong, and T. Gray. London: Springer.

Dalai Lama XIV, His Holiness & Tutu, Desmond with Abrams, Douglas. 2016. *The Book of Joy*. Penguin.

Ehrenfeld, John. 2024. Conversation 20 December.

Gunn-Wright, Rhiana. 2020. A green new Deal for all of us. In *All we can save: Truth, courage, and solutions for the climate crisis*, ed. Ayana Johnson and Katharine Wilkinson, 92–102. NewYork: One World.

Hickman, Caroline, Elizabeth Marks, Panu Pihkala, Susan Clayton, R. Eric Lewandowski, Elouise E. Mayall, Britt Wray, Catriona Mellor, and Lise Van Susteren 2021. "Climate Anxiety in Children and Young People and Their Beliefs About Government Responses to Climate Change: A Global Survey." The Lancet, 5, December. https://www.thelancet.com/journals/lanplh/article/PIIS2542-5196(21)00278-3/fulltext

Hawken, Paul. 2021. *Regeneration: Ending the climate crisis in one generation*. Penguin.

hooks bell, 1994. *Teaching to transgress: Education as the practice of freedom*. [Harvest] Routledge.

Johnson, Ayana. 2024. *What if we get it right? Visions of climate futures*. New York: One World.

Johnson, Ayana, and Katharine Wilkinson. 2020. *All we can save: Truth, courage, and solutions for the climate crisis*. New York: One World.

Kimmerer, Robin Wall. 2013. *Braiding sweetgrass: Indigenous wisdom, scientific knowledge, and the teachings of plants*. Milkweed Press.

Latour, Bruno. 2018. *Down to earth: Politics in the new climatic regime*. Paris: Polity.

Lorde, Audre. 2007 [1984]. "Poetry is Not a Luxury." *Sister Outsider*. Berkeley: Crossing Press.

Orr, David. 2016. *Dangerous years: Climate change, the long emergency, and the way forward*. Yale UP.

Rose, Jeff, and Adrienne Cachelin. 2018. Critical sustainability: Incorporating critical theories into contested Sustainabilities. *Journal of Environmental Studies and Sciences* 8:218–525.

Weissbourd, Richard, et al. 2024. "On Edge: Understanding and Preventing Young Adults' Mental Health Challenges." *Making Caring Common Project*. Harvard University Graduate School of Education. https://mcc.gse.harvard.edu/reports/on-edge

PART I

Being a Sustainability Educator: Pursuing Questions and Shaping Values

CHAPTER 2

What Are we Hoping to Sustain? A Personal Ecology

Eric Keeling

I'm listening to Icelandic pianist Vikingur Ólafsson play his arrangement of "The Arts and the Hours" by the eighteenth-century French composer Jean-Philippe Rameau. And I'm thinking about sustainability.

To be honest, I've gone through several phases of rethinking sustainability over the years I've been teaching. When I started, I was concerned that the whole topic was bogged down by fuzziness. Under the spell of Ólafsson's playing, my treatment here will be somewhat unconventional. But I will try to outline one clear (narrow) idea of sustainability before moving to broader thoughts.

In my ecology courses, we work through primordial questions: Why do we eat food? Why do we breathe air? What do plants "eat and breathe"? Where does the biomass of trees come from? Where does this mass go when a tree dies and decomposes on the ground? The answers are surprising to most college students. Energy and carbon flow through all these questions.

E. Keeling (✉)
New Paltz, US
e-mail: keelinge@newpaltz.edu

S. M. Wyman and J. Korenblat (eds.), *Sustainability Learning for Action and Community Engagement*,
https://doi.org/10.1007/978-3-032-11698-7_2

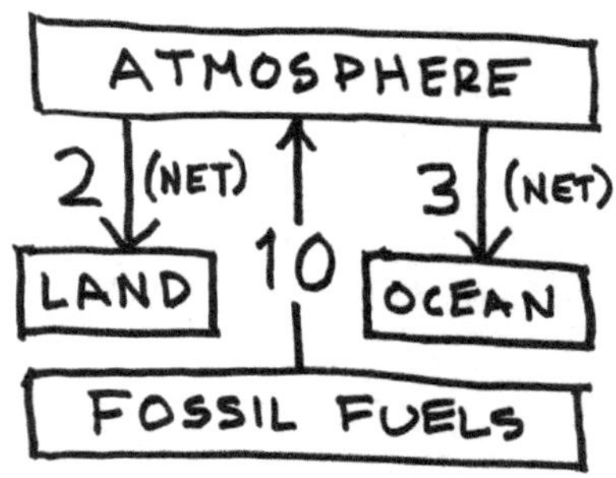

Fig. 2.1 Sketch of the global carbon cycle. Arrows represent "net fluxes" (transfers) of carbon in gigatons of C per year. Flux values are rounded averages for the period 2014–2023

When I get to global ecology, we sketch a simple model of the global carbon cycle (Fig. 2.1)[1] and cover the science of climate change that is closely tied to that cycle. The main evidence for human-caused global warming can be summed up as follows: CO_2 is a heat-trapping gas; it is increasing in the atmosphere; the increase is due to humans burning fossil fuels; the planet is warming; the warming is only explainable by human-caused emissions of CO_2 and other greenhouse gases.[2]

Fig 2.1 is also the gateway into a quantitative understanding of climate change solutions. Over the past decade or so, we have been emitting CO_2 into the atmosphere at an average rate of about 10 gigatons of carbon per year, while land ecosystems and oceans have captured about 2 and 3 net gigatons of carbon per year, respectively. It is easy to see that the rate of emissions is about double the rate of carbon capture, which means CO_2 is increasing in the atmosphere every year. So, our planet has an unstable atmosphere. We are living in a gradually warming world.

Now for one succinct, unfuzzy articulation of sustainability, at least in the narrow context of carbon and climate: Rearrange the numbers in the diagram. There are an infinite number of ways to make *sources* (emissions to the atmosphere) equal *sinks* (total net capture by land and oceans). Sustainability, applied to our atmosphere, requires a balance of sources and sinks, which would produce a stable atmosphere.

The sooner we reach a point where atmospheric greenhouse gases are stable or decreasing, the sooner global temperatures will stabilize.

[1] *Source* https://essd.copernicus.org/articles/17/965/2025/essd-17-965-2025-discussion.html.

[2] See: https://www.ipcc.ch/report/ar6/syr/summary-for-policymakers.

Looking at Fig. 2.1, we can see that stabilizing atmospheric CO_2 involves either increasing the total net uptake rate of CO_2 by land and oceans (sinks) or reducing emissions from fossil fuels (sources). Increasing sinks will likely become more difficult as the planet warms, so a baseline goal is to maintain sinks at current rates. With that approach, Fig. 2.1 shows that global emissions need to be cut roughly in half, probably more if sinks weaken over time. If we hope to draw CO_2 concentrations down, even greater cuts will be needed. With this basic quantitative understanding at the global scale, we can start to evaluate specific strategies and scale our actions down from the global to the national, to the community, and to the personal.

But now, as a matter of equity, things may get uncomfortable. At the national level, we should compare our average per-capita emissions to other nations. Alternatively, we could ask all high-emitting individuals around the world to cut emissions, regardless of their nationality (Chakravarty 2009). If you live in the United States, it doesn't matter which approach you take. Our average per-capita emissions are roughly four times the global average, twice the average for a European or Chinese person, and more than eight times the average for a citizen of India or Bangladesh.[3] Almost all of us in the United States are high-emitters no matter how you look at it. Accounting for global climate justice, our US national goal must be to reduce emissions by substantially more than 50%.

How can we do this? To understand the relative benefits and burdens of various actions, frameworks like Drawdown (Hawken 2017) or the Wedges (Pacala and Socolow 2004) are helpful. Climate solutions are at hand, and some are rapidly being deployed. Almost all these strategies will require government action so that burdens are shared fairly across society. In 2022 in the United States, we passed a climate bill that provided incentives for lowering emissions. In this context, debates about collective vs. personal actions present a false dichotomy. Voting for government action on alternative energy is a personal action *to affect* a collective decision. Taking advantage of those incentives is a personal action *enabled* by a collective decision. The time for personal action in the United States is now.

I believe a scientific and quantitative grip can instill hope and action on the climate problem. But science, on its own, is not always a motivator. Likewise, appealing to long-term risks or international justice might not

[3] Data available online: https://ourworldindata.org/grapher/co-emissions-per-capita.

work. Climate-related emergencies are now a regular occurrence worldwide, but there is still no shared sense of urgency. In the United States, we are promised inalienable rights: life, liberty, and pursuit of happiness. The shared burdens of acting on climate can always be postponed. Because of this, we live in cognitive dissonance, vaguely aware that our current patterns of energy and resource use are unsustainable, but unwilling or unable to change course.

In 2024, in a talk I gave to teachers in the New York State Master Teacher Program, I showed a video of a childhood friend and professional surfer, Rick Lasch, riding a wave. His movements are graceful, not flashy. In long flowing lines, he coaxes energy out of the wave, carving, rising, coming off the lip, ducking and turning through multiple breaking sections—a long ride. Consider this image from the perspective of human-environment interaction. The wave exists as an independent force of nature. The surfer adds a human dance on top of that—a dance with the pre-human world. The surfer rides the wave, but without exhausting its energy, adding beauty without dominating or diminishing its source. Perhaps this metaphor can hint at new ways to think about our relationship with energy in a time of environmental challenge.

I first saw someone riding an electric surfboard in 2023. No muscle energy required, no need to judge the waves, paddle for the wave, go from prone to standing, or feel the energy of the wave and tap into its wondrous and mysterious power. Tear it up, dude! Running on renewables, an electric surfboard emits no greenhouse gases. It is compatible with the goal of a stable atmosphere and climate. One might say, *it is sustainable*. But for me, it has no soul and threatens the sustainability of my surfing metaphor.

There are other more direct trade-offs between different visions of sustainability. A recent proposal in the New York Catskills would allow solar farms to produce electricity and use that energy to pump water uphill into reservoirs during the day. The potential energy of the elevated water could then be converted back into electricity again via turbines during the night when the sun is not shining. This clever system would provide renewable electricity 24 hours a day. But it would also entomb several valleys in the Catskills under water and behind concrete dams. The project was killed by popular outcry almost as soon as it was proposed. A complementary example is a proposal to mine the headwaters of the Bitterroot River in Montana to explore for rare-earth minerals needed in many of our computer, electronic, and alternative energy technologies, including batteries for electric vehicles.

I mention these trade-offs not to start a debate, but to point out possible hidden assumptions about what we are hoping to sustain. In the name of sustainability, there is still a risk that we end up doubling down on an unexamined life of consumption: more energy, more technology, more conveniences, more stuff. Personally, I want to keep the valleys of the Catskills undammed and the headwaters of the Bitterroot unmined. I'm willing to use less electricity, even green renewable electricity, to protect places like these and the opportunities they provide for connection to wild and natural things and processes.

I listen again to Ólafsson playing Rameau. On the face of it, music does not seem like part of the natural world. Jazz guitarist Pat Metheny has observed that music seems like something we weren't supposed to know about, as if there are cracks in the universe letting in something forbidden.[4] But we *do* know about music. It comes into the world through our heads, hands, and hearts.[5] Its primary channel is the human voice. There is language for describing music, just as there is language for describing ecosystems. But music expresses a truth with none of the ambiguity of words. Ólafsson's phrases entreat me to feel something deeper and older than modern concerns. Music is not an answer to current environmental problems. But it does address a more eternal question. How do we relate to the mystery of the world? Music's answer: *Add to the Mystery*.

While working on this essay, I decided to take a break, clean up the kitchen, and take out the compost. With the quantitative approach outlined earlier, I could figure out the benefit of composting orange peels instead of tossing them in the trash. Composting might reduce methane emissions from the landfill. It might reduce carbon dioxide emissions by decreasing the waste transported to landfills. But when I head to the compost bin, I realize that what matters is the practice itself. By gathering up the food waste, getting out under the open sky, and spreading and covering the scraps, I'm also stepping outside our lifestyle of economic productivity, where time is money. I pay the compost with orange peels, and I get back time for fresh air, infused briefly with smells of rotting organic matter and buzzing flies.

[4] See *Dialogues Between Neuroscience and Society: Music and the Brain* available online: https://www.youtube.com/watch?v=yhAbNv1gJT8&t=2292s.

[5] The phrase comes from guitarist Peter Sprague's album of the same name.

Like surfing, composting works with the forces of nature. Decomposition is a natural ecosystem process. Of course, it's happening in landfills also. But the landfill is an abstraction. It is a form of de-consciousness, as is our habit of throwing things into the trash. In contrast, composting is a ritual engagement with decomposition. Similarly, owning an EV or putting solar panels on my roof would have the obvious benefits of decreasing carbon emissions but would also increase my awareness and engagement with energy. These climate solutions are attractive for ritual, for mindfulness and interactions, and for instilling a personal ecology.

As I listen to the piano, I wonder whether there can be a personal ecology related to music. This is more difficult, so I start with a simpler question: What's allowing me to experience this music in the first place? I'm hearing it through Bluetooth speakers, connected to a handheld electronic device made by a company called Apple, running an application called Spotify. I don't understand this technology, which lends itself to the methodology of descriptive ecology. Where am I? What is this stuff? What is going on? I don't think I had ever heard the piece before now. An algorithm suggested it because other people who like this piece also like Bach, Chopin, Elgar, Grieg, and other composers I searched for using the app.

So, did the internet (or AI) enable this experience? Well, yes and no. No, in the sense that this connection was rooted by growing up in a house where this kind of music was played on a piano in our living room. More specifically, this affection was perhaps activated by hearing my sister play at my father's memorial. I can hear my father's piano playing in my sister's piano playing. And I can hear my sister's playing in this pianist from Iceland. The piece itself is reminiscent of Bach, and my father was fond of Bach. Maybe I *have* heard it before, somewhere deep in my childhood. Many other musical experiences in my life likewise play a part in this response.

Going much further back, could anatomically identical humans 40,000 years ago produce a Rameau or a Bach? Bach was part of a huge extended family of musicians. He learned from his uncles. He was eventually supported by nobility and the church. He benefitted from very sophisticated musical instruments that took thousands of years to develop. Somehow, I doubt he ever killed an animal with his bare hands or painted on the walls of caves. But he was indebted to that past, as we all are. There is a cultural stream of consciousness that brings music out of a deeper human history.

This descriptive ecology of music is necessarily exploratory, inductive, and unfinished. But it suggests that connecting with a piece of music grows from a complex personal and cultural history. The experience is only superficially enabled by the digital technology we're immersed in today. I value having access to this music, but more important is the consciousness of its presence and origins—the phrasing and touch of the pianist, the creative vision of the composer, the cracks in the universe adding to the mystery. The act of making and recording these observations about music is also a manifestation of consciousness. We don't want to become unconscious, non-conscious, or even less conscious. Should consciousness be a sustainability goal?

Other observations follow from these meditations. In 2024, the Apple company released an advertisement depicting a massive hydraulic press crushing an artist's studio that includes a piano, a drum set, a guitar, paints, and sculpting materials, including a clay sculpture of a human figure. After it's all crushed, an iPad emerges, and a singer intones "All I ever need is you." By the same reasoning, I suppose I could crush an iPad under the wheels of my car to liberate all the gifts of human creativity it has appropriated. Sorry, I *don't* need you; I have a guitar.

In fall 2022, I saw a weird triangular thing, lit up and moving across the night sky. A news report said it was a satellite, just one of thousands scheduled to be launched over the next few years, to bring the internet to all the corners of the world. "Don't worry, it's safe!" said the breathless newscaster. I've seen strange lines of lights crossing the night sky three more times since then: In Bryce Canyon in Utah around 4:00 AM. Again, ironically, during an outdoor presentation on Native American astronomy. And, finally, when I went out with my kids to say goodbye to the winter constellations—Orion, Sirius the Dog Star, and Gemini—setting in the West.

More recently, in the local market, a picture and headline on the front page of the *New York Times* caught my eye. It read, "The Internet's Final Frontier: Remote Amazon tribes of Brazil." The picture showed a young girl, a member of the Marubo tribe, in front of a grass-roofed hut. Looking at the hut and pictures of the village, I feel these people have a lot to teach us about how to live. A grass hut is familiar to me mostly as an image. I'd like to know how to build one. As exotic as the hut is, the young girl strikes a pose that is very familiar—head down, looking at a handheld device. At the time this article appeared, the tribe had had Internet access for eight months. The article ends with a statement from a tribal member: "… going back is no longer an option. … We can't live without the Internet."

After reading this, I found myself wondering, "Does *my* survival depend on the internet?" I wasn't sure of the answer. So, I googled it.

As we lose our awareness of the natural and cultural systems that sustain us, our trajectory becomes one of maximum technology for maximum convenience, with little discussion. But perhaps the bland acceptance that there is "no going back" can raise the question of what we hope to sustain. That question can't be answered until we recognize that adopting new technologies almost always involves benefits that are obvious, and losses that are not.

When it comes to early twentieth century technologies such as cars, paved roads, and chainsaws, we *did* have this discussion (Sutter 2009). With the passage of the 1964 Wilderness Act, our country decided to keep these mechanized uses out of some areas of the landscape. The decision not to max out on roads and automobiles was a moment of anti-hubris, a recognition of the deeper implications of Euro-American expansive land use. While wilderness designation does not prevent people from owning a car, it limits where cars can be used by keeping some places roadless. Thus, it preserves freedom to have cars, but also freedom to get away from cars.

A surfer using a surfboard on a wave exhibits the beauty of both using and limiting technology in the context of modern energy use. A hiker with a backpack in a wilderness area does the same for traveling across land in the context of mechanized land use. Are there analogous practices for a world that is always a mere click or swipe away from the ubiquity of cyberspace?[6] The Marubo people have decided to turn off the internet for their entire village during certain times of the day and night. They are hoping to sustain something that doesn't depend on the internet.

[6] The phrase "ubiquity of cyberspace" comes from Borgmann 2017.

In my talks and courses, I often finish by showing a picture of Biosphere 2. Most people don't recognize the building, and many don't know the story. It was an attempt to build a self-sustaining, enclosed ecosystem that humans could live in. Why it failed is an interesting story, but the short answer is: it was a carbon cycle problem.[7]

I like to say there are two lessons that we learn from Biosphere 2. The first is self-evident. For the time being we depend on Biosphere 1 (planet earth). This illustrates the importance of environmental sustainability—"life support"—at the broadest level. But the second lesson comes via a thought experiment. Suppose we *are* someday able to build an artificial environment capable of sustaining human life, an environment maximally designed and controlled by humans for human purposes. Is that the world we want to live in?

I vote against a world that is maximally controlled for human purposes. Instead, I envision sustainability as a dance with natural forces, forces that gave birth to human consciousness and creativity, a stream reaching back to pre-human origins and onward into the future, bountiful with the gifts of stars and waves and self-sustaining wildness, something to revere and be grateful for, a ritual engagement with the beauty and mystery of the earth.

Yes, we must act now to lower our greenhouse gas emissions. But the science of carbon and climate doesn't have to be a message of fear and compulsion. Behind the steady rise of CO_2 in the data from Mauna Loa (Fig. 2.2), there is another, more primordial pattern, the undulating annual cycle of the seasons, photosynthesis in play in spring and summer, respiration unmasked in fall and winter. Driving this pattern is Darwin's entangled bank of plants, singing birds, flitting insects, crawling worms, damp soils, meadows, mountains, forests, deserts and oceans. In all of this, I hear music. And I see my father's hands and fingers at the piano keys.

[7] For a non-technical summary see: https://dartmouthalumnimagazine.com/articles/biosphere-2-what-really-happened.

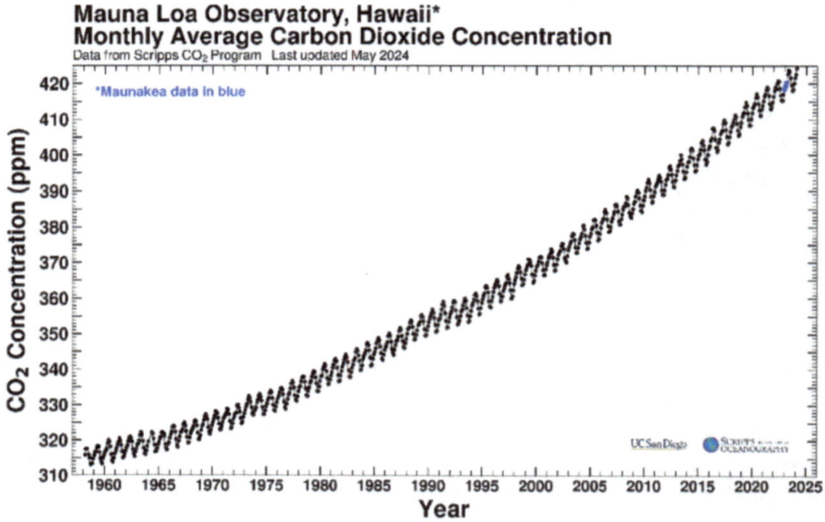

Fig. 2.2 The "Mauna Loa Record" of carbon dioxide measurements, aka the "Keeling Curve"

References

Borgmann, Albert. 2017. The force of wilderness within the ubiquity of cyberspace. *AI & SOCIETY* 32:261–265.

Chakravarty, Shoibal, Ananth Chikkatur, Heleen De Coninck, Stephen Pacala, Robert Socolow, and Massimo Tavoni. "Sharing global CO2 emission reductions among one billion high emitters." Proceedings of the National Academy of Sciences 106, no. 29 2009: 11884–11888.

Hawken, Paul. 2017. *Drawdown*. Penguin.

Pacala, Stephen, and Robert Socolow. 2004. Stabilization wedges: Solving the climate problem for the next 50 years with current technologies. *Science* 305 (5686): 968–972.

Sutter, Paul S. 2009. *Driven wild: How the fight against automobiles launched the modern wilderness movement*. University of Washington Press.

CHAPTER 3

Eddy: A Living and Learning Lab for Co-Creation

Andrea Frank

Introduction

Eddy is a co-creative lab for regeneration, for experimenting within a broad notion of sustainable living and learning at the local level. To render the Eddy concept of holistic education visible, we work with the metaphor of a whirlpool or countercurrent within a river: one that circles back iteratively, revisits, slows, and resists the rush forward. Located in the Hudson Valley of New York, US, our collaborative of SUNY New Paltz students, faculty, and allies often visit the streams and the mountains for their wisdom. We think of our different academic disciplines and life experiences flowing together, swirling, looking at each other, going deep together, and becoming creative and generative in that *eddying* space.

Eddy thus provides a framework for prototyping sustainability education through embodied engagement and experimental modalities. Our interactive research generates new knowledge and new relationships between humans and their own learning experience, between each other,

A. Frank (✉)
Art Department, Photography Program, SUNY New Paltz, USA
e-mail: franka@newpaltz.edu

S. M. Wyman and J. Korenblat (eds.), *Sustainability Learning for Action and Community Engagement*,
https://doi.org/10.1007/978-3-032-11698-7_3

and with Earth. Eddy collaborated with the City of Newburgh Health, Wellness, and Environmental Justice Collective to generate a pedagogical orientation we call *STREAM*. This methodology builds on the existing Science, Technology, Engineering and Math framework (STEM), integrated with Arts education (STEAM). Our approach involves interactions based on care and transdisciplinary collaboration with a focus on justice and Earth connection.

Eddy work is inherently collaborative and responsive to the current campus climate at our university. We host events that center the whole person in every aspect and hold space for listening and creative engagement. To listen, to create, and to engage with others is *to learn*. Projects are often integrated into classes across disciplines or are extracurricular. We believe in and empower each person's potential as an imaginative force, a systems thinker, and a change maker, aspects that coincide with established sustainability competencies (Brundiers et al. 2021). As we meet to plan an activity, we invite the extended community into our circle.

We are driven by the urgency of the growing polycrisis and related climate, biodiversity, and social justice tipping points. The Kogi of the Colombian Sierra Nevada de Santa Marta, who are deeply in tune with the rhythms and spirits of nature, warn us that Mother Earth is dying. We understand that we will be either passively swept into this disaster or join the *great turning*, a shift from Industrial Growth Society to a life-sustaining civilization (Macy 2009). We and our students can learn to live *with* Earth as Tiokasin Ghosthorse would say (Ghosthorse 2024). We can do the vital *inner and outer work* to co-create sustainable ways of living rather than contribute to an otherwise accelerating course toward planetary and species destruction. We encourage participants to respond and reflect on their individual experiences and values and on the practical work of seeking solutions.

Place is key. We define the *environment* as where one is situated: urban, suburban, rural, or so-called *wilderness*. This encompasses how our campus looks and supports biodiversity, how we use resources here, and how we engage with beings that live in this ecosystem. Located within a university, we are critically aware of the history of the unceded Munsee Lenape land on which we stand. We acknowledge that, through our collective institutionalized mindsets, we contribute to rushing the relentless train of progress, competition, and extraction toward an abyss of environmental destruction and unspeakable suffering. Without immediate intervention, this way of life will eventually render Earth, our dwelling place, uninhabitable. It is mind-boggling that this terrifying prospect is widely shared and discussed, but so far, we are collectively unable to integrate these words in meaningful ways (Wallace-Wells 2019). As Bruno Latour (2017) insists, "either we deny the existence of the problem, or else *we look for a place to land*" (5). For us, for the moment, this place is Eddy, where we can resist this denial and reclaim the *shared practice* that, Latour agrees, is imperative if we are to move into a new era where Earth, as a terrestrial force, plays a significant role in politics and human/creature futures (25).

As educators, students, and community champions, we actively question the mental models that have been normalized. Must we perceive ourselves as isolated individuals, as voracious and entitled consumers and extractors, as disconnected from others and Earth? In response, we examine closely how we teach and learn with each other and how these interactions can transform our ways of being to help us grow.

A sustainability perspective through an Eddy lens looks at the social dynamics and the health of all living beings that are here to learn and thrive. We acknowledge that human societies are far from homogenous or uniform; there are hierarchies and degrees by which certain people do or do not have access to opportunities and choices. We lean into *not knowing* and explore together how we can prototype a new paradigm, a culture of listening, care, and connection, even within an institutional context. We investigate how to put on the brakes, take the train of *progress at all cost* off those cliff-bound rails, and steer it toward places and pathways of collective well-being.

Genesis of the Eddy Concept and Community

For nearly a decade, I have been bringing together groups of people to design embodied sustainability pathways in a transdisciplinary context using creative think tank formats. In 2020, while COVID kept us in our different spaces at home, Michael Asbill and I started creating collaboration modalities at a distance via Zoom. A community of interested folks, mostly faculty, some students, and administrators, came together for experiential, embodied activities that incorporated System Drawing (Frank 2020) and practices adapted from Indigenous wisdom traditions. We worked with the Four Directions (van Kampenhout 2008) designed to promote reflection and connection and build trust through cooperation.

The whirling stream image came about as a potent metaphor during one of these online System Drawing think tanks using a co-creative digital whiteboarding application. Looking together at a photograph of flooded New Orleans, Louisiana, US, after Hurricane Katrina in 2005, the image of a busted pipe emerged. We considered the celebrated institutional model—*school as pipeline*—designed to produce graduates at great rates, seamlessly moving from school to graduation to workforce. We saw this phenomenon as a short-sighted production model that drives and feeds the larger, profit-bound systems. This mindset undermines the learning experience and pushes us toward the abyss of climate catastrophe and the extreme disenfranchisement of those *left behind* (Latour 2017, 22).

The fact that disaster can prompt creativity may be what saves us. We felt that, with COVID and the swelling of the Black Lives Matter movement, the pipeline had cracked, and suddenly the water once rushing through it into that sanitized pool of predetermined jobs with prescribed ideas, values, and ideals was leaking, *eddying*, bursting forth. Academic institutions could no longer facilitate the illusion of business as usual, of a growth model without mental health impacts, long-term social equity consequences, and the acceleration of ecological damage. With the pipeline shattered and the stream daylighted, we could reimagine education as flowing into places, as generative and regenerative energy, unbounded by the traditional seated classroom. That image of a living stream continues to help us visualize a goal for how we want to be in the world and *eddies* as places where anyone can slowly explore, spill, go deep, and experiment, knowing they are part of an ecosystem of creativity and care.

STREAM Methodology

With the image of the river eddy in mind, we came up with the STREAM education model, one that expands upon the familiar STEM framework, more recently developed into STEAM with the infusion of Arts Education. Eddy STREAM methodology encourages students to take an active role in their learning and to work collaboratively with others in embodied ways that acknowledge the whole person and the whole system. We emphasize attention to ecosystems and to the social, environmental, and economic dimensions of sustainability education and action. STREAM prepares learners to develop their gifts and potential contributions so we might live in a sustainable world that meets the needs of the present without compromising the ability of future generations to meet their own needs (Brundtland 1987, 16). To this end, we define the STREAM acronym to reference an expanded set of values and concepts, including:

- *Sustainability, Systems-thinking, Social justice, Science*
- *Transformative learning, Transdisciplinary collaboration, Technology*
- *Relevance, Relationship, Rigor, Recreation, Regeneration*
- *Earth connection, Embodied learning, Engineering*
- *Awareness, Arts*
- *Mindfulness, Math*

This educational framework informs all Eddy work.

I think of traditional STEM education as the engine that relentlessly drives the metaphorical train described earlier toward the cliff, the tipping point from which we cannot recover. Simply adding the "A" of the arts will not change the train's direction, as the underlying values and mental models remain the same. We need to not only put on the brakes and steer elsewhere but have the courage to actually disembark onto safe soil so we can tune in with the intricate functioning and carrying capacity of Earth. On the ground, there is freedom to re-invent and re-think what is regenerative for Earth and life. We can re-purpose that engine; we can package the powerful traditional STEM tools in a protective cocoon so that we can harness them toward the long-term well-being of all beings, toward life and thriving for thousands of years to come. Such purposeful activity brings deep satisfaction; it is a potent antidote to fear. This is how we may become true change agents and enter a time of transformation while taking care of each other rather than dodging disruption.

A STREAM orientation reminds us that we have much to learn from nature: how to keep our systems in balance. We have the capacity to review our technological tools to mimic nature and to celebrate interdependence (Buchholz 2024). This approach is joyful, collaborative, horizontally organized, and driven by communities. It can happen on campus or in intentional engagement contexts that bring universities and their local neighborhoods together.

Rendering Visible: Art's Contribution to Academia

As an artist, I strive to see beneath the surface. I look for patterns, whether they be energetic, emotional, or resonating with metaphor. I look for short-and long-term harm and complexity underneath what I perceive and experience around me, such as pesticides permeating the soil of our monocultured campus quadrant. And I imagine the unthinkably complex network of intelligent life that acts according to an ancient script inside the mossy root stump of a fallen tree. My creative research and my studio practice nourish each other and fuel my teaching and Eddy contribution (Fig. 3.1).

As we observe CO2 levels that drive global warming, and watch them climb on the Keeling curve way beyond what can keep our Earth systems in balance, many in the West seem immobilized. We are held captive by a sense of entitlement and our addiction to stuff, habits, systems, and ideas

Fig. 3.1 *Untitled (hole)* (2016), Scratched archival pigment print. 28" x 36", Andrea Frank

that drive our current paradigm. And many are overwhelmed by climate anxiety. There are potent modeling tools that help us understand what's required and what is at stake, e.g., M.I.T.'s En-ROADS Climate Solutions simulator (Climate Interactive 2014) and other frameworks. Can we imagine the Keeling curve flattening out and bending down again through a shift to ways of living and cooperating that radically draw down the atmospheric greenhouse gas concentration? (Hawken 2017). If not, we are in a full-blown, addiction-fueled crisis of imagination.

Art, unburdened by religious or political associations, can help dislodge us from this disturbing emotional state so we can move toward reconnection with each other, our emotional and imaginative dimensions, and Earth. This is a time to set aside the ideas of the lone artist genius and the professor-expert and lean into trusting and sharing spaces to learn and expand our creative potential together. We use our hands and minds, and trust our bodies' inherent wisdom and creativity as we dare to think of Earth, the plants, and other-than-humans as alive and intelligent.

Pouring Love and Care into the Local

As a German-born artist and educator who lives between North America and Europe, the COVID crisis made me realize that it is time to radically direct my energies locally. This required a conscious retreat from the push toward international engagement and exposure. For professional review in our Art Department, for example, the trajectory of success is measured in local, then national, and then international exhibitions. Yet, the real challenge is to affect the local through deep involvement and understanding, to take action in the communities where we live. Often separated by false walls around the university, this potential connection and exchange with neighbors and region takes listening and careful work, which can lead to deep societal transformation and collective resilience. We must be nimble and aware in our immediate spheres to nurture a culture of care and commitment that allows for genuine thriving. From here, we can serve regions and people we may never see or know.

My international connections feed my thinking: I value global exchange, collaboration, and the possibility to leverage ideas across continents. Nevertheless, for a public university such as ours, one that commits to serving the region and its people, we need to invert this value system and start privileging the local above the glamor or exoticism of the international, the abstracted concept of the *global* that privileges profit margins of the elite above the vast majority of the world's people (Latour 2017; Nussbaum 2011). We do this by experimenting and modeling site-specific ways of being in relationship at a fractal scale that can be adapted and multiplied (Brown 2019). A focus on the localized, ecological, and social conditions right here and now, a manifestation of Latour's terrestrial-informed "way of worlding" (Latour 2017, 54), keeps our attention where we can genuinely problem-solve and have an immediate impact through life-enhancing thought and action. The local understood in this way is not a rejection of the global but rather leverages the best of both worlds.

Strategy, Tools, and Practical Impacts

Eddy has a palette of engagement modalities, creative materials, collaborative drawing techniques, and embodied processes that we use as needed. Formats are adapted to the situation and context with a focus on holding a space where individuals can contribute without feeling constricted and collaborate where they can. This may happen in the campus museum, on the main green outside, or in a café in a neighboring city where we challenge the established division between our suburban university and the urban and rural spaces that surround us. We hold weekly meetings online, maintain an active website, and use our large, publicly available Eddy cart to deploy red metal folding chairs, a sound system, seating pillows, art supplies, a rain canopy, snacks, and other materials for creative and regenerative engagement sessions.

Our most important tool is being present for the realities of others by listening from a place of *not knowing,* without the usual need to come up with solutions. We are informed by a wide range of transformational strategies that help us generate innovative and embodied processes and experiences, such as Arawana Hayashi's Social Presencing Theater (SPT), designed to expand our intuitive capacities (Hayashi and Scharmer 2021) and Edward Lawson's theory of community engagement (Lawson and Wyman 2026). Our work also includes aspects of systemic constellation and Theory U processes (Scharmer 2018) such as 3D and 4D Mapping. 3D mapping involves arranging everyday objects on a table, each symbolizing a particular aspect of the local community, then interpreting from four directional and conceptual perspectives. 4D mapping is a type of performative, embodied constellation process where people represent and enact key players (students, President, birds, food service provider, alumni, mayor, Earth, etc.) to visualize our interactive social system. These systems-view activities allow us to understand where our unseen zones and barriers might be and to investigate leverage points and feedback loops (Meadows 2008; Sterman 2012). As we meet our silence, truth, and intention through these processes, we can intuitively sense the space between us where the opportunity for our highest potential future exists and physically shift the system toward how it wants to evolve. This creates a condition where we can prototype rapidly, with clarity and purpose (Scharmer 2018). We feel that education and action for sustainability work best when they are generative in this way, when they develop through a collaborative and embodied creative process informed by a STREAM pedagogy to imagine, envision, and shape situations and inner landscapes.

Our arts-infused projects foreground gratitude and amplify underrepresented voices as we attempt to celebrate joy, promote a culture of care, and address harm (Frank 2024). The impact is palpable when we gather in the spaces of our institutional buildings, outside on the ground, or in another campus corner where our collaborators and participants enter with their whole selves. Individuals share their experiences as they offer their creativity and strength in deeply touching ways and pull others into our process in an inclusive, welcoming manner. By enabling such episodes of collective cultural engagement, our students have opportunities to experience authentic connection, mitigating the sense of isolation that is so pervasive today.

Eddy in Action Using STREAM Methodology: Projects in Brief

Sound Your Truth

Eddy facilitates a student-run, recurring campus event, Sound Your Truth, born of a need to amplify underrepresented voices, particularly those of Black students and students of color. Contentious campus events and transnational reactions to George Floyd's death at the hands of police in 2020 demonstrated the need to attend to views and experiences that had been neglected, ignored, or denied. On or around Floyd's birthday, October 14, students create a space of creativity and listening to each other's truths or stories. This event is held at the heart of our campus, in front of the Sojourner Truth Library. Core questions posed for our first iteration included: *How should the heart of our campus beat? What needs to be said right now? Who needs to be heard?* The next year, students named the event *Emergence/y*, and in 2023, *Courage* (Fig. 3.2).

To plan this community engagement evening for approximately 200, students and faculty integrate the core aspect of the collaboration—active listening—into our classrooms in ways appropriate to their respective disciplines. In 2021, Art faculty assigned students the task of creating an accessible and reusable stage framework and held dialogues about listening to deep truths in a hands-on way. In response, Sculpture students created a low, round sound stage and structures to anchor three long paper scrolls where audience members could draw with homemade carbonized wood charcoal. Other students created molds and cast colored chalks in the shape of human ears to represent deep listening. Participants used these to record what they heard or wanted to say on outdoor blackboards, some of them on the sides of student-built Habitats for

Fig. 3.2 Sound your truth (Jordan Hennix). *Photo credit* Sarah Wyman

Artists (Draper 2024). These temporary shed structures were also used as photo booths and art exhibition spaces. The event is live-streamed with closed captions so people with hearing impairments can participate fully. Black Studies professor Anthony S. Dandridge challenges his students to speak their truths, to sing, show treasured family artwork, or creatively contribute in this context of sharing rather than explicit advocacy. Audience members demonstrate support by expressing visually, drumming, humming, or listening to digest the information, so a sense of trust and energy grows within the creative atmosphere. Students not registered in advance are often emboldened to approach the microphone and speak. A drumming circle hosted by Music Therapy students closes the event, which is documented by Photography and Digital Media & Production students.

Circle Back—Give Thanks

Another simple but impactful event we hold annually the week before US Thanksgiving is *Circle Back-Give Thanks*. We collectively create handmade cards using woodblocks, linoleum prints, or giant group watercolors cut into pieces. Then, we host open card-writing sessions in the library lobby where all members of our campus community are invited to express gratitude to those who help us learn, clean our rooms, share our living spaces, serve us food, coach our athletic teams, transport us to campus, help us get through the day, etc. These cards of appreciation are delivered via campus mail and foster a sense of care and connection.

Listening In: Mother Language Day

This annual event, held around the date of UNESCO's International Mother Language Day, provides an opportunity to celebrate the cultural richness that exists within our campus walls. Invisible linguistic abilities are made visible when students, administrative assistants, faculty, and other campus members read a poem or short text in a language they learned early. Each person explains why this bit of literature is important to them, allowing us to see and know each other more deeply. Instead of an explicit effort to engineer multiculturalism on our campus, this feels like a process of revealing the abundance that already lies within our community.

Signs of Harm—Attention to Discourse

Signs of Harm was an impromptu event set up on our main campus quadrangle to address a recent incident. A dog leash had been left hanging in a tree in a way that resembled a lynching noose. We circled our signature red folding chairs and student-sewn Eddy pillows and offered clay for haptic engagement as we gathered, processed, and responded. We invited the campus community, including students from specific classes and colleagues from our Psychological Counseling Center, to come and share their feelings and reactions to this happening. Participants discussed the non-negotiable truths and implications of how we each see and read objects and information based on our personal experiences and cultural backgrounds. For some, a possibly coincidental combination, a leash in a tree, evoked the historical phenomenon of mass lynching (extra-judicial killing) of African Americans that reached epidemic rates between the Civil War and the early years of the twentieth century. This event allowed participants to recognize and honor individual perceptions and grow as people by honoring the rich worlds of others.

Conclusion: Future Work and Play

Education and Action for Sustainability, which includes the Eddy STREAM view, insists that the practices and mindsets that have brought us here cannot take us further. And the way to bring deep, lasting improvements to our current crisis depends on including the perspectives of many, interweaving the genius of all involved and affected, and listening to the space between us. So, Eddy provides a generative space where each of us shows up with curiosity and where we listen to what can carry through us. This is a wondrous, emergent process that could not happen individually, one that needs collective holding and carrying and an openness to let that unfold.

One goal of Eddy is to render visible challenges and opportunities within the present moment that are hard to conceptualize (such as mass extinction, global instability, and local effects of the polycrisis). Creative responses using poetry, music, movement, and the visual arts have a great impact on communication, understanding, and discovery. We are in a continuous state of emergency with extreme weather events and violent conflicts happening at ever-increasing speed. We need to learn to calmly navigate the river that is rushing more furiously every day and develop the presence to let the currents take us to the edge where we can *eddy* with others, co-create and support each other, and do the necessary work upstream to calm and slow the flow.

We all can look at what and how we teach—is it ultimately destroying or supporting life? One step at a time, carefully and slowly, we question whether our thoughts and resulting actions are, as the Kogi ask, based on life force (Buchholz 2024). Once aligned with life and a thriving future, we can make our choices in ways that become creative, innovative, and joyful. This is where I see art's most important contribution. Through the wonder that comes from emergent co-creation while engaging the whole person, we wake up and address the inner lives and dimensions of our collaborators.

We are all invited to learn to adjust our old-paradigm expectations of life as we knew it, to be deeply grateful for Earth's gifts and share them, and to feel our connection with all that is in ways that make us yearn to shift and transform. Let us set the stage for a society that actively stewards the drawing down of atmospheric CO2 in a way that honors Earth, and embraces response-ability for social injustice, climate-caused suffering, and displacement. We can learn to do this from a place of genuine care and emotional intelligence so that greed, apathy, denial, or willed ignorance in the face of this urgent work can soften and melt away.

Eddy core collaborators include Michael Asbill, Ripley Butterfield, Lucas Jackson Peterka, Anthony S. Dandridge, Andrea Frank, Emilie Houssart, Wren Kingsley, Joshua Korenblat, Edward Lawson, Jr., Emily Puthoff, and Sarah Wyman. https://hawksites.newpaltz.edu/eddy/.

References

Brown, Adrienne Maree. 2019. *Emergent strategy: Shaping culture, changing worlds*. Anarchist Press.

Brundiers, K., et al. 2021. Key competencies in sustainability higher education–toward and agreed-upon reference framework. *Sustainability Science* 16:13–29.

Brundtland, G. H. 1987. *Report of the world commission on environment and development: Our common future*. UNESCO/Oxford UP.

Buchholz, Lucas. 2024. *A book of balance: Kogi wisdom for a good life and thriving earth*. Harper One.

Climate Interactive. 2014. En-ROADS Climate Solutions Simulator. https://www.climateinteractive.org/en-roads/

Draper, Simon. 2024. Habitat for Artists. https://simoncpdraper.com/page/1-Habitat%20for%20Artists.html

Frank, Andrea. 2020. Interview "heeding impulse: Michael Asbill in conversation with Andrea Frank". In *Lossy: Toward a sensory ecology*, ed. Modesta Di Paola, 1st ed. Canterano: Arachne.

Frank, Andrea. 2024. Eddy–collaborating across disciplines toward a regenerative culture, on campus and beyond. In *Expecting reality–connecting minds. Anticipators in Arts & Education*, ed. Anna Maria Loffredo, Rainer Wenrich, and Wanja Kröger. Waxmann.

Ghosthorse, Tiokasin. 2024. Sustain ability to live with earth. In *Sustainability education and action for learning and community engagement*. Palgrave/Springer.

Hawken, Paul. 2017. *Drawdown: The Most Comprehensive Plan Ever Proposed to Reduce Global Warming*. Penguin.

Hayashi, Arawana, and Otto Scharmer. 2021. *Social Presencing Theater: The art of making a true move*. Cambridge: PI Press.

Kampenhout, Daan van. 2008. *The four directions*. Berlin: Sine Causa.

Latour, Bruno. 2017. *Down to earth: Politics in the new climatic regime*. Cambridge, UK: Polity.

Lawson, Edward, and Sarah M. Wyman. 2026. STREAM: Community-led engagement for health, wellness and healing. In *Sustainability Education and Action for Learning and Community Engagement*. Palgrave/Springer.

Macy, Joanna. 2009. The Great Turning. Center for Ecoliteracy. https://www.ecoliteracy.org/article/great-turning

Meadows, Donella. 2008. *Thinking in systems: A primer*. Chelsea Green.

Nussbaum, Martha. 2011. *Creating capabilities: The human development approach*. Harvard UP.

Scharmer, Otto. 2018. *The essentials of theory U: Core principles and applications*. Berrett-Koehler.

Sterman, John D. 2012. Sustaining sustainability: Creating a systems science in a fragmented academy and polarized world. In *Sustainability science*, ed. Michael P. Weinstein and R. E. Turner. Springer. https://doi.org/10.1007/978-1-4614-3188-6_2.

Wallace-Wells, David. 2019. *The uninhabitable earth: Life after warming*. 1st ed. New York: Tim Duggan Books.

CHAPTER 4

Peace, Sustainability, and Educators

Samrat Pathania

Some mornings I wake up from a beautiful dream that is set in a landscape where snow-capped mountains pierce the heavens, the forest echoes with sounds of wind whistling in the pines, and streams of sparkling clear water bounce off rocks rounded by eons of erosion. This landscape is no dream. It is the place I was born—the state of Jammu and Kashmir in India. A famous Persian poet captures its beauty perfectly: *Agar Firdaus bar ru-ye zamin ast, Hamin ast o hamin ast o hamin ast*: "If there is a paradise on earth, it is this, it is this, it is this."

Tragically, this region is the scene of a seemingly endless conflict between the countries of India and Pakistan, the most militarized region in the world. I have never visited the Kashmir side of the state because my parents considered it too dangerous. Most people who have not experienced war or conflict cannot imagine the scale of the human tragedy

S. Pathania (✉)
Wallkill Central School District, Wallkill, NY, USA
e-mail: samratpathania@protonmail.com

Mid-Hudson Teacher Center, Educator Collective for Environmental Justice, Clean Energy Coaching Program, New Yorkers for Clean Power, New Paltz, USA

S. M. Wyman and J. Korenblat (eds.), *Sustainability Learning for Action and Community Engagement*,
https://doi.org/10.1007/978-3-032-11698-7_4

there. Tens of thousands of people have lost their lives in this decades-long conflict, unemployment remains high, and there's little hope of a peaceful resolution anytime soon.

I say this with trepidation, knowing that if we do not get our act together as a civilization and limit global warming to as close to 1.5 degrees Celsius as possible, conflicts like India-Pakistan might seem like child's play in the near future. The prospect of much bigger climate-change-driven conflicts keeps me up at night and fuels my passion for climate work. I am a software programmer turned high school educator. I firmly believe that climate solutions-focused education is necessary in our classrooms and communities.

We have already seen the devastating effects of floods and heatwaves in the Indian sub-continent over the last few years. A tenth of Pakistan was inundated, and over thirty million Pakistanis were displaced by climate change-induced floods in 2022. As the climate crisis intensifies in the coming years, more and more people will be forced to leave their homes across the world as a result of everything from desertification to rising sea levels. How will countries cope with this influx of refugees? We won't be surprised when increased resource scarcity leads to conflict. There are up to three-and-a-half billion people living in hotspots of high vulnerability to climate change. The world could see up to one billion climate migrants by 2050, according to the respected Lancet Commission. If you think of all the immigration problems in Eastern Europe, the United Kingdom, and the United States, with only a few million people, can you imagine what a billion would mean? In the United States alone, twelve million people could lose their homes due to sea level rise with a warming of 2 °C.

When the problem of climate change is stated thus, most people feel overwhelmed, and their usual response is something like—What can I do about this? I am just one person. I didn't create this problem—how can I solve it?

This may seem like a rational response, but I believe that it is mostly a mistaken one. People do not fathom the extent of their power—their power to drive fundamental institutional and cultural change. All the technologies needed to achieve the necessary deep cuts in global emissions by 2030 already exist (see Hawken 2017; Project Drawdown), and the policies that can drive their deployment are already proven. What we need is a cultural revolution, one that meets the needs of all of our fellow human beings within the ecological bounds of our living planet (Raworth 2017).

And who better to lead this revolution than teachers? Teachers understand community—they know how to build communities and nurture them; they know how to envision a future, and they know how to turn this vision into reality. To achieve a sustainable future within the next couple of decades, citizens of democratic societies across the world will have to ask tough questions of each other. They will have to do this in a spirit of community, a spirit of generosity—with love, compassion, and care. And teachers know how to do all of that. That's why I firmly believe that if we are to prevent the worst of climate change if we are to bring nature back from the brink, teachers across the world will be at the heart of our collective effort.

References

Hawken, Paul. 2017. *Drawdown: The Most Comprehensive Plan Ever Proposed to Reduce Global Warming*, Penguin; Project Drawdown. https://drawdown.org

Raworth, Kate. 2017. *Doughnut Economics: Seven Ways to Think Like a 21st-Century Economist*. White River Junction, VT: Chelsea Green.

CHAPTER 5

An In the Meantime, In Between Time, Type of Love

Demond M. Hill

Y'all, are we dreaming…because my inner child is screaming for joy.

With love bubbling within my soul, this form of love requires me to name the state of our world for our most marginalized children. As we dream and attempt to create a liberated world, many of our most precious babies are caught living amidst a deeply oppressive state. Their playfulness, joys, laughs, smiles, and dreams deferred.

They say liberation is hard. I agree. They say liberation requires some real patience. I agree.

So, this love that bubbles within requires me to dream of what to do in the meantime, between time: something birthed out of love, something radical.

As we experience a shared existence, we share fears, anxieties, and worries. In this sharing, there is a reckoning.

D. M. Hill (✉)
Department of Child Study and Human Development, Tufts University, Medford, MA, USA
e-mail: dhill03@tufts.edu

S. M. Wyman and J. Korenblat (eds.), *Sustainability Learning for Action and Community Engagement*,
https://doi.org/10.1007/978-3-032-11698-7_5

You can see/feel/communicate the everyday realities of living on the margins of the world. This empowers marginalized folks to respond to the effects of their marginalization and transform the "unlivable moments" into "livable moments" through love.

In the meantime, in between time is the underground railroad, trans youth housing each other, children imagining themselves as superheroes, a letter to an incarcerated family member, a warm hug from your grandmother, and the ladies in the community handing out food. It is necessary, unpopular, unseen, and essential. It is the past, present, and future. It is a form of radical love.

So, what is radical love?

As sung with fierce beauty, bell hooks (1994) huddles us together to consider,

> Without love, our efforts to liberate ourselves and our world community from oppression and exploitation are doomed. As long as we refuse to address fully the place of love in struggles for liberation, we will not be able to create a culture of conversion where there is a mass turning away from an ethic of domination. Without an ethic of love shaping the direction of our political vision and our radical aspirations, we are often seduced, in one way or the other, into continued allegiance to systems of domination—imperialism, sexism, racism, classism. (289)

So, in honor of bell hooks, *Radical love* isn't just a fleeting emotion or sentiment; it's a revolutionary act, a commitment to honoring our collective liberation. If a radical love bubbles within you, allow that love to manifest into action—a transformational and liberatory change. A radical form of love transcends boundaries, runs through barriers, and builds the foundation for profound healing and transformation in our lives and the world around us.

Radical love is a courageous act of defiance against oppression, injustice, and hatred. It is a pure refusal to accept the suffering of others. Instead, it is an unwavering commitment to standing in solidarity with the marginalized and advocating for their liberation.

But radical love isn't just about outward action; it starts with an inward journey of self-discovery and self-compassion. It requires us to confront our anxieties, worries, and insecurities and embrace ourselves with grace and care. No matter how disconnected the world may seem, we are connected. By recognizing our collective connectedness, we cultivate spaces for empathy, compassion, joy, and belonging.

But radical love isn't without its challenges. It requires us to confront our privilege and acknowledge the ways we've benefited from systems of oppression. It asks us to be engaged.

So, how do we cultivate radical love in our lives? Importantly, how do we love our inner child?

Within each of us resides a younger version full of innocence, curiosity, and boundless imagination.

Our inner child represents the essence of who we truly are before the world taught us to conform, doubt, and fear. Yet, amidst adult responsibilities and societal pressures, we often neglect this precious part of ourselves. Reconnecting with and loving our inner child can unlock a life of joy, fulfillment, and authenticity. It means honoring our dreams, passions, and desires. We must forgive ourselves for past mistakes and embrace our worthiness. It is a journey I am currently on. While on this journey, I am reminded that my behaviors and those of others are simply an attempt to protect, care for, and love our inner child in ways others could not.

When we love our inner child, we become more compassionate, empathetic, and connected to others. We learn to approach life with curiosity and courage and celebrate the beauty of *just being*. This journey begins with awareness of our inner child, continues with remembrance, self-care, and grace, and deepens with embracing the possibilities of belonging.

This essay is for those often unseen who deserve to be seen as the authors of strength, beauty, and joy. This paper honors and cherishes those here and those no longer here with us in the flesh.

As we navigate the complexities of our world, may we remember that radical love, born from the depths of our being, is not merely a concept to ponder but a call to action to uplift the marginalized, challenge injustice, and cultivate healing and transformation. In loving our inner child, we embark on self-discovery and connection. This journey reminds us of our inherent worthiness and capacity to create spaces of belonging and joy. So, let us, in the meantime, between time, dream. Let us create. Let us radically love.

References

hooks, bell. 1994. *Outlaw Culture: Resisting Representations*. Routledge.

Sharpe, C. E. 2016. *In the Wake: On Blackness and Being*. Duke University Press.

CHAPTER 6

The Way of the Real Human Being

Ilarion Merculieff

One day, when I was about five years old, I went out with my grandfather before sunrise to pray near the shore of the Bering Sea. It was an incredibly beautiful day, magical, with sunshine and no wind. As we walked home along the beach, our feet fell into a slow, rhythmic pace along the dark-colored sands. Small waves crested in rapid succession. We could hear the seagulls calling and the seals bellowing in a distant rookery. The sea air smelled fresh, and everything was alive and intense. The sky and sea were expansive and a myriad of blue hues. I could see the horizon for a hundred and eighty degrees.

Ilarion Merculieff—co-founder and/or co-director of the Indigenous Peoples' Council for Marine Mammals, the Alaska Oceans Network, the International Bering Sea Forum, the Bering Sea Council of Elders, and the Science Working Group for Snowchange.

I. Merculieff (✉)
Global Center for Indigenous Leadership and Lifeways, Aleut Nation, Homer, Alaska, USA
e-mail: lmerculieff@gmail.com
URL: https://www.gcill.world/contact

S. M. Wyman and J. Korenblat (eds.), *Sustainability Learning for Action and Community Engagement*,
https://doi.org/10.1007/978-3-032-11698-7_6

"Oh I love this day!" I proclaimed. "It is really good!"
"*Anaan eestahnaan Laikaiyax*," my grandfather said softly. "*Tututhaax*."

Even as a five-year-old child, I understood what he was saying. He was teaching me how to be a real human being through one of the age-old ways that Unangan (Aleut) people have used to survive and thrive along the Bering Sea for ten thousand years. He was saying that words are unnecessary, that they diminish the fullness of meaningful human experience. He was telling me to stop talking and to experience the world around me without words.

St. Paul Island, where I was born and raised, is one of five islands in the archipelago known as the Pribilofs. It is a magical and mystical place, a windswept outpost of volcanic rock twelve miles long and five miles wide, in the middle of the Bering Sea, about 300 air miles west of the mainland of Alaska and 800 miles from Anchorage. My people—the Unangan people—have lived in and along the Bering Sea for ten thousand years, and we live there still. In my childhood, St. Paul was home to some 1.4 million fur seals, 2.5 million seabirds, a thousand reindeer, an untold number of arctic foxes, and six hundred Ungangan Aleut people, including those in our sister village of St. George, about forty miles away by boat.

I was blessed to have a fully traditional upbringing, by which I mean I was raised by my entire village in the ways of the real human being. I was always welcomed into everyone's home and treated as if I were a member of the family. I was never scolded for anything and had the freedom to roam the island anytime day or night without restriction. Basically, I was free to explore my world inside and out without interference by adults. Of course, none of this would have been possible without the agreement of my parents. Most children in the village were raised this way.

Between the ages of four and six, I lived with Papa (my grandfather, Paul Merculieff) twenty-four hours a day, as was the custom at that time. I went to work with him, hunted with him, prayed with him, and slept with him. This is a tradition that goes a long way back. It was a way for my grandfather to get to know me and for me to get to know my grandfather.

At the same age, I was invited into many circles within the village. The Elders would invite me to go camping with them, and I would hear their stories. The men would take me out hunting and fishing. The women would take me berry-picking and later invite me to be with them while they prepared and cooked our wild foods. The villagers were my teachers. They taught me what they did, and they shared what they knew.

I found my *Aachaa* at the age of five. An *Aachaa* relationship develops between an older person and a child based on energetic and spiritual connections rather than family ties. *Aachaa* relationships are not proscribed or planned; they just happen. When that feeling of special connection occurs, the two become *aachaas* and call each other "*Aachaa*."

The word can't begin to encapsulate the relationship. Attempts to define it only diminish it. But for the next eight years of my childhood, my *Aachaa* took me under his wing. He took me out hunting and fishing with him. We hunted Steller sea lions, eider ducks, and whatever other migratory ducks came through. We fished for halibut. The men in my village were consummate hunters and fishers.

From him, through our experiences together, I learned much of what I know about relationship and reciprocity, the Unangan ethics and values of sharing, my relationship to myself and the natural world, and what it means to be an Unangan and a man.

The whole time he had me under his wing, my *Aachaa* spoke no more than about two hundred words to me. He gave no verbal instructions or explanations; he did not encourage me to ask questions. We were together in silence most of the time. He expected me to watch, listen, and learn on my own, using my inherent intelligence.

This is typical of our people. Words are not only superfluous, but they also constrain our intelligence. Adults never presumed any limitation to my intelligence or ability to learn, nor tried to tell me what I should learn. Instead, they provided learning opportunities. The adult's responsibility was simply to create a big open space in which the young one can learn.

The beauty and deep wisdom of that kind of learning is that it allowed me to reach my maximum potential. What I learned depended totally upon my own interests, initiative, experiences, interpretations, discernment, and intelligence. Western-cultured adults often begin with fixed ideas about what children should and should not know. They teach to that presumed and predetermined standard of knowledge, sometimes at the expense of the child's creativity, sense of self, and natural ability to learn on his or her own.

By contrast, nothing was held back from me. Anything I wanted to do, learn, or know, I could, without concern about my age. The only times adults intervened were for reasons of safety. The elders would gauge their responses by the questions I asked. They would answer at a level I could understand.

In Western culture, we are taught to give the "right" answers, or at least try to. And the "right" answers, conveniently, are what they told you they are, things you have to learn by rote. That's a reverse from the way I was raised. None of the things I learned about being Unangan came from books, and there were no wrong answers, only better or different ones. The Unangan learning process helped me to think creatively and critically, without judgment…something that has helped me immensely in my life and career.

Aleut people know that human intelligence exists and operates not simply in the mind but in the body and spirit as well. We learn with all of our senses: hearing, feeling, smelling, intuition, gut responses, thinking, emotions, "heart sense," and body signals. Intelligence is a system, synthesizing information from both sensory and non-sensory inputs. Underneath is a knowing that is profoundly connected to All That Is. Ultimately, this is the basis for our spirituality. The way of a real human being is to understand and feel this connection.

Pribilof Unangan are People of the Sea Lion. Steller sea lions are as central to our way of life as bowheads are to the Inupiat and bison were to the peoples native to the Plains. They have provided for us through good and hard times. In return, we preserve their habitat from disruption and protect them from wanton killing. We honor each one we kill by eating or using every possible part of the body and sharing the animal with as many families as possible.

As young hunters, we would sit on volcanic basalt boulders next to the sea for six, seven, even ten hours at a stretch, waiting for a sea lion to come by. We learned to be quiet inside and out and to pay attention to everything going on around us. Conversation was a distraction we could not afford. It might prevent us from sensing the sea lion out in the water, five or even ten miles away.

I sat in silence with my *Aachaa* and the other hunters for hours. I watched the sea, listening to the waves and feeling the rhythm of the ocean. Sheww, sheww, sheww. Our island is called the Birthplace of the Winds. I listened to the rhythm of the winds. I listened to the seals barking along the beach. I breathed deeply in the fresh sea air. The sun might shine through the clouds. Everything around me would be moving. Everything would be alive.

Sometimes, I let the rhythm of the wind and the waves and sounds of boisterous seals lull me into a serene stupor. The background of sound in a wild environment is full of rhythm and redundancy; it can easily carry you away into a dreamlike state. Once in a while, I might even doze off, having been up since three in the morning to be at the shore by six for a full day of hunting that typically lasted late into the afternoon. After all, I was only five years old.

But I noticed that the men never once dozed off or fell into daydreaming. Their awareness never slipped for a second. And then suddenly, a hunter would proclaim, "*Cowax ukukohx*!" A sea lion is coming!

Instantly, without anyone pointing, all the men would look to the same spot out in the open Bering Sea. Uncannily, the hunters would know a sea lion was coming even before anyone could physically see it. To a child my age, this seemed truly magical.

I came to understand it through an experience that at first might seem unrelated. About six miles from my village is a place called Tolstoi, where tens of thousands of seabirds nest on cliff ledges of rugged volcanic rock and raise their young on outcrops overlooking the sea. There are many different kinds of birds filling many different niches. Puffins have burrows near the top of the cliff. Kittiwakes make nests of tundra grass on the main body. Murres lay their eggs on the bare rock. Cormorants prefer the promontories sticking out from edges of the cliff face. Auklets nest in crevices underneath basalt boulders at the base.

Beginning at age six, I made regular trips to Tolstoi, leaving the village in the late night darkness so that I could be at the cliffside before sunrise when the seabirds stirred. I would later learn names for their forage: sand lance, tiny sea creatures called copepods, and oil-rich fish such as herring, capelin, and pollock. But at this age, I just loved the experience of them.

Near sunrise, birds began to slip off the cliff ledges and circle around in front of the dark volcanic basalt cliffs. Soon, thousands of birds would be flying in every direction in loops around the face of the cliff, passing just inches over my head as I sat below. The murres flew quickly; kittiwakes were slower. The air was filled with the sound of thousands of fluttering wings and bird calls. They were rich and happy sounds, and I loved them.

I watched the birds maneuver around each other, repeatedly landing and taking off from the tiniest of ledges. With my child's mind, I marveled at how thousands of birds could fly up-down, right-left, down-up, left–right, and diagonally, all at the same time, without any bird hitting another or even clipping a wing. How did they do this? I wondered. It looks like chaos, but they never even clip a wing.

From this expanse of deep observation, I asked myself a question. What made them different from me? I could think of several things right away. They don't use words or thoughts. They don't worry about tomorrow. They don't think about yesterday. They are full of life and intensely present in the moment. And somehow, they know where all the others are too.

I thought about the hunters, who were never lulled as I was. Like the birds, a good hunter has to be fully present at all times, experiencing the rhythms without being lulled to sleep by them. This quality of awareness makes the hunter more successful. With it, there is a better chance he will get the sea lion and a better chance he will not cause undue suffering when he kills it.

In my child's mind, I decided to be like the birds. I had been doing it all along as I watched them, taking in their movements without interference of thought, just like they did. And now, my grandfather's teaching about experiencing the world without words came suddenly and clearly to life for me.

It is logically impossible to put into words an experience that goes beyond words; such are the limits of language. In the wordless state, one becomes nothing but pure awareness, not attached to anything, not even thought. As soon as I go into my head, I get lulled again, in much the same way as I was once lulled into a stupor after hours on the rocks. But whenever I can be without thought and in awareness, my experience is enhanced, and new dimensions of human capability are opened to me.

As my awareness deepened I, too, learned to stay fully alert out on the rocks, to feel the sea lion before I could see it. More consistently, I could feel a halibut before it hit my jig fishing line. I could tell how the fish was hooked (by the lip, jaw, or torso); I could determine size and weight; frequently, I would be able to tell if it was male or female. That kind of inner knowing is inexplicable by empirical standards, but it is nevertheless quite real. It is a mark of manhood to me and my people.

Inupiat, Yup'ik, Sugpiaq, Dene, Unungan…these are all different code names for the real people, real human beings, or people. Virtually every Indigenous culture in Alaska refers to itself as *the people*, or *the human beings* or *the real human beings*. Real human beings are those who are profoundly connected, in spirit and with themselves. They are individuals who feel they are part of a whole and not separated.

You can recognize real human beings by how they inhabit their bodies. Real human qualities include patience, gentleness, soft-spokenness, observation, consideration for people and wildlife, cooperation, non-aggression, the ability to be present in the moment, and a deep reverence and respect for all living things. In Western-dominated cultures, these qualities are often associated with the feminine and dismissed as somehow of lesser or even negative worth in the fight for survival. In the Unangan worldview, however, they are the mark of a true person and a complete human being.

The way of the real human being is a proven pathway to living in long-term sustainable ways on our shared land. It can—and should—help us all deal more successfully with the daunting issues facing humankind.

CHAPTER 7

Drawing Matters

Andrea Kantrowitz

It feels as if we are living in uniquely precarious times. Of course, this has always been true for some of us. Today, our lives and our children's futures are being shaped by unprecedented fires, floods, pandemics, and political upheaval. Both human and non-human forces wreak havoc on our expectations for stability. We need to use all the prodigious capacities of the human imagination to meet these challenges. Yet, in trying times, the calm needed for creativity and innovation to flourish is difficult to come by. But we can seek out quiet moments and spaces where we open our eyes and hearts to encounters with unknown possibilities from which solutions might emerge. Drawing with paper and pencil or whatever is at hand, is a simple, accessible means to become more mindful, balance our inner and outer worlds, and nurture our imaginations. It provides a protected, sheltered place to reflect on and communicate our experiences, ideas, and observations. We set aside our worries and fears about the future and engage deeply with the present moment. We can draw ourselves into stillness. We record what we see or images and stories that emerge from our minds. In the virtual space of the drawing, we can

A. Kantrowitz (✉)
Department of Art, SUNY New Paltz, New Paltz, NY, USA
e-mail: kantrowa@newpaltz.edu

S. M. Wyman and J. Korenblat (eds.), *Sustainability Learning for Action and Community Engagement*,
https://doi.org/10.1007/978-3-032-11698-7_7

discover fresh perspectives and lines of inquiry that might help address the present crises that loom before us. Unforeseen possibilities and solutions can emerge. Drawing can help us visualize alternative futures.

Drawing is not just for artists; it's for everyone. Very few five-year-olds will say they can't draw. Pretty much everyone starts out drawing. But many adults will tell you that they can't even draw a straight line. Sure, when you are just trying to get from point A to point B as quickly as possible, such a mark can come in handy. But if you want to end up somewhere new, take the scenic route. Most young children scribble for the pure joy and pleasure of seeing the traces of their actions on paper. Sometimes, they invent a subject or story as they go along on their own or when prodded by adults. The young child knows, and the older child or adult sometimes must relearn that confusion and disorder can be opportunities for surprise and discovery. Messy lines are fun to make but sometimes lead you somewhere new. Drawing helps us harness the tendency of the human mind to perceive order, pattern, meaning, and agency in seemingly random events. Ambiguous rough sketches might suggest possibilities that couldn't be imagined without them (Fish and Scrivener 1990). As many children get older, they feel intimidated, lack confidence, and get frustrated that they can't hit the target, rendering an image that already exists in the world or in their minds. And K-12 schooling, at least in the United States, is all about hitting targets. We need to show, or maybe remind learners of any age, that drawing can be much more: it's about creating opportunities to observe, discover, or invent something never seen or imagined: a phenomenon or feeling that could not be seen or imagined any other way.

Drawing is well suited to leverage the complexity of human cognition, the sometimes fuzzy, perhaps messy, and ambiguous thoughts that can't easily be put into words (Goel 1995). This activity engages and coordinates all our senses and makes visible thoughts and perceptions that might otherwise slip past our conscious minds. And it is universally available. Even during the COVID-19 pandemic, when we were trying to figure out how to teach art online, most children had access to a pen or pencil and paper. Drawing leaves a trace of our movements to inspect, revise, and reconfigure those thoughts and perceptions in unforeseen and sometimes surprising ways. It makes thinking visible— when we are not ready or not able to put our thoughts into words, we can sometimes put them down in arrangements of lines and marks. It is about making connections: between

Fig. 7.1 *Drawing Hands* (2022) Andrea Kantrowitz

ideas and perceptions, between past and present, and between individuals (Fig. 7.1).

Drawing helps us take advantage of the ways we use our hands to think and communicate. The complexity of our brains and hands evolved in tandem. Our abilities to communicate symbolically through language and to make and use of tools arose from the same parts of our brain. Human hands are capable of an astounding range of deliberate and reflexive actions. According to the brain, when we draw, the pencil becomes a physical extension of the hand that holds it. The hand, pencil, and mind are one (Maravita and Iriki 2004). Hand gestures help us think and communicate in important ways (Goldin-Meadow 2005). Drawing integrates tool use with our abilities to communicate symbolically through gesture. When an experienced drawer holds a pencil, the tip of the pencil is mapped onto the area of the brain that controls the hand, as part of the body. As we move a pen or pencil across the drawing surface, we leave the traces of our gestures behind.

There is no basis in science for the idea that creative people are right-brained and logical people are left-brained (Wager et al 2003). Brain research has shown experienced drawers' brains are more connected and coordinated than others, especially when actively drawing (Bhattacharya and Petsche 2005). We use more of our brains, and our brain areas are more connected when we are engaged in creative activity, especially when challenged (Lindell 2011). Drawing connects our eyes, hands, and hearts: exteroception (awareness of the external environment), interoception (awareness of our internal bodily processes), and proprioception (awareness of our bodies in space.) The importance of these connections has been demonstrated recently in studies on neuroplasticity: the use of psychedelic drugs, psilocybin, and MDMA for treatment-resistant depression and post-traumatic stress disorder encourages the brain's capacity to reorganize and form new connections between neurons (De Vos et al. 2021).

Thought is action-oriented. Our human capacities for thought evolved out of the need to make sense of and take action in order to survive. Recent advances in cognitive and learning sciences demonstrate how we think and learn with all our senses, including the movement of our bodies in space, as they interact with our surroundings. If we understand thinking as arising from interactions between our bodies and the world, the medium where that happens is space. Spatial reasoning includes the ability to integrate different views of a single object and to imagine it from a variety of viewpoints. It also includes understanding relationships between objects and between multiple arrangements of items within a larger spatial context. The ability to imagine yourself in someone else's shoes, seeing the world from their perspective, involves spatial reasoning. This effort empowers learners of any age to engage with their environment and to think in generative, empathetic, potentially world-changing ways (Tversky 2019). It would be helpful for the development of spatial reasoning that students have frequent opportunities to create in three-dimensions in a variety of media. Because of its accessibility and immediacy, I focus on drawing as a means of developing spatial ability to improve communication, problem-solving, and imagining possibilities.

The blank page can be understood as a physical extension of our minds. We draw out two- and three-dimensional models of real and invented objects, actors, and scenes, pull them apart, and put them back together in new ways. Drawing is all about the relationships and movement of forms in space. We make marks in two dimensions on the drawing surface and play with them in the imaginary three-dimensional space of the drawing. In daily life, spatial analogies permeate our thinking so completely that we usually don't notice them. We sometimes say we need to "turn things over in our minds" when we want to analyze a subject thoroughly. We try to get "on top" of a situation, "cover" a topic, "uncover" the facts, or put something "to the side."

As with so many talents, some people seem naturally gifted, but everyone can improve in the area of spatial reasoning, which is linked with innovation and creativity across disciplines (Uttal et al. 2013). You use spatial reasoning when you try to shop efficiently at the supermarket, wrap a birthday present, pack the car for a long trip, or park the car once you get there. You also use this type of thinking to imagine (and sketch) a new layout of the furniture in your living room before you try picking up that heavy sofa. A sculptor might imagine the form of the final work in a hunk of marble before starting to chip away at it. A surgeon uses spatial reasoning to plan a repair to a torn knee ligament, explains her plan to the patient on a whiteboard, and does the surgery. High performance in spatial reasoning is crucial in most science, technology, engineering, and math (STEM) fields as well as in art and design.

Spatial reasoning is not adequately recognized or taught in K-12 education in the West. Children of higher socioeconomic status consistently surpassed their peers of lower socioeconomic status in spatial reasoning and verbal and standard math tests. Simple, one-time educational interventions, such as playing particular kinds of board games, have been shown to level the playing field for low-income children in spatial tasks, such as correctly placing the number seven on an unmarked number line (Booth and Siegler 2008; Siegler and Ramani 2009). Jonathan Wai and David Uttal, in their paper "Why Spatial Reasoning Matters for Education Policy" (2018), write,

> Spatial reasoning is linked to positive educational outcomes in STEM and is a strong predictor of success in the visual arts and vocational, manufacturing, and technical careers. Students with this ability are not being identified or developed effectively, and in fact seem to fall behind academically when their talents are not recognized. Consequently, society is missing out on innovation from a larger pool of spatially talented individuals who could be engineers and inventors if their abilities were more fully developed.

In their meta-analysis, Lakin and Wai found that unidentified yet spatially gifted students are even more likely to face academic challenges than those who are not spatially gifted (2020). But we can change the trajectory of a young person's life by appreciating and developing new ways of thinking through active engagement with the material world. New possibilities arise through the process of making. We can develop and enhance cognitive skills through drawing that are valuable within and beyond the art studio classroom. Visuo-spatial reasoning is important in exactly those domains critical to building alternative, sustainable futures and can be taught through drawing (Uttal et al. 2013; Wu and Rau 2019). Schools need earlier and more systematic "spatial talent identification and educational development," Lakin and Wai argue, because when students receive training and experience to develop spatial skills, research shows these abilities improve consistently (2020).

Males, on average, have a slight advantage over females in spatial reasoning, particularly mental rotation tasks, one of the only areas of human cognition where there is a statistical difference between male and female performance (Lauer et al. 2019). This gender gap emerges during the early elementary school years (after kindergarten) and grows larger through adolescence, which may suggest that environmental and sociocultural factors (rather than innate biological differences) play a significant role in driving the gender gap in spatial skills. The difference in spatial abilities may contribute to the gender disparity observed in some STEM disciplines, such as engineering and physics. Learning to draw real and imagined three-dimensional objects, characters, and worlds could help all children develop their spatial reasoning skills (Wu and Rau 2019). Culture does seem to be a mediating factor. Globally, the gender gap in mathematical reasoning (closely related to spatial reasoning) seems to correlate with measures of gender equity more broadly (Else-Quest et al. 2010). In the United States, gender differences in mathematical achievement are narrowing, perhaps as attitudes about these differences shift (Ramírez-Uclés and Ramírez-Uclés 2020).

Whether drawing changes the way your students think in a comprehensive and lasting way depends on the context and being explicit about what kinds of thinking you are aiming to teach through drawing. Most people do not transfer learning from one disciplinary domain to another unless they are explicitly instructed to do so (Perkins and Salomon 1992; Day and Goldstone 2012). Psychologists have found artists' perceptual advantages persist even when they are not drawing: drawing teaches us to see. This activity generally seems to improve mental focus and executive function, at least while we draw (Chamberlain et al. 2021). Studies suggest that drawing might also improve mental flexibility and creative idea generation in more durable ways (Drake et al. 2021).

We each have our own individual concerns, curiosities, and points of view. As we look out toward the horizon of possibility, we each see something different. Drawing helps us get to know our perspectives better. We become more aware of the limits of what we notice from where we sit. Where can the process of drawing itself take us? We can follow a suggestion, a squiggle, shadow, or smudge, and see where it leads. When we think through drawing, the drawing surface becomes a laboratory for our minds to generate and test ideas and perceptions. Heidi Edmundson, an emergency medicine consultant in the United Kingdom's National Health Service, conducted weekly wellness sessions for her medical team during the height of the pandemic. She writes, "Drawing often enables people to express emotions that are difficult to say. For some, drawing lets them acknowledge or accept feelings that they were unaware that they had. Drawing together also created a shared experience. It was a fun task and there was always a lot of laughter." For those on the front lines or during the long lockdowns, so much we had taken for granted was no longer available. But paper and pencils were still at hand, ready to help us process and share difficult experiences. They connected us when we felt disconnected; they provided an arena for introspection and communication, where we could learn about ourselves and share thoughts and feelings with others (BMJ 2020).

It would be wonderful to draw like Leonardo or even like the folks who draw Marvel Comics. But drawing benefits even people who don't draw like experts. In fact, those who DON'T have strong visuo-spatial skills to begin with seem to benefit more from drawing than those who DO have high preexisting skills and strong internal mental models. The process of externalizing thought helps to support those who need it most. Students who are not good visuo-spatial thinkers may benefit the most from drawing out their thoughts (Bobek and Tversky 2016). Through drawing, students can learn to correct innate perceptual distortions and develop the ability to notice, question, and direct their attention and internal processing. They build their capacities not only for spatial reasoning but also for self-regulation and metacognition. They might become better equipped to observe, discover, invent, and solve problems in the arts, sciences, and everyday life. Integrating sustainability education into art curricula can be an effective way to develop students' spatial reasoning abilities, environmental literacy, and engagement. This interdisciplinary approach holds promise for promoting educational equity across STEM and the arts.

Drawing to learn… may be more important than learning to draw. These days we have so many other ways to capture images that already exist. Digital photography, 3D drawing tools, and AI are as ubiquitous and handy as our cell phones; learning to render a preconceived scene or idea by hand may not justify all the time and hard work it requires. On the other hand, drawing to think and learn can be valuable. Even doodling may enhance listening and learning (Andrade 2010). If learners are confident that they have (or can figure out) the drawing tools and techniques they need for the task at hand, drawing can increase emotional engagement and motivation. Drawing offers everyone a way to discover that there is much more to observe than initially meets the eye. We come to realize that we know and understand less than we think we do, but at the same time, we gain access to ideas and perceptions otherwise hidden from our conscious minds. We are surprised by our own inventiveness. Through regular practice and reflection, we can develop habits of mind, heart, and eye that help us focus, slow down, and keep curiosity alive.

Drawing, as a practice, takes time and patience to develop. Over time, it becomes a habit that can help us be attentive and make meaning out of otherwise random, disconnected experiences. We learn to find beauty in unexpected places. We can use drawing as a tool of thought to enhance our abilities to observe, discover, and invent (Kantrowitz 2022). In the face of global pandemics and ecological disasters, everything we can do to cultivate and nurture human resilience, ingenuity, and understanding is of value. Our continued existence and perhaps all life on earth depend on how well we can think and work together. We must imagine and build a future world we all want to live in. Drawing together, metaphorically and literally, could play a part.

References

Andrade, Jackie. 2010. "What Does Doodling Do?" *Applied Cognitive Psychology: The Official Journal of the Society for Applied Research in Memory and Cognition* 24 (1): 100–106.

Bhattacharya, Joydeep, and Hellmuth Petsche. 2005. "Drawing on Mind's Canvas: Differences in Cortical Integration Patterns between Artists and Non-Artists". *Human Brain Mapping* 26 (1): 1–14.

BMJ. 2020. "Covid-19: How Drawing Helped Us to Deal with the Pandemic—the BMJ". *The BMJ*. August 6, 2020. https://blogs.bmj.com/bmj/2020/08/06/covid-19-how-drawing-helped-us-to-deal-with-the-pandemic/.

Bobek, Eliza, and Barbara Tversky. 2016. "Creating Visual Explanations Improves Learning". *Cognitive Research: Principles and Implications* 1 (1): 27.

Booth, J. L., and R. S. Siegler. 2008. "Numerical Magnitude Representations Influence Arithmetic Learning". *Child Development* 79: 1016–1031.

Chamberlain, Rebecca, Aaron Kozbelt, Jennifer E. Drake, and Johan Wagemans. 2021. "Learning to See by Learning to Draw: A Longitudinal Analysis of the Relationship Between Representational Drawing Training and Visuospatial Skill". *Psychology of Aesthetics, Creativity, and the Arts* 15 (1): 76–90.

Day, Samuel B., and Robert L. Goldstone. 2012. "The Import of Knowledge Export: Connecting Findings and Theories of Transfer of Learning". *Educational Psychologist* 47 (3): 153–176.

De Vos, Cato MH, Natasha L. Mason, and Kim PC Kuypers. 2021. "Psychedelics and Neuroplasticity: A Systematic Review Unraveling the Biological Underpinnings of Psychedelics". *Frontiers in Psychiatry* 12: 724606.

Drake, Jennifer E., Ariana Riccio, Rebecca Chamberlain, and Aaron Kozbelt. 2021. "Artists Have Superior Local and Global Processing Abilities but Show a Preference for Initially Drawing Globally". *Psychology of Aesthetics, Creativity, and the Arts*.

Goel, Vinod. 1995. *Sketches of Thought*. MIT Press.

Goldin-Meadow, Susan. 2005. *Hearing Gesture: How Our Hands Help Us Think*. Harvard University Press.

Else-Quest, Nicole M., Janet Shibley Hyde, and Marcia C. Linn. 2010. "Cross-National Patterns of Gender Differences in Mathematics: A Meta-Analysis". *Psychological Bulletin* 136 (1): 103–127.

Fish, Jonathan, and Stephen Scrivener. 1990. "Amplifying the Mind's Eye: Sketching and Visual Cognition". *Leonardo* 23 (1): 117–126.

Kantrowitz, Andrea. 2022. *Drawing Thought: How Drawing Helps Us Observe, Discover, and Invent*. Cambridge, Massachusetts: The MIT Press.

Lakin, Joni M., and Jonathan Wai. 2020. "Spatially Gifted, Academically Inconvenienced: Spatially Talented Students Experience Less Academic Engagement and More Behavioral Issues than Other Talented Students". *British Journal of Educational Psychology* 90 (4): 1015–1038.

Lauer, Jillian E., Eukyung Yhang, and Stella F. Lourenco. 2019. "The Development of Gender Differences in Spatial Reasoning: A Meta-Analytic Review". *Psychological Bulletin* 145 (6): 537.

Lindell, Annukka K. 2011. "Lateral Thinkers are not So Laterally Minded: Hemispheric Asymmetry, Interaction, and Creativity". *Laterality: Asymmetries of Body, Brain and Cognition* 16 (4): 479–498.

Maravita, Angelo, and Atsushi Iriki. 2004. "Tools for the body (schema)". Trends in cognitive sciences 8 (2): 79–86.

Perkins, David N., and Gavriel Salomon. 1992. "Transfer of Learning". *International Encyclopedia of Education* 2: 6452–6457.

Ramírez-Uclés, Isabel M., and Rafael Ramírez-Uclés. 2020. "Gender Differences in Visuospatial Abilities and Complex Mathematical Problem Solving". *Frontiers in Psychology* 11: 498272.

Siegler, Robert S., and Geetha B. Ramani. 2009. "Playing Linear Number Board Games—but Not Circular Ones—Improves Low-Income Preschoolers' Numerical Understanding". *Journal of Educational Psychology* 101 (3): 545.

Tversky, Barbara. 2019. *Mind in Motion: How Action Shapes Thought*. Hachette UK.

Uttal, David H., Nathaniel G. Meadow, Elizabeth Tipton, Linda L. Hand, Alison R. Alden, Christopher Warren, and Nora S. Newcombe. 2013."The Malleability of Spatial Skills: A Meta-Analysis of Training Studies". *Psychological Bulletin* 139 (2): 352–402.

Wager, Tor D., K. Luan Phan, Israel Liberzon, and Stephan F. Taylor. 2003. "Valence, Gender, and Lateralization of Functional Brain Anatomy in Emotion: A Meta-Analysis of Findings from Neuroimaging". *NeuroImage* 19 (3): 513–531.

Wai, Jonathan, and David H. Uttal. 2018. "*Why Spatial Reasoning Matters for Education Policy*". American Enterprise Institute.

Wu, Sally P. W., and Martina A. Rau. 2019. "How Students Learn Content in Science, Technology, Engineering, and Mathematics (STEM) Through Drawing Activities". *Educational Psychology Review* 31 (1): 87–120.

CHAPTER 8

Sustain Ability to Live With Earth

Tiokasin Ghosthorse

Please note that the word *Indian* does not originate from Christopher Columbus, who mistakenly believed he had arrived in "India." In 1492, India was not even called by that name; it was known as Hindustan. Furthermore, it is a misconception that Columbus discovered Indians. Indians, in fact, discovered Columbus lost and sick on the shores.

It is believed that Columbus coined the term *Indian* when he discovered the people of the land. Columbus, being an Italian and not fluent in Spanish, described the natives as *Una gente in Dios*, meaning "A people in God." Due to his pronunciation, the word *Indians* was derived, and it was considered a noble and respectable term.

I take umbrage when *Indians* is used derogatorily against Native peoples of the Western Hemisphere.

Here is a little story of how one man—Eduardo—learned about Western education.

Eduardo, a cultural anthropologist, wanted to arrive two weeks before a conference in a remote mountain to get a feel for the place and prepare

T. Ghosthorse (✉)
Itazipcola and Mnicoujou Lakota of the Sakowin Oyate & First Voices Radio, Lakota Nation, USA
e-mail: tiokasin@gmail.com

S. M. Wyman and J. Korenblat (eds.), *Sustainability Learning for Action and Community Engagement*,
https://doi.org/10.1007/978-3-032-11698-7_8

his thoughts on delivering a cultural paper to other academic scholars for possible publication.

Eduardo rented a car, carefully mapped a route to the remote village, and began to travel. But there was a breakdown, and the car rental agency said they could "send a mechanic in five days at the least due to the roads and scheduling." Eduardo thought this would indeed interrupt his preparation days and travel time. He asked the local people of the village if there was a car agency or someone he could hire to take him to the remote mountain. He found no one going in that direction, let alone anyone with a vehicle. They all walked!

Now, this really put a kink in his plans, but he had to get there as soon as possible or at least arrive a few days earlier than five. He would plan to work on his way there on his laptop computer or handwrite if need be. He was sure there was someone in the next village who had a car and began to walk without listening to the locals who said, "No one owns a car the rest of the way."

As Eduardo walked, he came upon an old woman who was going to the same conference but had started two weeks earlier from down the mountain with her dog and a bundle on her back. The old woman invited Eduardo to walk with her and said that she would show him the way. He couldn't take the chance to wait and see whether there was another car going in his direction on the main highway. Eduardo really was in a predicament and resisted before accepting that at least he might arrive a few days earlier and have some semblance of an intelligent paper to deliver with apologies.

Along the way, he often thought "if he could just get there ahead of time." So he would try to hurry the old woman and her dog along. Day after day, the old woman would walk from village to village and stop to eat and visit with the locals. Eduardo began to think he would never get to the conference on time and would ask every person for a car or at least someone younger who could walk faster than the old woman. Still there was no luck, no one who could do so, but for an elderly man who was going to visit relatives a few days later.

So now, Eduardo was in for a long trek to the conference. Still, he wished he had stayed at the place where his rental car failed, and at least he could have caught a ride and gotten there faster and saved time… saved time… saved time. He soon resigned himself, and his Western thoughts of *time is money and money is time*.

Day after day, he did the same thing, awakening while the old woman was doing a ceremony to greet the new day and give thanks, then eating as another ceremony, and even drinking a cup of water became a ceremony. Another ceremony for strength and guidance while walking. And giving thanks to all the trees, flowers, birds, sky, plants, and even the little ants along the way became a ceremony. At the end of each day, they would arrive at a different village and the same visiting, dancing, stories, eating, and ceremony.

He thought he would never get there. His computer battery had run out of power, and along the way, none of the villages had any electricity, so he threw that idea out and began to write down what he had learned at the university about culture. What frustrated Eduardo the most was that the old woman would hardly talk with him except to say "this way" as he followed, more doubtful than ever that they would get there. He had lost his watch somewhere along the way, had no idea what day or time it was, and had practically given up on the idea of ever presenting a quality paper if any. Then, one morning, after the same awakening ceremony and other ceremonies, the old woman said, "We are here," and they walked over the hill a day before the conference began.

Eduardo stood behind the podium and began to offer apologies and describe the circumstances of why the expected scholarly paper was not prepared. Then something happened, and he began to describe his journey along the way and what he had done for the last two weeks with the old woman. He began to understand that he learned more about culture than if he had taken the fast road in the rental car and had arrived two weeks earlier.

All of the songs, dances, food, language, ceremonies, and relationships with the people he unintentionally made along the way: even despite his textbook resistance to culture, he did not need a paper. He would be able to deliver something alive, someone living, people living reality and not using reality as he had learned through words on paper.

Eduardo now lives among the people. Eduardo now lives with the people.

Eduardo now lives.

CHAPTER 9

From (Teaching About) Existential Threats to (Teaching) Existential Resilience and Sustainability

Christine Wamsler

Invited to write on the theme of "Education for Learning and Community Engagement," this poem, "One Story," came to mind. I wrote it while reflecting on my experiences with education—going through Western modern society's educational system.

> As a child, I grew up with strong feelings
> A deep sense of who I am and what is important in life
> I felt one with the world.

The original version of the chapter has been revised. A correction to this chapter can be found at
https://doi.org/10.1007/978-3-032-11698-7_46

C. Wamsler (✉)
Lund University Centre for Sustainability Studies (LUCSUS), Lund University, Lund, Sweden
e-mail: christine.wamsler@lucsus.lu.se

S. M. Wyman and J. Korenblat (eds.), *Sustainability Learning for Action and Community Engagement*,
https://doi.org/10.1007/978-3-032-11698-7_9

Talking to the wind in autumn
Dancing with snowflakes in winter
Listening to flowering in spring
And jumping with grasshoppers in summer
Living and caring were one.

As I became older, culture took over
The stories I heard began to color my life.

I went to school
I learned that adults and experts know—not children
I learned that knowledge is external—not internal.

Kicked out of the world of wonders and wisdom
I started filling my brain with knowledge
To find a new place in the world.

I went through academia
Learned that the truth is complicated
Learned to express myself in complicated ways
Over time, I ceased to understand myself
Hiding under layers and layers of "othering" to feel protected.

At one point, my heart and body became rebellious
My voice refused to repeat others' stories
My thoughts refused to ignore my feelings:
Feelings of partiality, imperfection… of separation
I lost my voice.

Then I saw the separation in the world
How relentlessly we exhaust ourselves
Others and the planet around us
How we seek comfort in material distractions.

Consumerism, injustice, climate change
It's me. It's us. It's one story.

Searching for a new voice is scary
And a strong headwind whirls out there
But if you understand my silence, you may understand my words
Delving into stillness, you may hear my call.

I have been working in the field of sustainability for around 25 years now, in theory and practice. In simple terms, sustainability science, education, and practice are about finding solutions to complex, global challenges that threaten the existence of humanity, other creatures, and the planet, such as climate change, biodiversity loss, pollution, and global health pandemics.

These existential threats are generally framed as external to us, and therefore, we try to address them primarily with external, technical measures (Leichenko and O'Brien 2020). This is simply because the way we define problems automatically determines our responses to them.

The narrative is changing now, in that we understand that these threats are a reflection of an inner human crisis, a kind of alienation, an integral element of modern life forms (Berry 2003; Figueres and Rivett-Carnac 2020; Osberg et al. 2024; Rosa 2021; Stibbe 2015; Wamsler and Bristow 2022). Today's interconnected global threats result from our dominant narrative or paradigm of separation, which assumes our thinking mind is separate from our feelings and bodily emotions, that we are all separate from each other, that some humans are superior to others, and that we humans are separate from and superior to the rest of the natural world (Scott et al. 2021).

This situation manifests in our daily lives, where we relentlessly and increasingly exploit and exhaust ourselves, others, and the planet. The acceleration of our pace of life, increasing levels of stress, mental illness, and ecological catastrophe are examples of the same polycrisis (Rosa 2021; Wamsler and Bristow 2022). People have never been this anxious and mental distress is a key obstacle to human development (Helliwell et al. 2024). More people die because of suicide than war, terrorism, natural disasters, and murder together (Our World in Data 2022, 2023; UN 2023; Haines 2024; WHO 2023; Alves et al. 2024). In Europe, one-fifth of all animals are currently in danger of extinction (Davis 2023). 80% of insects are already gone (Davis 2023; Janicki et al. 2022). 90% of big fish are already gone (Christensen et al. 2014; The World Counts, n.d.).

This polycrisis has different names: a crisis of separation, a crisis of disconnection, a crisis of perception, a meaning crisis, a spiritual crisis, or an existential crisis (Clarke and Mcphie 2014; Ives et al. 2023; Vervaeke and dir 2020).

The question becomes, therefore, *how can we move toward a more caring and thus sustainable society*? This question requires us to ask and address existential questions in research, education, and praxis. In other words: questions about our identity, how we relate and are connected to ourselves, others, and the world.

Accordingly, existential resilience and sustainability also aim to address the inner human dimensions that underlie today's polycrisis, to ultimately increase individual, collective, and planetary wellbeing, regeneration, and flourishing. The human inner dimensions refer to people's individual and collective minds or mindsets, including their beliefs, values, worldviews, and associated inner qualities and capacities (cognitive, emotional, relational) (Wamsler 2020; Wamsler et al. 2020b; 2021). These mindsets are at the root of today's polycrisis because they tend to govern the underlying -isms in our society, such as individualism, dualism, materialism, consumerism, racism, extractivism, etc.

I believe that you, and all other educators, can see close connections to your own work in what I describe here. I would hope that you question and challenge our current understanding of identity and interconnection, of happiness, of living a good life and the ways we want to achieve it through our teaching and community engagement.

Today, we can see the outcome of our modern societies' dominant paradigms and associated values. Related priorities, beliefs, and values are reflected in our everyday lives, our culture, our social norms, education, and the institutional and political systems we have created (Osberg et al. 2024; Wamsler et al. 2022). This means that on the one hand, in modern societies, we are living in very good times: we are probably inhabiting the utopia our grandparents imagined, with modern healthcare, democracy, and human rights for many (Rosling et al. 2018). At the same time, the advancements that we enjoy today also have a downside, an unintended side effect: we are seeing ecological catastrophe, mental health crises, wealth inequality, and political turmoil at unprecedented levels (Suzman 2017).

Our universities and educational systems are part of this picture, a cause, a driver, and a place where resultant impacts become highly visible. We see an increasing number of students who have lost hope and belief in their own agency (Hickman et al. 2021; Ogunbode et al. 2022; Wamsler

2019; Wamsler et al. 2024; Wullenkord and Ojala 2023). We have systematically stripped away the role of inner qualities and capacities such as presence, self-reflection, care, and compassion, in education, institutional, and political systems (Wamsler and Bristow 2022).

If we trace the roots of today's sustainability crises through a culturally entrenched story of separation, we can clearly see the potential role of inner dimension capacities in creating transformation (Uhl 2003; Wamsler 2018; Wamsler et al. 2018; Wamsler and Bristow 2022). These affordances entail an exploration of how we relate–to ourselves, others and the world at large–in reflective, meaningful, ethical and responsible ways (Rifkin 2009). And this exploration nurtures and gives rise to what I call *active hope*, or a mature expression of hope, one that includes the aspiration to manifest our full potential, trust that this is possible, and an energy that is engaged in serving this unfolding (for individual, collective, and planetary regeneration and wellbeing).

Simply providing more information about today's polycrisis, more facts and figures, will not make any difference! In the preface to his book *About this Life: Journeys on the Threshold of Memory*, the late naturalist Barry Lopez (1999) advised the mother of a young, aspiring writer, "If she wishes to write well, she will have to become someone. She will have to discover her beliefs, and then speak to us from within those beliefs. If her prose doesn't come out of her belief, whatever that proves to be, she will only be passing along information, of which we are in no great need."

Similarly, in the context of education, we must discover our beliefs and explore how they relate to our (unsustainable) behaviors, mindsets, cultural norms, systems, and structures so we can adequately respond, resist or adjust if necessary. This requires transformational learning (Mezirow 2018) to adapt, understand, and, ultimately, address the root causes, symptoms, and impacts. Such learning involves practice-based, social, emotional, creative, and ethical components that can nurture transformational capacities, which can be clustered under the headings of awareness, insight, connection, purpose, and agency (Wamsler et al. 2021). Contemplative, nature- and arts-based approaches and storytelling provide powerful paths for transformational learning (Wamsler et al. 2022a). But nurturing certain inner capacities, is unfortunately not enough. Transformation can only be achieved if we systematically support integrated measures that link inner and outer dimensions of sustainability across all scales—at the individual, the collective, and the system levels (Ives et al. 2023; Wamsler 2018; Wamsler et al. 2021).

Measures at individual and collective levels involve linking "outer" approaches with activities that can help us tap into our inner potential, help us discover our personal thinking and internalized cultural messages of separation, superiority, and instrumentalization, develop alternative approaches, and create fields of change (Ives et al. 2023; Wamsler et al. 2021). This can take the form of creating related educational opportunities, working environments, networks, communities of practice, exhibitions, or festivals to balance our cognitive and empirical knowledge, for instance, of nature, with our pleasure in nature's beauty (Wamsler et al. 2022a). Empiricism is only vicious when it is deployed as the only way to understand and dominate the forces of nature or tell just one side of the story of nature's meaning (Brusslan 2020). Balancing our empirical knowledge is thus crucial.

Measures at the system level aim to mainstream and institutionalize the consideration of inner dimensions of transformation into existing institutional and political structures to foster more integrative approaches across all levels (Ives et al. 2023; Wamsler et al. 2021; Wamsler et al. 2020a). This entails, for instance, modifying organizations' vision statements, policies, regulations, and working structures, and their tools for project management, resource allocation, communication, monitoring, and evaluation. A concrete example would be to change communication structures and mechanisms by moving away from the common meta-messages that promote feelings of powerlessness and anxiety toward communicating a new theory of change that supports agency and hope and recognizes that inner development, collective and systems change are never separate, but are co-created.

From an ethical standpoint, it is crucial to highlight that the work across these three levels does not demand that we fix other people's views. This would turn people into objects to be changed rather than seeing them as agents of change (Bentz et al. 2022; Leichenko and O'Brien 2020). Instead, it is about creating integrative measures, spaces, structures, and conditions that can nurture a culture of inner development, mutual support, and engagement from a place of shared, universal values and interconnection to ultimately rattle unsustainable norms and systems (Ives et al. 2023; Sharma 2017; Wamsler et al. 2021.

It is thus vital to shift the conversation and undertake a critical analysis of the potential benefits of attending to inner dimensions and transformation. We must also consider their limits, what they are not, and how we can provide understanding and action that supports more integrative approaches that cut across the individual, collective, and system levels (see Earthrise Commons; earthrisecommons.org).

As a growing number of courses on sustainability and transformation are offered to diverse audiences in and outside academia, it is time to critically re-assess their potential to contribute to sustainability outcomes across different scales. Work conducted in the context of our Contemplative Sustainable Futures Program has addressed this gap. Related studies have looked into programs that aim to integrate inner and outer dimensions of change, nourish inner development to support culture and system change, and accelerate work toward the United Nations Sustainable Development Goals (SDGs).[1]

On the basis of our results, we identified the key elements that such education programs require to become vehicles for transformation across individual, collective, and system levels (Wamsler et al. 2023). We can steward this change by adapting and combining inner development methods in ways that support participants as they challenge current unsustainable social paradigms and understand how they can systematically co-integrate the human dimension of sustainability within existing cultures, mechanisms, and structures. These key elements are:

- **Context and understanding**: how we see the world

  Provide a comprehensive understanding of the entangled nature of today's sustainability crises in a complex. We need to understand that today's sustainability problems are deeply connected and always changing, and that we are part of those systems. This means examining how our thinking and our actions affect sustainability. (Figs. 9.1 and 2). Related transformative capacities of educators and learners are, for instance, interdependency, complexity, and systems thinking.
- **Learning approach**: how we learn

  Cultivate transformative spaces and integrative methods for exploring inner dimensions and nurturing transformative capacities on a continuous basis. Related transformative capacities of educators and learners are, for instance, presence, self-reflection, perspective-taking, and compassion.

[1] For an overview of related studies, please see: www.contemplative-sustainable-futures.com/general-3-1

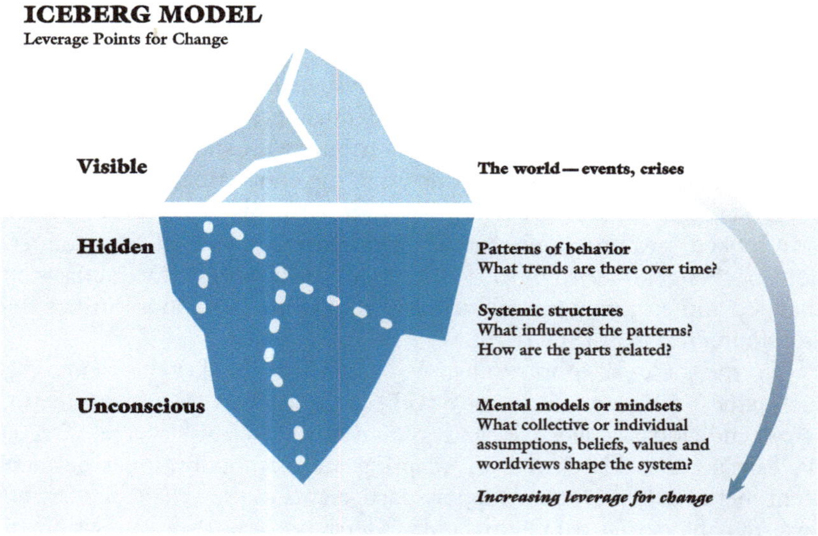

Fig. 9.1 Iceberg model. Adapted from Meadows (1999). Iceberg credit: CC-BY 3.0, Created by Michael Appleford, the Noun Project

- **Practical guidance and solutions**: how we engage

  Give practical guidance on how to design and implement measures that link inner and outer dimensions across individual, cultural, and systems change. Related transformational capacities of educators and learners are, for instance, emotional regulation, sense of agency, co-creation, creativity, active hope, and courage to act.
- **Quality assurance: how we ensure quality and ethics considerations across all aspects**

  Ensure quality education through the explicit consideration of ethics, the role of facilitators, and adequate monitoring and evaluation. Related transformational capacities of educators and learners are, for instance, humility, intrinsic value orientation, and equitable thinking (Fig. 9.3).

Together, these four essential pieces, or key ingredients, support individual, collective, and by extension, planetary wellbeing and flourishing by

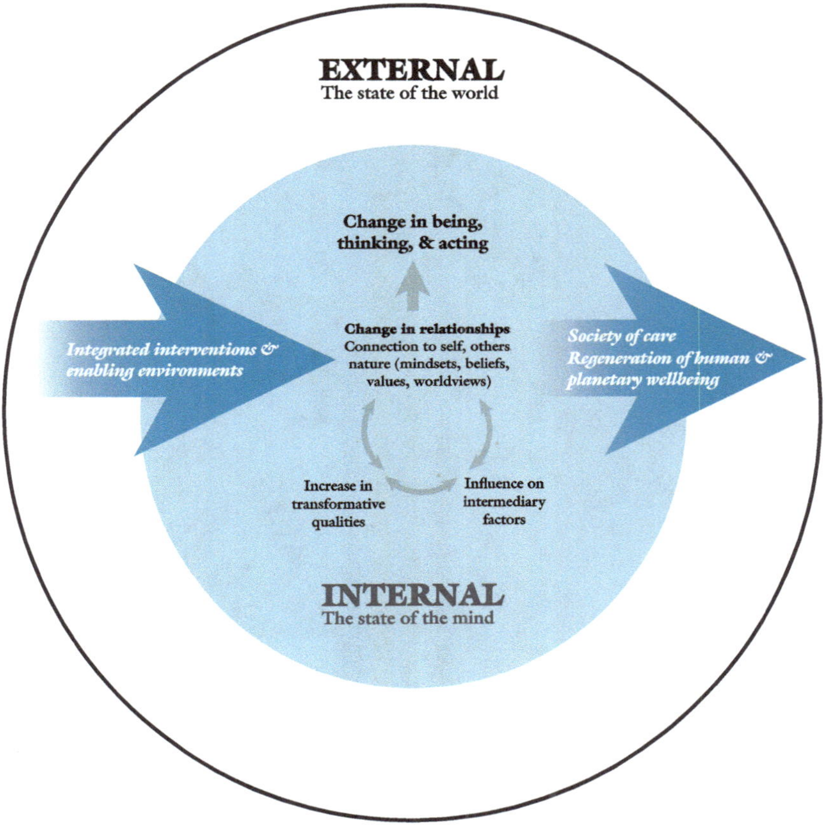

Fig. 9.2 Inner-outer transformation model. Adapted from Wamsler et al. (2021)

covering all key aspects of inner-outer transformation (ontology, epistemology, praxis, ethics). Self-awareness, personal growth, and the capacity for love and service on behalf of the common good become the ultimate goals (Uhl 2011, Fig. 9.4).

Applying and, at the same time, advancing knowledge on the complex intersection between sustainability, inner development, and transformation must become an integral part of our education and related cultures, mechanisms, and structures. As scholars, researchers, and educators, we have a choice of how we position our work with respect to dominant,

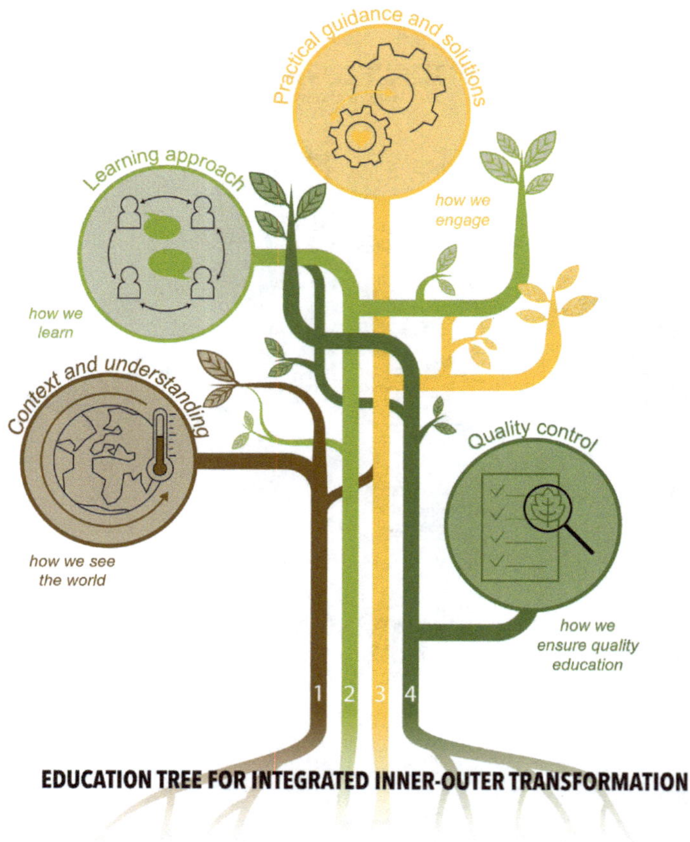

Fig. 9.3 Educational tree: four key ingredients for holistic learning and understanding to accelerate sustainability transformation. Adapted from Wamsler et al. (2023)

unsustainable institutional norms, growing resistance to these norms, and more integrative approaches—even if this means throwing ourselves into the humility of Not Knowing.

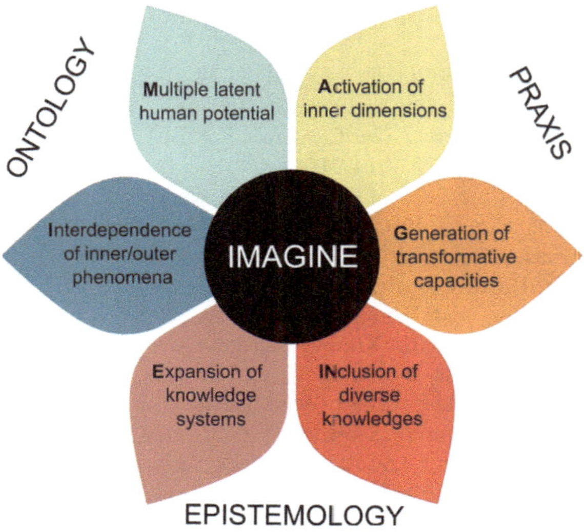

Fig. 9.4 Imagine framework: key contributions of the emerging field of inner-outer transformation for sustainability. Adapted from Ives et al. (2023)

References

Alves, Flávia Jôse, Erika Fialho Oliveira, Azevedo Paiva Jacyra, John A. de Araújo, Maurício L. Naslund, Vikram Patel Barreto, and Borges Machado Daiane. 2024. The rising trends of self-harm in Brazil: An ecological analysis of notifications, hospitalisations, and mortality between 2011 and 2022. *The Lancet Regional Health–Americas* 31. https://doi.org/10.1016/j.lana.2024.100691.

Bentz, Julia, Karen O'Brien, and Morgan Scoville-Simonds. 2022. Beyond "blah blah blah": Exploring the "how" of transformation. *Sustainability Science* 17 (2): 497–506. https://doi.org/10.1007/s11625-022-01123-0.

Berry, Thomas. 2003. The new story. In *Teilhard in the 21st centrury: The emergin spirit of earth*, 77–88. Arthur Fabel and Donal St. John.

Brusslan, Elizabeth Millán. 2020. Learning to see nature, learning to love nature: lessons from Alexander von Humboldt. *Center for Humans and Nature* (blog). 9 March 2020. https://humansandnature.org/learning-to-see-nature-learning-to-love-nature-lessons-from-alexander-von-humboldt/.

Christensen, Villy, Marta Coll, Chiara Piroddi, Jeroen Steenbeek, Joe Buszowski, and Daniel Pauly. 2014. A century of fish biomass decline in the ocean.

Marine Ecology Progress Series 512:155–166. https://doi.org/10.3354/meps10946.

Clarke, David A. G., and Jamie Mcphie. 2014. Becoming animate in education: Immanent materiality and outdoor learning for sustainability. *Journal of Adventure Education and Outdoor Learning* 14 (3): 198–216. https://doi.org/10.1080/14729679.2014.919866.

Davis, Josh. 2023. One Fifth of All Species in Europe Threatened with Extinction. *Natural History Museum London*, 9 November 2023. https://www.nhm.ac.uk/discover/news/2023/november/one-fifth-of-all-species-in-europe-threatened-with-extinction.html.

Figueres, Christiana, and Tom Rivett-Carnac. 2020. *The future we choose: Surviving the climate crisis*. Knopf Publishing Group.

Haines, Julia. 2024. The Deadliest conflicts across the globe in 2023. *U.S. News*, 18 January 2024. http://www.usnews.com/news/best-countries/articles/countries-with-the-deadliest-conflicts-of-2023.

Helliwell, John F., Richard Layard, Jeffrey D. Sachs, Jan-Emmanuel De Neve, Lara B. Aknin, and Shun Wang. 2024. *World happiness report 2024*. University of Oxford: Wellbeing Research Centre.

Hickman, Caroline, Elizabeth Marks, Panu Pihkala, R. Susan Clayton, Eric Lewandowski, Elouise E. Mayall, Britt Wray, Catriona Mellor, and Lise van Susteren. 2021. Climate anxiety in children and young people and their beliefs about government responses to climate change: A global survey. *The Lancet Planetary Health* 5 (12): e863–e873. https://doi.org/10.1016/S2542-5196(21)00278-3.

Ives, Christopher D., Niko Schäpke, Christoph Woiwode, and Christine Wamsler. 2023. IMAGINE sustainability: Integrated inner-outer transformation in research, education and practice. *Sustainability Science* 18 (6): 2777–2786. https://doi.org/10.1007/s11625-023-01368-3.

Janicki, Julia, Simon Scarr, and Jitesh Chowdhury. 2022. Insect populations are declining at an unprecedented rate. *Reuters*, 6 December 2022. https://graphics.reuters.com/GLOBAL-ENVIRONMENT/INSECT-APOCALYPSE/egpbykdxjvq/

Leichenko, Robert, and Karen O'Brien. 2020. *Climate and society: Transforming the future*. Wiley.

Lopez, Barry. 1999. *About this life: Journeys on the threshold of memories*. Vintage Books.

Meadows, Donella. 1999. *Leverage points: Places to intervene in a system. The sustainability institute papers*. Hartland: The Sustainability Institute.

Mezirow, Jack. 2018. Transformative learning theory. In *Contemporary theories of learning: Learning theorists … in their own words*, ed. Knud Illeris, 2nd ed., 114–128. London: Routledge.

Ogunbode, Charles A., Rouven Doran, Daniel Hanss, Maria Ojala, Katariina Salmela-Aro, Karlijn L. van den Broek, Navjot Bhullar, et al. 2022. Climate anxiety, wellbeing and pro-environmental action: Correlates of negative emotional responses to climate change in 32 countries. *Journal of Environmental Psychology* 84:101887. https://doi.org/10.1016/j.jenvp.2022.101887.

Osberg, Gustav, Mine Islar, and Christine Wamsler. 2024. Toward a post-carbon society: Supporting agency for collaborative climate action. *Ecology and Society* 29 (1): art16. https://doi.org/10.5751/ES-14619-290116.

Our World in Data. 2022. *Natural disasters*. Our World in Data. https://ourworldindata.org/natural-disasters.

Our World in Data. 2023. *Terrorism*. Our World in Data. https://ourworldindata.org/terrorism.

Rifkin, Jeremy. 2009. *The empathic civilization: The race to global consciousness in a world in crisis*. New York: J.P. Tarcher.

Rosa, Hartmut. 2021. *Resonance: A sociology of our relationship to the world*. Polity Press.

Rosling, Hans, Ola Rosling, and Anna Rosling Rönnlund. 2018. *Factfulness: Ten reasons we're wrong about the world—And why things are better than you think*. Hodder Stoughton.

Scott, Brian, Elise Amel, Susan Koger, and Christie Manning. 2021. *Psychology for sustainability*. 5th ed. Routledge.

Sharma, Monica. 2017. *Radical transformational leadership: Strategic action for change agents*. Berkeley: North Atlantic Books.

Stibbe, Arran. 2015. *Ecolinguistics: Language, ecology and the stories we live by*. Routledge.

Suzman, James. 2017. What the kalahari Bushmen can teach us about a 15 hour work week. *Quartz*, 14 December 2017. https://qz.com/work/1156937/what-hunter-gatherers-understand-about-ai-automation-and-our-economic-future.

The World Counts. n.d. *Overfishing statistics*. https://www.theworldcounts.com/challenges/planet-earth/oceans/overfishing-statistics. Accessed 26 March 2024.

Uhl, Christopher. 2003. *Developing ecological consciousness: Path to a sustainable world*. Lanham, Md: Rowman & Littlefield Publishers.

Uhl, Christopher. 2011. *Teaching as if life matters: The promise of a new education culture*. John Hopkins University Press.

UN. 2023. Global homicides hit record high in 2021 as post-lockdown stress set in. *United Nations News*, 8 December 2023. https://news.un.org/en/story/2023/12/1144392.

Vervaeke, John, dir. 2020. *Awakening from the meaning crisis [Lecture Series]*. https://johnvervaeke.com/series/awakening-from-the-meaning-crisis/.

Wamsler, Christine, Björn Wickenberg, Helena Hanson, Johanna Alkan Olsson, Sanna Stålhammar, Helena Björn, and Hjalmar Falck et al. 2020a. Environmental and climate policy integration: Targeted strategies for overcoming barriers to nature-based solutions and climate change adaptation. *Journal of Cleaner Production* 247 (February): 119154. https://doi.org/10.1016/j.jclepro.2019.119154.

Wamsler, Christine. 2018. Mind the gap: The role of mindfulness in adapting to increasing risk and climate change. *Sustainability Science* 13 (4): 1121–1135. https://doi.org/10.1007/s11625-017-0524-3.

Wamsler, Christine. 2019. Contemplative sustainable futures: The role of individual inner dimensions and transformation in sustainability research and education. In *Sustainability and the humanities*, ed. Walter Leal Filho and Adriana Consorte McCrea, 359–373. Cham: Springer International Publishing. https://doi.org/10.1007/978-3-319-95336-6_20.

Wamsler, Christine. 2020. Education for sustainability: Fostering a more conscious society and transformation towards sustainability. *International Journal of Sustainability in Higher Education* 21 (1): 112–130. https://doi.org/10.1108/IJSHE-04-2019-0152.

Wamsler, Christine, and Jamie Bristow. 2022. At the intersection of mind and climate change: Integrating inner dimensions of climate change into policymaking and practice. *Climatic Change* 173 (1): 7. https://doi.org/10.1007/s10584-022-03398-9.

Wamsler, Christine, Laureline Simon, Gwen Ducros, Gustav Osberg. 2024. *Transformative climate resilience education for children and youth: From climate anxiety to resilience, creativity and regeneration,* literature review conducted for the ERASMUS+ project 2023–1-SE01-KA220-SCH-000158705. Lund University. http://lucris.lub.lu.se/ws/portalfiles/portal/197488523/A2.Literature_Review.Framework.FINAL.10.10.2024.pdf

Wamsler, Christine, Jamie Bristow, Kira Cooper, Gretchecn Steidle, Sara Taggart, Lene Søvold, Jessica Bockler, and Tom Oliver. 2022a. *Theoretical foundations report: Research and evidence for the potential of consciousness approaches and practices to unlock sustainability and systems transformation.* Report of the UNDP Conscious Food Systems Alliance (CoFSA), United Nations Development Programme UNDP. https://www.contemplative-sustainable-futures.com/_files/ugd/4cc31e_143f3bc24f2c43ad94316cd50fbb8e4a.pdf.

Wamsler, Christine, Johannes Brossmann, Heidi Hendersson, Rakel Kristjansdottir, Colin McDonald, and Phil Scarampi. 2018. Mindfulness in sustainability science, practice, and teaching. *Sustainability Science* 13 (1): 143–162. https://doi.org/10.1007/s11625-017-0428-2.

Wamsler, Christine, Gustav Osberg, Jeroen Janss, and Liane Stephan. 2023. Revolutionising sustainability leadership and education: Addressing the human dimension to support flourishing, culture and system transformation. *Climatic Change* 177 (1): 4. https://doi.org/10.1007/s10584-023-03636-8.

Wamsler, Christine, Gustav Osberg, Walter Osika, Heidi Hendersson, and Luis Mundaca. 2021. Linking internal and external transformation for sustainability and climate action: Towards a new research and policy agenda. *Global Environmental Change* 71:102373. https://doi.org/10.1016/j.gloenvcha.2021.102373.

Wamsler, Christine, Gustav Osberg, Anna Panagiotou, Beth Smith, Peter Stanbridge, Walter Osika, and Luis Mundaca. 2022a. Meaning-making in a context of climate change: Supporting agency and political engagement. *Climate Policy* 23 (7): 1–16. https://doi.org/10.1080/14693062.2022.2121254.

Wamsler, Christine, Niko Schäpke, Carolin Fraude, Dorota Stasiak, Thomas Bruhn, Mark Lawrence, Heike Schroeder, and Luis Mundaca. 2020b. Enabling new mindsets and transformative skills for negotiating and activating climate action: Lessons from UNFCCC conferences of the parties. *Environmental Science & Policy* 112:227–235. https://doi.org/10.1016/j.envsci.2020.06.005.

WHO. 2023. *Suicide*. https://www.who.int/news-room/fact-sheets/detail/suicide.

Wullenkord, Marlis C., and Maria Ojala. 2023. Climate-change worry among two cohorts of late adolescents: Exploring macro and micro worries, coping, and relations to climate engagement, pessimism, and well-being. *Journal of Environmental Psychology* 90:102093. https://doi.org/10.1016/j.jenvp.2023.102093.

CHAPTER 10

Teaching the Wild Within and Around Us

Joshua Korenblat

Introduction: The Fox and the Little Prince

One summer evening over dinner with a friend, I tried to explain a personal, emergent sustainability teaching philosophy. This approach to learning combines careful, hands-on observation of nature with the abstract concepts and data typically presented in the classroom. I recognized that my students and I needed a more lively relationship—even a dialogue and phenomenal engagement—with the sustainability-related source material in our visual communication design courses: *Designed World, Data Visualization*, and *Information Design*. As scholar Dalia Nassar puts it: "We know about climate change, mass extinction, pollution, the rising oceans—but we don't seem to act. Or better: we know but we don't really know. We know through graphs and numbers, but we don't know in an embodied, visceral way. We know only intellectually, not as whole, embodied beings. This reveals that there is a problem with our way of knowing" (Chao n.d.).

J. Korenblat (✉)
Department of Design, State University of New York, New Paltz, NY, USA
e-mail: korenblj@newpaltz.edu

S. M. Wyman and J. Korenblat (eds.), *Sustainability Learning for Action and Community Engagement*,
https://doi.org/10.1007/978-3-032-11698-7_10

Ruminating on the truth in her words, I found myself thinking back to my early career in the Art Department at *National Geographic Magazine*, where we illustrated topics that couldn't be photographed, such as paleontology, natural history, archeology, and other scientific concepts. Rather than relying only on abstract charts and maps, we created artistic portraits. Over six years as a junior staff member right out of college, I contributed in small ways to projects photo editors assigned to the Art Department to illustrate stories: dinosaurs and pterosaurs; the evolution of whales; the collectivist societies of ants; a portrait of King Tut; the broken-legged boy pharoah behind the golden mask and cobra-vulture crown; Xibalba, the Mayan Underworld accessed through cenotes and caves; climate change.... We dove into subjects like these, beyond the reach of photography alone.

Even knowing these roots, my teaching philosophy still felt fuzzy, burgeoning, and hard to articulate as I munched falafel with my friend under the restaurant's bright artificial light. "Oh," the artist, educator, and cultural sustainability ambassador Ibrahim Khazzaka said, "like the fox and the Little Prince!"

Aha, I thought, returning to one of my favorite books. In Antoine de Saint-Exupéry's classic children's story *The Little Prince*, a wise fox teaches a life-changing lesson about the importance of attention and care. Saint-Exupéry drew the source material for this lesson from his time as a professional airplane pilot. In his lyrical memoir *Wind, Sand and Stars* (originally published in 1939), Saint-Exupéry recounts how, in 1935, while chasing a speed record from Paris to Saigon, he and his mechanic, André Prévot, crashed their Simoun aircraft in the Libyan desert after heavy clouds forced them to a perilous altitude. For five days, they survived on grapes, crackers, and wine, hallucinating and dehydrated. Navigating the vast dunes on foot, he and his mechanic encountered some mysterious three-toed tracks in the scorching sand. A fennec fox. This curious creature, native to North Africa, is the smallest fox in the world—the size of a rabbit—and has the largest ears of any canid relative to their size. On a practical level, the tracks suggested the presence of life-sustaining water. On a spiritual level, the discovery of another living creature, his "little friend" whom he deliciously imagined "trotting blithely along at dawn and licking the dew off rocks" (Saint-Exupéry 2002, 118), helped Saint-Exupéry and Prévot overcome an overpowering sense of desert isolation until they met a Bedouin who rescued them.

Botanist Robin Wall Kimmerer might call Saint Exupéry's marooned yearning for other signs of life in the desert a type of "species loneliness"—"a deep, unnamed sadness stemming from estrangement from the rest of Creation, from the loss of relationship" (Kimmerer 2015, 208).

We don't need to be stranded in the Sahara to feel a sense of species loneliness. Walk through a suburban neighborhood where patchwork lawns replace native forests, meadows, and other richly variegated ecologies, where we can easily identify brands but not the leaves and trees all around us. Here and in many other settings, humans have only a fragmentary experience with their surroundings and lack meaningful encounters with the wild—both outside and within—the wild, the places beyond our agency and control. Drawing from her Potawatomi heritage and the traditions of Western New York State's indigenous peoples, Kimmerer laments this separation from the more-than-human world (Abram 1997): "As our human dominance of the world has grown, we have become more isolated, more lonely when we can no longer call out to our neighbors" (Kimmerer 2015, 208).

For Saint-Exupéry, the presence of the fennec fox enlivened the otherwise desolate landscape and the inner landscape of his imagination, inspiring the famous relationship that would become the one between the Little Prince and the fox. In his Saharan encounters with these creatures, he had discovered what would become a central theme of *The Little Prince:* meaningful bonds can transform the most traumatized and solitary landscapes into sacred places of beauty, connection, and wonder.

In the Sahara Desert of Saint-Exupéry's fictional world, the pilot of a crashed plane meets the Little Prince, an enigmatic, otherworldly boy who shares stories of his interplanetary travels. One such story describes the moment he became disillusioned upon discovering thousands of roses seemingly identical to the one rose he tended to on his home planet. Sadly, his special rose was now just an ordinary one. Then, a wise fox appears and soothes him like a teacher and companion: "If you tame me," the fox says, "then we shall need each other. To me, you will be unique in all the world" (Saint-Exupéry 2000, 61). The Little Prince realizes his single rose is special after all because of the care he took tending to it; such a relationship between humans and other animals is a process called *taming*.

For the fox, taming requires patient observation and attention to create a deep understanding and bond. Indeed, the English translation of "to tame" misses the true meaning Saint-Exupéry sought to express in the original French. While *dompter* and *apprivoiser* mean "to tame," *dompter* refers to subduing a strong force, whereas *apprivoiser* suggests a gentler, patient approach of becoming acquainted with an animal. Saint-Exupéry chose *apprivoiser*.

The themes and conversational storytelling style of *The Little Prince* share elements with folklore. Each art form pursues truth through understanding reality and inquiring into what makes a life ethical and well-lived. The German word for folklore, *Wundermärchen*, translates as "Wonder Tale." And wonder, though seemingly magical, is actually quite simple: it sustains our engagement with the world while renewing it. As philosopher Abigail Tulenko writes, folklore "preserves ideas *softly*. It's putting flowers in a vase rather than drying them—watery narrative is always moving, always in flux, so the ideas stay green and do not become brittle" (Tulenko 2024).

The fox and the Little Prince speak in this conversational, folkloric manner to me. The fox elaborates that taming creates an intimacy between creatures: "I shall know the sound of a step," presumably the Little Prince's, "that will be different from all the others" (Saint-Exupéry 2000, 62). Now consider my small dog, a mini-golden doodle, who barks at strangers but knows I'm present, even without seeing me. Hearing me approach, recognizing my inimitable footfall on the floorboards, sighing, he perks his ears and rolls over to have his belly softly scratched.

"The tame animal is perhaps the most natural of all," muses writer and animal welfare advocate Matthew Scully, "displaying qualities that human kindness can elicit" (Scully 2002, chap. 3, para. 101). The wild is everything in nature beyond human control, but here, taming that wildness is a dynamic not of power, but kinship. A special kind of kinship that relies upon human kindness, and more. As the Little Prince and the rose foreshadow, and the fox notes, we are forever responsible for what we come to know well. Through study, careful listening, and sustained imaginative engagement, we discover what makes each being special, even wonderful, in all the world.

We also learn about our humanity by becoming patiently acquainted with other animals through an imaginative engagement that includes storytelling and art, just as Saint-Exupéry did in *The Little Prince*. Scholar Tim Morton goes further, moving beyond human/non-human and tame/wild distinctions to embrace our embodied entanglement with other beings in ordinary, liminal places. These human-modified spaces lack dramatic appeal and are rarely considered worthy of conservation, yet

Morton sees them as sites of strange, meaningful, and beguiling ecological relationships (Morton 2010, 273).

How do we teach this kind of experienced and embodied knowing to our students when people rarely model such relationships with other animals beyond their kinship with family dogs and cats? After all, we're living through what scientists call the Sixth Great Extinction, due to human activity that sees us as separate from the rest of nature.

Conservation efforts often seek to codify regulations and enforce compliance, a fear-based approach to enlisting support for animal welfare. While advocacy for such efforts requires emotion to animate and amplify them, including fear, compassion, and outrage, the most renewing emotion of all often remains dormant: Wonder.

Wonder offers unique advantages as a guide to what we value in our educational objectives, especially for creating a deep sense of ecological awareness. Wonder helps lead us out of narrow self-interest toward the original joy of life itself, "like an infant cast into the shores of light" (Nussbaum 2001, 189). This capacity to transcend self-interest makes wonder expansive and replenishing compared to other emotions harnessed in environmental education, conservation efforts, and advocacy work for peace and justice. Unfortunately, compassion has limited tolerance for otherness and distance; it's proven to be like a flame that can only light a small circle around it. Outrage is like a firebrand that might feel righteous but also has the capacity for volatility and burnout, and a dark side that can lead to retributive anger rather than what philosophy professor Martha Nussbaum calls "Transition-Anger" for improving the world (Nussbaum 2023, loc. 9 of 372).

Wonder, meanwhile, provides a renewable source of energy for change. Wonder complements traditional reason with imaginative thinking and embodied learning, inviting us to suspend immediate reactions and see beyond our preconceived notions. Through sensory engagement with our surroundings and narrative imagination, a student who experiences wonder can no longer feel indifferent and alienated from their learning material; instead, they experience an intimacy that's not possible in a fact-based education model alone.

These transformative qualities of wonder remind me of the educational vision of Nobel Laureate Rabindranath Tagore, the Bengali polymath who advocated for learning in *tapovan*, or forest schools outdoors, that preserve "the perfect symphony between one's life and the world outside" and help students "realize at every step how our training and knowledge have an organic connection with our surroundings" (Thriveni 2018). Wonder may be the most persuasive emotion available to educators seeking to guide students toward this model of ecological education, which focuses on our interplay with the world around us.

To cultivate wonder in practice, we can make our conservation-focused and related classes more place-based, as Tagore modeled in the outdoor classrooms of his school in northeast India. This means getting to know the animals around us as individuals, not just as representatives of their species as they're portrayed in textbooks. We also need to attempt to understand the worldviews and lifeworlds of these beings, as naturalist Jacob von Uexküll called them, their *umwelt*. However, we can't focus solely on animals and other beings in isolation. We need to situate them within their ecosystems, seeing their role as homemakers, not just habitat dwellers. This intimate approach to understanding individual creatures opens pathways to a broader understanding that braids together science, the study of nature; the humanities, the study of human experience; and art and design, which interleave all these fields. As Scully analogizes, in getting to know one animal, we can wonder over many: "In a single drop of rain can be seen the colors of the sun" (Namier, as quoted by Scully 2002, chap. 8, para. 146).

Getting to know one animal in an ecosystem is like observing a single raindrop in a rainbow. Just as a single raindrop refracts all the colors of sunlight within itself but only reveals one color to our limited human eye, one animal embodies many roles and connections within an environment, yet only reveals a part of the ecosystem's complexity when studied alone. To truly understand the whole *rainbow*, we need to see how animals, plants, and other organisms interact. Each species, like each raindrop, contributes a part to the whole; the true beauty and function of the ecosystem emerge only when considering all the parts working together. Though in 1820, the poet John Keats lamented the Enlightenment era's focus on rationality and the scientific precision of prisms, by unweaving a rainbow, we can reach a form of narrative understanding that at once enlivens our imagination and adds to our archives of knowledge (Shermer 2005).

For example, in rainforested Costa Rica, naturalist Craig Holdrege observes how sloths embody the rainforest's dappled qualities of constancy and slowness, warmth and coolness. The forest's stable conditions are evident in the sloth's slow movements. In her fluctuating body temperatures, she embodies the forest's sunlit and rainy ambience of heat and coolness. The sloth, whose algae-encrusted fur blends into the green leaves she munches upon, is a microcosm of the rainforest ecosystem. She carries a languid yet lively habitat for other organisms in her fur, and she contributes to the forest's nutrient loops through the gradual decomposition of her waste, left behind on the forest floor after a once-a-week trip down from the trees. In this way, the sloth is the rainforest; her beings and doings can only happen there (Holdrege, n.d.).

This chapter reminds educators that each of us is part of an ecosystem comprised of human and more-than-human worlds. After all, despite human cultural designs and messages that can induce a sense of species loneliness, humans make up only 0.01% of the biomass of life on Earth (Ritchie 2019). We can teach students to understand themselves not as separate from other people and nature, but connected through their relationships with others, their environments, and the non-humans among us whom they study and imagine. All of these entities, those perceived and beyond perception, co-create our world and shape who we are.

Drawing upon the fox's lesson in *The Little Prince*, I propose an accessible pedagogical approach called ***nature journaling***. Regardless of the subject area or grade level, nature journaling is an educational form of stewardship that relies on attention, curiosity, creativity, and, yes, wonder. Nature journaling can become a meaningful collaboration between knowledge seekers and the lifeworlds they study. It's a go-to activity for learning about and taking action on sustainability, one that promotes community engagement. And nature journaling helps us see that nature is not somewhere else; nature is here, at our fingertips, in our minds and hearts, and even in the mosaic of landscapes all around us in the designed world.

We Need a New Story for Teaching Conservation, Biodiversity, and Animal Welfare

If fear, compassion, and outrage inadequately guide future change agents toward pro-biodiversity and conservation decision-making and goals, ***wonder*** might offer a more effective path. Indeed, research indicates that positive emotional engagement motivates conservation-minded ecological practices more effectively than abstract ethical arguments or a pure reliance on scientific information (de Lange et al., 2022, p. 1). Nussbaum offers a more effective alternative to compassion or outrage-based appeals to preventing animal suffering and the myriad injustices committed against the more-than-human world. Drawing upon the Capabilities Approach, an economic framework first built by Nobel Laureate Amartya Sen and then further shaped into a social justice framework by Nussbaum, she asks: *What does each species need to flourish according to their nature?* First, they need conditions that meet a minimum threshold for the pursuit of life, liberty, and happiness. For a wolf, this might mean living in a pack and having territory to roam. For a sea turtle, this includes the ancient capability to migrate and build beachside nests. And for their thumb-sized hatchlings, the opportunity to waddle back to the moonlit sea without the disorienting interference of nearby artificial lighting.

This approach transforms conservation education. Instead of inducing students to feel sorry or angry on behalf of animals that have long been overlooked or exploited in human plans, we can invite students to understand what diverse forms of life require to live in fulfillment of their needs, and to live well, and then work to make those conditions accessible. Here, cultivating wonder requires not just a set of questions about what other creatures want to be and do, but also an attendant set of observations.

We also need a fundamental shift in how we talk about animals because discourses shape what we think about in the first place. In ecological discourse, animals are often reduced to resources that live in "habitats," not homes. Questions of human utility and economic development dominate this resource-focused discourse. The United Nations Global Goals exemplify this mindset, frequently referring to animals as mere resources (Stibbe 2007, 2). Meanwhile, animal rights discourse often portrays animals as passive victims rather than beings that actively strive to fulfill their purposes, feeling what is good or bad for them, just as we do (Weber 2016, Introduction, para. 3).

Some hope emerges in conservation education, where documentaries like *Planet Earth* and *March of the Penguins* offer intimate views of individual animal lives that inspire wonder and beckon us to take action for conservation. Yet even here, a troubling pattern persists: media focused on conservation or stewardship often privilege species that already appeal to us, such as the cuddly or rare (Stibbe, 2007, p. 2), like penguins, snow leopards, and mini-golden doodles.

This selective attention to certain species troubles me deeply. Recently, I attended a sustainability conference where a speaker made a statement as if it were common knowledge: *Polar bear images don't work for climate advocacy*, she claimed. *Community health messaging is much more effective at motivating action.* Chilled, I glanced around the room to see if anyone else seemed to feel the same unease. When I see images of polar bears on shrinking ice, I feel a profound sense of sadness and injustice. Nussbaum defines injustice as "significant striving blocked by not just harm but also wrongful thwarting, whether negligent or deliberate. Often, thwarting includes the infliction of pain, which impedes almost every ordinary activity of an organism (perceiving, eating, moving, loving)" (Nussbaum 2023, loc. 8 of 372). That wrongful thwarting of a valued life aspiration motivates me to learn more and take action. Don't other people feel the same way? What about the polar bears of the Hudson Bay, who might go extinct by 2030, "because sea ice is becoming too thin to support them as they hunt, according to new research"? (Gaffney 2024). What about *biophilia* as a catalyst—biologist E.O. Wilson's term for humans' innate love of living beings ("Biophilia Hypothesis" 2025)?

But then I looked at some more data. The speaker was right! Studies show that framing climate change as an immediate public health crisis *does* motivate people more than images of distant Arctic animals. This pragmatic approach makes strategic sense for short-term advocacy and policy change. Yet, something feels woefully wrong about abandoning those polar bear images in our discourse. To illustrate the point with an admittedly stark contrast: If we care about melting ice because it threatens the value of Miami beachside developments, have we really developed a deeper, more loving relationship with the rest of nature that makes long-term sustainability possible? And suppose climate advocates have learned that showing images of polar bears on melting ice doesn't motivate people to act. Instead, public health messages about threats to human communities are more effective. In that case, this approach perpetuates the very human-centered thinking that contributed to our ecological crisis in the first place. One problem with the human-centered approach: it tends to separate us from the rest of nature. Not only that, this fragmentary way of thinking shapes conservation education and discourse in other ways.

Consider the United States Endangered Species Act (ESA). For fifty years, the ESA has tried to protect nature through laws and regulations that target specific species. This approach creates compliance (and implicit fear) rather than love. While the ESA is a landmark piece of legislation that has notably helped save bald eagles in the United States, fewer than 3% of its listed species have recovered since the ESA's passage (Wright 2023). Even worse, it creates "perverse incentives," where landowners sometimes destroy habitats that endangered species call home to avoid the presence of these beings on their property regulated by government restrictions.

"The golden-cheeked warbler, a beautiful bird that breeds only in a small area of oak-juniper scrub in Central Texas, provides a chilling example of the dark side of human nature, as well as the ineffectiveness of the Endangered Species Act," Doug Tallamy, an entomologist and wildlife professor at the University of Delaware, writes. "Golden-cheeked warblers were never abundant because of their restricted breeding range, but increasing development in Central Texas in the 1980s triggered a formal review of the warbler for protection under the Endangered Species Act. The reaction of many private property owners was immediate and unfortunate: instead of taking action to support warbler flourishing, they cut down the oak-juniper scrub required by the warbler"(Tallamy 2020, loc. 33). Tallamy argues that the golden-cheeked warbler would do better if property owners could take care of their oak-juniper habitat through government incentives.

Our ecological crisis demands another approach that diverges from traditional top-down compliance strategies that rely on fear rather than promoting genuine engagement and collaboration. As Tallamy argues, we can't address environmental challenges through expert-driven initiatives alone. The scale and urgency require the broader participation of everyday people. We need to raise the inner-conservationist within each of our students, extending their environmental consciousness beyond formal education into their daily lives and personal spaces. This doesn't mean training everyone to become professional environmentalists, incentivizing them, or indoctrinating a particular viewpoint. Instead, we need to nurture *amateurs* who can inquire into the capabilities of other animals in the word's truest sense: those who act from love rather than obligation. It turns out that when it comes to the more-than-human world, such a love just might be an innate part of our heritage—and the stories we tell and images we share that can render these realities visible.

Why Children Love Animals (And What That Teaches Us)

Non-human animals help us understand ourselves and the world around us through a process of reflective engagement. We make meaningful analogies that illustrate our similarities and differences with them, and the sense of wonder this creates is distinct from our interaction with mere objects (Bendik-Keymer 2021, 162). In his 1980 essay, "Why Look at Animals?" art critic John Berger observed that humans first engaged with wild animals through gesture and symbolism in cave art, seeking to understand their place within a non-anthropocentric ecological order. Humans aren't alone in these acts. Other creatures also express creativity to co-create the world from sea to sky: pufferfish inscribe circles in the sandy plum of seabeds, male bowerbirds craft elaborate nests with blue bottle caps to woo mates, and mockingbirds mimic car alarms (All About Birds 2025). Humanity's earliest art was often placed over marks left by cave bears, creating a palimpsest of human and animal expression. Early human structures show beaver-influenced design, with archeological sites indicating that humans repurposed beaver-cut timber (Newman 2025). This challenges the narrative that humans distinguish themselves solely through cultural achievements.

Berger notes this reverent relationship evolved as humans transitioned to farming 12,000 years ago and domesticated animals, from caves to carts. Yet even domesticated animals were viewed as sources of meaning and ritual, not mere resources. Mythic heroes like Orpheus in Ancient Greece could communicate with non-human animals with music; notably, it was humans who were deemed to lack this gift of relating, not other animals. Today, while so many animals live in fragmented landscapes, polluted seas, or in cages or factory farms, the magic of ancient cave walls persists in children's worlds through animal characters in consumer culture. As philosopher Andreas Weber observes:

> Children are fascinated by animals. They crave to be nourished with animal images of all kinds...cuddly toys, plastic dinosaurs and dribble bibs embroidered with teddy bears stride ahead while their wild counterparts vanish. Nearly every kid's first picture book is crowded with animals. And this is a fact across all cultures. (Weber 2016, chap. 6, para. 41—42)

This universal fascination suggests something deeper than cultural preference. If "children help us to mediate between the ideal and the real" (Kinchin et al. 2012, 15), then we can identify an opportunity to continue an education about caring for and protecting interspecies relationships well past childhood. A 2022 study found that 97% of award-winning children's books featured animals, with 80% of these books having animals as their main characters (Hooykaas et al. 2022, 673). Children who play with stuffed giraffes, elephants, and lions might be forgiven for not knowing how precious few exist in the wild. Unlike our early ancestors, children and adults encounter diverse animals primarily through the passive consumption of media, which shapes how these creatures are culturally perceived and how they might interact with them. The story we need, then, is at once new and very old: it lives in humanity's deep ecological past and each of our earliest memories of reading and playing, when we discovered our kinship with the more-than-human world through books, stuffed animals, and toys.

How might we tell this new (yet very old) story in education? Anthropologist David Abram says a story must make sense in the original meaning: to enliven the senses (Abram 1997, 158). This points toward what Weber calls an *ecology of feeling that leads us to a new ecological ethics* (Weber 2019). "We should protect other beings because we love them. We love them because we are a part of them, and even more because they are part of us." (Weber 2016, Introduction, para. 26).

In holistic education, we complement the teaching of reason and its useful tools with imaginative modes of thinking, feeling, saying, and doing. This transformative education model includes sensory engagement and embodied learning: an embrace of sound, sense, taste, visual patterns, touch, and the sixth sense: intuition (Hauk et al. 2015, para. 1). A holistic pedagogy for interspecies relationships rekindles wonder for students learning about sustainability in secondary and post-secondary schools.

An Accessible Pedagogical Approach: Nature Journaling as Loving Attention

To rediscover a sense of wonder, educators can give students of any age, studying any subject, the opportunity to give the world their full, compassionate attention. How? By keeping a nature journal. Why? Our students aren't just nature journaling for the sake of wonder; they're observing to understand better what each creature needs to flourish. Nature journaling is an archetypal activity of learning, a practice of observing animals as individuals within their unique *umwelt* and ecosystem. Imagine Leonardo Da Vinci, amidst his backward writing, journaling the patterns of whirlpools and seeing the same swirls in the braids of a young woman's hair, as he notes the details of a flight of birds over Florence. Author Michael Gelb identifies qualities in Da Vinci's notebooks that we can all practice, including: *Curiositá*, or curiosity and asking questions; *Connessione*, seeing relationships between different fields and ideas; *Sensazione*, developing perception through observation and sketching;

and *Dimostrazione & Sfumato*, learning from experience while embracing ambiguity. *Sfumato* is a painting technique of showing the smokiness of atmospheric perspective (Gelb 2000).

In one of my courses, a liberal arts course about the world we design called *Designed World*, I teach my students these practices. Nature journaling works exceptionally well as part of their applied study of biomimicry. This design practice listens to nature as teacher and companion, as biomimicry expert Janine Benyus puts it (Benyus n.d.). Nature and culture are deeply interconnected; we experience nature through our feelings, imagination, and art, all while also engaging with science. Journaling is a way to do all of this with that sustained, compassionate attention that nature journaling expert John Muir Laws calls *love*.

Students can adopt the nature journaling approach of Laws by engaging in regular outdoor observation sessions. There's an open access treasure trove of resources on his website, as well as at the Wild Wonder Foundation, an educational non-profit that focuses on nature journaling and which he helps lead. One place to begin is with a free, downloadable zine that illustrates the most accessible, primary approaches for beginning nature journalers. Using his "I notice, I wonder, it reminds me of" framework, they practice attention, curiosity, and creativity (Wild Wonder Foundation, n.d.). Learners record changes in a dedicated place over time, whether in a wild, suburban, or urban setting. The practice brings together science and art through what Goethe called "tender empiricism." Rather than distant analysis, students engage all their senses in respectful dialogue with nature.

Laws also uses a simple three-beat inquiry sequence, "Let's see, could it be, let it be," as a trampoline for scientific thinking and artistic, narrative imagining (Laws 2020). By studying a nature journal entry of their own or a peer, a student can say, "Let's see…" and engage in inductive reasoning, observing and gathering examples to answer questions about *what*, *when*, *where*, and *how*, while reflecting on the sample size of their collected observations and their own biases and perspectives. "Could it be," meanwhile, leads to deductive reasoning: hypothesizing and testing one's assumptions. Science offers three virtuous learning outcomes for sustainability learners. The first virtue: the practice of being true to nature by accurately observing and representing the world; the second: trained judgment founded on the deliberative interpretation of data; the third: mechanical objectivity that relies upon instrumental measurements and standard procedures (Daston and Galison, as summarized in Christiansen 2023, 128). Yet students also need to appreciate the virtue of “Let it be”—an organizing idea for all journal entries that can’t be engaged with or reflected upon through those informal scientific procedures. This is the source material for stories, poetry, art, philosophy, and spirituality, as seen in works like *The Little Prince*.

According to Laws, pictures, words, and numbers comprise the three languages of nature journaling. These three languages enable students to engage in dialogue with the more-than-human world, which can’t express itself through human language but communicates through other gestures and signs. The very act of walking enhances this creative process, as research demonstrates that movement increases novelty in thinking (Oppezzo & Schwartz 2014). Nature journaling cultivates this embodied knowledge. Through sustained attention to specific places—sketching, questioning, walking, wondering—students develop what the fox in *The Little Prince* teaches: we become responsible for what we come to know through patient care. This responsibility, chosen through relationship rather than imposed through regulation, offers our best foundation for conservation. The practice transforms how we think, feel, and act—creating the kinship with other species our ecological moment demands (Fig. 10.1).

Fig. 10.1 *Nature journal example: mallards (by the author).* Reflecting on these sketches, the author coined a personal term, *mallardine*, to describe the faithfulness of ducks when they bond anew as a pair each year. The example illustrates how nature journaling draws on observation and imagination alike

Conclusion

Many conservation efforts react to a crisis. The Endangered Species Act, for instance, protects animals only after they're on the brink. By then, we're too late to fix the real normative problem, the narrative of separation between humans and the more-than-human world. Educators need a more proactive and sustaining, wonder-based approach. From a scientific perspective, Tallamy argues we should focus on preventing the crisis by turning toward our backyards and ordinary landscapes, planting native plants that create wildlife corridors and healthy ecosystems from the outset. From a philosophical perspective, Nussbaum offers her proactive approach: becoming patiently acquainted with other beings and understanding what they want to be and do in their own homes.

Nature journaling helps students not only practice the three languages of pictures, words, and numbers, but also practice these proactive dispositions. Nature journaling draws upon careful observation of natural processes and their interplay with our senses, reason, and imagination. A student who observes which caterpillars eat which plants, when birds need the most caterpillars, and how these processes connect becomes more effective at stewarding landscapes. A chickadee needs an environment rich in edibles for caterpillars: "An incredible 6,000 to 9,000 caterpillars are required to make one clutch of chickadees" (Tallamy 2015). A seemingly prosaic backyard bird suddenly becomes a way of understanding abundant needfulness and a vast ecosystem. This approach prevents problems instead of just responding to them, building lasting commitment based on knowledge rather than temporary emotional highs that burn out.

An effective pedagogy for conservation may come from allowing students to rediscover their childhood wonder and our heritage of kinship with other species. Rather than overwhelming them with problems to solve, we invite students into experiences that awaken their sense of responsiveness and belonging within the community of life. Through direct observation and sustained engagement with particular places and creatures, students develop the caring foundation requisite for all conservation efforts. When education begins with wonder and relationship, students don't need to be convinced to care—they discover they already do. This kinship, rooted in wonder and sustained by love, offers our best hope for responding to ecological challenges that threaten all creatures whose fate is, in truth, our own. Keep in mind the fox in *The Little Prince*: we become forever responsible for what we tame, for what we come to wonder over, imagine, and know well.

References

Abram, David. 1997. *The Spell of the Sensuous: Perception and Language in a More-Than-Human World*. New York: Vintage Books.

All About Birds. 2025. Why Do Some Birds Mimic the Sounds of Other Species? Cornell Lab of Ornithology. Accessed June 30, 2025. https://www.allaboutbirds.org/news/why-do-some-birds-mimic-the-sounds-of-other-species/.

Bendik-Keymer, Jeremy. 2021. The Other Species Capability & the Power of Wonder. *Journal of Human Development and Capabilities* 22 (1): 154–179.

Benyus, Janine. n.d. Janine Benyus – Co-Founder of the Biomimicry Institute. Biomimicry Institute. Accessed June 30, 2025. https://biomimicry.org/janine-benyus/.

Berger, John. 1980. *About Looking*. New York: Pantheon.

Biophilia Hypothesis. 2025. Wikipedia. Last edited June 23, 2025. https://en.wikipedia.org/wiki/Biophilia_hypothesis.

Chao, Sophie. n.d. An Interview with Dalia Nassar. Morethanhumanworlds. Accessed June 30, 2025. https://www.morethanhumanworlds.com/mthm-dalianassar.

Christiansen, Julia. 2023. *Building Science Graphics: An Illustrated Guide to Communicating Science Through Diagrams and Visualizations*. Boca Raton: AK Peters/CRC Press.

de Saint-Exupéry, Antoine. 2000. *The Little Prince*. New York: Clarion Books.

Wild Wonder Foundation. n.d. I Notice, I Wonder, It Reminds Me Of. The Wild Wonder Foundation. Accessed June 30, 2025. https://www.wildwonder.org/activity-iniwirmo.

Gaffney, Austyn. 2024. Missing a Global Climate Target Could Spell Disaster for These Polar Bears. *New York Times*, June 13, 2024.

Gelb, Michael J. 2000. *How to Think Like Leonardo da Vinci: Seven Steps to Genius Every Day*. Revised ed. New York: Dell.

Hauk, Marna, Madeline Zimdars, Kadin Westermann, Jessica Soderberg, Heather Schmidgall, Rocco Presicci, Daisy Perez, et al. 2015. Senses of Wonder in Sustainability Education for Hope and Sustainability Agency. *Journal of Sustainability Education*, December 3, 2015. https://www.susted.com/wordpress/content/senses-of-wonder-in-sustainability-education-for-hope-and-sustainability-agency_2015_12/.

Holdrege, Craig. What Does It Mean to Be a Sloth? The Nature Institute. Accessed June 30, 2025. https://www.natureinstitute.org/article/craig-holdrege/what-does-it-mean-to-be-a-sloth.

Hooykaas, Martin V., Marloes G. Holierhoek, Joris S. Westerveld, Menno Schilthuizen, and Ionica Smeets. 2022. "Animal Biodiversity and Specificity in Children's Picture Books." *Public Understanding of Science* 31 (5): 671–88

Keats, John. 1884. "Lamia." In The Poetical Works of John Keats. Bartleby.com. https://www.bartleby.com/lit-hub/poetical-works/36-lamia/.

Kimmerer, Robin Wall. 2015. *Braiding Sweetgrass: Indigenous Wisdom, Scientific Knowledge and the Teachings of Plants*, 1st ed. Minneapolis: Milkweed Editions.

Kimmerer, Robin Wall, Gavin Van Horn, and John Hausdoerffer. 2021. Kinship Is a Verb. *Orion: Nature and Culture* 40, no. 4: 22–29.

Kinchin, Juliet, Aidan O'Connor, Tanya Harrod, and Medea Hoch. 2012. *Century of the Child: Growing by Design, 1900–2000*. New York: Museum of Modern Art.

de Lange, Elizabeth, William Sharkey, Sofia Castelló y Tickell, Julien Migné, Robyn Underhill, and E. J. Milner-Gulland. 2022. Communicating the Biodiversity Crisis: From 'Warnings' to Positive Engagement. *Tropical Conservation Science* 15: 1–19.

Laws, John Muir. 2020. The process of Science (with Inference). YouTube video, 1:03:00. December 9, 2020. https://www.youtube.com/watch?v=nSp5LSxiatA.

Morton, Timothy. 2010. Guest Column: Queer Ecology. PMLA 125, no. 2 (March): 273–82. https://doi.org/10.1632/pmla.2010.125.2.273.

Nassar, Dalia. 2022. *Romantic Empiricism: Nature, Art, and Ecology from Herder to Humboldt*. Oxford: Oxford University Press.

Nassar, Dalia. Interview by morethanhumanworlds.com. An interview with Dalia Nassar. morethanhumanworlds.com. Accessed February 26, 2025. https://www.morethanhumanworlds.com/mthm-dalianassar.

Newman, Sarah. 2025. Did Animals Provide the Blueprints for Human Culture? *Aeon Essays*, June 13, 2025. https://aeon.co/essays/did-animals-provide-the-blueprints-for-human-culture.

Noddings, Nel. 2005. *The Challenge to Care in Schools: An Alternative Approach to Education*. New York: Teachers College Press. Kindle.

Nussbaum, Martha C. 2001. *Upheavals of Thought: The Intelligence of the Emotions*. New York: Cambridge University Press.

Nussbaum, Martha C. 2010. *Not for Profit: Why Democracy Needs the Humanities*. Princeton: Princeton University Press.

Nussbaum, Martha C. 2023. *Justice for Animals*. New York: Simon & Schuster. Kindle.

Oppezzo, Marily, and Daniel L. Schwartz. 2014. Give Your Ideas Some Legs: The Positive Effect of Walking on Creative Thinking. *Journal of Experimental Psychology: Learning, Memory, and Cognition* 40 (4): 1142–1152.

Popova, Maria. 2017. Why Look at Animals: John Berger on What Our Relationship with Our Fellow Beings Reveals About Us. The Marginalian (blog), January 2, 2017. https://www.themarginalian.org/2014/04/01/why-look-at-animals-john-berger-about-looking/.

Ritchie, Hannah. 2019. "Humans Make Up Just 0.01% of Earth's Life — What's the Rest?" Our World in Data. Accessed February 18, 2025. https://ourworldindata.org/life-on-earth.

Saint-Exupéry, Antoine de. 2002. *Wind, Sand and Stars*. San Diego: Harcourt. Kindle.

Scully, Matthew. 2002. *Dominion: The Power of Man, the Suffering of Animals, and the Call to Mercy*. New York: St. Martin's Press. Kindle.

Shermer, Michael. 2005. Unweaving the Heart. *Scientific American*, October 1, 2005. https://www.scientificamerican.com/article/unweaving-the-heart/.

Snyder, Gary. 2020. *Practice of the Wild*. Berkeley: Counterpoint Press. Kindle.

Stibbe, Arran. 2007. Haiku and Beyond: Language, Ecology, and Reconnection with the Natural World. *Anthrozoös* 20 (2): 101–112.

Stibbe, Arran. n.d. Part-3. Stories We Live By. Accessed February 18, 2025. https://www.storiesweliveby.org.uk/part3.

Tallamy, Douglas W. 2020. *Nature's Best Hope: A New Approach to Conservation That Starts in Your Yard*. Illustrated. Portland, OR: Timber Press. Kindle.
Tallamy, Douglas W. 2015. The Chickadee's Guide to Gardening. *The New York Times*, March 11, 2015. https://www.nytimes.com/2015/03/11/opinion/in-your-garden-choose-plants-that-help-the-environment.html.
Thriveni, C. 2018. Rabindranath Tagore's Philosophy on Indian Education. *Bhoomi Magazine*, February 6, 2018. https://bhoomimagazine.org/2018/02/06/rabindranath-tagores-philosophy-on-indian-education/.
Tulenko, Alicia. 2024. Folktales Like Philosophy Startle Us into Rethinking Our Values. *Aeon*, February 26, 2024. https://aeon.co/essays/folktales-like-philosophy-startle-us-into-rethinking-our-values.
Weber, Andreas. 2016. *The Biology of Wonder: Aliveness, Feeling, and the Metamorphosis of Science*. Gabriola Island, BC: New Society Publishers. Kindle.
Weber, Andreas. 2019. *Enlivenment: Toward a Poetics for the Anthropocene*. Cambridge, MA: MIT Press. Kindle.
Wright, Katherine. 2023. The Endangered Species Act at 50. PERC. October 17, 2023. https://www.perc.org/2023/10/17/the-endangered-species-act-at-50/.

PART II

Creating Transformational Learning Experiences

CHAPTER 11

Environmental Leadership for Those Not in Charge: Pragmatism, Critical Inquiry, and the Workable Good

Paul Hirsch

Introduction and Overview

What does it take to provide leadership in the context of society's most pressing environmental issues? What does environmental leadership look like, and what forms can it take? Can anyone be an environmental leader, or does this role require a specific set of knowledge and skills? I've explored these questions for over two decades: as a student and scholar in environmental sustainability, philosophy, and policy, and as a facilitative environmental leader, working in the areas of affordable housing, urban forestry, green manufacturing, protected areas, wildlife management, and the relationship between water and energy. For the past thirteen years, as a faculty member in the Environmental Studies Department at SUNY College of Environmental Science and Forestry (ESF), I've guided undergraduate

P. Hirsch (✉)
Department of Environmental Studies, SUNY Environmental Science and Forestry, Syracuse, NY, USA
e-mail: pahirsch@esf.edu

S. M. Wyman and J. Korenblat (eds.), *Sustainability Learning for Action and Community Engagement*,
https://doi.org/10.1007/978-3-032-11698-7_11

and graduate students to explore these questions. For the past eight years, I've had the opportunity to engage a diverse range of non-traditional students and working professionals in these inquiries through an online graduate certificate in Environmental Leadership. In this chapter, I draw on these experiences to share key ideas, conceptual frameworks, and pedagogical strategies for teaching present and future environmental leaders.

Below, I define the concept of *facilitative environmental leadership* that undergirds the approach. Next, I summarize the key elements of the Environmental Leadership curriculum. Then, I offer reflections about teaching in an online format. I conclude by discussing the opportunity for facilitative environmental leaders to think of their work not only as analytical, critical, and facilitative, but also as transformative.

Facilitative Environmental Leadership, or Leadership for Those Not in Charge

When teaching environmental leadership to graduate students, community leaders, and working professionals, one of the first things I say is that *addressing our most serious and pressing environmental issues requires learning to work in situations in which one is not in charge*. What do I mean by this?

First, I mean that environmental concern is not always the driver of social and political change that we might wish it to be. Generally, across the political spectrum, people care about clean water and air; they care about the beauty and integrity of nature; they care about the places where they live and recreate. These concerns are usually (although not always) uncontroversial. At the same time, however, they are not everyday priorities for many people. They are not seen as pressing issues related to daily survival, to livelihood, to health, or to economic vitality. When decisions are made at individual, organizational, or political levels, other seemingly more immediate issues can push environmental concerns to the margins.

Second, environmental leaders must learn to work in situations in which they are not in charge because environmental issues, both urban and rural, spill across boundaries that structure civic life and political decision-making. While there are departments and divisions dedicated to environmental protection at all scales of government, real-world problems rarely if ever fit neatly into their administrative purview. Therefore, any meaningful progress toward solving these problems must be made by people working both within and across the boundaries of discipline, culture, and/or organization. This within- and across-work, furthermore, is rarely structured hierarchically: rather, it is "poly-centric" in the sense of Nobel Laureate Elinor Ostrom's characterization of multiple overlapping centers of governance (2017). Within a poly-centric system, power is distributed and, in some cases, shared.

How, then, can we understand environmental leadership as a role that is explicitly distinct from forms of leadership associated with "being in charge?" Environmental leadership is *facilitative leadership*: leadership that functions through the inspiration, engagement, and organization of multiple actors, at various scales, around shared issues. The role of a facilitative environmental leader is to support reflection, dialog, and decision-making that recognizes the diverse ways in which our shared biophysical environment sustains the values we hold dear. Facilitative environmental leaders work at all levels of government and in businesses and other organizations where there are departments and divisions dedicated to environmental protection. They also work in communities, organizations, and parts of government and other organizations not explicitly structured around environmental protection.

In the Environmental Leadership curriculum, working with a notion of facilitative environmental leadership, we learn how to engage key partners in planning and decision-making, how to ask the right questions, how to critically reflect on how problems are being framed and by whom, and how to help illuminate appropriate pathways forward, while attending to the complex trade-offs involved. Below I describe the elements of the curriculum which is divided into three components: *Ethical and Philosophical Foundations*, *Thinking Together*, and *The Art of Facilitative Environmental Leadership*. The details offered below are not focused on specific readings or assignments; rather, I provide the conceptual background for the courses and programs I teach in the hope that others can adapt, remix, and be in dialogue with the ideas presented.

The Environmental Leadership Curriculum

Ethical and Philosophical Foundations

The purpose of philosophical and ethical inquiry is to provide a foundation—or, perhaps, a dynamic flow—that can undergird our work as facilitative environmental leaders. Philosophical inquiry helps us understand our commitments, enriches our learning and the exercise of our practice, and can help guide the paths we choose in our lives and careers. We begin with the concept of *place*. The concepts of place, and sense of place, have been developed and used by writers across disciplines to understand how people develop particularized relationships to the environments in which they live, work, and find meaning (Tuan 1975; Lewis 1995; Hiss 1998; Massey 2008; Seamon 2022). The concept of place also presents a challenge for many of us, whether because we have moved a lot, feel displaced by the dynamics of life, confront issues of safety, or for other reasons. Either way, the concept of place is uniquely suited to help us make connections between ideas and experiences and to help us tell the story of who we are and what we are up to as environmental thinkers and problem-solvers. Rather than introducing ourselves to our classmates as students in a particular discipline, or professionals in a particular organization, in the Environmental Leadership courses we introduce ourselves to each other with the concept of place as a background, using the prompt: *when someone asks you where you are from, how do you answer and why?* We then read scholarship on place and sense of place, including critical reflections on these concepts. Authors with diverse perspectives are assigned, and students can find and share other research and relevant concepts.

From this personalized grounding in place, we begin to engage in broader philosophical and ethical inquiry about the environment, with the aim of understanding the different perspectives that have evolved in relation to environmental issues. This prepares us to jump into current environmental challenges and debates, which—whether apparent on the surface or not—are always philosophically and ethically laden. While there is not enough time for a comprehensive overview of all the ethics and philosophies that can be applied to environmental issues, there are two parallel goals that I hope to achieve by exposing students to a select range of them. First, I try to give students a sense of the various sources and streams that can be and are being drawn on in environmental thinking and decision-making. We cover virtue ethics, kin-centered ethics (Kimmerer 2011), emerging ideas about the rights of nature, and approaches that account for the role of the environment in human flourishing. Students are also encouraged to explore and share additional philosophical and ethical perspectives that have intrigued and moved them. Second, because the work of facilitative environmental leadership always occurs in contexts where people are working with diverse value orientations, I expose students to a set of philosophical tools that can help them navigate ethical diversity. The overarching aim is to help students clarify their own views and values and train them to be agile in different contexts and situations.

By *philosophical tool*, I mean a concept that we can use to listen for what matters in a given situation and draw on what we hear to find words and concepts that help us communicate with others who also care (although perhaps in different ways). As an example, when I worked as a graduate student with the philosopher Bryan Norton, I was exposed to the philosophy of *environmental pragmatism*. Environmental pragmatism, as Norton and others have developed it, draws on the tradition of American pragmatism—a philosophical orientation that emphasizes the importance of developing practical, context-sensitive solutions to pressing problems. Pragmatism rejects the notion that there is a single, universally correct answer to these problems and instead advocates for adaptable, flexible approaches that take into account the specific circumstances and values of the communities involved. I've found pragmatist philosophy to be an extremely helpful approach to figuring out, in the context of real and ongoing environmental issues, how to be pluralist without being relativist. That is, how to embrace diversity without giving up on the ideas of truth and right and good as meaningful concepts altogether. In the Environmental Leadership curriculum, we learn to apply the insights of environmental pragmatism to transform the complexities and uncertainties associated with many pressing environmental issues into openings for listening, for dialog, and for joint decision-making across people and groups with diverse values and points of view.

Along with pragmatism, another important tool for facilitative environmental leaders is critical inquiry: into the ideas and commitments one may take for granted; into one's positionality as someone interested in preserving nature and the privilege this often entails; and into the social structures that shape what counts as an environmental problem and whose views are counted (see Hirsch 2018). Seen through a critical lens, the way we talk about the environment and environmental problems is a reflection of who we are and how we have been taught to think, as well as the groups and organizations with which we associate and align. Does the term *environment* evoke leafy trees, sandy sweeps, or urban landscapes, for example? Critique involves learning to pay careful attention to the role of language in shaping how we think about and relate to each other and the natural world, and to the role of knowledge—including what we know, how we know it, and who gets to speak.

In a deeper sense, critical inquiry can lead us to confront a tension that is built into academic scholarship on environmental ethics and problem-solving. Namely, to the extent that environmental ethics is understood as an extension of primarily Western philosophical traditions, there is a tendency, in the words of indigenous scholar and philosopher of science Kyly Whyte, to "privilege certain dimensions of ethics and exclude other dimensions, such as the exclusion of the importance of differences in models of environmental stewardship, conceptions of the history of environmental ethics and the procedural justice of acts of translation" (Whyte 2015, 15). The point of confronting these tensions, and of critical inquiry more broadly, is not to romanticize a particular culture or forego the idea of a canon of concepts altogether, but rather to learn to listen for alternate voices that may have other stories to tell. Critical inquiry also helps us problematize the various ways in which we sometimes simplify and flatten alternative perspectives, for example, through efforts to translate, reconcile, and integrate them. Ultimately, critical inquiry into values, language, knowledge, and power is important in working through contemporary issues related to environmental problem-solving, including the role of scientific understanding in an increasingly polarized society.

Paying attention to the role of power and privilege in the flow of ideas opens both danger and possibility, however. The danger is that we fall off a cliff into an infinite regress—we become relativists who can only offer critiques of critiques, with no way to put a stake in the ground. Or, falling off the cliff on the other side, we hold fast to this or that idea and become fundamentalists of a sort—in thrall to a religion, a philosophical perspective, or even a scientific worldview. The challenge for facilitative environmental leaders is to navigate the terrain between these two cliffs. By introducing students to the complementary ideas of pragmatism and critical inquiry, I hope to support them in making and defending claims, reflecting on their own positionality in doing so, and working to continually expand their own and others' perspectives.

Thinking Together Within and Across Boundaries

One answer to how we can find a middle path between relativism and fundamentalism is democracy, and more specifically, deliberative process. That is: hashing things out in a public setting. Our multi-layered democratic system offers a variety of opportunities to unfold collective thinking in the context of pressing environmental issues and problems. Of course, a quick look at our democracy in practice raises some questions: what do we do in this world not only of diversity and difference but also of division? What do we do in the case of increasing polarization in the domain of public values? What can we agree on as a basis for environmental values that could underpin our collective survival, our obligation to care for others, our notion of kin, and our quest to live a good and virtuous life? How do we combat the various forms of fundamentalisms—including scientism—without throwing out the possibility of making shared progress, of increasing knowledge and understanding of the world, of protecting things we mutually care about and on which we rely?

The challenge for a facilitative environmental leader is not to answer these questions alone but to find ways to support people with diverse views in thinking together about shared problems, making decisions together, and learning from those decisions. This, in turn, calls for three sets of capacities, each of which is necessary in order to support meaningful processes of thinking—and acting—together in the context of pressing environmental issues. First, because environmental problems transcend the boundaries of the counties, states, and countries that must respond to them, facilitative environmental leaders must learn to support people in thinking *across* existing boundaries. This requires metaphors, models, concepts, and tools that can support our own and others' abilities to think at the large scales of space and time necessary to respond to pressing environmental concerns. Second, in order to effect meaningful change, processes of thinking together must occur *within* the boundaries of functioning institutions and in relation to existing and emergent policies. A key part of the Environmental Leadership certificate program thus entails learning the factors that make environmental problems so challenging to govern; being able to describe and discuss the institutions and policy instruments available for responding to those challenges; and seeking leverage points within the policy system. Third, we must develop capacities for supporting thinking together by people and groups who not only have diverse values and value systems, but who have distinct interests and cherished identities.

Supporting thinking together across boundaries, within boundaries, and by people with diverse—and sometimes divided—values, interests, and identities calls for both pragmatism and critical inquiry. Applying the tools of pragmatism, we learn to seek points of convergence wherein people with diverse viewpoints can agree on what to do even if they cannot agree on the underlying *why* (Norton 1994). Applying the tools of critical inquiry, we learn to ask whether there are ways in which problems are being framed that elide or ignore key aspects of the history and context of the issues involved and, ultimately, if there are ways that current environmental challenges can be reframed that are more responsive to the range of values, interests, and identities involved. We also learn the power of grounding the process of thinking together in people's experiences in and relationships to specific places. Through grounding in place, we can search for pathways forward that are *appropriate* with respect to the values, interests, and identities of the people whose lives will be most affected (see Raymond et al. 2023).

The Art of Facilitative Environmental Leadership

It is helpful to think of the practice of facilitative environmental leadership as an art as much or even more than as a science. There is an art to navigating the terrain between a relativism that rejects the possibility of shared meaning, and various fundamentalisms that reject difference and diversity. As we have seen, language matters. The words and concepts we use shape who participates and how; the way we describe problems shapes the tools and solutions we apply to address them. In designing and teaching a curriculum for facilitative environmental leaders, I draw on my own experiences to offer specific language for acknowledging *complex trade-offs*, for seeking the *workable good*, and for selecting modes of engagement from a *facilitative leaders' palette*.

The language of complex trade-offs comes from my participation in an initiative after completing my graduate studies called "Advancing Conservation in a Social Context: Working in a World of Trade-offs" (see McShane et al. 2011; Hirsch et al. 2011). The initiative included researchers and nongovernmental leaders from Vietnam, Tanzania, Peru, England, and the United States. We looked at environmental problem-solving contexts characterized by multiple issues, including struggles for sustenance or livelihood, degradation of ecosystems, cultural diversity, and divisiveness based on political and other commitments. Our key finding was that, when a multiplicity of interests, values, and identities are intermingled, "win–win" solutions are in many cases rhetorical fantasies, and at worst, win–win language is a means of obscuring losses and silencing marginalized perspectives. We therefore sought to develop an approach to the analysis of environmental problems that takes trade-offs seriously. Our working premise was that more explicit recognition of trade-offs—and the hard choices they entail—may help avoid the disenchantment and disenfranchisement that come with inevitable losses. In some cases, recognizing losses may allow them to be appropriately grieved, and efforts can be directed toward healing and repair. Ultimately, this should lead to more resilient and sustainable outcomes in the long term, the only term in which environmental protection matters.

As the initiative unfolded, it became clear that algorithmic solutions to calculating and mitigating trade-offs, in particular those based on quantitative approaches of counting costs and benefits, may be helpful but would not be sufficient for dealing with the multi-dimensionality of the most pressing issues. We therefore refined the concept to *complex trade-offs* and sought to develop a narrative-based approach that can be adapted and applied to problems across a variety of contexts (see Hirsch et al. 2013; Russo and Hirsch 2023). In the Environmental Leadership certificate program, the concept of complex trade-offs is taught to students as a way to develop an expansive view of a problem—and of various solutions—that takes both gain and loss seriously and does so in a way that is inclusive of a range of values, interests, and identities.

Acknowledging complex trade-offs, of course, only gets us so far. The next question is how to move forward—how to develop workable and appropriate pathways in the midst of complex trade-offs. A workable good, as presented in the Environmental Leadership curriculum, is not the "right" or "best" or "optimal" or "win–win" or "scientifically justified" or "objective" decision. Rather, *workable goods* are pathways for continued development anchored by a set of principles that are narratively meaningful and functionally applicable in light of the diverse values, interests, and identities at stake. These principles should be articulated and refined through ongoing dialog with those who are affected by decisions. Ultimately, a workable good is a decision that can live and breathe and be adapted in the context of its application as the future unfolds.

Keeping a workable good living and breathing while adapting it in light of a changing future is part of the practice of facilitative environmental leadership, and where the art of it comes most into play. Everybody engages in this work from a different set of experiences and a unique set of facilitative leadership capacities. Some people are excellent researchers and analytical thinkers; others are consensus builders; others are skilled in the art of negotiation or conflict resolution; some are seasoned advocates and champions of environmental and other causes; some of us are attuned to bias and injustice, and know how to be allies and accomplices in making the necessary changes. By asking ourselves and each

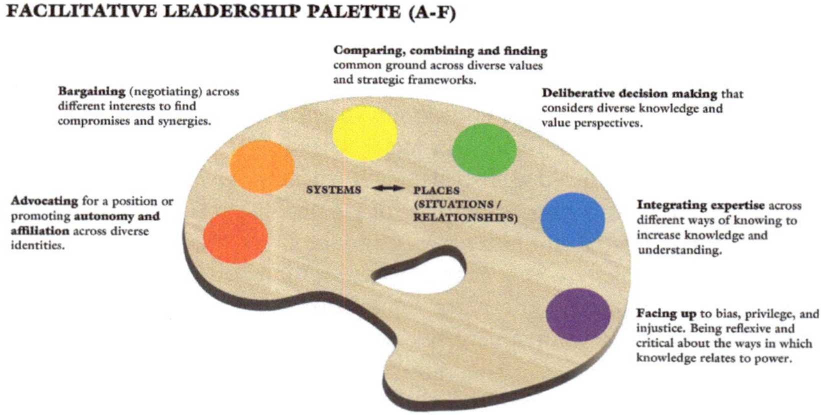

Fig. 11.1 Facilitative leadership palette

other what our strengths in these areas are, where we want to grow, and where we could use teammates and support, we begin to develop a more full-spectrum approach to environmental leadership. We also develop the ability to select an appropriate *palette* of engagement approaches for a given context and set of interconnected issues. Below is an illustration of the facilitative leaders' palette we work with in the Environmental Leadership curriculum (Fig. 11.1).

Strategies for Online Curriculum Delivery

Over the past eight years, I've taught courses and designed curricula, based on the above, for traditional and non-traditional students working to earn their graduate certificate in Environmental Leadership. All the courses have been fully online, which, if you had asked me beforehand, would have been an undesirable and infeasible way to teach this material. Nevertheless, I have found the experience to be invigorating, and it has become very clear that expanding existing classes into the online space allows people to participate in the academic community who did not previously have access to it. Since the courses I teach were taken online, our more traditional residential graduate students have been joined in the virtual classroom by:

- Environmental professionals working throughout New York at the local, state, and regional scales;
- People working in corporate settings including public relations, human resources, supply chain management, and the automotive industry;
- Staff of regional non-profits focused on affordable housing, biocultural restoration, and food sovereignty;
- Environmental advocates, community leaders, teachers, and librarians;
- Faculty and staff from other SUNY institutions.

The fact that my experience in teaching online started several years before the global pandemic has been instructive. As many have come to learn since then, being forced to teach online with no preparation, no support, and no intrinsic motivation is a very different experience than teaching online when both the students and instructor are online because they choose to be. I was motivated to teach online and fortunate to have both preparation time and adequate support. And with each passing year, with the opportunity to layer my semester-length efforts atop the infrastructure accreted through my efforts in previous years, I have developed more appreciation for and understanding of the medium. As I told people who asked me at the time of the pandemic, and continue to do so today, I have come to learn that online teaching is *teaching*: each instructor does it differently, and each finds a way to reach their students (or not) given the constraints and opportunities of the learning environment, whether two dimensional or three, synchronous or asynchronous, or some combination.

That being said, there are unique aspects of teaching in the online space, and there are specific strategies one can employ to orient oneself. Two pedagogical interventions, made by ESF's Open Academy teaching and learning specialists, were particularly helpful to me. First, I was encouraged to reorient the in-person curriculum, which had been developed with traditional ESF graduate students in mind, by asking myself the Backwards Course Design question: *what do I want the students to be able to DO after they complete the program?* Before this stage, course planning had centered more on: *what will students READ?* This action-oriented, student-centered approach stimulated a profound (and still developing) shift in my thinking about what it means to teach undergraduate and graduate students (Wiggins and McTighe 2005). Secondly, and relatedly,

I was encouraged by the same pedagogical expert to pay close attention to the courses' learning objectives and how they aggregate into program-wide learning objectives. Just as clarifying principles is key to good, facilitative environmental leadership, specifying learning objectives—and how they will be achieved—has been essential to the development of a robust online learning experience.

It may be that these two orientational shifts represent standard good pedagogical practice. For me, they were invaluable in making the transition into the online teaching space. The capabilities I developed through classroom experience, including the abilities to deliver a lecture and to facilitate conversations, remain essential in the online space, but they are not sufficient. As I came to understand—and am still grappling with—online teaching is as much or more about being an architect of a structured learning experience as it is about delivering lectures and facilitating dialog.

Conclusion: Cultivating a Capacity for Transformation

Facilitative environmental leadership requires analytical, facilitative, and critical capacities. It also requires what we might call a *transformative* capacity—that is, a capacity to change the terms of reference, the overarching metaphors, by which we generate solutions and evaluate options.

We can cite the experience of Aldo Leopold in his essay "Thinking like a Mountain" as illustrative of what is meant by a transformative capacity (see Hirsch and Norton 2012). Early in his career as a forester in the Southwest, United States, Leopold set out to eradicate wolves to enhance deer populations and improve deer hunters' opportunities. But these efforts had unfortunate consequences as deer populations exploded and vegetation suffered. In the words of Leopold, "in the end, the starved bones of the hoped-for deer herd, dead of its own too-much, bleach with the bones of the dead sage, or molder under the high-lined junipers" (1949, 140). When Leopold saw the green fire go out in the eyes of a wolf he had killed, he knew at a visceral level that he had failed to think across the boundaries of his narrow interests and point of view. Through witnessing the compounding of his error, he learned more about what it means to "think like a mountain," to value wolves as essential to maintaining a healthy system, and to embrace values relative to the integrity of natural systems rather than their mere instrumental exploitation.

While much went into Leopold's transformation, and care must be taken not to overgeneralize (see Whyte 2015), we can draw from his narrative the insight that it is a profound mistake to think that one can be "in charge" of the complex mountainside dynamics that occur between wolves, deer, and plants. If we add in the dynamics of soil, water, air, and other forms of life, including other people and the various roles they play, the idea of being in charge becomes even more unreasonable. Yet that doesn't mean that the actions of individuals, groups, organizations, and political units don't matter. They matter greatly, and thus it is also incumbent on Leopold, and all of us, not only to think *across* and outside the boundaries to which nature does not conform, but also to think and act *within* the boundaries that structure human activity. In doing so, we will have to live and work with the messiness that comes with the full diversity of human experiences, wants and needs, aspirations, and senses of self and other. At times, the art of facilitative environmental leadership may lie in recognizing unbridgeable gaps, and in finding pathways to coexistence in the absence of the power that comes with being in charge. At other times, not being in charge can serve as the basis for facilitative environmental leaders to be bridge builders and connectors.

References

Brosius, J. P., A. L. Tsing, and C. Zerner. 1998. "Representing Communities: Histories and Politics of Community-Based Natural Resource Management." *Society & Natural Resources* 11 (2): 157–168.

Hirsch, P. D., W. M. Adams, J. P. Brosius, A. Zia, N. Bariola, and J. L. Dammert. 2011. "Acknowledging Conservation Trade-Offs and Embracing Complexity." *Conservation Biology* 25 (2): 259–264.

Hirsch, P. D., and B. G. Norton. 2012. "Thinking Like a Planet." In *Ethical Adaptation to Climate Change: Human Virtues of the Future*, edited by A. Thompson and J. Bendik-Keymer. Cambridge, MA: MIT Press.

Hirsch, P. D., J. P. Brosius, S. O'Connor, A. Zia, M. Welch-Devine, J. L. Dammert, A. Songorwa, T. C. Trung, J. L. Rice, Z. R. Anderson, S. Hitchner, J. Schelhas, and T. O. McShane. 2013. "Navigating Complex Trade-offs in Conservation and Development: An Integrative Framework." *Issues in Interdisciplinary Studies* 31: 99–122.

Hirsch, P. 2018. "Sustainable Change in a Fractured World." In *A Sustainable Philosophy: The Work of Bryan Norton*, edited by B. Minteer and S. Sahotra, 87–97. Basel: Springer International Publishing.

Hirsch, P. 2015. *Painting Participation: A Full Spectrum Guide to Selecting Tools, Methods and Approaches for Fostering Productive Dialog about Complex Problems*. Washington: American Geophysical Union Thriving Earth Exchange. https://thrivingearthexchange.org/wp-content/uploads/2015/01/Painting-Participation.pdf

Hiss, Tony. 1998. "The Experience of Place." *Aperture* Winter. https://issues.aperture.org/article/1998/1/1/the-experience-of-place

Kimmerer, R. 2011. "Restoration and Reciprocity: The Contributions of Traditional Ecological Knowledge." In *Human Dimensions of Ecological Restoration: Integrating Science, Nature, and Culture*, 257–276.

Leopold, A. 1949. "Thinking like a Mountain." In *A Sand County Almanac*, 129–133. New York: Ballantine Books.

Lewis, E. 1995. "Connecting Memory, Self, and the Power of Place in African American Urban History." *Journal of Urban History* 21 (3): 347–371.

Massey, Doreen. 2008. "A Global Sense of Place." In *The Cultural Geography Reader* (pp. 269–275). Routledge.

McShane, T. O., P. D. Hirsch, T. C. Trung, A. N. Songorwa, A. Kinzig, B. Monteferri, and S. O'Connor. 2011. "Hard Choices: Making Trade-offs Between Biodiversity Conservation and Human Well-Being." *Biological Conservation* 144 (3): 966–972.

Mouffe, C. 2005. *On the Political*. Psychology Press.

Nersessian, N. J., Newstetter, W. C., Kurz-Milcke, E., and Davies, J. 2002. "A Mixed-Method Approach to Studying Distributed Cognition in Evolving Environments." In *Proceedings of the International Conference on Learning Sciences* (pp. 307–314). Hillsdale, NJ: Lawrence Erlbaum Associates.

Norton, Bryan G. 1994. *Toward Unity Among Environmentalists*. Oxford University Press.

Norton, B. G. 2020. *Sustainable Values, Sustainable Change: A Guide to Environmental Decision Making*. University of Chicago Press.

Ostrom, E. 2017. "Polycentric Systems for Coping with Collective Action and Global Environmental Change." In *Global Justice* (pp. 423–430). Routledge.

Raymond, C., P. Hirsch, B. Norton, A. Scott, and M. S. Reed. 2023. "Rethinking Appropriateness of Actions in Environmental Decisions: Connecting Interest and Identity Negotiation with Plural Valuation." *Environmental Values* 32 (6): 739–764.

Russo, F., and P. Hirsch. 2023. "Navigating Complex Trade-offs in Public Health Interventions." *Journal of Evaluation in Clinical Practice* 29 (3): 430–437.

Seamon, D. 2022. "Sense of Place." In *International Encyclopedia of Geography: People, the Earth, Environment, and Technology*.

Tuan, Yi Fu. 1975. "Place: An Experiential Perspective." *Geographical Review*, 151–165.
Whyte, K. 2015. "How Similar Are Indigenous North American and Leopoldian Environmental Ethics?" Available at SSRN 2022038.
Wiggins, G. P., and McTighe, J. 2005. *Understanding by Design*.

CHAPTER 12

My Place in the World: Designing a Regenerative, Sustainable Education Foundation for Students in the Twenty-First Century

Andrea Varga

Introduction

I started down the path of teaching sustainability in the 2010s, motivated by a dissatisfaction with what was going on in our federal government in terms of inaction about climate instability and biodiversity loss. I also felt frustrated with the state of higher education, how our disciplinary-specific, rigid approaches are not serving us well. I noticed the connection between ecoanxiety and increasing mental health challenges in our community. I wondered what I could do to be more involved without a degree in sustainability and worried, "Who am I to say anything? I am a climate activist over there, and I'm an artist over here." And I realized this

A. Varga (✉)
Department of Theatre Arts, SUNY New Paltz, New Paltz, NY, USA
e-mail: vargaa@newpaltz.edu

S. M. Wyman and J. Korenblat (eds.), *Sustainability Learning for Action and Community Engagement*,
https://doi.org/10.1007/978-3-032-11698-7_12

separation wasn't cutting it for me, that I wanted to create intersections between sustainability, climate action, and creativity.

I became a member of our interdisciplinary Sustainability Faculty Fellows Program, and this was revelatory because I felt welcomed. No one seemed to think, "What's the weirdo artist doing in a room with all the smart scientists?" It was a little scary for me at first, and I worried, "Am I going to be criticized for doing this?" But then I think of my STEM friends who claim, "Oh, I'm not an artist." In response, I insist, "No, you are an artist, and your creative work is welcome and needed." We all have something to offer in this effort to confront our planetary and human/creature challenges. I realized early on: *I can do this thing*. I can apply sustainability to everything I teach. I may not have a PhD, but I can effectively integrate science into my teaching; I can bring my empathy; and I know what I'm talking about. My own research question is, *How can we make sustainability studies digestible and approachable?*

My classes have become a laboratory for sustainability education in various ways. Learners at any age can use research and their own knowledge and find unique ways to communicate through visual arts, creative and persuasive writing, zines, and videos. Drawing exercises engage students no matter their artistic experience, which helps them care more about the topics. If they're observing beaver dams in their nature journals, for example, they make the connections themselves between self and environment, and empathy follows. Rather than lecturing prescriptively about what students "should" do for sustainability, I seek to create a space where they can explore what they're genuinely curious about through conversation, creativity, and playfulness. This replaces feelings of guilt and occasional academic resistance with a sense of empowerment as they productively engage based on their own interests. There is something for everyone—learners, parents, educators, and civic leaders—some form of creativity that resonates and provides an outlet to constructively explore sustainability concepts. When learners feel autonomy over their growing awareness, development, and self-expression, they become much more invested and empowered to take positive action.

I have the joy of teaching a variety of learners, from incoming students to graduating seniors in interdisciplinary general education cohorts, as well as art students in studio classes. I integrate climate science and data into art classes and artistic expression and art practice into Humanities seminars. The key is engaging curiosity. I find expressive arts, which include creative writing, to be a great incubator for curiosity. Instead of thinking, "What should I do?" Students are empowered to ask, "What do I want to do?"

Everyone needs basic climate science to understand climate change and other challenges that are exacerbated by it: poverty, migration and displacement, health issues, trauma, and biodiversity loss. You don't have to be a climate scientist to teach these topics; in fact, we all have an obligation to understand and address these issues. Likewise, students can utilize climate science without being scientists and have confidence that this foundation will allow them to expand their disciplinary approach to climate/sustainability education without fear of criticism, no matter their discipline.

My Place in the World (And Yours)

To identify my own positionality, I'm a mom, an artist, an educator, a gardener. I grew up in an arts and science household in the Intermountain West of the US, in a dynamic environment, a space with my dad as a plant scientist and my mom as a visual artist, with no separation between the two fields of inquiry.

I always joke that my dad was eager to have me learn botanical Latin at the age of five, but I was too busy drawing flowers. Now, I'm trying to learn botanical Latin and climate change science, and I need to explore sustainability topics with my students. It's harder in your 50s, but I care more about it now. It's probably better that I simply learned to draw flowers back then, and now I can decide whether the classification matters or not.

I teach art and sustainability at SUNY New Paltz, US, and I happen to be in the School of Fine & Performing Arts. I also teach in our interdisciplinary Honors program, where I have been able to incubate my interdisciplinary, Sustainability-focused seminars—*Intro to Sustainability* and *Ethical Fashion*. I now teach sustainability in all my classes, from *Costume Design* to *History of Fashion*, *History of World Dress*, and *Mixed Media Art for Storytelling*. I approach sustainability differently for each class, but through all of them, we explore the intersections of science, art, society, and economics, as featured in the *Hechinger Report* (https://hechingerreport.org/how-colleges-can-become-living-labs-for-combating-climate-change/) (Preston 2024). My mission is to engage students across disciplines in sustainability education and action, empowering young people to lean into sustainability in ways that make sense for them in their professions and lives.

Student Mental Health: A Look at the Data

Once I moved on from being a climate activist in my free time to being a full-time sustainability educator, I had to seriously confront the anxiety crisis that is plaguing our children and higher education students. We have data on the depth of this matter: "Climate change has important implications for the health and futures of children and young people, yet they have little power to limit its harm, making them vulnerable to climate anxiety" (Hickman et al. 2021). This was the first large-scale investigation of climate anxiety in children and young people globally and its relationship with perceived government response. The researchers surveyed 10,000 children and youth aged 16 to 25 in 10 countries in 2021, post-pandemic. "Respondents across all countries were worried about climate change, with 59% very or extremely worried, and 84% at least moderately worried. More than 50% reported each of the following emotions: sad, anxious, angry, powerless, helpless." More than 45% said their feelings about climate change negatively affected their daily life and functioning, and many reported a high number of negative thoughts. 75% said they think the future is frightening, and 83% said people have failed to take care of the planet. Respondents rated governmental responses to climate change negatively and reported greater feelings of betrayal than reassurance (Hickman et al. 2021).

We know this level of anxiety is neither appropriate nor healthy for our children and our university students, typically aged 18–25. As adults, we have a sense of urgency about the environmental and social challenges we face. This anxiety is probably appropriate for us, but we must figure out how to channel these emotions productively. We know most youth don't have those skills yet. That is what I lean into as an educator: how to take the alarming information and then, in age-appropriate ways, pull that apart and help students understand it, confront it, digest it, and act upon it. That's where creativity comes in (Fig. 12.1).

Getting Started: An Entry into the Material of Creative Sustainability

I tend to start my classes gently with a creative activity, no matter the topic, to bring everyone together. For example, learners may have an

Fig. 12.1 *Overlook* (2023) Andrea Kantrowitz

index card at their station, pencils or black pens to create a little piece of mini-art and to share a thought or reflection. The following is an activity borrowed from artist Lynda Barry (2019), an educator of educators.

Take the index card and draw a little frame. It can be a line, it can be dots, and it can be zigzaggy. Everybody can make a frame, and it gets you started. Put your name up in one of the corners and the date, and I will provide a prompt. It might be a bug that represents you or a woodland creature, for instance. Here's the elegance: We can all do it, it's a bug. This is not about creating some artistic masterpiece or personal expression. It's just about getting your hands moving and your brain flowing and having a little bit of fun and playfulness at the beginning of class. Barry believes we're all artists because we all express emotion. We all can communicate with each other in one way or another. When we're little, we don't question whether we can make marks on a page to express ourselves. In fact, it's one of the first things that we do innately: before language develops, we're making marks on surfaces.

Somewhere along the way, we dismiss creative expression and creative mark-making as a skill some people possess and some people don't. Barry believes we should re-embrace the three-year-old in ourselves and do more. I tell my students, "I don't mind doodling at all." This casual activity can help a student focus on what's being said and seen. It isn't a sign of inattention; it's a way to calm down and to focus on being in this space—being you right here and now, then we'll see where we can go together.

Using Reflection Exercises to Measure Progress and Increase the Sense of Shared Experience

I started doing a poll on my index cards at the beginning of the semester: on a scale of zero to ten, how worried are you about climate change and social injustice? 10 being "The planet's on fire, we're doomed, there's nothing we can do" and zero being "It's all just propaganda, nothing's going on, we're fine." Typically, scores fall between eight and nine. That's an unbearable level of intensity, and that isn't helping learners engage. I deploy the poll again at the end of the semester, and students tend to come down closer to six, giving nuanced responses like, "Some days, I get up to a ten, but I try to maintain more of a five or six." Progress is being made, and we're starting to learn how to modulate our anxiety and channel our emotions into actions.

We need to help our children, our university students, and other community members share, exchange, and ask questions. There are approaches that won't freak them out; they're already stressed enough. Let's give them space to process, express, be creative, and have conversations. That's the feedback I receive, too: "Thank you for allowing us to talk about this." I share my own feelings as well: "Yeah, I am up to a nine or ten sometimes too." Looking at each other and realizing you're not alone is incredibly helpful.

The words that we use matter to students. They know climate change is happening and that social unrest, injustice, and racial and economic inequality exist around the world. One of the messages they're getting in this media-rich world is that "Gen Z needs to save the planet." That message is disingenuous because the planet is not going anywhere—changes created by our anthropogenic systems need to be addressed. Instead, ask, "Let's look at your own community, your own life: what do you care about?" Students tend to respond to that question in a positive way instead of shutting down.

In our sustainability-infused or focused classrooms, we are providing a new place to turn, a place to metabolize these emotions and uncertainties about the future and the state of our human family through creativity, productivity, community, and connection, instead of self-isolation, consumerism, guilt, and doom scrolling on social media.

Teaching Creative Critical Thinking with the UN Sustainable Development Goals

I use the UN Sustainable Development Goals (SDGs) as a framework for students to select an area they're interested in learning more about. I'll say, "Pick one UN Global Goal that resonates with you and tell me why." I never have students who refuse, though sometimes they say it's overwhelming to pick just one since the Goals are all interconnected—which is a great place to start a conversation on interdependence.

Students select one area to research in-depth: if someone picks Goal 10: *Reduced Inequalities*, they explain why it matters to them. Then, we discuss how that Goal connects to others. As they dig in through research, they start detecting the intersections between goals. We grapple with definitions, approaches, and solutions through their interests and experiences and integrate climate science with environmental justice as we talk about our biospheres and biodiversity. I often use specific topics, such as *where fibers come from* and *what materials make our clothes,* alongside plant science and chemistry, as well as the labor and animal rights issues involved in apparel fabrication, to demonstrate the applicability of this knowledge to their own lived realities. I share my own experiences with animals, growing up on a farm and working as a professional designer. I try to guide students to see that you can use many approaches to understand a problem and work toward solutions—and that your own experiences, interests, and expertise matter tremendously.

Meanwhile, alongside the research, we are drawing and doing observation exercises, so learners engage their hands and bodies, which I find helps them care more about the ecosystem they are living in. We take walks on campus and in a nearby nature preserve where the beavers are always redesigning the landscape and paths. Students start thinking about our local watershed, how important clean water is for life, and then how they use and interact with these resources. It's a great opportunity to talk about interconnected ecosystems, the importance of beavers, and how they link SDGs14: Life Below Water and 15: Life on Land, with our own local environment.

Walking, talking, observing, and drawing relieve some pressure the students experience from the media. The goal of my classes is not to figure out the big picture right now but to start to see parts of it with curiosity and care. I begin with, "Let's examine things you're observing and you're caring about." And often, it's not nature or the climate crisis that they're bringing in. One might comment, "You know, my best friend is suffering from a mental health crisis, and I'm really worried about them." This is an opportunity to talk about SDG 3: *Good Health and Well-Being,* and how critical this is to meet the challenges of our times. If you have a friend who can't get out of bed in the morning to take a walk with you, what do you do about that? I use creative approaches to problem-solving, student-centered learning, and critical thinking about complex systems to address these issues.

For my interdisciplinary seminar classes, students engage in topical modules related thematically to a grouping of Global Goals or issues such as Consumerism and Marketing. After engaging in the materials for the module, they each select a topic where they want to take a deeper dive, and they research, then write a short-form research article that is shared on the discussion board. All students must read and respond to several classmates for each learning module. This teaches them agency and helps them understand that their communication style matters—plus, they learn from each other and discover the power of peer engagement—as well as the seriousness of doing research before you come to conclusions. For their final project, they have full agency to select a topic that matters to them. They research it and then come up with a creative communication project to share the issue and some solutions or introduce change makers in that area, and how people can get involved on a local level. They are encouraged to examine the ways they like to be creative and then figure out how to incorporate that into a communication strategy.

Nature Journaling to Develop Observation and Wonder

As an incubator for curiosity, creative projects help us move from the struggling space of "What should I do about this problem?" to ask, "What do I want to do to help with these issues?" We can deploy these methods to engage wonder, joy, and playfulness. Then, we have more space in our brains to deal with seemingly intractable problems. Nature journaling is an accessible place to begin an artistic and observational journey to unlock creative problem-solving. John Muir Laws writes

> Children need nature. Contact with the natural world improves health and reduces stress. Nature is also a rich and meaningful place to learn. There is no computer program that can replicate the excitement of seeing a squirrel up close, the intrigue of studying leaves, or the calm of watching clouds. (Laws and Lygren 2020, 4)

This process allows us to take time to slow down and look at something in front of us deeply and draw parts of it, not worrying about the artistic expression part but just putting marks on the page. Then, add some words. "What is going on here? What do I see? What is the name of that tree? Why are the blossoms pink?" Learners at any level enter interconnected modes of curiosity—and they want to do more and more of this. When I'm with students outside, unexpected things happen all the time. For example, there are often individuals who are afraid of bugs and not happy about something buzzing around their heads. They'll say, "I hate bees!" That's an opportunity to say, "That's a yellow jacket, that's not a bee. Let's talk about the difference, the threat level, and how to respond." Learners can take that experience and draw out their feelings in the nature journal (Fig. 12.2).

Nature journaling has the immediate benefit of making people slow down and connect the brain with the body (hand). It has the longer-term benefit of helping learners observe their lived spaces, which naturally leads to conversations, "Are those trees supposed to be blooming in February?" They become more deeply invested through their own inquiry and emotional senses. I use this practice with my art students to encourage them to think about science through the lens of art. It is also useful for my non-art scientists, social scientists, business, or humanities students to relax, sense, and slow down.

Student Research: Use Creativity to Increase Exploration and Personal Investment

I had a Biology major student collecting tiny bits of data on crickets—lots of scientific observation. He took *Costume Design* for fun and came up with such wonderfully fresh, sweet ideas in the form of storytelling. At

Fig. 12.2 *Ashokan: Walnut ink study*, Ashokan center (2023) James Fossett

the end of class, he said, "Thank you so much for letting me be in your class, because it's like I have this other part of my brain that I wasn't even using. And I needed to go into that part of my brain, the part that did this creative work, so when I went back to the lab, that other part of my brain was ready to go again" (*used with permission*).

This is key: we can engage other sections of our students' brains to let parts that are overworked rest a little bit. And when they return to the task or disciplinary methodology they're doing daily, they say, "I'm refreshed. I can see things a little differently."

Michael, a senior in Geology doing research on soils, came up to me struggling to select a final topic for the *Intro to Sustainability Seminar*, saying, "I know probably everybody's expecting me to just do it on soils." I said, "I doubt any of the other students are thinking about what your final project is." He said, "Don't you expect me to do it on soil?" I said, "Not at all. It's great if that's what you want to do, but maybe you need a break from soils." I asked, "What do you do for fun?" The look on his face did not compute. Then, he replied, "Well, I guess hiking, biking." I said, "Great, there's your final topic: do it on biking. That's one of our community sustainability solutions: micro-mobility." He immediately realized he could talk about the Bike Friendly Campus initiative. All he needed was a little nudge to personalize the research project (*comments used with permission*).

This magic happens often in my classes. Michael did a zine project over a couple nights, staying up till 2:00 a.m. because he was having so much fun making mini folded magazines. He observed, "That's flow." He then asked whether we could print out copies and make a sustainability zine literature display for orientation week, finding ways to share out further with our community.

Teaching with Mini-Zine-Making

One of the simplest modes for creativity in the classroom works across all ages. A zine is the opposite of a magazine, a glossy, published, usually for profit, publication. Zines are very personal, handcrafted, and quirky. They should be easily replicable and distributed. That's part of the fun. This process takes personal drawing and observation to the next level, where someone feels confident to say, "I can do this, and now I can share it."

Anybody can make a zine. And they don't have to be able to write; words aren't necessary. Roy, a Geology and Earth Science Education Honors student, was very interested in our local watershed. He was activated when I started talking about synthetic fabrics, the plastic microfibers they shed into our waterways particularly from doing laundry, and the implications of plastics pollution. His brain was exploding because as a native of the Hudson Valley, he cares very much about our ecosystem health. He said, "Wait, you're telling me that there are invisible plastics in our water that are potentially poisoning all of us?" He was very animated and interested in understanding more, but apprehensive about doing something creative like a zine to address this topic for his final project. He created a mini-zine that could be photocopied or printed over and over, in tandem with his final presentation on synthetic microfiber pollution in our local waterways. This kicked off a multi-semester project and a series of on-campus and community presentations. With a small undergraduate research grant, we purchased external microfiber filters for clothes washing machines. Roy's research on a simple sustainability issue engaged people throughout the community in everyday solutions.

In addition to reducing and ultimately eliminating synthetic textiles from our wardrobes, installing filters is something we can do immediately.

The most inspiring development from Roy's interest, engagement, and activation on this issue was that he inspired other students to work on it too. We developed a mini-workshop on understanding synthetic textiles and how to do laundry in more sustainable ways. This included reducing synthetic microfiber waste, reducing water and energy usage, eliminating harmful chemicals and other byproducts that end up in our wastewater, and raising awareness of SDG 14: Life Below Water. Roy's energy and enthusiasm to share this information led him to become a campus Sustainability Ambassador during his senior year, and he drew in the Beyond Plastics team to focus on synthetic microfiber pollution. Today, Roy is a high school Earth Science teacher, and he continues to educate about microfiber pollution in his own classroom and community. On our campus, the Sustainability Ambassadors in cooperation with our Campus Auxiliary Services, Office of Residence Life, and Office of Campus Sustainability have installed a pilot program of microfiber filters on dormitory washing machines.

Additional Resources for Research and Teaching

Paul Hawken's *Regeneration,* a publication from Project Drawdown, incorporates the social and economic spheres of sustainability along with the focus of drawing down carbon in the atmosphere. I utilize the book, website, and other media to engage students with folks and organizations around the world that are multi-solving, creatively deploying solutions across sectors and effectively confronting our most wicked problems. The Regeneration approach is particularly effective because of its clear message that we all have a role to play in the work of addressing climate change and systemic social issues. I also recommend the documentaries *Life on our Planet* (Spielberg 2022) and *Breaking Boundaries: The Science of Our Planet* (Clay 2021) to help students engage with climate science and understand natural systems.

Conclusion

When we present learners, even adult ones, with a pencil, a piece of paper, and a prompt, they're ready. They're eager to be asked what they're thinking, how they're feeling, what's stressing them out, or what's fun.

These questions and activities become playful, and then learners share with their peers. This builds self-confidence: "Oh, I'm a person who's worthy to be in this space with other people, with paper and scissors and books." There's almost a hunger; they didn't realize that they're starving from a lack of working with materials and each other in the real, physical, lived world.

Ultimately, the more students do this kind of work, the less anxiety they express and the more wonder and humor they end up sharing as they build community. They are not turning away from the big issues; these pedagogies and approaches constitute a different way of framing sustainability challenges that allows for more possibilities. As educators, we can guide students into futures-thinking and systems-change, visioning a healthy, just, and thriving future. This orientation helps learners feel excited that the way forward can be different—rather than spinning out with anxiety in a doom-cycle.

If you give learners a front row seat to the natural world by engaging them with creative exposure...walks, nature journals and zines, school gardens, tiny terrariums, worm composting bins, and then allow them to express how they feel through words, art, music, cooking, and movement, they are unstoppable! They come up with solutions that make sense in their own lives and communities. There is tremendous power in this approach. I believe creative problem-solving taps into the very essence of what it is to be human, as part of nature, ourselves. Artists out there: *bring out your inner scientist*, and scientists out there: *nurture your inner artist!* We need everyone in this effort.

References

Barry, Lynda. 2019. *Making Comics*. Drawn & Quarterly.

Clay, Jon. Dir. 2021. *Breaking Boundaries: The Science of Our Planet*. Silverback Films.

Hawken, Paul. 2021. *Regeneration: Ending the Climate Crisis in One Generation*. Penguin Books.

Hickman, C., E. Marks, P. Pihkala, S. Clayton, R. E. Lewandowski, E. E. Mayall, B. Wray, C. Mellor, and L. van Susteren. 2021. *Climate Anxiety in Children and Young People and Their Beliefs About Government Responses to Climate Change: A Global Survey*. Lancet Planet Health. Volume 5, Issue (12):e863–e873. December 2021. https://pubmed.ncbi.nlm.nih.gov/34895496/

Laws, John Muir, and Emilie Lygren. 2020. *How to Teach Nature Journaling*. Berkeley: Heyday.

Preston, Caroline. 2024. "How Colleges Can Become 'Living Labs' for Combating Climate Change." *The Hechinger Report*. July 30. https://hechingerreport.org/how-colleges-can-become-living-labs-for-combating-climate-change/

Spielberg, Steven. 2022. *Life on Our Planet*. Amblin Television & Silverback Films.

CHAPTER 13

Teaching and Learning for Sustainability: with Hope, Love or Rage?

Roland Tormey

Introduction

University instruction is often framed as an entirely cognitive activity, and so it may seem provocative to entitle a chapter in a way that places emotion as central to teaching and learning about sustainability. The title is indeed doubly provocative as it implies the possibility that a 'negative' emotion (such as *rage*) be part of teaching and learning. While many may well accept that *hope* and *love* are part of sustainability education, we may instinctively shy away from incorporating *anger*—an emotion that is seen by many as the antithesis of rational thought. But the title of this chapter takes its lead from the way the Brazilian educator Paulo Freire described his own book *The Pedagogy of Hope*: 'written in rage and love, without which there is no hope' (1992/2014, 4). When it comes to sustainability education in universities, we may need to make space for the right amount and type of rage, and some compassion, if we are to have hope.

R. Tormey (✉)
Teaching Support Center (CAPE), Ecole polytechnique fédérale de Lausanne (EPFL), Lausanne, Switzerland
e-mail: roland.tormey@epfl.ch

S. M. Wyman and J. Korenblat (eds.), *Sustainability Learning for Action and Community Engagement*,
https://doi.org/10.1007/978-3-032-11698-7_13

A focus on emotion may seem out of place in a discussion about university, yet a little reflection reveals that a wholly cognitive institutional landscape probably does not correspond with our experiences as educators: in my own case, each week after teaching three hours in a row, I generally feel exhausted. This is not because of the physical effort I put into teaching (I move around the class, but not enough to account for such exhaustion) but rather because of the emotional effort that goes into it. My sense from talking to colleagues over three decades is that others feel the same. As much as I like to think of what I do as rational work, it involves considerable emotional labor. As the renowned Andy Hargreaves notes, 'Good teachers are not just well-oiled machines. They are emotional, passionate beings who connect with their students and fill their work and their classes with pleasure, creativity, challenge and joy' (1998, 855). While teaching is an emotional experience, so too is learning. As this chapter will explore, understanding learning as an emotional experience is important to sustainability education.

In reality, emotion is already integral to university education, and ignoring it won't make it go away (Kotluk and Tormey 2023). Emotion is also central to how people respond to questions of justice, care, and fairness which affect ourselves, other people, and entities other than humans. Addressing this aspect of human experience is, therefore, an important part of learning for a sustainable future and a key aspect of addressing the inner dimensions of transformative sustainability action and education (Lewis et al. 2024). To make this argument, I will briefly define what emotion is; look at how emotion enters into questions of 'right' and 'wrong' (as well as how it is systematically excluded from such inquiries); and finish by highlighting processes that can help students to learn with the right kind of rage and love (compassion), and therefore, with hope.

An emerging practice in justice scholarship is the inclusion of a positionality statement. This allows readers to understand how the researcher and their perspective are part of the research (see Secules et al. 2021). I am a sociologist and learning scientist who worked for over a decade in Ireland on questions of educational inequity (particularly those linked to gender and ethnicity, Sustainable Development Goals (SDG 5 Gender Equality and 10: Reduced Inequalities), as well as on integrating sustainable development into education both in Ireland and more widely in Europe (SDG 4: Quality Education), (UNECE Expert Group on Competences in Education for Sustainable Development 2011). I came to appreciate the importance of emotion in how people built relationships with others or built systems for excluding others. Recognizing the importance of teachers' emotional competence, I therefore worked on how student teachers can develop their own emotional capacities (Corcoran and Tormey 2012) to be better able to work with the emotions that are part and parcel of educational inequality issues. In the last decade, my focus has shifted to Switzerland and the education of engineers, but my concerns with issues of inequity, of sustainable development, and of emotion (Lönngren et al. 2023; Lönngren et al. 2024; Kotluk and Tormey 2023; Tormey 2024) have remained central.

What Is Emotion, and Why is it Important in Sustainability Education?

There is not a single accepted definition of *emotion* (Frijda 2008, 68; Bellocchi 2019). Indeed, concepts of emotion have shifted over time such that the definition used by many researchers today differs in important ways from the popularly accepted understanding of the term. Barbalet (2001) for example, identifies what he calls a 'conventional' approach to defining emotion which sees emotion as distinct from reason and as a bodily disturbance that undermines reason. This approach can be framed as a 'Cartesian' approach, building on the conception of the philosopher René Descartes that the mind and the physical body are distinct and separate. Emotion has long been, in the words of Solomon (2008), conceptualized in Western cultures as needing to be held in check by reason because it was seen as 'more primitive, less intelligent, more bestial, less dependable, and more dangerous than reason' (3). This view emphasizes emotion as non-cognitive, a bodily reaction to a stimulus, and dangerous. This view probably remains the most common (implicit) perception of emotion and guides people's thinking about it. It would hardly be surprising then, that universities where people implicitly accept this view of emotion would want to have little to do with it.

Although this conventional definition of emotion seems deeply imbedded in the common-sense and everyday concept used by many people in the West, emotion researchers today tend to employ one closer to what Barbalet calls a 'critical' definition that differs in a number of important ways. For example, one commonly used scientific definition of emotion sees it as a multi-component episode in which a person is prepared to respond to a perceived event. The different elements include a cognitive component involving the evaluation of a situation, a neurophysiological component in which the body changes (such as increased heart rate or secretion of particular hormones), a motivational component in which the person is predisposed toward particular ways of thinking and acting, a motor expression component often seen in non-conscious facial expressions, and a subjective feeling component (Shuman and Scherer 2014). Thinking is not something that happens in response to emotional changes in the body (as in the conventional approach); rather the emotion *includes* cognitive, biophysical, motivational, and subjective elements. Other, more sociological accounts also see social and cultural factors and organizations that shape many of those components (Bericat 2016).

This reintegration of mind, body, and the social, which has become fundamental to contemporary emotions research, reconnects people to others and their planet, which is integral to sustainability education. The 'conventional' view of emotion as bodily and distinct from reason (and dangerous), is arguably part of a wider legacy of Descartes; it lays the foundations for a mechanistic view of science and a sense of human separateness from their physical environment. Insofar as sustainability education is frequently concerned with reconnecting the human intellect with biological, environmental, and social systems, it makes sense that emotion, seen in this more critical way, will be part of how we link our thinking to our world and our own bodies.

The change in how emotion is understood was part of a wider 'emotional revolution' (Sutton and Wheatley 2003, 328) in academic research that grew from the 1990s onwards. Part of this revolution was another change in how emotions were perceived by scholars: while the traditional account of emotion sees it as a dangerous and as 'disturbing' our cognition, researchers since the 1990s have increasingly focused on the idea that emotions do not simply *disturb* thinking but rather, they *facilitate particular kinds* of thought. This more nuanced view of emotion was perhaps best seen in the introduction of the concept of emotional intelligence (Mayer and Salovey (1993); Nussbaum 2001) to account for the way that thinking with emotion is integral to solving complex everyday problems, as well in as neuroscience perspectives which emphasize that emotion is integral to 'rational' judgements (Damasio 1994). More recently, this research has also included how a range of emotions is essential to ethical and pro-social behaviors (Haidt 2001, 2003; Hoffman 2008; Nussbaum 2016; Cherry 2021). To be clear, the change from 'conventional' to more 'critical' views is not a switch from seeing emotion as dangerous to seeing emotion as uncritically positive: instead, the critical perspective sees different emotions as facilitating contrasting patterns of thinking and acting. Whether or not these are positive may well depend on the context in which they occur, the goals of the people involved, and on the perspective of the observer.

The emotional revolution included a massive growth in interest in emotion in teaching and learning, and there is now a large body of research showing that emotions 'profoundly affect students' and teachers' engagement, performance and personality development' (Pekrun and Linnenbrink-Garcia 2014a, 659). Thinking about emotions in education is often categorized in terms of their object focus, described as (i) topic emotions, (ii) epistemic emotions, (iii) achievement emotions, and (iv) social emotions (Pekrun and Linnenbrink-Garcia 2014b).

Topic emotions refer to the material being studied. This might include an empathetic response to a character in a work of eco-fiction for someone studying literature, or a sense of awe at the elegance of a mathematical proof. For example, in a study of the emotions experienced by young people (aged 16 to 25 years old) while learning about sustainability topics, Dunlop and Rushton (2022) found that students expressed fear of being judged by others who might see them as being out of touch with other students' lives and concerns (fear, embarrassment, shame), as well as a sense of being held individually responsible for addressing sustainability challenges (guilt). Climate anxiety or *ecoanxiety* could also be seen as a topic emotion (Clayton 2020). Dunlop and Rushton (2022) also found students experienced some emotions that felt more positive—such as hope—in their learning around sustainability. Topic emotions like these can impact upon a person's willingness to engage with the material: content which gives rise to emotions that feel bad may encourage people to disengage.

Epistemic emotions refer to those linked to processing of information in a learning situation. For example, a person who finds out something they did not previously know may experience surprise, while someone who finds themselves unable to solve a problem that they expected to solve will perhaps feel frustration. Epistemic emotions may well be particularly important for students in sustainability learning, since sustainability education requires engaging with problems that are often of quite a different type and systemic complexity from those they typically face in traditional academic settings. Lönngren, Ingerman, and Svanström (2017), for example, observed engineering students grapple with a wicked sustainability problem involving access to water in Jordan, Israel, Lebanon, and Syria. They noted that students used various approaches to try to solve the problem, with some students using methods more suited to the kinds of routine or well-structured problems commonly used in their education programs. They note that these students experienced high levels of frustration, a sense of being 'stuck,' and this contributed to giving up. This kind of emotion, felt in response to a recognition that one's method isn't working, would typically be seen as an epistemic emotion. In the case of sustainability challenges, the probability of such a reaction is high, and may further contribute to (eco-)anxiety about the 'solvability' of such problems.

Achievement emotions refer to the judgements that are made about how well educational outputs meet standards of quality, whether or not a learner feels they have done something well. Common achievement emotions include pride, shame, and anxiety. Some of the key features of wicked sustainability problems such as those studied by Lönngren et al. (2017) are that (i) the time frame is open ended and so there is no moment at which the solution can be definitively evaluated and (ii) there are competing value systems in place, so no solution can be unambiguously 'good' for all stakeholders (see also Seager, Selinger and Wiek 2012). Since these problems are not open to unambiguous evaluation which can lead to a sense of achievement, they may be less likely to lend a sense of pride or achievement when compared to other domains of learning.

Social emotions refer to the interaction students have with teachers and with each other. This interaction can give rise to a range of emotions including those linked to warmth (which enables collaboration), a sense of emotional security (which allows them to try, fail, get feedback, and improve), and a sense of respect for their status and that of others. This can include emotions like anger when they feel disrespected or admiration when they see others who deserve their respect (Tormey and Isaac 2022, 157+). These social emotions are probably evident in all learning situations but are particularly prominent in group projects (Kotluk et al. 2023; Germanier et al. 2023). Such collaborative activities are often part of sustainability learning, and this can have a significant impact.

Emotions, then, are present in all learning and affect how students engage and how much and what they learn. As the next section will explore, emotions can contribute significantly to sustainability learning goals.

Emotions and Sustainability Learning

The dominant way of thinking about emotion in Western cultures as a problem to be removed from rigorous, ethical problem-solving is built mainly on the work of developmental psychologist Lawrence Kohlberg (Rest et al. 2000, 382). Kohlberg's work focused on the development of cold, rational moral judgements and explored how people apply 'universal' principles. Perhaps the best-known critique of Kohlberg's work comes from moral psychologist Carol Gilligan (1982) who argued that ethical reasoning could equally be grounded in a sense of care in context within social networks. This ethics of care approach provides a 'compelling foundation for environmental ethics,' according to Powys White and Cuomo (2017, 235).

Influenced by the work of writers Carol Gilligan and the philosopher Martha Nussbaum, Sabine Roeser (2012a, 2012b) has argued that emotions are crucial to ensure people develop a full understanding of the nature of risk; 'rather than being biases that threaten objectivity and rationality in thinking about acceptable risks, emotions contribute to a correct understanding of the moral acceptability of a hazard' (2012a, 107). If we take an environmental concern such as climate change that affects people who are often distant from us in space or time (in the future): 'altruistic emotions can help us care about the needs and rights of people who are far away, even if it means that we have to make personal sacrifices' (2012b, 1036). Work by Pfattheicher et al. (2016) has added empirical support to this proposition that compassion plays a role in pro-environmental values, pro-environmental intentions, as well as reported pro-environmental actions.

Other scholars have looked at emotions that may be relevant to motivation to act. Rees, Klug, and Bamberg (2015), for example, have explored how different ways of framing the same sustainability educational content can induce guilt and lead to either greater or lesser motivation to act. Higgs et al. (2020) looked specifically at how guilt and shame impact people's perceptions of ethical cases. They found that participants who felt shame, for example, reported the highest levels of personal responsibility while those who felt guilt saw the problem as more pressing. Similarly, Samantha Stanley and colleagues have found eco-anger to be associated with pro-environmental personal behavior (Stanley et al. 2021). Emotion may not only motivate or inhibit action, but it may also improve cognitive learning (Thiel et al. 2013).

The impact of emotion on learning is not always positive, however. While some degree of compassion may increase people's engagement with material, too much may become distressing for learners and cause them to withdraw from learning—a process known as *empathic overarousal* (Hoffman 2008). The meta-analysis on ethics learning by Logan Watts et al. (2017), for instance, found that ethics cases with low to moderate realism have more impact than realistic cases that may involve highly emotive content such as multiple deaths. Similar issues with disengagement may also arise with ecoanxiety (Stanley et al. 2021). Similarly, while anger may have a positive effect on motivation to act, it may have a more negative impact on the quality of moral reasoning that goes into that action: feeling angry is associated with a less thorough evaluation of ethical dilemmas and an increased focus on punishment and retribution (as compared to more restorative actions) (see Thiel, et al. 2011).

Political philosophers in recent years have been divided on the role of anger in pro-social action. Nussbaum (2015; 2016) has discussed the risk that anger motivates an inordinate focus on punitive retribution. For her, the only productive form of anger is what she calls *Transition-Anger*, anger addressed at injustice but without seeking retribution (2015, 54). Others have argued for a larger role for a kind of anger that targets beliefs, policies, behaviors, and social systems rather than targeting individual people (Cherry 2021). The question for *rage* then, may be what kind of rage and how much, rather than whether or not to feel rage within the teaching and learning context.

It may be useful here to think of emotion as being like complexity in a mathematical problem: if there is insufficient complexity in the problem, the student is not stretched beyond their existing ability; they get bored, and they do not learn. If the complexity is too great, the student is overwhelmed, unable to see how to respond and—again—shuts down. But the level of complexity of problem a student can face will change over time as they develop a set of thinking skills that enables them to navigate such complex problems. If, however, the complexity is of the wrong type (such as a student having to cope with an error the teacher made), the student will again, not learn.

Teaching Strategies to Integrate Emotion into Sustainability Education

This final section outlines some strategies for university teachers to actively embrace emotion as fundamental to the learning experience, even those educators who feel that they themselves do not have much expertise.

Develop emotional awareness and language: Although there are many different typologies that reflect ways in which emotional competences are understood (for example, as an intelligence and as a learnable skill), there are shared features that tend to be found across a range of different models. Hoemann et al. (2021) identify a common feature of many accounts of emotional expertise: experts 'are *consciously aware* of their [emotional] experiences and... experts use specific language to *verbally represent* them' (2021, 1174). Since emotions are already present in teaching and learning, a starting point could be in helping students become more consciously aware of the emotions they are feeling and develop language to describe them. Giving people a space to make their emotions explicit through self-reflection contributes to developing such competence. This may be as simple as inviting students to 'check-in' at the outset of class by sharing a word that describes their current emotion. In contexts of artistic forms of expression, this too could be used since emotions are often linked to colors, textures, or patterns. During the COVID pandemic, many in higher education developed a practice of inviting students to write in the chat how they were feeling at the beginning of a synchronous online class. After the return to face-to-face teaching, the use of classroom response software allows students to share anonymously how they are feeling at the start of class and attend to the feelings of others. The recognition that others are / are not feeling the same as themselves in that moment may also contribute to students' capacity for emotional empathy (intra-personal competency). Understanding the diversity of students' feelings is also very helpful information for the teachers. If the teacher is concerned that the emotional language used is insufficiently precise, displaying a typology of emotions (such as Plutchik's Wheel of Emotions) on a slide can support students in deciding which terms to use.

Explicitly include emotion in examples: Since there is evidence that (some) emotions can aid cognitive learning and improve student motivation in relation to sustainability and ethical issues, it would make sense for sustainability educators to intentionally add emotional information (about, for example, the feelings of those affected by events or the nature of the effects they experienced) to case studies or examples studied as part of sustainability learning. Indeed, as my colleague Nihat Kotluk and I found (Kotluk and Tormey 2024), even when examples that involve potential harm to others are framed in emotionally neutral terms, students identify emotions within these cases. Rather than leaving it to chance to have such emotions recognized, it would make sense for teachers to be deliberate about this.

We identified a number of strategies for embedding *compassion*, for example, in the curriculum. These included (i) *implicit induction*, through identifying someone who is affected by an event, by defining potential harm to them, and by including information that may induce a feeling of connection with them, (ii) by *explicitly expressing* that others feel compassion or sympathy for the person and (iii) by *clarifying negative consequences* to that person due to the event. We found that all three methods increased students' rating of the intensity of compassion in the case without impacting on other emotions (such as anger). Equally, it is worth considering that specific features of case studies may reduce compassion. Nussbaum (2013) has argued that compassion requires a judgement that something bad has happened to someone else, that the person does not deserve their suffering, and a judgement that the observer is themselves similarly vulnerable. This suggests that if a case does not identify anyone who might suffer individually, if blame is attributed to that person, or if the differences between the life of that person and that of the reader are emphasized, the case may actually *diminish* the student's responsive compassion.

Making problem solving strategies explicit: Johanna Lönngren et al. (2017) identified that one source of frustration and ultimately withdrawal for students is that the problem-solving approaches needed to address wicked sustainability problems are not the same as the problem-solving approaches more commonly used in university education (which are variously described as 'tame' or 'well-structured'). They concur with Hattie (2009) that explicit teaching of problem-solving strategies or foregrounding pedagogical strategies for the students tends to be more effective than simply exposing students to different strategies. Furthermore, given the ambiguous nature of 'success' in addressing the vast complexity of such wicked sustainability problems, including the *polycrisis*, teachers may need to be very explicit with students about the criteria against which student success is measured (assessment rubric or exemplification). Being as explicit as possible with students about what success looks like can increase student success and, simultaneously, reduce unnecessary stress.

Addressing Social Emotions in Complex Problem Solving: Since solving sustainability problems requires the input of many perspectives, it often requires that students work in teams or groups. These groups may be within a discipline or interdisciplinary. Social emotions can play an important role in the success of student's learning. In particular, there is evidence that, in gender diverse groups, women tend to be assigned less technical roles and may not have their opinions taken as seriously as those of male colleagues (Aeby et al. 2019). This can be associated with emotions such as shame, anger, and anxiety. Women students may also be required to perform additional emotional labor to fit into such teams (Germanier et al. 2023). This can impact both the learning of women in such groups and on the success of the project. Similar issues can arise for other minoritized students. Groupwork can also be particularly challenging for neurodiverse students who may find the need to imagine the perspective of others to be difficult or to impose an additional cognitive load. One way to address this is to provide diversity and implicit bias training before students work in teams and strategies to ensure a more balanced participation (Isaac et al. 2023).

Address ecoanxiety: While empathic overarousal and ecoanxiety may lead students to withdraw from engagement with sustainability issues, educational interventions that include the possibility of taking some collaborative action—('hands'), alongside work on concepts and ideas ('head') and emotional processing ('heart')—appears to be more successful in reducing anxiety (Olsen et al. 2024). Olsen et al. (2024) have found that, while there is little research evidence on effective educational strategies to address ecoanxiety, there is a consensus that environmental education should 'provide space for learners and/or educators to process their emotions regarding climate change, including their eco/climate anxiety' (20).

Conclusion

While universities often imagine themselves to be emotion-free spaces for rigorous intellectual pursuits, emotions are not simply present in universities; they are important to learning in general and particularly in educational settings that address human, planetary, and other creature thriving. It makes sense to engage openly with feelings so they can positively contribute to and guide students' learning. Emotions like *compassion* play a positive role in learner engagement with sustainability issues, but so too may emotions that have historically been regarded as 'negative' such as *guilt* and even *rage*. While the evidence about *anger* is more nuanced, it clearly has a role in sustainability education. Numerous pedagogical strategies are accessible to those who wish to recognize and benefit from the emotion that is already present in their teaching and learning.

References

Aeby, P., R. Fong, M. Vukmirovic, S. Isaac, and R. Tormey. 2019. The impact of gender on engineering students' group work experiences. *International Journal of Engineering Education* 35 (3): 756–765.

Barbalet, J. M. 2001. *Emotion, Social Theory, and Social Structure: A Macrosociological Approach*. Cambridge University Press.

Bellocchi, A. 2019. Emotions and teacher education. In *The Oxford research encyclopaedia of education*, ed. G. Noblit. Oxford: Oxford University Press.

Bericat, E. 2016. The sociology of emotions: Four decades of progress. *Current Sociology* 64 (3): 491–513. https://doi.org/10.1177/0011392115588355.

Cherry, M. 2021. *The Case for Rage*. Oxford: Oxford University Press.

Clayton, S. 2020. Climate anxiety: Psychological responses to climate change. *Journal of Anxiety Disorders* 74:102263. https://doi.org/10.1016/j.janxdis.2020.102263.

Corcoran, R., and R. Tormey. 2012. *Developing Emotionally Competent Teachers: Emotional Intelligence and Pre-service Teacher Education*. Oxford: Peter Lang.

Damassio, A. 1994. *Descartes' Error: Emotion, Reason and the Human Brain*. New York: Putnam and Sons.

Dunlop, Lynda, and Elizabeth AC Rushton. 2022. Education for environmental sustainability and the emotions: Implications for educational practice. *Sustainability* 14 (8): 4441. https://doi.org/10.3390/su14084441.

Freire, P. 1992/2014. *Pedagogy of Hope, Reliving Pedagogy of the Oppressed*. Bloomsbury.

Frijda, N. 2008. The psychologists' point of view. In *Handbook of Emotions*, ed. M. Lewis, J. M. Haviland-Jones, and L. Feldman Barrett, Third ed., 68–87. New York: The Guilford Press.

Germanier Rachel, Darioly Annick, Kotluk Nihat, Tormey Roland. 2023. "But of course I'm going to look happy" or "he needed to know I was angry"?

Comparing use of emotional labour in teamwork in engineering and hospitality students *EAPRIL 2023 conference, Belfast, Northern Ireland (UK), November 22–24, 2023.* https://infoscience.epfl.ch/record/310305?ln=en&v=pdf.

Gilligan, C. 1982. *In a Different Voice; Psychological Theory and Women's Development*. Cambridge MA: Harvard University Press.

Haidt, J. 2001. The emotional dog and its rational tail: A social intuitionist approach to moral judgment. *Psychological Review* 108 (4): 814. https://doi.org/10.1037/0033-295X.108.4.814.

Haidt, J. 2003. The moral emotions, in *Handbook of Affective Sciences,* edited by R.J. Davidson, K.R. Scherer & H.H. Goldsmith, H. H., 852–870. Oxford: Oxford University Press.

Hargreaves, A. 1998. The emotional practice of teaching. *Teaching and Teacher Education* 14 (8): 835–854. https://doi.org/10.1016/S0742-051X(98)00025-0.

Hattie, J. 2009. *Visible Learning: A Synthesis of Over 800 Meta-analyses Related to Achievement*. London: Routledge.

Higgs, C., T. McIntosh, S. Connelly, and M. Mumford. 2020. Self-focused emotions and ethical decision-making: Comparing the effects of regulated and unregulated guilt, shame, and embarrassment. *Science and Engineering Ethics* 26 (1): 27–63. https://doi.org/10.1007/s11948-018-00082-z.

Hoemann, Katie, Catie Nielson, Ashley Yuen, Jacob W. Gurera, Karen S. Quigley, and Lisa Feldman Barrett. 2021. Expertise in emotion: A scoping review and unifying framework for individual differences in the mental representation of emotional experience. *Psychological bulletin* 147 (11): 1159.

Hoffman, M. L. 2008. Empathy and Prosocial behaviour, in *Handbook of Emotions*, ed. M. Lewis, J. M. Haviland-Jones, and L. Feldman Barrett, Third ed., 440–455. New York: Guilford Press.

Isaac, S., N. Kotluk, and R. Tormey. 2023. Educating engineering students to address bias and discrimination within their project teams. *Science and Engineering Ethics* 29:6. https://doi.org/10.1007/s11948-022-00426-w.

Kotluk, N., and R. Tormey. 2023. Compassion and engineering students' moral reasoning: The emotional experience of engineering ethics cases. *Journal of Engineering Education* 112 (3): 719–740.

Kotluk, N., and R. Tormey. 2024. The impact of different methods of increasing the intensity of compassion in engineering ethics cases. *European Journal of Engineering Education.* https://doi.org/10.1080/03043797.2024.2341758.

Kotluk, N., R. Tormey, R. Germanier, and A. Darioly. 2023. Emotional labor experienced in team-projects: A comparison of engineering and hospitality students. In *European Society for Engineering Education (SEFI) 51st annual conference 2023.* Dublin. https://doi.org/10.21427/B0JC-X967.

Lewis, A. L., I. Providoli, A. Sundermann, A. Frank, D. Fischer, P. Frank, M. Studer, R. Tormey, L. J. Trechsel, and C. Wamsler. 2024. *Addressing inner dimensions for sustainability in higher education; A facilitator guide*. Bern.: Centre for Development and Environment (CDE), University of Bern, with Bern Open Publishing (BOP). https://doi.org/10.48620/76349.

Lönngren, J., Å. Ingerman, and M. Svanström. 2017. Avoid, control, succumb, or balance: Engineering students' approaches to a wicked sustainability problem. *Research in Science Education* 47:805–831. https://doi.org/10.1007/s11165-016-9529-7.

Lönngren, J., I. Direito, R. Tormey, and J. Huff. 2023. Emotions in engineering education, in *The International Handbook of Engineering Education Research*, ed. A. Johri, 156–182. London: Routledge.

Lönngren, Johanna, Alberto Bellocchi, Maria Berge, Pia Bøgelund, Inês Direito, James L. Huff, Khairiyah Mohd-Yusof, Homero Murzi, Nor Farahwahidah Abdul Rahman, and Roland Tormey. 2024. Emotions in engineering education: A configurative meta-synthesis systematic review. *Journal of Engineering Education* 113 (4): 1287–1326. https://doi.org/10.1002/jee.20600.

Mayer, J. D., and P. Salovey. 1993. The intelligence of emotional intelligence. *Intelligence* 17 (4): 433–442. https://doi.org/10.1016/0160-2896(93)90010-3.

Nussbaum, M. 2001. *Upheavals of Thought, the Intelligence of Emotions*. Cambridge: Cambridge University Press.

Nussbaum, M. 2013. *Political Emotions*. Cambridge MA: Harvard University Press.

Nussbaum, Martha C. (2015). Transitional anger. *Journal of the American Philosophical Association* 1 (1): 41–56. https://doi.org/10.1017/apa.2014.19.

Nussbaum, M. 2016. *Anger and Forgiveness, Resentment, Generosity, Justice*. Oxford: Oxford University Press.

Olsen, E. K., D. F. Lawson, L. R. McClain, and J. D. Plummer. 2024. Heads, hearts, and hands: A systematic review of empirical studies about eco/climate anxiety and environmental education. *Environmental Education Research*. https://doi.org/10.1080/13504622.2024.2315572.

Pekrun, R., and L. Linnenbring-Garcia. 2014a. Conclusions and future directions, in *International Handbook of Emotions in Education*, ed. R. Pekrun and L. Linnenbrink-Garcia, 659–675. New York: Routledge.

Pekrun, R., and L. Linnenbring-Garcia. 2014b. Introduction to emotions in education, in *International Handbook of Emotions in Education*, ed. R. Pekrun and L. Linnenbrink-Garcia, 1–10. New York: Routledge.

Pfattheicher, S., C. Sassenrath, and S. Schindler. 2016. Feelings for the suffering of others and the environment: Compassion fosters Proenvironmental tendencies. *Environment and Behavior* 48 (7): 929–945. https://doi.org/10.1177/0013916515574549.

Powys White, K., and C. J. Cuomo. 2017. Ethics of caring in environmental ethics: Indigenous and feminist philosophies, in *The Oxford Handbook of Environmental Ethics*, ed. S. M. Gardinier and A. Thompson, 234–248. Oxford: Oxford University Press.

Rees, J. H., S. Klug, and S. Bamberg. 2015. Guilty conscience: Motivating pro-environmental behavior by inducing negative moral emotions. *Climatic Change* 130:439–452. https://doi.org/10.1007/s10584-014-1278-x.

Rest, James R., Darcia Narvaez, Stephen J. Thoma, and Muriel J. Bebeau. 2000. A Neo-Kohlbergian approach to morality research. *Journal of Moral Education* 29 (4): 381–395. https://doi.org/10.1080/713679390.

Roeser, S. 2012a. Emotional engineers: Towards morally responsible design. *Science and Engineering Ethics* 18:103–115. https://doi.org/10.1007/s11948-010-9236-0.

Roeser, S. 2012b. Risk communication, public engagement, and climate change: A role for emotions. *Risk Analysis* 32:1033–1040. https://doi.org/10.1111/j.1539-6924.2012.01812.x.

Seager, T., E. Selinger, and A. Wiek. 2012. Sustainable engineering science for resolving wicked problems. *Journal of Agricultural Environmental Ethics* 25:467–484.

Secules, D., C. McCall, J. A. Mejia, C. Beebe, A. S. L. Masters, M. Sánchez-Peña, and M. Svyantek. 2021. Positionality practices and dimensions of impact on equity research: A collaborative inquiry and call to the community. *Journal of Engineering Education* 110 (1): 19–43.

Shuman, V., and K. R. Scherer. 2014. Concepts and structures of emotions. In *International handbook of emotions in education*, ed. R. Pekrun and L. Linnenbrink-Garcia, 23–45. London: Routledge. https://doi.org/10.4324/9780203148211.

Solomon, R. C. 2008. The philosophy of emotions, in *Handbook of Emotions*, ed. M. Lewis, J. M. Haviland-Jones, and L. Feldman Barrett, Third ed., 3–16. New York: The Guilford Press.

Stanley, S. K., T. L. Hogg, Z. Leviston, and I. Walker. 2021. From anger to action: Differential impacts of eco-anxiety, eco-depression, and eco-anger on climate action and well-being. *The Journal of Climate Change and Health* 1:10003. https://doi.org/10.1016/j.joclim.2021.100003.

Sutton, R. E., and K. F. Wheatley. 2003. Teachers' emotions and teaching: A review of the literature and directions for future research. *Educational Psychology Review* 15 (4): 327–358. https://doi.org/10.1023/A:1026131715856.

Thiel, C. E., S. Connelly, and J. A. Griffith. 2011. The influence of anger on ethical decision making: Comparison of a primary and secondary appraisal. *Ethics & Behavior* 21 (5): 380–403. https://doi.org/10.1080/10508422.2011.604295.

Thiel, C. E., S. Connelly, L. Harkrider, L. D. Devenport, Z. Bagdasarov, J. F. Johnson, and M. D. Mumford. 2013. Case-based knowledge and ethics education: Improving learning and transfer through emotionally rich cases. *Science and Engineering Ethics* 19:265–286. https://doi.org/10.1007/s11948-011-9318-7.

Tormey, R. 2024. Teaching ethical decision making to students as 21st century professionals, in *The Routledge Handbook of Global Sustainability Education and Thinking for the 21s Century*, ed. M. John. London: Routledge.

Tormey, R., S. Isaac, and with Hardebolle, C. and Le Duc, I. 2022. *Facilitating Experiential Learning in Higher Education; Teaching and Supervising in Labs, Fieldwork, Studios and Projects*. London: Routledge.

UNECE Expert Group on Competences in Education for Sustainable Development. 2011. *Learning for the future, competences in education for sustainable development*. Utrecht: Programma Leren voor Duurzame Ontwikkeling.

Watts, L. L., K. E. Medeiros, T. J. Mulhearn, L. M. Steele, S. Connelly, and M. D. Mumford. 2017. Are ethics training programs improving? *A Meta-Analytic Review of Past and Present Ethics Instruction in the Sciences, Ethics & Behavior* 27 (5): 351–384. https://doi.org/10.1080/10508422.2016.1182025.

CHAPTER 14

Climate Change and Mental Health: Students' Use of Poetry to Express Ecoanxiety

Nosipho Faith Makhakhe

Introduction

Little research has been done to explore how young people from the Global South experience and mitigate *ecoanxiety*, any form of mental stress and unrest induced by the ecological crisis (Pihkala 2020). This chapter seeks to shed light on this oversight by discussing climate change as a threat multiplier with significant impacts on mental health. I present a transformative creative writing learning opportunity explored by students from the Durban University of Technology (DUT) in the city of Durban, South Africa, as a means to learn about human-induced climate instability and ecoanxiety and address these feelings. This group of twenty-one university students in their second year engaged in a Collaborative Online International Learning (COIL) experience with students from Sheffield

N. F. Makhakhe (✉)
Centre for General Education, Durban University of Technology, KwaZulu-Natal, South Africa
e-mail: nosiphom8@dut.ac.za

S. M. Wyman and J. Korenblat (eds.), *Sustainability Learning for Action and Community Engagement*,
https://doi.org/10.1007/978-3-032-11698-7_14

Hallam University in England to explore and exchange students' knowledge, understandings, and experiences of climate change and mental health in their respective unique contexts. Prior to this COIL engagement, students from DUT admitted to having a limited understanding of climate change and its impact. They viewed the climate intensified weather patterns as an act of God unrelated to any human-induced activities. As a lecturer, I found the use of poetry a compelling means to convey complex emotions in response to natural disasters and develop the requisite *eco-literacy* to describe their experiences and feelings. The term eco-literacy was first used by US educator David Orr and Fritjof Capra in 1990 to introduce environmentally friendly lifestyle values through education, principles of responsibility and love for the earth.

As a learning model, I focus on the aspect of our virtual exchange project which involved students being asked to explore their feelings regarding climate change. This session was significant because, for the first time, the DUT students reflected on the catastrophic flooding event that took place in Durban and the impact that it had on their mental health. Durban experienced severe flooding and mudslides in April 2022, which caused widespread damage to infrastructure. The floodwaters destroyed roads and bridges, making it difficult for emergency responders to reach affected areas. At least 435 people were reported dead, fifty-four were missing, and thousands more were displaced from their homes (Joubert 2023). Considering this climate-induced event, I asked my students to reflect on their lived experiences during and after the flood using art as a means of expression. I discuss students' poems as well as provide my own interpretations as someone who also lived through the 2022 Durban floods, historically, the greatest climate change-induced disaster in South Africa.

Climate Change as Mental Health Threat Multiplier

According to the United Nations, climate change refers to long-term shifts in temperature and weather patterns exacerbated by human activities such as the burning of fossil fuels. The burning of coal, oil, and gas generates harmful greenhouse gases that envelop the earth, trapping the sun's heat and bringing about a rise in temperatures. This warming has resulted in unfavorable outcomes such as intense droughts, water scarcity, severe fires, rising sea levels, flooding, and many more catastrophic weather-related disasters. People worldwide are negatively impacted by the adverse environmental effects of climate change. This impact is not only physical but emotional. Research has shown that this has been the most neglected health condition. It is only in recent years that the awareness of mental health as an important aspect of health and well-being is slowly gaining traction with research evidence that mental health challenges affect an estimated billion people worldwide (Lawrance et al. 2021). The World Health Organization defines mental health as a state of well-being that enables people to cope with the stresses of life, realize their capabilities, and be active members of society through building relationships, learning, and working for the greater good (WHO 2022). Various social and structural determinants may protect or undermine our mental health. For example, exposure to unfavorable social, economic, geopolitical, and environmental factors puts people at risk of experiencing adverse mental health conditions, and climate change is one such phenomenon (WHO 2022). According to Lawrance et al. (2021), climate change and mental health are two of the most significant challenges facing societies currently. Yet, the nexus of these two problems has not been adequately explored or addressed.

Climate change and rising temperatures have a *threat multiplier* impact on mental health in the forms of increased social violence, post-traumatic stress disorder (PTSD), and suicide rates. A South African study predicts that aggression, violent behavior, and homicides could increase by 18% with temperatures that go higher than 30 degrees Celcius / 86 degrees Fahrenheit (Subramaney et al. 2022). Flooding events have been shown to increase the risk of PTSD, depression, or low mood (Lawrance et al. 2021). In South Africa, it is further predicted that with every one-degree Celsius increase in warming, displacement of people due to flood events will rise by 50%. This disruption or forced migration contributes to economic and psychological challenges as it affects livelihoods and access to essential services (Subramaney et al. 2022). Studies conducted in various countries such as India, Finland, Italy, the United States of America, and Mexico have all found an association between higher temperatures and increased suicide rates (Charlson et al. 2021). A specific statistic shows that the continued rise of temperatures could result in an additional 21,770 suicides by 2050 in the United States and Mexico alone. While in Finland, temperature changes contributed to an estimated 60% variance in suicide rates (Dumont et al. 2020).

The pervasive effects of climate change are most likely to contribute to compounding vulnerabilities particularly among those with existing mental illnesses and those living in poorer countries. Research shows that the impacts of climate change will severely affect the developing world far more than it will impact developed countries due to the limited infrastructure and ability to respond to natural disasters within these contexts (Lawrance et al. 2021). Young people have been found to be affected by climate change and feel a deep sense of despondency due to a perception that they can do very little to mitigate climate change since they do not occupy positions of power (Lawrance et al. 2021). Studies have found that young people between the ages of 18 and 35 reported higher levels of climate-related anxiety which affected their ability to function optimally in everyday life. Furthermore, females experienced higher stress levels than males of a similar age group (Coffey et al. 2021). This sense of hopelessness and dread is evidence of *ecoanxiety*. This set of negative emotions manifests both mentally and physically. For example, people feel physically sick, develop panic attacks, irritability, depression, helplessness, guilt, frustration, and anger as well as a host of other emotions associated with the ecological crisis (Coffey et al. 2021). A review by Coffey et al. (2021) found studies that explored the concept of ecoanxiety were conducted in Western countries and study populations were mostly Caucasian. The recommendation is that more studies are needed that explore the concept of ecoanxiety among Black people or people of color, as well as a need for more studies conducted in the Global South (Coffey et al. 2021). Hence the focus of this chapter on Black students in a South African university and their conceptualization of climate change and mental health, particularly ecoanxiety.

Learning Model: Poetry as a Means to Express and Mitigate Ecoanxiety

Our COIL collaboration on creative responses to climate change and ecoanxiety ran once a week for three weeks in March 2024. The program was structured into three broad themes; session one, “Our cities” started off with an icebreaker to establish rapport between the collaborating South African and English students. Thereafter, students were given a chance to introduce their respective cities, showcasing any historical facts, tourist and cultural attractions, as well as speak about a recent consequential climate event that took place. During session two, “Our feelings," students were encouraged to use artistic modalities including poetry and drawings to explore their feelings regarding climate change and ecoanxiety. Students reflected upon a range of emotions such as hope, fear, anger, justice, and compassion regarding the environment. The last session had students focus on climate change policies that exist within their countries and engage in a critical discussion regarding issues of political will, implementation, accountability, and the proposed policy targets and deadlines.

Upon learning about *ecoanxiety* for the first time, the DUT students expressed a range of emotions, from fear to a desperate appeal for climate change to stop. There was also a deep sense of hopelessness, a loss of control, and sense of being at the mercy of the environment. When we began our conversations about our experience during these floods, students expressed emotions that they felt but could not quite name. They would describe their sensations and emotive responses to shifts in the weather: *whenever it starts to rain, I would feel nervous, or I would want to hurry and get home because I'm scared that this rain is going to be a violent storm*. They reported a sense of fear, anticipation and apprehension whenever the weather changed. During one of our classes, the clouds started gathering, and I asked my students to check their emotional response, "How do you feel? Look at what's happening outside." A heavy rain was beginning, a literal hailstorm. Next time we met, I asked, "What happened that day? How did people get home? How did people feel?" Students reported (to paraphrase): *I was feeling anxious. I was afraid. I was nervous. I didn't know what to expect and now whenever it rains, I feel that feeling*.

The concept of ecoanxiety was important to students because it gave them a means to articulate and express their emotions regarding the changing weather patterns, intensified by climate change. During class discussions, some students said that they felt a sense of fear and dread during heavy rains, however, they did not understand these feelings. Upon learning about ecoanxiety, they have developed a language to understand and express traumatic emotions related to climate change-induced weather conditions, which makes them aware of the significance of their feelings.

For this creative exercise, some students chose to narrate their experiences through storytelling or drawing, some put together collages, and a few students wrote poetry to express their emotions related to ecoanxiety. Their *eco-literacy* increased as they observed their own emotional responses as a shared experience with their peers. They were able to identify, name, and own the phenomenon of *ecoanxiety*. If you can't name an experience, it's hard to describe and share it; you're not sure what it is. Now, they could explore their emotional responses because they were given a language to do so. For some, this was a moment of epiphany. Our next tasks as educators were to (1) provide creative learning experiences as opportunities to express and share these feelings through art and (2) to extend the concept of ecoanxiety to a global context with attention to politics and policy.

Using Poetry to Reflect and Address Mental Health Challenges

Poetry writing has been used by researchers as a process to facilitate reflective or reflexive thinking. According to Luce-Kapler (2009), writing poetry enables one to convey complex experiences in a meaningful way, and this process can shed light on new and emerging insights. Poetry as a mode of reflective writing can enhance self-understanding (Owton 2017). As a vehicle to process and express difficult emotions this mode of writing can have a therapeutic or cathartic effect.

In this context, DUT students expressed sensitive emotions regarding their experiences of the Durban floods. One student wrote a poem where she personifies climate change and appeals to its "human nature" (*used with permission*).

> Our beautiful earth is where we live,
> Where we were supposed to be happy and comfortable.
> We are powerless, hopeless, angry and nervous
> Because of you climate change….
> Can you stop your performance, climate change?
> People are dying, animals are dying, houses are floating,
> And some people are lost, can't you get tired, climate change?
> Our future is unclear, our next generation will never see the
> Beauty of our environment because of you, climate change.

Another student expressed a deep sense of fear and hopelessness induced by climate change (*used with permission*).

> Dark it is, so scary it was,
> Rain, lightning,

Oh Rain, can you stop now!
There are floods,
Flooded hope,
Drains gurgle,
Leaves break,
I check the forecast,
These four walls closing in on my brain,
I am suffocating and I am losing hope,
Will these floods ever end?

These poems take a passive approach to the issue of climate change. There is an inherent failure to recognize that as people, we can take action to reduce damage to the environment. A US study by Geiger et al. (2021) found that increased feelings of hope were important in determining action against climate change. Furthermore, the extent to which people felt anxious or helpless was not strongly correlated with their intentions to act. This evidence shows that while it is important to educate people about climate change, it is equally important to emphasize the incremental actions that they can take to contribute positively to the environment and establish a sense of hope.

Another aspect that emerged from the poems and class discussions was that students did not make an immediate connection between the Durban floods as the consequence of climate change. This skepticism is further expressed in the poem below, which ends with the question "Could it truly be climate change?" (*used with permission*).

Will it ever end?
Oh rain, can you stop now!
There are floods, flooded hope.
All that grassy, green, warm weather
Doesn't just mean heatwaves and fire.
And while daily predictions and conclusions are inaccurate,
We experienced weather that wasn't predicted.
Could it truly be climate change?

This poem expresses a sense of oblivion without any interrogation as to what causes floods and the role that humanity plays in influencing changing weather patterns. With that said, it is important to acknowledge that engaging with issues of climate change is itself a privilege. When most people in poorer countries are faced with a lack of basic needs, the preoccupation is with daily survival instead of thinking about an unknown future. Research by the United Nations confirms that rich countries place a higher priority on climate change than poorer countries. The people in these countries tend to worry more about necessities such as food, housing, and employment despite the fact that climate change has a direct bearing on these issues (Plumer 2014).

At the end of the COIL project, I questioned students about how they felt about the future now that they had learnt about climate change and its consequences. Most students expressed a combination of dread, hopelessness, and uncertainty as captured in the poem below. They based these feelings on the fact that climate change is largely due to man-made actions and bad decisions. This renders the battle against climate change difficult because people are inherently selfish and are bound to act in ways that benefit their individual interests above the need to save the environment. Furthermore, students acknowledged and advocated for a worldwide, concerted effort to address climate change issues.

> The world is changing and I'm scared,
> The future is exposed, the world is in pain.
> Forests burn, animals suffer,
> Trees are broken, their beauty is lost.
> And all around the planet yearns.
> My moods are like climate change,
> When I'm sad it glooms like winter,
> When I'm happy it blooms like spring.
> My smile is presented by the sunshine.
> I try to stay positive, to hope for change
> To save the earth, our mother land.

The last few lines of the poem demonstrate a slight shift in emotion from a sense of fear to a budding sense of hope. It is this flicker of hope that needs to be nurtured and transformed into sustainable action.

Discussion

The slight shift in the DUT students' perspective was demonstrated in the students' final COIL reflections, where they discussed what they learned in the project. One student spoke about how they have started implementing daily changes, such as carrying reusable bags when going to the store to avoid buying plastics. They also seek out and purchase ethically produced products. Another student spoke about avoiding fast fashion outlets and looking into thrifting and buying good quality used clothing. Others expressed a need to educate those around them about climate change so that they would be more conscious of the environment. This shows the importance of taking the conversation about climate change a step further than academic focus on the problem by working with students to explore existing solutions that may inspire action rather than emotional and mental paralysis induced by a state of hopelessness.

Students were also encouraged to critically engage with climate change policies so that they could explore strategies and solutions as proposed by their respective governments. The DUT students looked at the South African climate change strategy, which aims to achieve zero carbon emissions by 2050 (Joubert et al. 2022). However, the major critique of the bill is that it lacks clear accountability measures that will ensure that all sectors contributing to major carbon emissions and pollution are monitored and held responsible. The DUT students were also surprised to learn from the Sheffield Hallam students about the lack of political will demonstrated by the UK government that has not taken enough decisive action against climate change. Both student groups concluded that reaching zero emissions by 2050 for both countries sounds like wishful thinking.

COIL, as an intercultural exchange for students, provides an important learning opportunity. Prior to the exchange, the DUT students assumed that the United Kingdom, being a first-world country, would demonstrate great strides in the reduction of behaviors that harm the environment; however, students noticed the laissez-faire approach to climate change in both countries' political contexts. Another assumption that the DUT students made was that they would not have any information of value to share with the Sheffield Hallam students since those students come from a more technologically and economically advanced country. The DUT students battled an underlying inferiority complex, coming from a "third world" country. After the first session, the DUT students realized that they shared more similarities with the Sheffield Hallam students than their perceived differences. The icebreaker was an important exercise that highlighted these similarities, as students shared interesting unknown facts about themselves and their countries including direct experience with a climate-induced disaster. Thus, before engaging students in COIL, it is important for facilitators to explore their students' assumptions about a particular country and groups of people and find ways to have those assumptions questioned and challenged as part of the learning experience.

Conclusion

Upon engaging in the COIL project, my DUT students' perspectives on climate change and its causes developed significantly. This demonstrates the importance of *eco-literacy*, which is defined as the ability to understand the surrounding environment and how human behavior has an impact on the ecosystem (Setiawati et al. 2020). The ability to understand how the human system is interlinked with the natural system helps foster sensitivity and empathy for the environment. It creates a conscious awareness of the harm that humans can cause to the environment and how to mitigate that existing or potential damage. This understanding may be cultivated in schools to help learners develop a sense of hope and optimism through engaging in protective behaviors that contribute positively to the environment (Setiawati et al. 2020; Bennett 2012; McBride et al. 2013).

Our COIL collaboration provided such an opportunity for students to compare their understandings of ecoanxiety and other climate-related mental health challenges. By developing their eco-literacy and practicing self-expression through the creative arts, they gained the means to address and process their feelings around their own personal experiences with environmental instability, especially the Durban floods and mudslides of 2022. Through their international exchange with Sheffield Hallam students, DUT students encountered varied perspectives on mental health challenges and on the way countries address or fail to address climate change through policy and practice.

References

Bennett, Lisa. 2012. https://www.ecoliteracy.org/article/what-does-it-mean-be-ecoliterate. Accessed 19 June 2024.

Charlson, Fiona, Suhailah Ali, Tarik Benmarhnia, Madeleine Pearl, Alessandro Massazza, Jura Augustinavicius, and James G. Scott. 2021. "Climate Change and Mental Health: A Scoping Review." *International Journal of Environmental Research and Public Health* 18 (9): 4486.

Coffey, Yumiko, Navjot Bhullar, Joanne Durkin, Md Shahidul Islam, and Kim Usher. 2021. "Understanding Eco-Anxiety: A Systematic Scoping Review of

Current Literature and Identified Knowledge Gaps." *The Journal of Climate Change and Health* 3: 100047.

Dumont, Caroline, Elizabeth Haase, Trygve Dolber, Janet Lewis, and John Coverdale. 2020. "Climate Change and Risk of Completed Suicide." *The Journal of Nervous and Mental Disease* 208 (7): 559–565.

Geiger, Nathaniel, Janet K. Swim, Karen Gasper, John Fraser, and Kate Flinner. 2021. "How Do I Feel When I Think About Taking Action? Hope and Boredom, Not Anxiety and Helplessness, Predict Intentions to Take Climate Action." *Journal of Environmental Psychology* 76: 101649.

Joubert, F., J. Rushton, and M. Kadodia. 2022. "Key Aspects of South Africa's Much-Anticipated Climate Change Bill." Key Aspects of South Africa's Much-Anticipated Climate Change Bill (polity. org. za) (22 August).

Joubert, Leonie. 2023. https://www.dailymaverick.co.za/article/2023-05-10-a-perfect-storm-how-the-deadly-2022-durban-floods-hold-crucial-lessons-for-the-future-of-the-city-and-others-like-it/. Accessed 18 June 2024.

Lawrance, Emma, Rhiannon Thompson, Gianluca Fontana, and Neil Jennings. 2021. "The Impact of Climate Change on Mental Health and Emotional Wellbeing: Current Evidence and Implications for Policy and Practice." Grantham Institute Briefing Paper 36.

Luce-Kapler, Rebecca. 2009. "Serendipity, Poetry and Inquiry." In*Poetic Inquiry*, 75–78. Brill.

McBride, Brooke B., Carol A. Brewer, Alan R. Berkowitz, and William T. Borrie. 2013. "Environmental Literacy, Ecological Literacy, Ecoliteracy: What Do We Mean and How Did We Get Here?"*Ecosphere* 4 (5): 1–20.

Owton, Helen, and Helen Owton. 2017. "Poetry as Reflective Writing." *Doing Poetic Inquiry*, 85–102.

Pihkala, Panu. 2020. "Anxiety and the Ecological Crisis: An Analysis of Eco-Anxiety and Climate Anxiety." *Sustainability* 12 (19): 7836.

Plumer, B. 2014. https://www.vox.com/2014/9/23/6835285/why-rich-countries-worry-more-about-climate-change-than-poor-ones. Accessed 6 May 2024.

Setiawati, Eka, Ayi Teiri Nurtiani Usmaedi, Agustinus Tandilo Mamma Nuryati, Yusdiana Sirjon, Suci Aprilyati, Ayu Fajarwati Ruiyat, and Y. N. S. Hartuti. 2020. "Develop Green Behaviour Through Ecoliteracy for Early Children." *Systematic Reviews in Pharmacy* 11 (11): 1551–1558.

Subramaney, Ugasvaree, Samantha Iyaloo, Tejil Morar, Tanesha Nayager, and Indhrin Chetty. 2022. "Climate Change and Mental Health Implications in South Africa." *Wits Journal of Clinical Medicine* 4 (3): 151–156.

World Health Organization. 2022. https://www.who.int/health-topics/mental-health#tab=tab_1. Accessed 18 June 2024.

CHAPTER 15

Physics 101: Understanding the Fuel of Our Modern Civilization

Samrat Pathania

> Our planet is a lonely speck in the great enveloping cosmic dark. In our obscurity, in all this vastness, there is no hint that help will come from elsewhere to save us from ourselves.
> —The one-and-only Carl Sagan

In his 1985 congressional testimony, Dr. Sagan made a case for urgent action on climate change with his usual eloquence and clarity (2021). What is striking about Dr. Sagan's testimony (that did *not* result in the requisite governmental policy and collective national action) is that though he is absolutely clear that we urgently need to stop burning fossil fuels, he makes no mention of most of the technologies that we now consider crucial to solving the climate crisis. The reason for that omission is rather simple: those technologies either did not exist or were not considered scalable in 1985. It is tragic indeed that during the decades

S. Pathania (✉)
Wallkill Central School District, Wallkill, NY, USA
e-mail: samratpathania@protonmail.com

S. M. Wyman and J. Korenblat (eds.), *Sustainability Learning for Action and Community Engagement*,
https://doi.org/10.1007/978-3-032-11698-7_15

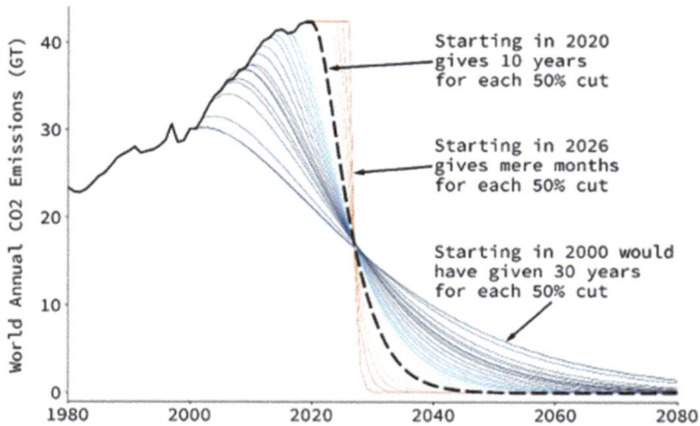

Fig. 15.1 Mitigation curves required to hit 1.5 °C targets (Griffith 2022)

since the testimony of Dr. Sagan, the amount of greenhouse gases has continued growing in the Earth's atmosphere even though humans have developed better and newer technologies to eliminate the use of fossil fuels. So here we are in 2025, woefully behind our climate targets, as this graph succinctly demonstrates (Fig. 15.1).

If one takes a few moments to reflect on the implications of this graph, one may be left with a sense of overwhelming despair, especially if they are unaware of the state of existing climate solutions and those in the pipeline (pun unintended). That despair is unwarranted and dangerous. There is no doubt that it is time for us to wage World War Zero—to paraphrase Saul Griffith, author of *Electrify: An Optimist's Playbook for Our Clean Energy Future*—if we are to bring our fossil fuel-dependent civilization to Net-Zero emissions. Luckily, we have the right weapons to do this: heat pumps, electric vehicles, induction cooktops, and renewable energy to power them all. As a teacher of Physics, I find it rather exciting—even though the overall situation is grim—that introductory Physics courses provide the perfect ground for exploration of the aforementioned climate solutions. These **Beneficial Electrification** solutions replace "direct fossil fuel use" (for example, propane, heating oil, and gasoline) with electricity in a way that reduces overall emissions and energy costs (EESI).

The content of General Physics courses 1 and 2 has been stable for several decades. Due to relative invariability in the foundational concepts, the value of the course to the student is largely determined by the experience created by the instructor in the classroom. The importance of context and applicability of course content to student learning and retention is well-established (Hulleman and Harackiewicz 2009; Hulleman et al. 2008; Lau and Roeser 2002; Newmann 1992). Traditionally, General Physics courses tend to be fast-paced as educators seek to cover fundamental concepts to satisfy the needs of a variety of different STEM majors. Instructors strive to impart an understanding of a range of physical phenomena by encouraging mastery of the mathematics of physical laws. Unfortunately, this appeal to the beauty and symmetry of mathematics and its explanatory power can fall flat for the student who is not convinced of its aesthetic value, sees no connection to their future, or is simply trying to swim against the strong current of new abstractions. Taking these factors into account, incorporating climate solutions into introductory Physics is a win-win on multiple fronts. If students can see the application of simple, physical concepts with regard to current technologies at the forefront of the climate fight, then their struggles with the course content will seem justified. In fact, they might be willing to invest greater effort since application and outcomes are intimately tied to their shared future. This has been my experience in teaching College Physics with a focused exploration of *beneficial electrification*.

In the last few years, I have had hundreds of conversations with people about climate solutions as they pertain to people's personal lives, especially through my work as a clean energy coach with New Yorkers for Clean Power. A lack of understanding of the technological and economic aspects of a solution is one of the biggest impediments to their adoption. It is very unlikely that someone will buy an electric vehicle (EV) or a heat pump if all they hear/know is uninformed noise in their social circles or on digital media. What makes matters worse is that electricity seems fundamentally mysterious to many people despite the fact that our modern civilization would cease to exist without it. While I have productive conversations on this topic at the local farmers' market and other public venues, I find no better place to address this problem than in an introductory Physics course. This transformation in course content does not require a radically different curriculum from what exists currently. The strategy is largely about a difference of emphasis that ultimately renders the subject matter more relevant to student lives.

Below, I provide details about how I accomplish this in Physics courses and the positive impacts on student learning and confidence.

Why Climate Solutions in the Physics Curriculum?

During the last three years, I have organized multiple workshops for K-12 teachers to help them begin the process of incorporating climate solutions into their curriculum. The most common question I am asked is, *Do you provide a ready-to-go curriculum in your workshop*? There are essentially two parts to my response:

(a) Ready-to-go curricula demand context independence (focus on global rather than local) and can, therefore, fail to directly illuminate the solutions most relevant to students' lives.
(b) Authentic learning demands authentic teaching, which can only come from lived experiences of the teachers themselves.

This does not mean that I do not have curricular examples to offer to teachers or encourage them to share their own experiences. This chapter aims to be exactly such an exemplar. Also, I am not claiming that every teacher must implement all solutions that they bring into their classrooms. However, human beings, especially youth, are often most inspired by actions of those they know personally (Team 2014). Therefore, the climate solutions curriculum in my Physics classes is centered around **Beneficial Electrification**, what this looks like in my household, and the possibilities for my students' homes, given the current state of technology, understanding, and legislation.

In order to meet our climate goals over the next 20–25 years, the US will need to replace approximately one billion fossil fuel-based machines with their efficient, electric counterparts and power them all with renewable energy (Rosenberg 2021). Most of these machines are in our households and power what we have come to expect and enjoy as the American way of life. So, it should come as no surprise (and yet it does to many) that the fossil-fueled machines in our homes, including our vehicles, collectively produce about **40% of US energy-related carbon emissions** (Rosenberg 2021). This single fact energizes me to teach climate solutions the way I do; I can truly help my students become sustainability ambassadors in their communities. By understanding the basic working principles of these technologies and, consequently, their benefits, students can demystify them for their family members. This is crucial because, as a democracy, our government cannot force large-scale electrification on its citizens. We tried mask mandates during the COVID-19 pandemic—this unfolded in many parts of the US with resistance to the point of violence. However, this time around, we do not have to do it all at once—in fact, we cannot do it all at once. This makes public education and outreach crucial to achieving our climate goals. The Inflation Reduction Act (2022) made the single biggest climate investment in history by allotting $369 billion to clean energy provisions. Though the Trump administration has undone many of these incentives, our students can still help their families understand the importance of beneficial electrification so that we stay on track to meet our climate goals.

Necessary Deviation from Tradition

I teach College Physics 1 at a public high school where the curriculum requires me to cover the following units:

- Kinematics (One and Two Dimensions)
- Dynamics
- Impulse and Momentum, Work, Power, and Energy
- Statics and Rotational Motion
- Simple Harmonic Motion and Mechanical Waves
- Fluid Mechanics
- Thermodynamics

Obviously, the topic of electricity and/or electromagnetism is a glaring omission from this list. That said, the curriculum for introductory Physics courses, specifically Physics 1, was never intended to be built around electrification, so the omission is quite understandable. Also, the units have typically been taught in a certain order, as listed above. Consequently, the exploration of electricity usually happens in a College Physics 2 course. Most of my students do not plan to take another Physics course in their lives. Therefore, I am presented with a unique challenge: I need to incorporate beneficial electrification, but the traditional way of teaching physics does not offer me the possibility. To humbly follow in the footsteps of the giants who refused to follow tradition and gave us the current understanding of our universe, I teach Physics in a non-traditional manner, albeit not in a radically different manner. The curricular transformation is simply a matter of what, when, and how I emphasize concepts during the year.

Setting the Stage

Before beginning exploration of the solutions, I introduce my students to the problem of climate change by investigating some of the underlying essential concepts:

(a) Origin of fossil fuels
(b) Primary uses of fossil fuels
(c) The Greenhouse Effect
(d) The Keeling Curve
(e) Impacts of Climate Change and Consequent Warming

This is accomplished using reading assignments, videos, and open discussions modeled on the Socratic method (see Earthrise Commons; earthrisecommons.org for teaching materials and resources). My goal is to make sure that all my students, irrespective of their backgrounds, have a foundational understanding of the role of fossil fuels in our world and the consequences of their unbridled use. Each of these topics could be worth days of exploration and study. However, just as we choose how much depth to cover in any class, we can make similar choices with regard to introducing students to climate change and its causes. I aim for students to have a level of understanding so that they may explain the issue satisfactorily to their younger siblings or parents who may not have a background in science. I use a similar metric when determining the amount of technological detail I discuss regarding electric machines that will replace their fossil-fueled predecessors.

Framework for Solutions

In order to fulfill our beneficial electrification goals, we need to continue overhauling the existing infrastructure at both the residential and commercial levels in the following four categories:

- Energy Production
- Energy Storage

- Energy Distribution
- Energy Consumption

Stated thus, the problem might seem technical and probably inaccessible to the layperson. However, most people are already familiar with some of the technologies pertaining to these categories—solar panels, EVs, home battery storage solutions, heat pumps, etc., to name a few. Crucially, what I find missing is a coherent picture of how these technologies work and how they interact with each other to achieve our climate goals. I address this in the remainder of my class after I introduce my students to the problem of climate change. The culminating project is a three-week-long Net-Zero home project where students analyze the energy usage and emissions of their own homes and then "retrofit" their households to run exclusively on 100% renewable electricity (Fig. 15.2).

The Most Important Unit in the World (Energy and its Trade Unit)

Energy is fundamentally mysterious. Our physical models of the universe and our most fundamental laws, are attempts to demystify this entity of nature. The law of conservation of energy is one that most high school students know verbatim, for they learn it during middle school. Additionally, $E = mc^2$ is one of the most popular equations in the world. Yet, most people do not have much experience dealing with different units of energy except when they talk about how many calories are in their food or drink of choice. This lack of understanding is not surprising given that energy itself seems intangible; it is hard to have a good intuitive understanding of energy's units. In contrast, we do have a good intuitive comprehension of a gallon because of our evolutionary experience of dealing with a three-dimensional world of visible or tangible volumes. Given that our twenty-first-century technological civilization is the most energy-hungry civilization that the world has ever seen (see Fig. 15.3), and this hunger must be satiated with renewable electrical energy to maintain our planetary equilibrium, I strongly believe that every literate person should develop an intuitive understanding of the unit in which electrical energy is traded—the **kilowatt hour** (**kWh**).

One of the prominent ways I deviate from the Physics curricular norm is with an early focus on the **kWh** which increases the amount of time my

Fig. 15.2 Beneficial electrification in our homes (Rosenberg 2021)

students spend using this unit of energy. Instead of waiting to finish kinematics and dynamics and then begin with energy, I introduce energy via kinetic energy as soon as students are introduced to the law of inertia. I utilize these concepts to guide learners to uncover the conservation of mechanical energy via a lab assignment; then we define gravitational potential energy. The students' discovery of this *dance of energy* from one form to another sets the foundation for the whole cycle of energy production, distribution, storage, and consumption. Of course, we work with the SI unit of energy, the Joule, for many problems. But we start working with the **kWh** simultaneously instead of waiting for labs in the electricity unit in Physics 2, as done traditionally.

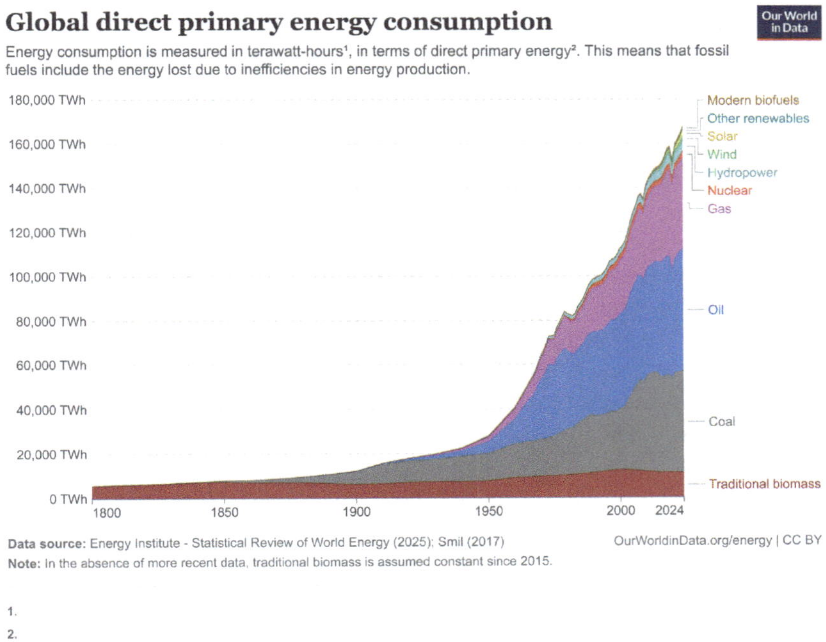

Fig. 15.3 Trends in global energy consumption

Energy Storage: Expanding the Definition of a Battery

Early classroom exposure to energy and the **kWh** creates an opportunity to dabble with energy storage in Quarter 2 of the course in the form of the **Gravitational Battery Assignment**. Students design a gravitational battery with a single block of material that can provide a 24-hour electrical backup for their home. They choose shape, dimensions, material, and height. Upon completing calculations, we compare aspects of practicality (cost, size, etc.) to a chemical home storage solution such as a Tesla Powerwall. We do not cover details of a chemical battery (for example, cathode/anode materials, etc., concepts more relevant to a chemistry course).

This assignment serves as the source of a key revelation to the students, which often causes quite a stir in the classroom: *a battery does not necessarily require some magic of chemistry*—it is simply a store of energy that can be harnessed as electrical energy in the future. Therefore, students are excited to learn about **pumped hydro storage**, a relatively older technology and currently the largest form of energy storage at the electric utility-scale (Hawken). We do not, however, go into the details of how a hydroelectric plant produces electricity at this stage of the course; we save this for the unit on electromagnetism.

We add one more exciting storage possibility later when we investigate thermodynamics: **Sand Battery Storage**. Though this technology is not yet commercially established (like lithium-ion batteries), if successful, sand battery storage might be a revolutionarily simple solution to the rather difficult problem of district heating. Students make a straightforward calculation using specific heat of sand to verify the claims of the company that has installed the first commercial sand battery for Vatajankoski's district heating network in Kankaanpää, Finland (Polar Night Energy, n.d.).

Energy Production: Getting the Basics Down

Electromagnetic Induction is at the heart of our current electricity production paradigm, with the notable exception of photovoltaics. It is also central to the consumption of electricity in one of the most amazing of human inventions: the electric motor. However, approaching electromagnetic induction course content via electrostatics, electric current, etc., is a process that usually unfolds over a few weeks in a Physics 2 course. Educators typically lack the luxury of time in a Physics 1 course. Therefore, once we have covered topics of energy, work, and power in the larger context of Mechanics, I chart quite a direct path to a foundational understanding of electromagnetic induction by covering the basics of electric fields, voltage, current, resistance, electrical power/energy, and Ohm's law. I do not try to cover every nuance of these concepts (for example, AC vs DC) but focus on helping my students develop a simple mental schema for how electricity is generated and the ability to quantify these transactions in terms of power and energy.

At the end of this exploration, I expect my students to be able to:

- broadly describe the similarities and differences between fossil-fueled power plants and non-photovoltaic sources of renewable electricity generation such as hydropower and wind turbines.

- use units kW/MW and kWh/MWh when discussing the capacity of power plants and energy storage (coupling renewables with batteries).

During the exploration, we revisit pumped hydro storage with this new understanding of electromagnetic induction and look at some of the most interesting power/energy-related numbers from the Blenheim–Gilboa Pumped Storage Power Station in the Catskill Mountains of New York. We also take a brief glance at the US wind industry by means of a few short videos that highlight the success of onshore wind in Texas—the wind capital of the US—while underlining the immense unexplored potential for offshore wind along the US coasts (WWF International 2020; Financial Times 2022). All of these forays generate rich, multi-faceted conversations in the class, and students often ask, *why don't we talk more about such things in our other courses?*

Energy Consumption: The Heart of Beneficial Electrification

Of the electrical devices central to our move away from fossil fuels in our daily lives, the most important ones are electric cars and heat pumps. Our school is located in a rural part of New York where public transportation is poor. Consequently, the families of most of my students have multiple cars. The national trend of a decline in teen interest in obtaining driver's licenses is not representative of our community (Highway Statistics, n.d.). Many of the juniors/seniors in our school report their success at driving tests with excitement. I am confident that the American car culture, for better or for worse, is thriving in our school community. Therefore, I introduce EVs very early in the year when we start with kinematics. Given that EVs generally have very impressive acceleration compared to their ICE (internal combustion engine) counterparts, they make a good subject for fun problems using equations of motion (see Earthrise Commons; earthrisecommons.org). The induction motor is the "engine" of the EV. However, it is a rather complicated piece of engineering for an introductory physics course. We do not cover the technology in any detail, but we do go over the working principle of the electric motor to give students a sense of what makes the wheels turn in EVs. Additionally, we cover the emissions reduction and fuel savings for EVs later when we work on the Net-Zero Home project.

Students begin the process of understanding electricity consumption by estimating annual electricity usage in their homes. They look at the technical specifications (voltage, amperage) of their most used electrical devices. Later, they compare this to their actual electricity usage by compiling electric bills from the previous twelve months. None of my students have ever looked at their electric bills before. Most do not even know where the electric meter is located at their homes. They often report colorful dinner table conversations with their parents as a consequence of their inquiry into this previously invisible facet of their lives.

In the thermodynamics unit, the students learn about the woeful efficiency of their beloved ICE vehicles and the limits that the laws of thermodynamics place on improving this efficiency. However, the most important thing that they study is probably the coolest (pun intended) of all the climate solutions—the humble **heat pump**. Most students have never heard of these devices and do not know their working principle. They are surprised that high-efficiency heat pumps are perfectly capable of handling our homes' heating and cooling requirements even when the temperatures fall into the single digits (Fahrenheit). We look at the nearby Mohonk Preserve's annual weather data (Mohonk Preserve 2023) to confirm that the current air-source heat pumps available on the market can handle all that the New York weather throws at them. Heat pumps are our last stop as far as curriculum is concerned. It is a wonderful stopping point before we jump into the final task of the year—the Net-Zero Home Project.

Bringing It All Together

The climate crisis affects the life of every single human being on our planet. However, among the group that stands to suffer the longest is our youth—the largest generation of youth in the history of our species (Youth Empowerment, n.d.). The last few years have seen a rise in support for youth empowerment. Organized action by young people is starting to bear fruit in a big way, such as the recent landmark climate case *Held Vs Youth of Montana* (Rott and Beyram, 2023). Sixteen youth residents of Montana filed a suit against the state of Montana, arguing that state policies worsened the effect of climate change on their lives and thus violated their rights as guaranteed by the Constitution of Montana.

District court judge Kathy Seeley ruled in favor of the youth. A ruling of this magnitude was unimaginable just a few years ago. We must keep this momentum going. For me, empowering my students is synonymous with providing them with skills and knowledge that can help them become sustainability ambassadors in their communities. The **Net-Zero Home** project, (described in detail on), creates a possibility for learners to begin this process and understand the fundamental transformation that is deeply personal given that fossil fuel use in our households for cooking, heating, transportation, etc., is collectively responsible for about **40% of US energy-related carbon emissions** (Rosenberg 2021).

Conclusion

The enormity of the climate crisis and the seeming insurmountability of the cultural and political forces that stand in the way of concerted climate action drive many people, especially the youth, to despair. However, the outcome is not yet decided. What is certain is that fossil fuels can no longer be at the heart of the engine that powers our civilization. To meet our climate goals, we must electrify our world and produce that electricity from non-fossil fuel sources. One cannot argue against this fact. Therefore, beneficial electrification must happen no matter what happens in the coming few years of technological advancements. We can argue, discuss, dissect, and disagree about how this will unfold. Should we take a sticks or carrots approach? What incentives should we provide to households and corporations? What is the financial responsibility of the Western democracies toward developing nations? How can we rein in the emissions from hard-to-electrify sectors like the aviation industry? However, it is inarguable that most of the technologies required for beneficial electrification are here, can be implemented at scale, and can save individuals, governments, and corporations money while generating employment and reducing pollution. This is why I am so passionate about incorporating beneficial electrification into Physics courses.

References

carlsagandotcom. 2021. Carl Sagan Testifying Before Congress in 1985 on Climate Change [Video]. YouTube, August 19. https://www.youtube.com/watch?v=Wp-WiNXH6hI.

Environmental and Energy Study Institute (EESI). n.d. Beneficial Electrification. EESI. https://www.eesi.org/electrification/be.

Financial Times. 2022. US Offshore Wind: Tapping into an Underused Resource | FT Energy Source [Video], November 7. YouTube. https://www.youtube.com/watch?v=h3WoOie3cKI.

Griffith, S., and S. Calisch. 2021. One Billion Machines. Technical White Paper https://content.rewiringamerica.org/reports/one-billion-machines-one-pager.pdf (Rewiring America, 2021).

Griffith, S. 2022. *Electrify: An Optimist's Playbook for Our Clean Energy Future.* The MIT Press.

Hawken, Paul. 2017. *Drawdown: The Most Comprehensive Plan Ever Proposed to Reverse Global Warming*. Penguin; Project Drawdown: https://drawdown.org/Hawken, Paul. 2021. *Regeneration: Ending the Climate Crisis in One Generation.* Penguin. https://regeneration.org/

Highway Statistics Series—Policy | Federal Highway Administration. n.d. https://www.fhwa.dot.gov/policyinformation/statistics.cfm

Household Electrification Savings Calculator. n.d. Rewiring America. https://www.rewiringamerica.org/app/ira-calculator

Hulleman, C.S., and J.M. Harackiewicz. 2009. "Promoting Interest and Performance in High School Science Classes." *Science* 326: 1410–1412. https://doi.org/10.1126/science.1177067.

Hulleman, Chris S., et al. 2008. "Task Values, Achievement Goals, and Interest: An Integrative Analysis." *Journal of Educational Psychology* 100 (2): 398–416. https://doi.org/10.1037/0022-0663.100.2.398.

Lau, S., and R.W. Roeser. 2002. "Cognitive Abilities and Motivational Processes in High School Students' Situational Engagement and Achievement in Science." *Educational Assessment* 8: 139–162. https://doi.org/10.1207/s15326977ea0802_04.

Mohonk Preserve. 2023. Weather Data—Mohonk Preserve, September 6. https://www.mohonkpreserve.org/what-we-do/conservation-programs/conservation-science/weather-data/.

Newmann, Fred M. 1992 *Student Engagement and Achievement in American Secondary Schools.* Edited by Fred M. Newmann. Teachers College Press.

Polar Night Energy. n.d. Sand Battery. Retrieved September 25, 2023, from https://polarnightenergy.fi/sand-battery.

Rewiring America. 2023. Pace of Progress. White Paper. https://www.rewiringamerica.org/pace

Rosenberg, J. 2021. Electrify Everything in Your Home. Technical White Paper. https://content.rewiringamerica.org/reports/electrify-home-guide.pdf (Rewiring America, 2021).

Rott, N., and S. Beyram. 2023. Montana Youth Climate Ruling Could Set Precedent for Future Climate Litigation. NPR, August 23. https://www.npr.org/2023/08/23/1194710955/montanayouth-climate-ruling-could-set-precedent-for-future-climate-litigation.

Team, B.I. 2014. *EAST: Four Simple Ways to Apply Behavioural Insights*. London: Behavioural Insight Team.

WWF International. 2020. How Texas Is Using Wind Power to Create a Cleaner Future [Video]. YouTube, January 8. https://www.youtube.com/watch?v=95eUlfPFbz0

Youth empowerment | United Nations Development Programme. n.d. UNDP. https://www.undp.org/governance/youth-empowerment.

CHAPTER 16

Teaching the UN SDGs in a Child Development Course: Empowering Educators

Andrea M. Noel

Introduction

Teachers are natural agents of change and effective models for developing and reinforcing children's attitudes and dispositions toward land stewardship and social equality. This is great news for sustainability education since Schools of Education can train teachers to be effective impact multipliers. At our current juncture in history, where humankind faces immense environmental challenges directly and indirectly related to children's learning, health, and general well-being, teachers' involvement and leadership in sustainability efforts are essential. Schools of Education will enjoy a powerful impact when they offer greater support to pre-service

A. M. Noel (✉)
Department of Early Childhood and Childhood Education, SUNY New Paltz, New Paltz, NY, USA
e-mail: noela@newpaltz.edu

S. M. Wyman and J. Korenblat (eds.), *Sustainability Learning for Action and Community Engagement*,
https://doi.org/10.1007/978-3-032-11698-7_16

teachers' knowledge development about the relationship between children's learning, health, and our sustainable future, and nurture their skills in how to bring about necessary change.

Schools of Education have not optimized this potential impact by prioritizing the development of teachers'-in-training skills and knowledge of sustainability, nor in advocacy and civic action. Integrating sustainability education into all students' General Education programs is important, but also into Education program requirements and specific Education courses.

This chapter discusses the integration of the United Nations Sustainable Development Goals (SDGs) into one required foundation course for the Education major, Child Psychology and Development. After a description of the context, including information about students' Education program and the course itself, an argument will be made for the appropriateness of connecting sustainability, particularly in relation to the UN SDG framework, to this course with modifications of course goals and objectives. A semester-long assignment designed to integrate and connect sustainability with content in Child Psychology and Development and students' informal feedback about the project provided by students after completing the project will be discussed. The chapter will close with my reflections on student learning, possible improvements to the project, and implications of the learning experience for pre-service teachers' future work with children, some of whom will live and work in communities with serious frontline sustainability challenges.

Context: The Students, Program, and Course

This Child Psychology and Development class is taught within the School of Education, and enrolled students are predominantly in the undergraduate Early Childhood and Childhood Education teacher education program, although some take the course for General Education (GE) social science credit. While several professors in the School of Education have been trained as Sustainability Fellows, to date, only one instructor has taught with consistent incorporation of sustainability concepts and links specific content to the SDGs.

Appropriateness: Why Add Sustainability to Child Development?

In Child Psychology and Development, students learn about a variety of theories of learning and personality development in educational contexts through scaffolded assignments and activities. Coursework and discussion focus on children's growth and development between the prenatal period and elementary school age. In addition to theories of bio-social, cognitive, and social-emotional development, some of the topics that have been traditionally addressed include the development of gender roles, the impact of ethnic/culture/language differences, child trauma, and children's learning differences. During initial class sessions, concepts essential to future discussions are introduced, including "theory," "plasticity," "learning," "epigenetics," etc. The concept of sustainability and how it relates to bio-social, cognitive, and social-emotional development is now also included as a key concept in my classes.

The study of sustainability in this course is structured around an introduction to the United Nations Sustainability Goals for several reasons. Many teaching materials are available for educators working in childhood education, and these are invaluable resources for practicing and pre-service teachers. The SDGs are especially credible for a college-age audience, as they were conceptualized and developed in 2015 by the United Nations, a highly prestigious international organization founded on collaboration and peacebuilding. Furthermore, the SDGs were approved by 193 countries and territories and based on surveys of young people aged 16–25 around the world.

Systems theories, which are important to sustainability education, have long been relevant to the field of child development and learning. Already in the 1970s, Urie Bronfenbrenner posited his bio-ecological systems theory, which helps us interpret and clarify the many levels of variables that interact to impact children's development, health, learning, and general well-being (Bronfenbrenner 1979; Bronfenbrenner and Morris 2006). This theory and its systematic way of studying children, their circumstances, and their experiences had already been an important theory in this course. As such, the expanded application of systems thinking fits well with existing content and a sustainability-informed focus on overlapping factors in a child's development.

In studying children's development, Bronfenbrenner encourages thinking beyond the variables in a child's immediate environment. Close variables, such as those related to children's homes and relationships, are situated in what is called the *microsystem*, and those further removed, such as culture and government policies, are in the *exosystems* and *macrosystems*. Variables in each of these systems act and interact to shape development. This view is especially useful in helping us make sense of how variables related to the 17 SDGs impact children's development. For example, children need to be well nourished for optimal development to occur. SDGs 1 and 2 ("No Poverty" and "Zero Hunger") both concern filling basic needs such as food, and these are closely related to the child's daily experiences. Interactions with schools and childcare facilities occur in the *mesosystem*, which can be influential in the provision of adequate nourishment. Further afield in the *exosystem*, are variables such as parents' employment status, which impacts a family's food supply through the amount of available income. Even further removed but still consequential is the *macrosystem* with influential cultural attitudes and government policies. For example, attitudes about taxing may result in laws that allow tax deductions for children. These deductions or credits might mean more money available for a family to buy food or to meet other basic needs that decrease the likelihood of poverty. Thus, many levels make up the system, and what happens in each sector will impact children's development.

Sustainability in childhood education is receiving increasing attention. Daelmans et al. (2017) argue that "it is the collective responsibility of governments, and all development health partners to guarantee that every individual starting life in every corner of the world is given the family care, education, health services, and nutrition to do so" (8). Samuelsson and Park (2017) state that "values related to sustainability promote a certain type of pedagogy" (273) and that teachers need opportunities to "think, reflect, and innovate" (280). They add that educated staff is a precondition for involving children in sustainability and that "all adults working and interacting with children have to be well educated … both when it comes to knowledge about children and awareness of sustainability" (282). Modifications to this Child Development course give pre-service teachers knowledge about the SDGs that will inform their future pedagogy, and experience in using that knowledge to promote change.

Modifications to the Course

Adding sustainability content requires changes in how the instructor thinks about the course and presents and applies content rather than numerous additions to the curriculum. Using the UN SDG framework is relatively easy, because several goals are so clearly connected according to the targets. Traditionally, the content is designed predominantly for teaching and working with individuals or small groups of children. It did not consistently link to broader world challenges, such as the impact of environmental issues on children, despite problems such as pollution and global warming, which have disproportionately high impacts on children (EPA 2014). Sustainability-centered content, however, links to current local and world circumstances and issues, making the course more relevant to learners' lived experiences. In addition to changes in approach and thinking, sustainability concerns frame concepts discussed at the start of the semester, and student expectations are clarified to reflect attention to the SDGs. A new sustainability statement, additional course objectives, and a semester-long, redesigned project serve as the main vehicles to reach these new objectives.

Sustainability is introduced early with the 1987 Bruntland sustainable development definition of "meeting the needs of the present without compromising the ability of future generations to meet their own needs." The three sustainability dimensions of environment, society, and economy are first discussed using a conversation prompt, which asks students to discuss and list some "present and future needs of children." Then, the instructor presents the 17 SDGs, and students are asked to link their previous answers to this information, which supports the pivot to a more specific discussion on how environmental factors impact children's psychological and physical development. A short reading from the United States Environmental Protection Agency (EPA 2014) entitled *Climate Change and Children's Health* ends the discussion with students enlightened about both the obvious and not-so-obvious ways children are impacted by climate change. Learners are typically not surprised by problems such as higher rates of asthma for children in areas with high ozone levels but are alarmed by risks posed by simply being a child. These include increased risks from spending more time outside than adults, to their developing immune systems, and putting more things into their mouths (EPA 2014).

The class syllabus is enriched with a sustainability statement and additional Student Learning Outcomes (SLOs). The sustainability statement offers a brief description of the SDGs based on the material on the UN website, and includes the 17 goals as well as the overarching objective to improve the lives of people everywhere (http://www.un.org/sustainabledevelopment/). The syllabus statement calls out Goals 3 (Health and Well-being) and 4 (High Quality Education for All) and informs students that class discussions, readings, and projects will consider the role of sustainability in child development to develop a fuller understanding of how many overlapping variables impact children's well-being. Two student learning outcomes (SLOs) are added to the course syllabus, including one that concerns students' ability to articulate the relationship between sustainability, the SDGs, and child development as well as its impact throughout the childhood period. The second is to be able to apply systems theory when interpreting intersecting factors that influence child development and learning.

Presentation of Sustainability: The Concept and Then the Project

A semester-long project offers an effective and practical way to connect sustainability with traditional child psychology content and topics. The project requires students to choose a contemporary issue related to children's development, research how the issue impacts psychological, physical, and/or emotional development, and link the issue with the SDGs. Students also brainstorm and eventually choose and implement an idea for a civic action designed to bring about positive change related to the topic of study. The focus of topics as well as the inclusion of a civic action, distinguish the project from ones in previous semesters and propel students to think about themselves as change makers rather than merely reporters of findings. In line with traditional projects, students must also analyze the problem from various theoretical perspectives, locate and read relevant current peer-reviewed research, and present findings to the class at the end of the semester. Student groups have successfully investigated social determinants that affect child psychology, health, and development, such as the impact of agricultural chemicals, air or water pollution, food insecurity, childhood obesity, chronic noise exposure, lack of universal pre-kindergarten, support for immigrant/refugee children, the state of children's mental health, and gun safety/danger.

The Action Project

The action project is introduced to students with materials from the UN SDG website, particularly through examples of "Invent, Innovate, and Campaign," which effectively present examples of young people around the world creating actions in these three ways. After this introduction, the meanings of "action" and "promoting change" are discussed. The conversation often revolves around recently famous or local leaders whose work relates to class content, such as Marley Dias (2018), a 10-year-old girl who created an initiative to bring books with diverse characters to her school. International youth leaders such as Greta Thunberg et al. (2020) have also proven interesting to students, who are then encouraged to brainstorm ideas about inventing, innovating, and campaigning to solve a problem associated with their research topic.

Clarification with an Example

Students understand specific project requirements when they work through an example with the instructor. During one class session early in the semester, a local issue is presented and discussed. The issue serves as a vehicle to learn content as well as an example of how this topic could be used to meet project requirements.

For example, just prior to the pandemic (2016–2018), several local New York State school districts identified water fountains with lead levels higher than state standards (Record Online 2016). Lead poisoning is a well-known example of how environmental pollution can impact children's physical and intellectual development and learning. This topic offers a solid opportunity to present content on children's intellectual development, link the SDGs, and explain project requirements, including actions taken to promote change.

Scientists were already aware of the dangers of lead poisoning in the late 1800s, and as summarized by Sorensen et al. (2019), about 12 million families in the US live in homes with problematic or toxic levels of lead. High blood lead levels are associated with lower IQ and behavioral disorders and directly impact the problematic achievement gap in US schools. Most of the impacted children are from lower socio-economic status (SES) families and are susceptible to lead poisoning through toxic water and dust from chipping lead paint.

The problem of lead toxicity in dynamic child development can be analyzed through a bio-ecological systems perspective (Bronfenbrenner 1979; Bronfenbrenner and Morris 2006). Lead toxicity may impact children at many levels within the bio-ecological system. Within the micro- and mesosystems, behavior changes caused by lead poisoning may impact family relationships and relationships with other children at school, in addition to academic achievement. Farther afield, in the exosystem, how lead dust is dealt with at a parent's place of work may impact how much dust is brought home on clothing. Furthermore, cultural, legal, and governmental norms and laws regarding lead abatement in the macrosystem interact with and more indirectly impact families and schools.

Students readily apply the SDGs to the problem of lead toxicity. Most of the goals relate, with the most obvious being 1: No Poverty, 3: Good Health and Well-Being, 6: Clean Water and Sanitation), 10: Reduced Inequalities, and 12: Responsible Consumption and Production.

While it may be disheartening to discuss lead toxicity, considering the many years this has been a known danger for children, the problem can also be viewed through a lens of progress. For example, legislation has resulted in the removal of lead from gasoline, and this has substantially decreased the number of children affected (Angrand et al. 2022). Lead is thus an example of an environmental pollutant with serious consequences for children, but it is also an issue that society is making progress to solve. Many civic actions have been undertaken to support change, and this success strengthens the argument that we can indeed make an impact. Pre-service teachers need to see past successes in addition to having opportunities to practice connecting content, research, and action for change.

Discussion/Evaluation/Closing

There are several notable findings to discuss about this semester-long project from both the pre-service teacher and the student perspective.

Students' knowledge of SDGs and how they relate to child psychology and development increased greatly over the course of the semester despite students' insecurities. Many students joined the class with no knowledge of the SDGs, as evidenced by initial class discussions, but they ended the semester with knowledge and the ability to relate the SDGs to important content, theory, and research in child psychology, as demonstrated by their successful completion of the project. Students gained an understanding of the SDGs as an umbrella of sorts under which healthy child development is supported and sustained on a global level, as well as the important impact many elements of a system have on children's psychology and development. For most students, needing to think about and create an action that could promote positive change was a new experience.

Finally, while this class integrates the SDGs into a course and students meet the learning objectives outlined at the beginning of the course, SDG information is not integrated into all Child Development courses. At the time of this writing, integration of sustainability at my institution is rather haphazard, and infused into courses primarily if the instructor has had training as a sustainability fellow. A literature search using the search terms "pre-service teachers" and "Global Goals" or "SDGs" gives reason to believe that this may also be the case in other institutions. For example, teacher educators at the University at Buffalo published a piece about pre-service teachers in a social studies methods class organizing a global literacy fair (Chicola and Ceprano 2009). Other publications were similar and reported on projects from a variety of countries. It could be that teacher education programs are starting to integrate sustainability more systematically into students' programs, but research related to whether this is actually happening appears to be unavailable. The process should start with the integration of the SDGs or another sustainability framework into the development or updating of Schools of Education strategic plans. Appropriate funding must be allocated to support this important effort to create an organized effort.

Conclusion

Pre-service teachers will begin their careers in diverse communities that vary in setting, wealth, the number of refugees, English fluency, the immigrant status of families, persons from historically disadvantaged groups, and many other characteristics. Environmental challenges disproportionately impact many of these communities. We know that people experiencing poverty, immigrants/refugees, and people of color are more likely to live and go to school in communities that are struggling with environmental challenges and are less likely to recover from weather disasters caused by global warming. Our future teachers need to be ready to understand and help address these challenges in partnership with community members so they can best serve the children and families where they work. Integrating content and skills into courses that help pre-service teachers develop the skills to do so isn't just appropriate, it's essential.

References

Angrand, R.C., G. Collins, C. Landrigan, J. Philip, and Valerie M. Thomas. 2022. "Relation of Blood Lead Levels and Lead in Gasoline: And Updated Systematic Review." *Environmental Health* 27 (1): 138. https://doi.org/10.1186/s12940-022-00936-x.

Bronfenbrenner, U. 1979. *The Ecology of Human Development: Experiments by Nature and Design*. Cambridge, MA: Harvard University Press.

Bronfenbrenner, U., and P. Morris. 2006. "The Bioecological Model of Human Development." In *Handbook of Child Psychology* (6th ed), edited by Richard M. Lerner, 793–346. Hoboken, NJ: Wiley.

Daelmans, B., G.L. Darmstadt, J. Lombardi, M.M. Black, P. Britto, S. Lye, T. Dua, Z.A. Bhutta, & M. Richter. 2017. "Early Childhood Development: The Foundation for Sustainable Development." *The Lancet* 389: 8–11.

Chicola, N.A., and M. Ceprano. 2009. "Preservice Teachers Collaborating with Families to Foster Global Literacy Connections." *International Journal of Learning* 16(8): 221–233.

Dias, M. 2018. *Marley Get's It Done: So Can You!* New York, NY: Scholastic Press.

Environmental Protection Agency. 2014. *Climate Change and Children's Health*. https://www.epa.gov/sites/default/files/2014-05/documents/ochp_climate_brochure.pdf.

Johnson D.W., and Roger T. Johnson. 1998. *Learning Together and Alone: Cooperative, Competitive, and Individualistic Learning* (5th ed.). Saddle River, NJ: Pearson.

Nelson, Loui Lord. 2013. *Design and Deliver: Planning and Teaching Using Universal Design for Learning*. Baltimore. MD: Brookes Publishing ISBN: 9781598573886.

Samuelsson, Ingrid Pramling and Eunhye Park. 2017. "How to Educate Children for Sustainable Learning and for a Sustainable World." *IJEC* 49: 273–285. https://doi.org/10.1007/s13158-017-0197-1.

Thunberg, G., S. Thunberg, M. Ernman, and B. Ernman. 2020. *Our House Is on Fire: Scenes of a Family and a Planet in Crisis*. New York, NY: Penguin Books.

Record Online. 2016. https://www.recordonline.com/story/news/2016/05/11/ulster-county-findshigh-lead/30434817007/ and https://www.ny1.com/nyc/all-boroughs/education/2017/04/28/nycschools-new-testing-water-elevated-lead-levels-more-than-1000-buildings.

Sorenson, Lucy C., Ashley M. Fox, Heyjie Jung, and Erika G. Martin. 2019. "Lead Exposure and Academic Achievement: Evidence from Childhood Lead Poisoning Prevention Efforts." *Journal of Population Economics* 32: 179–218.

CHAPTER 17

Weaving Climate Action Solutions Throughout General Chemistry and Other Core STEM Curricula

Megan A. Ferguson

Introduction

For decades, the United Nations has recognized education as a critical component on the pathway to building a sustainable society, and given the magnitude of our current climate crisis, incorporating more climate change material into curricula should be a particularly high priority (Meehan 2018). However, introductory college science classes for majors, as well as many high school core science courses, tend to have standardized curricula with very limited ability to introduce additional units or modules. In this chapter, we illustrate how to coherently integrate examples of climate *solutions* throughout an introductory science course to emphasize student agency and engage learners more deeply without sacrificing substantial class time to interact with the required content. While the problems society faces with climate change are extensive, so

M. A. Ferguson (✉)
Department of Chemistry, SUNY New Paltz, New Paltz, USA
e-mail: fergusom@newpaltz.edu

S. M. Wyman and J. Korenblat (eds.), *Sustainability Learning for Action and Community Engagement*,
https://doi.org/10.1007/978-3-032-11698-7_17

too are the applications of key science themes to known climate solutions, many of which can be regularly linked to topics covered in a conventional introductory STEM class. We use general chemistry as the primary example and provide details about how the same approach could be used in biology, physics, or engineering courses. This enhances students' educational experiences as they apply science concepts to positive actions toward solving a major societal and environmental problem.

Pedagogical Needs

Whether or not a college or university has made an institutional commitment to education and action for sustainability, introductory-level courses are an excellent way to introduce key concepts and foster a cross-disciplinary understanding of complex challenges. However, core science classes tend to be large, lecture-based, and have a high DFW rate (low grades of D, fail, or withdraw from the course). Moreover, student interest can be a challenge from the outset. For example, chemistry is required as a cognate course by nearly every major in the sciences and engineering, so the majority of students in a given class may not already be particularly interested in chemistry. This is also true for college physics and all high school core science courses. Choosing applications of course subject matter relevant to everyone—for instance, ameliorating climate instability—can enhance student interest in the subject and a feeling of belonging in the course. A longitudinal, interview-based study of computer science majors at Carnegie Mellon University (Margolis et al., 2002), for example, found that women's interest in computer science is often linked to societal improvements made possible by computing; after applying the study's recommendations, including the incorporation of computer science applications into introductory courses, female representation in the major increased from 7% to 42% in just five years.

Moreover, Michalopoulou et al. (2019) posit that the field of chemistry teaching needs to focus on making sure graduates understand how chemistry relates to other disciplines. They introduce a systems-based course revolving around the UN SDGs, but this is an optional MOOC (massive open online course) rather than a core content course. In the absence of resources available to offer such options, this approach can be modified using our strategy: intentionally incorporating applications focused on the multidisciplinary problem of addressing climate change.

Another advantage of the approach taken in this chapter stems from the fact that many students in introductory college science courses (or secondary education) do not typically have the flexibility in their schedule to take courses that address sustainability issues more deeply. Their majors often require that they take at least three math and science courses, often with a lab included, meaning they may be full-time students even before considering typical first-year college requirements like a writing/composition course. Even if they already have an interest in sustainability issues, they may not have the time to take elective courses outside of their major and graduation requirements. Incorporating sustainability themes such as climate solutions across the STEM curriculum allows greater exposure to many more students.

The Challenge—And a Solution

The key challenge associated with incorporating "new" material into core science courses is the class time that must then be taken from other topics. Curricula for college chemistry and engineering are clearly specified by the American Chemical Society and the Accreditation Board for Engineering and Technology, respectively. Biology and physics have no analogous institutions, but content in introductory college courses is effectively standardized by the available textbooks and the College Board's Advanced Placement exams. Thus, replacing a portion of course content with new material is not typically a viable solution.

Our approach involves *brief but clearly marked climate solution applications introduced regularly* throughout the course. These may require only five or ten minutes a week, thus not detracting significantly from concept-learning time. Yet, each has the same framework for introducing climate solutions—and drawing students' attention initially to that format—which magnifies the impact as students recognize more quickly how chemistry is being used to address a major societal problem. We also replace "conventional" problem-solving examples with ones more relevant to climate instability and ways to combat it.

To make sure students recognize the theme of incorporating climate change solutions, they must be introduced early. We recommend explicitly addressing this plan in the syllabus. For example, we included the following preface in a General Chemistry syllabus:

> Only a fraction of you plan to become chemists, but hopefully you all have an intention of creating a better world through your work, whatever that work may be. In this course, we will learn many chemical principles and solve a lot of word problems, but we will also focus on the power of chemistry for good in the world. Climate change is one of the biggest challenges currently facing humanity, and chemistry is deeply involved both with the physical processes involved in climate change and with many of its solutions. In fact, the American Chemical Society, which accredits chemistry programs at United States institutions, lists "Advance Sustainability and the Environment" as one of its four public policy priorities.

After going through the syllabus, we take about five minutes to show students the Drawdown website (www.drawdown.org) and accompanying book (Hawken, 2017), which detail more than eighty climate change solutions currently being used across multiple sectors (for example, energy, land use, building infrastructure). A few examples are given to pique student interest. Then, students are told that they will be seeing regular applications of chemistry topics to these Drawdown solutions once a week throughout the semester. These weekly examples not only help to achieve sustainability learning goals, but they also break up the lecture and infuse problem-based learning and interdisciplinary learning into the course.

Climate Change and the Impact of Knowledge and Perceptions on Action

By way of introduction to the connection between our course and the broader topic of environmental sustainability (with implications for the social and economic aspects as well), primary and secondary sources allow students to draw their own conclusions about the role of scientific research and innovation in the search for solutions. Rather than direct students toward prescriptive takeaways, we create learning experiences where they can apply concepts to a variety of possible solutions.

The synthesis reports published every five to ten years by the Intergovernmental Panel on Climate Change (IPCC) provide an excellent example. These reports have become increasingly stronger in language—not just in terms of how high the scientific confidence in a given statement is, but also in the tone of statements designed for policymakers. As climate change has intensified and the audience (policymakers) has been better targeted, figure captions have changed accordingly. For example, figures showing similar (updated) data in the 2007, 2014, and 2023 reports have the following captions:

- "Examples of possible impacts of climate change..." (IPCC, 2007)
- "The relationship between risks from climate change, temperature change, cumulative carbon dioxide (CO2) emissions and changes in annual greenhouse gas (GHG) emissions" (IPCC, 2014)
- "Risks are increasing with every increment of warming" (IPCC, 2023)

Moreover, the most recent IPCC report (2023) is the first to specifically use the phrase "net-zero CO_2 emissions;" in fact, it asserts that "Limiting human-caused global warming requires net zero CO_2 emissions." Substantially more text is devoted in the 2023 report to emphasizing that rapid adaptation of mitigation measures is necessary to limit adverse effects and that the costs to society only increase by postponing these measures.

While most college and high school students are not knowledgeable about the IPCC reports, virtually all are familiar with at least some of the risks associated with climate change, and many feel overwhelmed by the enormity of the crisis. Increased knowledge about climate change can foster hope (Li et al. 2019) and self-efficacy, defined as one's belief in one's own capacity to make a difference (Liu et al. 2022). Hope positively correlates with willingness to take actions that would mitigate climate change (Kerret 2016).

Examples of Climate Action Solutions in the Chemistry Curriculum

Approximately once per week, we show two to three slides to highlight climate change solutions. The first slide is always marked "Drawdown Solution" in the same recognizable format, accompanied by a picture from the Drawdown website with the title of the solution, an indication of how many CO_2 equivalents could be saved, and how much money would be spent or gained by applying the solution to the predicted extent. The second (and third, if applicable) slide relates the solution to one or more course topics that have been recently covered. In some cases, understanding the chemical concept is critical for describing how the solution works; in others, the solution can simply be used as an example of a broad topic. For instance, Solar Hot Water improves energy efficiency because heating water for showers and other household uses accounts for about 25% of household energy use worldwide. First, students learn the calorimetry equation

$$\Delta H = q = sm\Delta T,$$

where ΔH is change in enthalpy, equivalent in this application to the heat (q) added, s is the specific heat of a substance, m is the mass of the substance, and ΔT is the resulting change in temperature of that substance. Then this equation is applied to a solar hot water system to determine how sun-heated water could be used for a building's hot water system, reducing the need to heat it either by electricity or fossil fuel burning.

As an example of a Drawdown solution being used to illustrate a broad topic, Landfill Methane Capture presents a way to prevent methane, a potent greenhouse gas resulting from the decomposition of materials in landfills, from leaching into the atmosphere. Instead, it is collected and burned for energy. We provide the methane combustion reaction as an example of a redox reaction, and various concepts of redox chemistry (for example, oxidation numbers, species that are oxidized or reduced) are applied.

Additional example topics that touch on different aspects of the General Chemistry curriculum include the following:

Alternative Cement: Balancing Chemical Reactions and Stoichiometry

When manufacturing cement, limestone is heated in kilns, producing calcium oxide, a key component of cement also known as lime, as well as waste carbon dioxide:

$$CaCO_3(s) \rightarrow CaO(s) + CO_2(g)$$

Both the CO_2 produced by this reaction and the CO_2 emissions due to heating the limestone and other processing, as well as the enormous quantity of cement used in modern society, make the cement industry the third-largest contributor to anthropogenic CO_2 emissions (Andrew,

2019)."Alternative cement" refers to cement in which 50% or more of the CaO is replaced by other materials, including certain clays, ground glass, or untreated limestone, thus averting the CO_2 emissions that would have resulted from producing that CaO.

This topic, as well as nearly any other climate change solution that has a chemical reaction directly associated with it, can be used early in the General Chemistry curriculum at the high school or university level to demonstrate balancing reactions and stoichiometry. Moreover, there are promising techniques available at the pilot scale (thus not yet included in Drawdown solutions) that use microbes to produce CaO in a completely CO_2-neutral reaction (Jacoby 2023; Beatty et al. 2022). Students are often quite surprised to learn how much cement production contributes to global greenhouse gas emissions.

Nutrient Management: Limiting Reactants and Percent by Mass

Nutrient Management refers to the practice of applying fertilizers only at specific times and locations according to need. In addition to the indirect savings of greenhouse gas emissions associated with the production, transport, and application of excess fertilizer, nutrient management prevents soil bacteria from converting excess nitrate fertilizers into nitrous oxide, a greenhouse gas with a warming potential nearly 300 times higher than CO_2. Matching fertilizer choices with plant needs, in chemistry terms, involves considering the percent by mass of nitrogen, phosphorus, and potassium in a given fertilizer. There should be no clear limiting reactant of those three nutrients; that is, the nutrients should be balanced components that contribute to plant growth.

Alternative Refrigerants: Molecular Polarity

A majority of the refrigerant compounds currently in use are hydrofluorocarbons (HFCs), which are thousands of times more potent as greenhouse gases than CO_2. Alternative refrigerants include ammonia, carbon dioxide itself, and propane. All these molecules, as well as HFCs, have simple structural formulas for which Lewis structures, followed by molecular geometry and molecular polarity assessment, can be written and completed. While the magnitude of a molecule's dipole moment does not always coincide with its global warming potential, it is related to infrared absorption; correlations can be drawn between molecular dipole moment and strength as a greenhouse gas. In addition to some direct applications of tasks that students typically find highly theoretical, most students are taken aback by the magnitude of the climate impact of managing and replacing refrigerants.

Green Roofs: Enthalpy of Vaporization

Green roofs refer to buildings with rooftop gardens. Throughout the year, the soil in these gardens provides insulation to minimize heat transfer between the building and the outside air. Furthermore, rain is absorbed by the soil, and when that moisture evaporates (either directly or via plant evapotranspiration), it cools the building. Given the particularly high enthalpy of vaporization of water, this is a highly efficient cooling process.

High Efficiency Heat Pumps: Ideal Gas Law and Endothermic and Exothermic Reactions

Worldwide, buildings use nearly one-third of all electricity generated, and well over one-third of that energy is devoted to heating and cooling. Increased efficiency in heating/cooling could cut this energy use by 30–40%. Heat pumps use a relatively small amount of energy to transfer heat between the inside and outside of a building. In heating mode, an air-source heat pump uses a fan to bring cold outside air in contact with coils containing a liquid refrigerant. Heat from the air is transferred to the refrigerant, evaporating it (an endothermic reaction). That refrigerant then runs through a compressor, which increases the pressure of the system; by the ideal gas law, increasing the pressure without changing the volume or number of moles of gas causes the temperature to increase. That high-temperature gas is pumped inside the building, and as it releases heat, it recondenses to a liquid in an exothermic reaction and is pumped back outside. In cooling mode, the process is similar but reversed.

However, Fig. 17.1 maps several additional Drawdown climate change solutions to various topics covered in general chemistry. Figures 17.2–17.4 map climate change solutions to topics in general biology, general physics, and introductory engineering, respectively.

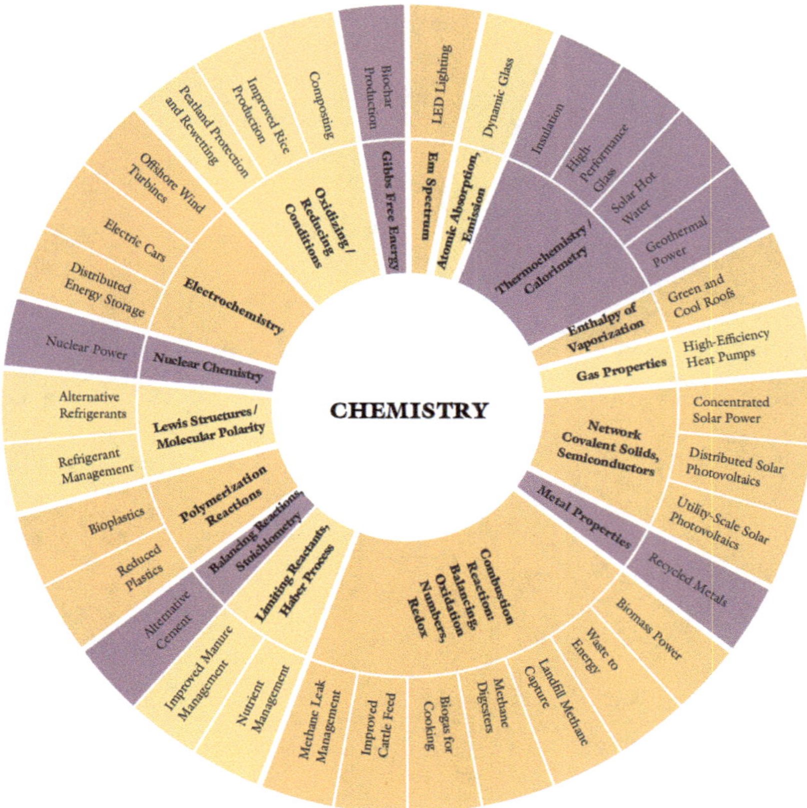

Fig. 17.1 Concepts covered in chemistry (inner ring) and the Drawdown climate change solutions that relate to those topics (outer ring)

Response in General Chemistry

Regular incorporation of climate solutions can increase students' engagement with chemistry topics while also inspiring them. In an end-of-semester evaluation, one student wrote:

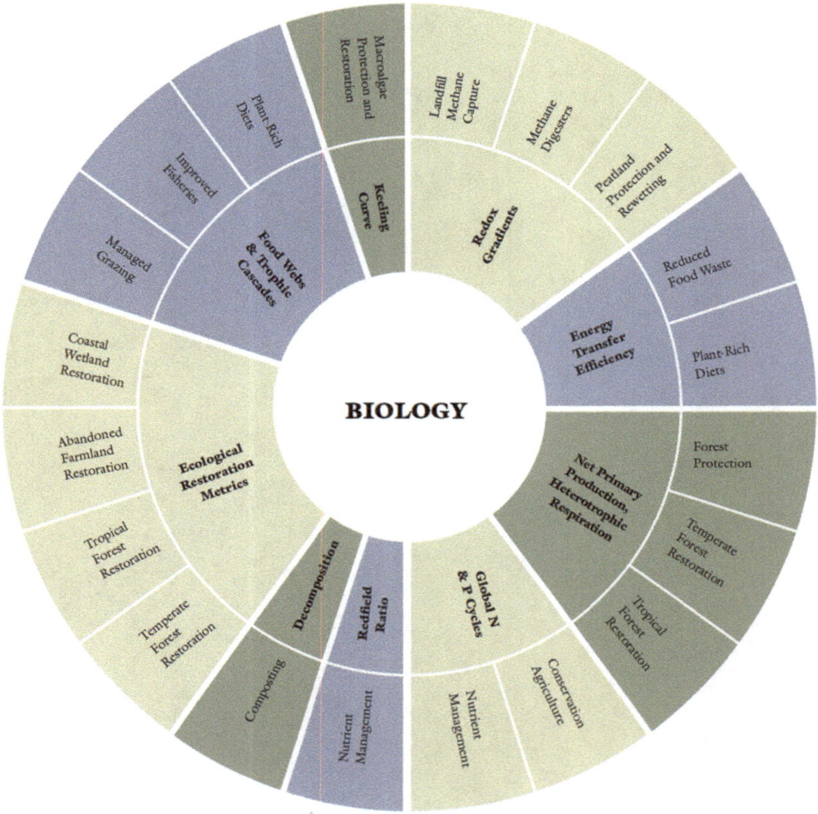

Fig. 17.2 Concepts covered in biology (inner ring) and the Drawdown solutions that relate to those topics (outer ring)

> My favorite part of the course was when you talked about how what we were learning related to the climate change solutions. It was very hopeful and inspiring, rather than how doom and gloom the information about climate change often feels. (*used by permission*)

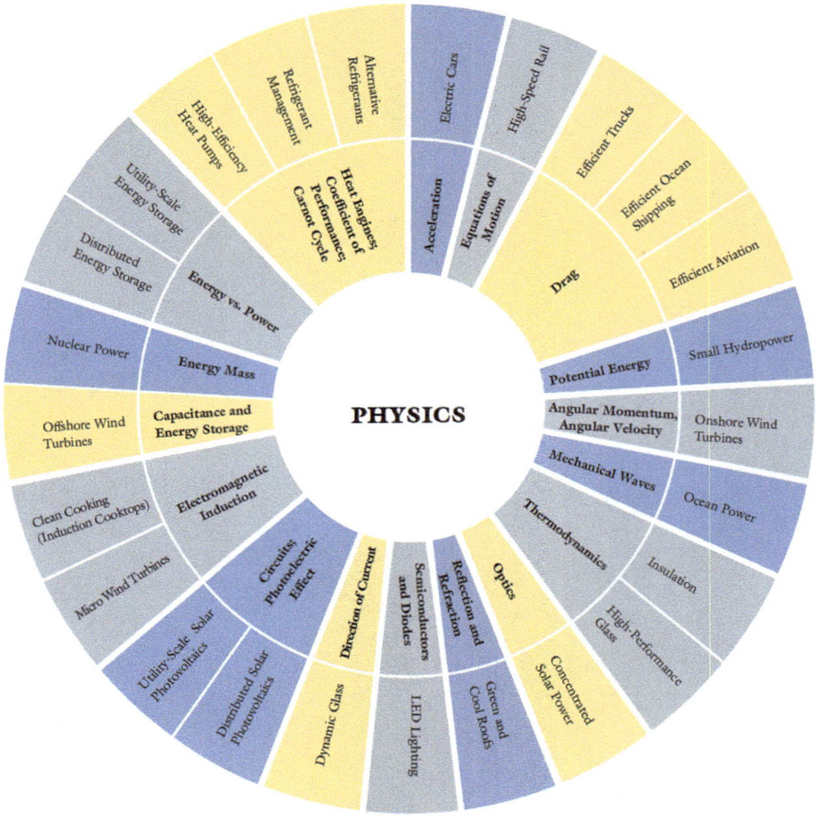

Fig. 17.3 Concepts covered in physics (inner ring) and the Drawdown solutions that relate to those topics (outer ring)

Broadening This Approach to Other Science and Engineering Courses

The approach of including short but regularly occurring climate change solution examples that use core disciplinary concepts can be mirrored in high school and college courses not only for chemistry, but also for biology, physics, and engineering. Introducing students to these applications may provide their first exposure to sustainability through a science lens. At the college level, these courses typically have high enrollments and

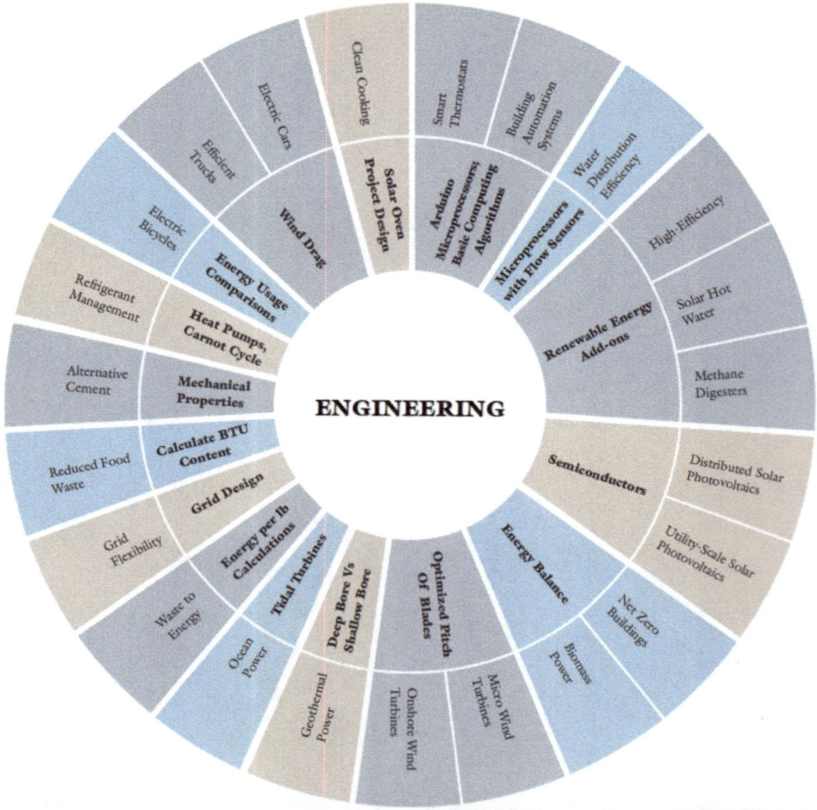

Fig. 17.4 Concepts covered in engineering (inner ring) and the Drawdown solutions that relate to those topics (outer ring)

contain mostly students who are taking multiple science and mathematics courses. Therefore, a large number of students who might not otherwise take sustainability-related courses can meet this learning objective.

Some of the "best-matched" Drawdown solutions are unique to a given discipline, whereas for other solutions, multiple science courses may relate different concepts to aspects of the same solution. See Earthrise Commons; earthrisecommons.org for how common topics in biology, physics, and engineering could be related to Drawdown solutions.

Conclusions

Directly incorporating sustainability issues into introductory science and engineering courses with heavily prescribed curricula can be difficult, but the potential for impact, both in terms of enhanced student learning and increased awareness of climate change and its solutions is considerable. With the approach described in this chapter, we are of course, only scratching the surface of each climate solution rather than delving deeply, but the net effects of this surficial treatment are many:

(1) Students understand how the concepts they are learning help to solve real-world problems and are thus reminded that there are reasons for learning the material beyond "making the grade." This enhances interest in the overall subject.
(2) Brief, picture-rich examples of content applications to climate solutions help to refocus students whose attention may wander in an otherwise uninterrupted lecture.
(3) Students learn a variety of approaches to mitigating climate change, which empowers them to have more hope and take an active role in solving the climate crisis.
(4) Students taking a course like this are less likely to be enrolled in other courses in which sustainability topics are given substantial time because science and engineering majors usually must take several subject courses with strictly defined curricula at once. Thus, this may be one of the relatively few exposures they have to sustainability topics.

More courses need to incorporate sustainability into their content, and given the limited time we have available in introductory STEM courses, we have chosen to focus on climate change solutions. Application within General Chemistry has yielded very positive student results, and similar opportunities exist in introductory courses across the STEM fields.

References

Andrew, Robbie M. 2019. Global CO_2 emissions from cement production, 1928–2018. *Earth System Science Data* 11 (4): 1675–1710.

Beatty, Danielle N., Sarah L. Williams, and Will V. Srubar. 2022. Biomineralized materials for sustainable and durable construction. *Annual Review of Materials Research* 52: 411–439.

IPCC. 2014. *Climate Change 2014: Synthesis Report. Contribution of Working Groups I, II and III to the Fifth Assessment Report of the Intergovernmental Panel on Climate Change*, ed. Core Writing Team, Rajendra K. Pachauri, and Leo Meyer. Geneva, Switzerland.

IPCC. 2023. *Climate Change 2023: Synthesis Report. Contribution of Working Groups I, II and III to the Sixth Assessment Report of the Intergovernmental Panel on Climate Change*, ed. Core Writing Team, Hoesung Lee, and José Romero. Geneva, Switzerlan.

Hawken, Paul, ed. 2017. *Drawdown: The Most Comprehensive Plan Ever Proposed to Reverse Global Warming*, 240. New York: Penguin Books. Updated data available at www.drawdown.org

IPCC. 2007. *Climate Change 2007: Synthesis Report. Contribution of Working Groups I, II and III to the Fourth Assessment Report of the Intergovernmental Panel on Climate Change*, ed. Core Writing Team, Rajendra K. Pachauri, and Andy Reisinger, 104. Geneva, Switzerland.

Jacoby, Mitch. 2023. Brewing building materials. *Chemical and Engineering News* 1010 (19): 20–25.

Kerret, Dorit, Hod Orkibi, and Tammie Ronen. 2016. Testing a model linking environmental hope and self-control with students' positive emotions and environmental behavior. *Journal of Environmental Education* 47 (4): 307–317. https://doi.org/10.1080/00958964.2016.1182886.

Li, Christine J., and Martha C. Monroe. 2019. Exploring the essential psychological factors in fostering hope concerning climate change. *Environmental Education Research* 25 (6): 936–954. https://doi.org/10.1080/13504622.2017.1367916.

Liu, Yujuan, Yeoeun Song, and Xisen Wang. 2022. Increasing preservice science teachers' climate change knowledge, hope and self-efficacy in an online chemistry course. *Journal of Chemical Education* 99 (7): 2465–2473. https://doi.org/10.1021/acs.jchemed.2c00074.

Margolis, Jane, Allan Fisher, and Faye Miller. 2002. *Caring about connections: Gender and computing*. Pittsburgh: Carnegie Mellon University, School of Computer Science.

Meehan, Casey R., and Brett L. M. Levy. 2018. Lauren Collet-Gildard Global Climate Change in U.S. High School Curricula: Portrayals of the Causes, Consequences, and Potential Responses. *Science Education* 102 (3): 498–528. https://doi.org/10.1002/sce.21338

Michalopoulou, Eleni, and Dudley E. Shallcross. ed. Atkins, Aisling Tierney, Nicholas C. Norman, Chris Priest, Simon O'Douherty, et al. 2019. The end of simple problems: Repositioning chemistry in higher education and society using a systems thinking approach and the United Nations' Sustainable Development Goals as a framework. *Journal of Chemical Education* 96 (12), 2825–2835. https://doi.org/10.1021/acs.jchemed.9b00270

CHAPTER 18

Using the UN SDGs to Teach Strategic Industry Research

Jack L. Harris

Strategic Industry Research (SIR) provides a significant opportunity to teach sustainability, multi-level thinking, systems theories, and social innovation to undergraduate and graduate students by illuminating functional connections between social systems and markets. More specifically, SIR provides a pedagogy and a management practice to connect the 17 United Nations Sustainable Development Goals (SDGs) as an interrelated set of communication and management problems to be solved by individuals in the classroom and practice. This chapter presents a method of teaching SIR through sustainability, using an example of a craft business with an international reach. Quercus Cooperage barrel makers are located in a rural region marked by significant economic and social inequality, the Hudson Valley, New York, US, that is under rapid socio-economic transition from pandemic-driven urban relocation.

J. L. Harris (✉)
Jack Harris, Department of Human Ecology, School of Environmental and Biological Sciences, Rutgers University, New Jersey, US
e-mail: jackharr@sebs.rutgers.edu

S. M. Wyman and J. Korenblat (eds.), *Sustainability Learning for Action and Community Engagement*,
https://doi.org/10.1007/978-3-032-11698-7_18

What Is Strategic Industry Research?

Strategic Industry Research, or simply *strategic research*, is a business, policy, and government decision-making approach that provides an expansive or multi-level view of the economic, social, environmental, and competitive trends and pressures that surround any organization. It places organizational activities within their ecological, economic, political, and social contexts. SIR is a process through which competitive pressures, economic needs, political dynamics, and social and cultural trends are assessed to provide perspective on an organization's economic viability and inter-organizational relationships over time. SIR is used to assess competitive pressures in business and industry for for-profit firms but is also vital for identifying broad economic, political, social, and cultural trends that may reshape any type of organization's operations or provide new opportunities (Porter 1980). Most importantly, SIR is needed to identify and engage stakeholders active in an organization's community, region, sector, or industry (Freeman 1984). SIR can also be used by non-profit and public sector entities to identify organizational and inter-organizational challenges, potential solutions, and decision pathways (Bryson 2011). The SDGs provide a framework through which to address the intertwining of strategy, economics, and stakeholder relationships in ways that address both strategic and ethical issues of the organization and community. Furthermore, the SDGs are particularly well-suited to highlight interconnections between business and non-profit stakeholders that impact community well-being or create shared value (Ansell and Gash 2008; Austin 2010; Porter and Kramer 2011; Reid et al. 2019; Wallace 2019).

Teaching Strategic Industry Research through Sustainability with the UN SDGs

The UN SDGs provide a framework for showing the interconnection between people, planet, and profit to students learning sustainability and SIR for the first time. For those studying research methods and case analysis, they also illuminate environmental, economic, and social impact problems as interconnected units of analysis. Research methods and case analysis using SIR approaches and the SDGs can be embedded in undergraduate classes on subjects such as Sustainability Studies, Corporate

Social Responsibility, Social Impact & Innovation, Stakeholder Engagement, Non-profit Management, Strategic Communication and Public Relations, Organizational Communication, and others. Modules that use the SDGs to frame research questions and connect seemingly disparate parts of a regional or industry ecosystem can also exist within standalone undergraduate research methods courses. My courses use the SDGs as a research and communication framework to help students learn how to do applied research that addresses the interrelated needs of communities, organizations, and people.

The SDGs enhance SIR studies by exploring the interconnection of organizational and inter-organizational processes that address grand challenges and wicked problems. They highlight and engage the relationships between ecological, economic, and social well-being, which can make the global local and render our teaching more relevant to students. Often seen as abstract or even separate from actual human and community experiences, the SDGs were established in 2015 with the goal to end poverty, protect the planet, and ensure social and economic well-being in terms first fully articulated in the Bruntland Report (Our Common Future, 1987). The SDGs provide a re-articulation of the triple bottom line: People, Planet, and Profit (Bode, Rogan, and Singh 2019) that continues to be prevalent in corporate social responsibility and social impact measures such as Environmental, Social and Governance (ESG) investing. As a multi-layered, interrelated set of processes, goals, and practices, the SDGs provide opportunities for students to grapple with complex and concrete examples of interconnected economic, environmental, and social problems.

By connecting the UN SDGs with SIR, we are not limited to problems of global industries and mass production. These skills and tools can be used to assess craft-based businesses like the New York barrel making business explored in this chapter. With an increasing focus on wicked problems (Weber and Khademian 2008) in our communities and on grand challenges in Business Schools and management research (George et al. 2016), there is a need for skilled business and communication professionals to more fully understand the business, socioeconomic, and ecological environments in which they are embedded. Grand challenges are complicated issues that cross organizational and disciplinary boundaries and require interdisciplinary solutions to bridge science, engineering, technology, management, communication, the social sciences, and the humanities divisions. Wicked problems are complex, collective, systems problems that typically require collaboration across the business, government, and non-profit sectors. Net Zero solutions to carbon emissions are a prime example of a wicked problem that becomes a grand challenge as it requires concerted action by the private, public, and non-profit sectors to ensure ecological, economic, social, and health considerations are well-balanced.

The SDGs can breathe new life into student understanding of the triple bottom line, acquaint students with concrete examples of stakeholder capitalism, and with the interconnections between business and social innovation. Stakeholder capitalism, with roots in Anglo-American stakeholder theory (Freeman 1984; Freeman et al. 2010) and in mid-twentieth century Nordic industrial democracy (Rhenman 1968; Strand and Freeman 2015; Strand 2015), balances economic, social, and political interests in ways that foster both community and organizational survivability over the long-term. Stakeholder thinking emphasizes that people, communities, and organizations with moral or legal rights to participation have a claim on, or obligation to, their communities or organizations and should be included in problem-solving and decision-making as a way to ensure all voices are heard and balanced (Harris 2022). Given its wide application, however, stakeholder thinking frequently defaults to managerial frameworks that ignore power, governance, and land use differences amongst stakeholder groups, often excluding local communities and indigenous perspectives (Banerjee 2000, 2002; Sarkki, Heikkinen, and Löf 2021).

Stakeholder thinking in SIR, however, teaches students to think beyond profit motives and economic incentives to consider interconnections between society, economy, and the environment for problem-solving and decision-making. Stakeholder framing is important for teaching inter-organizational relationships and for reversing the assumption commonly taught in mainstream economics courses that social relationships are embedded in markets rather than seeing markets as an outcome of social relationships built over time and that markets and *development* are ultimately only two tools of many for addressing sustainability challenges and the UN SDGs. With SIR connected to the UN SDGs, students learn that the core techniques of business and communication analysis are not separate from issues of sustainability and social impact. Organizational sustainability leaders emphasize the need for communication and analytical skills in managerial and other social contexts for students seeking jobs in sustainability-related sectors.[1]

The Strategic Industry Research Process

When analyzing organizations and industries, we begin with a macro view to explore their respective environments from a 50,000-foot level. This is especially true when working across sectors and with communities. All organizations work within a socioeconomic environment that influences their activities and provides opportunities and constraints. Understanding and including the perspectives of those who use or inhabit an environment can provide more focused analysis and offer more realistic solutions and recommendations (Xiao et al. 2018; Reid et al. 2019). SIR analysis can be used in communication, competitive, financial, and strategic analyses in the private, public, plural, or non-profit sectors. The examples in this chapter illustrate the broad interconnections between businesses, non-profits, and communities. This pedagogical framework was originally developed for a Strategic Communication class to highlight the shared value created through interconnections between a hyperlocal craft business and a sound understanding of a community's economic and social needs.

SIR involves a five-step process. In the communication analysis example below, we start by examining the overall industry environment (the 30,000-to-50,000-foot view) in which a particular business or non-profit organization operates. External industry analysis usually starts with publicly available sources found online and through academic and public library resources. Conversations and interviews with people knowledgeable about the industry and associated economic, social, cultural, and ecological trends impacting the industry help students gather information about organizations, cross-sector relationships, and partners that operate in the industry. Organizational analysis shares much with industry analysis, but in these activities, the focus is on the particular organization and key intra-organizational capacities and issues rather than broader relationships and trends. Organizations may be part of industries and impacted by those more general trends, but they are also unique, individual entities with

[1] Employer Panel, Rutgers University Faculty Workshop on Sustainability in the Curriculum, April 11, 2024.

their own sets of internal and external dynamics. Students and researchers may want to pay particular attention to identifying both organization-specific and industry or sector-wide information when accessing these sources, including conversations and interviews with experts, librarians, publicly available resources, associations and consulting reports. Since organizational analysis focuses on single organizations, students and researchers may also have access to organizational records through public sources and media reports or, more rarely, internal organizational records made available.

At the next level, students research and analyze a particular topic, issue, or project. In the communication analysis example below, the goal is to understand specific communication, messaging, and processes in order to design an effective media campaign that engages stakeholders. These may include suppliers and even competitors, but they expand to include the media, government civil society, and the non-profit sector more broadly. University relationships or partnerships with businesses and community organizations can also be considered stakeholder relationships (Freeman 1984; Freeman et al. 2010). Stakeholder theory connects business goals, such as profit and economic loss with broader concerns about ethics, society and the well-being of communities (Bonnafous-Boucher and Rendtorff Dahl 2016). Connecting stakeholder theories with SIR through the SDGs creates a powerful learning tool for showing students how theories, methods, policy frames, and applied business and community problems are interconnected, with ecological, economic, social, and political processes lying at the heart of sustainability.

We orient learners to SIR by explaining that their audience will want to know initial findings and critical interpretations, hypotheses, and expert recommendations and have the opportunity to provide feedback. In the communication analysis examples initially used in this case, the goal of the project was to assess how the focal organization could best communicate its value as a partner to non-profit organizations and local policymakers. Freeman (1984, 2010) argues that businesses and the leaders that manage these organizations actually create value for employees and communities for suppliers, customers, and investors. Despite the financialization of the economy (Ott 2011) over the last half-century, the stakeholder premise has held firm, culminating in the 2018 restatement of the purpose of a corporation by the Business Roundtable, which commits corporations to building value for all stakeholders.

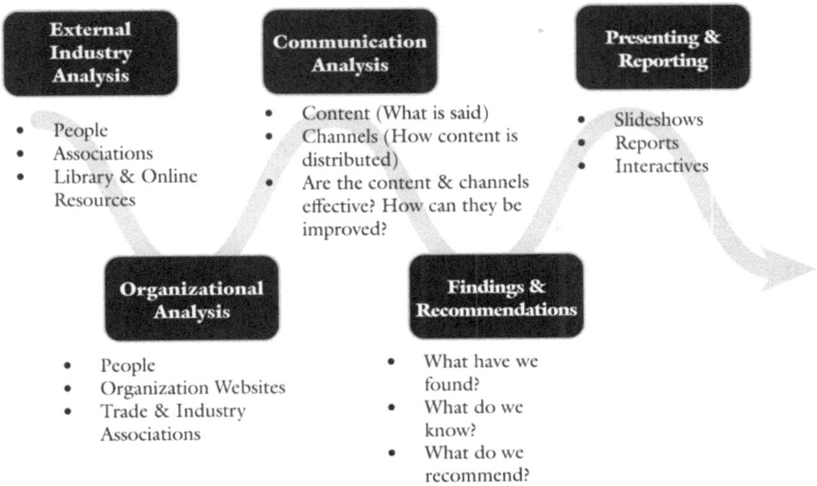

Fig. 18.1 The strategic industry research flow in communication analysis

Finally, students must make decisions as to how information will be communicated. We emphasize student choice and consequent ownership/responsibility of the SIR project by inviting learners to present their findings and analysis via a report or white paper, a presentation, through social media, or as a creative and interactive experience that allows the audience to engage with the information and findings directly. Inviting learners to communicate analyses, data, and source material across different textual, numerical, and visual formats allows each one to best utilize their own cognitive strengths, learning, and unique personalities as part of the project. Inquiry-based, self-directed projects help to individualize projects, render abstract concepts more concrete, and increase student buy-in(Fig. 18.1).

Case Study: Quercus Cooperage (for-Profit Craft Business)

The Quercus Cooperage case example demonstrates how an SIR approach to a relatively small business shows the interconnection of multiple SDGs in a regional, Hudson Valley context. Nearly all students are in-state at this university, 90 minutes from New York City and 90 minutes from Albany, the state capital. The region is rural, has a legacy of industry dating back to the American Revolutionary War, and was hard hit by rapid deindustrialization in the late twentieth century.

Quercus Cooperage is a hyperlocal craft business with a global market. John Cox and his team are barrel makers, a traditional industry in the region, but adapted to the twenty-first century with their emphasis on barrel making for the craft distillery market and sake vessels for export to Japan. By focusing on local for-profit and non-profit organizations that make the Mid-Hudson Valley their home, we can introduce students to the interconnections between local organizations, the larger regional community, and the socioeconomic realities that anchor them.

Using the development of a potential *earned media*[2] campaign to drive stakeholder engagement, this case demonstrates the connection of a small for-profit business to a broader industry: the American Craft Barrel Industry and the Global Barrel Making Industry, including Japanese companies in the distillery business operating at grand global scales. Quercus also intersects with adjacent industries, such as the New York Distilling industry and regional non-profits engaged in water and land conservation issues. The SDGs can be used to frame sustainability in New York's Mid-Hudson Valley region with emphasis on the interconnection between six different goals as an integrated system.

The Quercus Cooperage case and others like it, can be used to teach students an integrative trans-disciplinary approach to solving broad-level sustainability problems as well as more discrete problems related to business, communication, marketing, cross-sector and regional partnerships, stakeholder engagement, and messaging. This learning activity asks us to view sustainability at scale—from the hyperlocal to the global and how a small craft business is connected to larger environment, social, and governance issues.

While at first glance, a small private business such as Quercus would appear to be solely connected to SDG 9: Industry, Innovation and Infrastructure (Industry) and SDG 8: Decent Work and Economic Growth (Jobs/Careers), closer examination reveals that this business relies on the conservation of land (to maintain the unique type of oak wood used in their products) and also upon the protection and maintenance of quality

[2] Earned media is the industry term in Media and Strategic Communications for free media and social media coverage earned through organizational activities and strategic communication campaigns.

water resources in the region, SDG15: Life on Land (Environment) on which many of their distillery customers rely. Thus, this small business' competitive needs align with those of the tourism, distilling, and brewing industries that require open space, clean and protected waters, and healthy regional ecosystems and landscapes to support their varied needs. This also aligns with New York City's need for 1.1 billion gallons of fresh, clean drinking water every day (SDG 6: Clean Water and Sanitation). Most of New York's water supply is sourced by reservoirs and requires significant open space and wooded areas to act as buffer zones. Three reservoir systems serve nine million customers across the state. The Catskill reservoir system in which Quercus Cooperage is located provides 40% of NYC's water supply and is more than 100 miles north of the city. Examples like this demonstrate to our students symbiotic social and ecological relationships that encompass hyperlocal businesses like Quercus, their own lives, as (mainly) New Yorkers, metropolitan cities, and global markets for barrel and masu making.

Companies such as Quercus require a high level of skill to operate. In addition to the mastery of very specific, expert woodworking skills, barrel making also requires mathematical skills and knowledge to correctly understand the varied curves, shapes, and dimensions of the barrels and wood products produced in the business. Likewise, distilling, a key customer of Quercus, is a seemingly simple business that requires a vast knowledge of soil and grasses, chemistry, and math to ensure the correct measure and timing of proportion, flavor, and taste of each distilled product. These requirements place Quercus within a broad range of regional stakeholders across the private, public, and plural sectors that provide education, training, and youth development services. Quercus employees complete rigorous education and training programs to ensure a qualified workforce (SDG 4: Quality Education), which further aligns the company with SDGs 8, 9, and 11: Sustainable Cities and Communities (Jobs/Careers, Industry, & Communities). Thus, seemingly simple businesses can showcase complex interconnections among ecological and socio-economic processes in a given region.

Context and Regional Pressures

Socio-economic concerns often prove compelling to our students whose families tend to have first-hand knowledge of local history and present conditions. Deindustrialization hit the Hudson Valley hard over the last thirty-years with major employers such as IBM's non-commission of large facilities, which impacted both direct and indirect employment in the region. Inequality has been a significant marker of the Hudson Valley for decades (SDG 10: Reduced Inequalities). The pandemic-driven boom in real estate prices and people moving out of New York City to the region only exacerbated these trends by making housing scarcer and more expensive. Employment and job training opportunities remain significant challenges in the region (SDG 8). Effective economic development and job training require a concerted effort by stakeholders across the public, private, and plural sectors at community, organizational, and inter-organizational levels (SDG 17: Partnership for the Goals).

Reflection on the Practice

I have designed this case practice for use in capstone and senior seminar level courses, despite interruptions, diSIRuptions, and delays in course implementation caused by the pandemic and its aftermath. Consistent inclusion of such modules across courses can facilitate the assessment of student learning using Bloom's Taxonomy for Sustainability Learning through SIR (Table 18.1). While teaching at the University of Illinois in Urbana-Champaign, I adapted the module by adding information on the Kentucky and Illinois bourbon and craft distilling businesses to introduce students to their own local and global business environment and its associated ecosystems.

Expanding Project Scope for Action Plans

While we have not been able to build an actionable plan with students that an organization can implement so far, current projects address community food practices and systems in a small urban coastal community near my current university. This project has resulted in a report to two non-profits within the community food system and a research study on social impact networks relating to food access and security. In our senior practicum class, we are analyzing this urban coastal community's food system (SDGs 2: Zero Hunger, 3: Good Health and Well-Being, 11, 12: Responsible Consumption and Production, 17) from business (SDGs 8 and 9), cultural food practices (SDGs 5, 10, 11), and land use (SDG 15) perspectives. This integrates systems, socio-economic, and ecological thinking into the development of analysis and reporting on food

Table 18.1 Applying Bloom's taxonomy of learning to strategic industry research

Bloom's Taxonomy of Learning	*Applications to Strategic Industry Research for Sustainability*
Higher-order sustainability thinking skills developed through strategic industry research	• Systems Thinking (interconnection of social, ecological, organizational, and inter-organizational systems) • Futures Thinking (desired end states, creating vision, mission, purpose, goal setting) • Scenario Thinking (plausible, possible futures)
Creating	• Project plan and timeline • Organizational & Industry Reports • Policy briefs • White papers • Messaging strategy/strategies • Press releases • Social media communications • Communications and marketing collateral
Evaluating	• Strategies and scenarios at organizational & inter-organizational levels
Analyzing	• Industry ecosystems • Supply chains • Small & entrepreneurial businesses • Corporate social responsibility messaging and strategies • Social Impact Strategies • Stakeholder relationships & engagement • Communication & marketing strategies • Public & social policies • Economic Development • Workforce Development
Applying	• Academic reports & articles • Industry sources; written reports, information from human subject matter experts • Financial information • Government information • Social statistics

(continued)

Table 18.1 (continued)

Bloom's Taxonomy of Learning	*Applications to Strategic Industry Research for Sustainability*
Understanding	• The UN SDGs • Appropriate economic, ecological, organizational, and social context for the ClasSIRoom project • How to identify and use sources to create reports, messages, strategies, and briefing materials
Remembering	The types of sources to use in strategic research: • Library collections (databases, hard copy sources, research services) • Company & industry reports • White papers • Policy briefs • Company & organizational websites • Financial & statistical information • Human subject matter experts • Location of sources (libraries, company & organizational websites, people/subject matter experts, think tanks, consulting firms, academic centers & institutes)
Lower-order sustainability thinking skills reinforced through strategic industry research	Recall and explanation of definitions and key concepts for sustainability thinking. Examples of these concepts and definitions may include the UN SDGs, the Bruntland Report, sustainable development, the triple bottom line (people, planet, profit) and related economic frameworks, the planetary boundaries framework, Intergovernmental Panel on Climate Change (IPCC), climate neutrality, Net Zero, biodiversity, sustainable Food Systems.

access and security in the community. We have also begun exploring social innovation (SDG 9) possibilities within this community food system. Once the practicum project is completed, a new teaching module on this community food system will supplement the Quercus case in our Introduction to Sustainability course. A preliminary case study and teaching module on community-engaged teaching practices in this community food system has been accepted for presentation at Umbra Institute's June 2026 Biennial Food Conference in Perugia Italy.

Conclusion

A case like Quercus demonstrates the symbiotic and interconnected nature of business, environmental, and social goals. Teaching the SDGs in a business, economic, or non-profit context does not have to rely on brand-name organizations that students are familiar with or investigate FT1000 global corporations with sophisticated corporate social responsibility and social impact operations. Rather, by focusing on local or regional businesses, we can highlight how the SDGs can be used to understand concrete interconnections between people, communities, organizations, and economic needs at more granular levels in students' own backyards. Successful climate action requires the implementation of sustainability goals and practices at both the local and transnational levels. Future pedagogical goals with this approach include the development of experiential learning opportunities for students with case study partners and their related stakeholders by offering apprenticeships, internships, summer employment, the development of business, communication, or marketing plans, and even policy advocacy and stakeholder engagement around relevant, symbiotic SDGs. The SIR inquiry-based approach provides a method for teaching and learning that makes the SDGs come alive for students in relevant, clearly applicable ways.

References

Ansell, Chris, and Alison Gash. 2008. Collaborative governance in theory and practice. *Journal of Public Administration Research and Theory: J-PART* 18 (4): 543–571.

Austin, James E. 2010. *The Collaboration Challenge: How Nonprofits and Businesses Succeed Through Strategic Alliances*. San Francisco, CA: John Wiley & Sons.

Banerjee, Subhabrata Bobby. 2000. Whose land is it anyway? National Interest, indigenous stakeholders, and colonial discourses: The case of the Jabiluka uranium mine. *Organization & Environment* 13 (1): 3–38. https://doi.org/10.1177/1086026600131001.

Banerjee, Subhabrata Bobby. 2002. Contesting corporate citizenship, sustainability and stakeholder theory: Holy trinity or praxis of evil? *Academy of*

Management Conference. (August -13) Denver, Co, US. https://my.aom.org/ProgramDocs/pdf/AOM_2002_Annual_Meeting_Program.pdf

Bode, Christiane, Michelle Rogan, and Jasjit Singh. 2019. Sustainable cross-sector collaboration: Building a global platform for social impact. *Academy of Management Discoveries* 5 (4): 396–414. https://doi.org/10.5465/amd.2018.0112.

Bonnafous-Boucher, Maria, and Jacob Rendtorff Dahl. 2016. Stakeholder Theory*: A Model for Strategic Management*. Springer.

Bryson, J. M. 2011. Strategic planning for public and nonprofit organizations: A guide to strengthening and sustaining organizational achievement *(4th Edition)*. San Francisco, CA: Jossey-Bass. Business Roundtable. https://opportunity.businesSIRoundtable.org/ourcommitment.

Freeman, R. Edward. 1984. *Strategic Management: A Stakeholder Approach*. Marshfield Massachusetts: Pitman Publishing.

Freeman, R. Edward, Jeffrey S. Harrison, Andrew C. Wicks, Bidhan L. Parmar, and Simone de Colle. 2010. *Stakeholder Theory: The State of the Art*. Cambridge, UK: Cambridge University Press.

George, Gerard, Jennifer Howard-Grenville, Aparna Joshi, and Laszlo Tihanyi. 2016. Understanding and tackling societal grand challenges through management research. *Academy of Management Journal* 59 (6): 1880–1895. https://doi.org/10.5465/amj.2016.4007.

Harris, Jack L. 2022. *Hyperlocal Organizing: Collaborating for Recovery Over Time*. MD: Lexington Books.

Ott, Julia C. 2011. *When Wall Street Met Main Street*. Harvard University Press.

Porter, Michael E. 1980. *Competitive strategy: Techniques for Analyzing Industries and Competitors*. New York: The Free Press.

Porter, Michael E., and Mark Kramer. 2011. Creating shared value. *Harvard Business Review* no. January-February:4–17.

Reid, Amy, Michelle Abraczinskas, Victoria Scott, Morgen Stanzler, Jonathan Scaccia, Abe Wandersman, Rohit Ramaswamy, and Rohit Ramaswamy. 2019. "Using Collaborative Coalition Processes to Advance Community Health, Well-Being, and Equity: A Multiple–Case Study Analysis From a National Community Transformation Initiative." Health Education & Behavior 46 (1_suppl): 100S–109S. doi:https://doi.org/10.1177/1090198119838833.

Rhenman, Eric. 1968. *Industrial Democracy and Industrial Management*. London: Tavistock Publication Ltd.

Sarkki, S., H. I. Heikkinen, and A. Löf. 2021. Reindeer herders as stakeholders or rights-holders? Introducing a social equity-based conceptualization relevant for indigenous and local communities. *In Nordic Perspectives on the Responsible Development of the Arctic: Pathways to Action. Cham: Springer Polar Sciences. Springer.* https://doi.org/10.1007/978-3-030-52324-4_13.

Strand, Robert. 2015. Scandinavian stakeholder thinking: Seminal offerings from the late Juha Näsi. *Journal of Business Ethics* 127 (1): 89–105. https://doi.org/10.1007/s10551-013-1793-0.

Strand, Robert, and R. Edward Freeman. 2015. Scandinavian cooperative advantage: The theory and practice of stakeholder engagement in Scandinavia. *Journal of Business Ethics* 127 (1): 65–85. https://doi.org/10.1007/s10551-013-1792-1.

Wallace, Jennifer. 2019. "Understanding wellbeing and devolution in Scotland, Wales and Northern Ireland." In *Wellbeing and Devolution: Wellbeing in Politics and Policy*. Palgrave. https://doi.org/10.1007/978-3-030-02230-3_2.

Weber, Edward P., and Anne M. Khademian. 2008. Wicked problems, knowledge challenges, and collaborative capacity builders in network settings. *Public Administration Review* 68 (2): 334–349. https://doi.org/10.1111/j.1540-6210.2007.00866.x.

Xiao, Yu, Kai Wu, Donovan Finn, and Divya Chandrasekhar. 2018. "Community Businesses as Social Units in Post-Disaster Recovery." *Journal of Planning Education and Research*, October, 0739456X1880432. doi:https://doi.org/10.1177/0739456X18804328.

CHAPTER 19

Incorporating COIL and the UN SDGs for Global Learning for All

Nicole A. Simon

Introduction

In a collaborative online session, learners from international backgrounds shared snapshots of their respective waterways in Recife, Brazil (Capibaribe River) and Long Island, New York, US (Long Island Sound and Atlantic Ocean), highlighting pollution issues ranging from industrial runoff to agricultural contamination. They exchanged anecdotes of degraded environments and discussed the global implications of such challenges. Together, they explored grassroots initiatives and successful restoration projects, emphasizing the transformative power of cross-cultural collaboration to address urgent environmental concerns and pave the way for a more sustainable future on a global scale.

This chapter presents a Social Business Enterprise collaboration between students in Recife, Brazil, and Long Island, New York that uses the United Nations Sustainable Development Goals (SDGs) to develop skills in critical thinking, intercultural competency, systems thinking, and

N. A. Simon (✉)
STEM Department, Nassau Community College, New York, NY, USA
e-mail: nicole.simon@ncc.edu

S. M. Wyman and J. Korenblat (eds.), *Sustainability Learning for Action and Community Engagement*,
https://doi.org/10.1007/978-3-032-11698-7_19

global awareness. Using the COIL learning methodology, students who may not typically travel abroad or have access to international cross-cultural experience, learn new interpersonal and digital professionalism competencies, and see themselves as part of a global community. By leveraging online platforms and intercultural partnerships, COIL, and similar virtual exchange experiences contribute to building a more sustainable, accessible, and resilient educational ecosystem for all.

Collaborative Online International Learning

Collaborative Online International Learning (COIL) is a virtual exchange (VE) program that connects learners from around the globe to collaborate on a shared project using online platforms. COIL fosters digital literacy while lowering the physical barriers that separate learners from diverse cultural backgrounds. It also enables information, ideas, and perspective sharing across international, interdisciplinary groups. Successful COIL instructors make informed decisions in design and implementation so international teams from different cultures that may lack in-person collaboration experience, enjoy the full benefits of transnational exchange (Jie & Pearlman 2018). COIL implementations are becoming more common, partly due to the internationalization of education and the shift to online learning during the COVID-19 pandemic.

The SUNY COIL center was established in 2004 and has since brought together many disciplinary subjects and an international network of over 223 universities (Rubin 2017; SUNY COIL Center), while additional academic institutions have internationalized courses by following this model (Liu and Shirley 2021). COIL offers a promising avenue for fostering intercultural competency and curiosity among learners worldwide. Studies indicate that its benefits extend across diverse cultural backgrounds with the potential to equitably internationalize course materials. However, adopting COIL necessitates addressing potential challenges in language and communication across cultures.

Community College COIL Experience

Community college learners often miss opportunities to study abroad due to limited resources. However, their diverse campus populations offer a unique advantage. Connecting with peers from similar backgrounds in distant countries fosters cultural exchange and global awareness, enriching their educational experience without the need to travel abroad. They can connect deeply on a global level because they see similarities in their cultures, perspectives, and environmental challenges.

Our community college learners often juggle full-time work alongside their studies, and school is not always the top priority. However, our partners understand these challenges well. Over years of collaboration, we've built an understanding of these pain points, and prioritized flexibility. We aim to be as accommodating as possible, recognize the need to adjust deadlines and provide ample resources for learners who face difficulties that may involve time, financial resources, family and work obligations, and educational preparation. This aligns closely with four-year institutions' challenges, where student responsiveness may only sometimes meet expectations.

Learning Activity: Using the SDGs in Sustainable Social Business Models

Learners from SUNY Nassau Community College, US, and the Federal University of Pernambuco, Brazil, worked together synchronously online and asynchronously to understand the importance of international or global sustainability in social business enterprises by examining characteristics of ecological water use: costs, value, impact, and water quality models. They studied these features in the context of ethics, global perspectives, effective communication, decision-making, and partnerships. By the end of the COIL experience, learners gained experience communicating effectively through online interactions and developed skills in critical and analytical thinking by recognizing and respecting the validity of unfamiliar types of thinking and reflecting, coexisting, dialog, sharing, acting, and solving problems in learning environments marked by social and cultural diversity.

Sustainability Education and the United Nations Sustainable Development Goals

Sustainability topics have become more mainstream in education during the past two decades partly because diverse disciplines can commit to addressing these challenges from their bodies of knowledge and perspectives. *Sustainability* has transformed into a potent "boundary term" that acts as a unifying force, connecting disparate groups around a common agenda (Scoones 153). Sustainability work has catalyzed innovation in ideas, spurred political mobilization, and instigated policy changes. This transformative momentum was particularly pronounced during the UN Conference on Environment and Development (UNCED) in Rio in 1992. This is especially true for sustainable development, which some scholars consider a paradox because *development* in terms of productivity, GDP, and improved quality of life—all based on increased consumption—has destabilized our environment and contributed to the wealth gap. Globally, a curriculum that emphasizes Education for Sustainability (EfS) promotes sustainable development and thriving in all its forms: economic, social, and environmental.

Following the introduction of the UN Sustainable Development Goals (SDGs) in 2015, there has been a resurgence of interest in sustainability education in higher education institutions (HEI). The 17 SDGs comprise 169 targets set out by 193 UN member states to advance peace and prosperity by 2030. Each of the seventeen SDGs addresses a specific aspect of development, and they are all interconnected, emphasizing the need for holistic and integrated approaches to sustainable development. Achieving these goals requires concerted efforts from governments, civil society, businesses, and individuals worldwide. Overall, the SDGs represent a shared vision for a better future, where everyone can enjoy a decent standard of living, peace, and prosperity while safeguarding the planet for future generations. SDG 4: *Quality Education*, describes how learning is a fertile area of opportunity for improvement that makes other global goals more possible. Research has examined how university learners perceive and understand concepts related to sustainable development and the importance of imparting sustainability information and approaches (Armon et al. 2020; Barth 2015; Chang and Wi 2020; Cachelin 2019; Mossman 2018; Michel 2020).

I've seen firsthand at the two-year community college where I teach how aligning with the SDGs promotes equitable higher education. By embracing the SDGs, institutions address societal issues while ensuring access to quality education through the many resources available. Open Educational Resources (OERs) play a key role, offering free, adaptable materials that can be translated into multiple languages, to enhance accessibility for learners from diverse backgrounds. These initiatives create an inclusive learning environment where all learners with internet access can thrive and contribute to global sustainability goals using zero-cost materials.

When I COILed with a partner in Brazil, we completed a Social Business Enterprise collaboration with my Sustainable Energy course that allowed learners to work together on water and air pollution issues using the SDGs to address pressing social and economic instability issues. The shared course content was anchored in global concerns and activities with the SDGs as an overarching framework in both classrooms. *Climate Action* (SDG 13), *Sustainable Cities and Communities* (SDG 11), *Decent Work and Economic Growth* (SDG 8), *Life on Land* (SDG 15), and *Good Health and Well-Being* (SDG 3), are particularly pertinent to our COIL module, which analyzes the relationships between environmental, climatic, sustainability, and cultural heritage concerns.

Learners used simulations to look at what was going on in both countries so they could determine a list of similar problems that both countries faced. We also looked at ecological challenges particular to Recife and Long Island, respectively, and tried to come up with ideas to mitigate damage at the local level and start brainstorming about more significant repercussions at the global level.

Alignment of COIL with UN SDGs: Course Modifications

An embedded SDG-informed COIL project can integrate into an existing course without disrupting it, and in fact, COIL can enrich its curriculum with global perspectives. By partnering with international institutions, learners engage in cross-cultural collaboration on course topics, practicing critical thinking and intercultural competence. These projects can align with course objectives, enhance learning outcomes, and provide practical applications of theoretical knowledge in real-world contexts. Ultimately, embedding COIL can help learners develop global awareness, cultural humility, and empathy within the course framework (Anderson and Or 2023). How does one begin a COIL collaboration?

Begin by identifying which SDGs are relevant to the content or theme of your COIL/Virtual Exchange (VE) program. As an instructor, Integrating and applying course material using the SDG framework can be frictionless due to the clear connection between the COIL and SDG methodologies; each focuses on experiential learning and proactive attention to intended learning outcomes. Instead of numerous curriculum updates, incorporating sustainability topics requires a shift in perspective as to how to present and apply the existing course material. For instance, I employed the COIL format to integrate Alternative Energy and Social Business courses for cross-cultural exchange and understanding between students in Long Island and Recife. We modified our syllabi to include COIL modules, aligned learner expectations and course objectives, assignments, and assessments with the SDGs, and each introduced the triple bottom line of sustainability (people, planet, and profit) at the semester's outset. This approach deepened learners' comprehension of sustainable practices in academic and professional pursuits and prompted critical thinking about their personal roles in sustainable development.

Curriculum Design

As professors, my Brazilian colleague and I introduced our respective learners to our collaborative study, outlining its waterway conservation purpose and expected outcomes. We formed six teams, comprised of students from both universities. Prior to this initiative, many of our learners had experienced multicultural encounters within their own countries. However, engaging in this collaborative project allowed them to directly compare cultural perceptions, observe how their peers reacted in various contexts, and gain insights into their own cultural perspectives from different angles. We facilitated discussions and shared engaging content on a dedicated Padlet site, to enable interaction through icebreaker introductions that encouraged learners to reflect on their cultural experiences and views on the SDGs. We focused on developing entrepreneurial and communication skills to address real-world challenges and tasked learners with creating social enterprises that targeted specific SDGs. This approach aimed to promote intercultural collaboration, entrepreneurship, and digital literacy, recognizing their significance as twenty- first-century skills for an interconnected world.

Discussion

Results indicated that participants developed a critically sensitive lens and increased their understanding of the SDGs, which enhanced topic understanding. With a new perspective, learners could discover details about the SDGs, their schoolwork, their partners, and themselves. This allowed for a growth mindset and an awareness of local and global issues. The next iteration of this COIL course will include working with NGOs (non-government organizations).

I advocate for incorporating alternate pedagogical approaches such as Universal Design for Learning that outlines multiple means of engagement, representation, action, and expression (https://udlguidelines.cast.org/). This offers diverse avenues for learners to express themselves effectively and enhances their learning potential. I prioritize flexibility and user-friendliness across both campuses to ensure learners have access to unique learning opportunities. By integrating elements that may span the entire semester but offer a different perspective from traditional courses, we empower learners to collaborate with international peers, kindling teamwork and the development of self-driven projects. While providing some guidance, the emphasis remains on student initiative and collaborative problem-solving.

Looking ahead, I see areas for growth and expansion in our curricular design and places to address existing challenges. Exploring simulations to address transportation and environmental challenges, alongside integrating AI tools for translation assistance, would significantly enhance our curriculum and better support our students' learning experiences. Additionally, logistical factors like time zone disparities, variations in learning management systems (LMS), and differences in academic calendars can impact the degree of collaboration between international partners. While COIL presents significant opportunities, its successful implementation requires careful consideration of these factors (Rubin 2017; Liu and Shirley 2021).

Cultural disparities in learner expectations are also significant; they might include implicit differences in learning styles, variations in university academic conduct rules, cultural contrasts between interpersonal and productivity priorities, and instances of racism and microaggressions. In Mexican classrooms, for example, there's often a strong emphasis on respecting authority figures such as teachers, with learners typically using titles like *maestro, profesor*, or *profesora*, reflecting the cultural value of deference to authority. Like Mexico, Brazilian classrooms may also emphasize respect for teachers but with a more relaxed and informal tone. Learners often address teachers using their first names, reflecting the country's friendly and inclusive communication style. In Jordanian classrooms, in contrast, hierarchical structures may be more pronounced, with teachers occupying a position of authority and learners expected to show deference through attentive listening and minimal questioning during lessons. Instructors new to the COIL experience may be unfamiliar with the pluralism of classroom cultures in these settings. Research collaborations are widespread in international universities; however, teaching partnerships are uncommon (Rubin 2017; Liu and Shirley 2021).

Instructors' and learners' pedagogical methods have evolved because of the digital revolution in education (Adams Becker et al. 2017). For this reason, educating young adults to be twenty-first-century learners requires building digital literacy and intercultural abilities (Partnership for twenty-first-century abilities 2011; Rubin and Guth 2015). Value-added technology applications can enhance teaching and learning within and outside the classroom (Lenkaitis 2019, 2020). Both synchronous (real-time) online meetings and asynchronous (*not* time-specific or simultaneous) computer-mediated communication (SCMC) can connect parallel learners in various geographic and cultural contexts for Virtual Exchange (Belz 2003; Dooly 2017; O'Dowd 2018).

The SUNY OSCQR Rubric **(Online Course Quality Review Rubric)** sets high standards for online, hybrid, or high-flex courses tailored to diverse campus contexts globally. Given the varying access to instructional designers, it's vital to ensure partners understand these standards for effective collaboration and reduced logistical barriers. Cultural awareness enhances the global learning experience and fosters appreciation for diversity. Emphasizing teamwork and communication skills prepares learners for success in the workforce. Encouraging diverse communication skills is key for long-term growth. Integrating COIL + OSCQR principles, as I have done, aims to create a cohesive learning environment worldwide.

As an instructional designer, I facilitate interdisciplinary collaborations across different levels, drawing from my OSCQR certification. In COIL settings, I ensure courses are designed to suit both campuses by emphasizing co-creation with partners from the outset. I emphasize shared ownership to foster mutual investment. Accessibility is prioritized, leading us to establish a user-friendly, password-protected Google site as our primary platform, which ensures a safe environment for learners to share and access materials. Leveraging existing technology accounts minimizes costs for learners and partners and provides essential tools for project success.

Cross-Cultural Dialog

To craft democratic procedures vital for sustainability learning, use the SDGs as a framework for cross-cultural dialog. Encourage discussions about how global challenges and solutions vary across contexts and how international cooperation is essential for achieving the SDGs. By empowering learners to engage with global challenges and work toward shared goals, COIL/VE initiatives contribute to building a more inclusive, equitable, and sustainable world. Integrating the SDG of *Quality Education* (Goal 4) into COIL and VE initiatives is a powerful way to promote inclusion and equity, enhance cross-cultural understanding, and foster global citizenship. Here's how the SDG of Quality Education can be addressed through COIL/VE:

1. **Enhance Access to Education and Promote Cross-Cultural Understanding:** Community college learners, often without access to traditional study abroad opportunities, benefit from interactions with culturally diverse peers through online platforms. These learners, hailing from underserved communities and often as first-generation higher education attendees, engage in task-based language learning. These goal-oriented joint activities can spark a sense of belonging and augment intercultural communication and collaborative skills.

2. **Share Best Practices in Teaching and Learning:** COIL/VE initiatives serve as forums for educators to collaborate across borders, sharing expertise in curriculum development, instructional design, and pedagogical approaches. By drawing on diverse experiences, these collaborations enrich the quality and effectiveness of education delivery.

3. **Address Global Challenges in Education:** COIL/VE projects, aligned with SDG 4, tackle global educational issues, including access to quality education, teacher training, literacy, and technology integration. Through collaborative research and advocacy efforts, participants identify innovative solutions and contribute to positive changes in educational systems worldwide.

4. **Empower Learners and Educators:** Participation in COIL/VE experiences empowers both learners and educators to drive their learning and professional growth. By engaging in cross-cultural collaborations, they develop critical thinking skills, intercultural competence, and digital

literacy, which are crucial for navigating the challenges of the contemporary globalized world in personal, public, and professional realms. Students informally practice digital literacy through social media all the time. COIL helps formalize some of these 'folk literacies.'

Assessment of COIL Experience

The assessment dimensions of a COIL exchange need to inform the design of the learning activities and intended outcomes. They typically include evaluation objectives, formats, and assessment criteria. We required a team digital portfolio and presentation as scaffolded deliverables where we could provide formative and summative feedback on learners' understanding of key concepts throughout the project. We chose the types of evaluation aligned with the course's collaborative character and learning objectives to measure our learners' intercultural competency with an ensemble of techniques: peer evaluations, self-assessments, and cultural reflection essays. When grading, we considered the collaborative character of the course in relation to assessable individual intercultural proficiency criteria.

The COIL assessment design elements ensured that the deliverables represented learners' content knowledge while capturing and measuring the growth of their intercultural competency. Include reflection and evaluation activities that invite learners to assess the impact of their participation in COIL/VE programs on their understanding of the SDGs and their ability to contribute to sustainable development. Encourage learners to reflect on their learning experiences and identify areas for further growth and development.

Further Tips for COIL Curious Educators

Building a robust connection with your partner from the outset is the quintessence of successful collaboration. Give sufficient time for discussion and planning to address potential challenges and technological issues. Make sure everyone agrees on how that technology is used and that the language features in it are easy for everyone to use and understand. Tools such as Google Translate can help overcome language barriers and promote effective communication. Begin with small, manageable projects and incrementally scale up to prevent overwhelm and better secure a positive professional collaboration for yourselves and a learning experience for the students.

My courses have an array of majors in the sciences, and this disciplinary diversity enables us to remain adaptable and flexible when collaborating with partners on various projects. Over the years, I've engaged with numerous campuses, tailoring our focus based on their specific majors. Before each semester, I sit down with my partners to discuss project goals, using a flexible template that can be customized to suit different themes and topics. When incorporating the SDGs, I emphasize the concept of *thinking globally and acting locally*. Encouraging learners to develop small-scale projects on their campuses allows them to share their experiences and contribute to the greater good. This gives them practice in creating a sustainability culture that values global awareness and action. Encouraging learners to collaborate—especially during the pandemic—proved incredibly rewarding. Despite living miles apart, they discovered shared experiences highlighting our common humanity. Instead of focusing solely on our differences, they recognized that cultural diversity unites us. It's surprising but true: these differences serve as a meaningful bond that brings learners together.

Conclusion

During the pandemic, my learners and colleagues demonstrated remarkable resilience and solidarity amidst the chaos of adjusting to remote learning. Recognizing the shared nature of this global experience, they rallied together and deepened their understanding of worldwide issues. This collective perspective broadened their horizons beyond a solely US-centric view. One particularly remarkable instance was when a previously quiet student, initially hesitant about completing the course, surprised us all with an exceptional project. She teamed up with a partner to produce a captivating video illustrating a student's remote learning experience during the pandemic. Her intricate drawings, dynamic animations, and her partner's expertly crafted script brought the story to vivid life. This project stood out as one of the most impressive we had encountered, highlighting our learners' remarkable growth and adaptability during such challenging times. We can learn much about priorities during a crisis because we can't change much of the existing reality. We can only change what we attend to in our work and how we treat our partners. In

COIL, we prioritize sustainability competencies for working and learning together, thinking globally, and acting locally (Brundiers 2021).

Through these initiatives, we provided our learners with expanded access to global learning opportunities and prompted them to think critically about global issues and solutions to achieve SDGs. Our project-based collaboration on Social Business enterprise facilitated cross-cultural teamwork, leadership development, and presentation skills. In essence, leveraging virtual exchange programs with SDGs-driven social entrepreneurship challenges offers an engaging, impactful, and inclusive approach to cultivating learners' global competencies, digital skills, and collaborative problem-solving abilities. At the same time, this puts into pedagogical practice the sustainable development goals worldwide. Successful COIL experiences can play a key role in equipping learners from diverse backgrounds with the knowledge and tools to contribute meaningfully to achieving the SDGs.

References

Adams Becker, S., M. Cummins, A. Davis, A. Freeman, C. Hall Giesinger, and V. Ananthanarayanan. 2017. NMC horizon report: 2017 Higher education edition. The New Media Consortium.

Anderson, A., and Or, J. 2023. Fostering Intercultural Effectiveness and Cultural Humility in Adult Learners Through Collaborative Online International Learning. *Sage Journals*. https://journals-sagepub-com.libdatabase.newpaltz.edu/doi/https://doi.org/10.1177/10451595231182447.

Armon, J., S. Scoffham, and C. Armon. 2020. *Prioritizing Sustainability Education: A Comprehensive Approach*. London: Routledge.

Barth, M. 2015. *Implementing Sustainability in Higher Education: Learning in an Age of Transformation*. London: Routledge.

Belz, J. A. 2003. Linguistic Perspectives on the Development of Intercultural Competence in Telecollaboration. *Language Learning & Technology* 7 (2): 68–99. 10125/25201.

Brundiers, K., et al. 2021. Key Competencies in Sustainability Higher Education—Toward an Agreed-Upon Reference Framework. *Sustainability Science* 16: 13–29.

Cachelin, A., L. Ivkovich, P. Jensen, and M. Neild. 2019. Leveraging foodways for health and justice. *Local Environment*, 24 (5): 417–427. https://doi.org/10.1080/13549839.2019.1585771

Chang, Chew-Hung, Kidman, Gillian, and Wi, Andy, Eds. 2020. *Issues in Teaching and Learning for Sustainability: Theory into Practice*. Abingdon: Routledge.

Commander, N., et al. 2022. Virtual Exchange: A Promising High-Impact Practice for Developing Intercultural Effectiveness Across Disciplines. *Journal of Virtual Exchange* 5 (1–9). https://doi.org/10.21827/jve.5.37329.

Dooly, M. 2017. Telecollaboration. The Handbook of Technology and Second Language Teaching and Learning, ed. C. A. Chapelle and S. Sauro, 169–183. Wiley. https://doi.org/10.1002/9781118914069.ch12.

Forward, M. L., K. Ikeda, and B. LeSavoy. 2020. The Global Goals: Integrating the UN Sustainable Development Goals in COIL Collaborations [Pre-recorded conference session]. International Virtual Exchange Conference, September 14–16. Online.

Hackett, S., J. Janssen, P. Beach, M. Perrault, J. Beelen, and J. van Tartwijk. 2023. The Effectiveness of Collaborative Online International Learning (COIL) on Intercultural Competence Development in Higher Education. *International Journal of Educational Technology in Higher Education* 20 (5). https://educationaltechnologyjournal.springeropen.com/articles/https://doi.org/10.1186/s41239-022-00373-3. Accessed June 18, 2024.

Jie, Z., and A. M. G. Pearlman. 2018. Expanding Access to International Education Through Technology-Enhanced Collaborative Online International Learning (COIL) Courses. *Int J Technol Teach Learn* 14: 1–11.

Leal Filho, W., et al. 2022. Towards Symbiotic Approaches Between Universities, Sustainable Development, and Cities. *Scientific Reports* 12 (11433). https://doi.org/10.1038/s41598-022-15717-2.

Lenkaitis, C. A. 2019. Valuing Technology in the L2 Classroom: Student and Teacher Perceptions, Preferences, and Digital Identity. In *Embracing Digital Learners in an Age of Global Educational Change and Rapid Technological Advancements*, ed. M. Montebello, 173–189. Hershey, PA: IGI Global https://doi.org/10.4018/978-1-5225-9304-1.ch011

Lenkaitis, C. A. 2020. Teacher Candidate Reflection: Benefits of Using a Synchronous Computer-Mediated Communication-Based Virtual Exchange. *Teaching and Teacher Education* 92. https://doi.org/10.1016/j.tate.2020.103041.

Liu, Y., and T. Shirley. 2021. Without Crossing a Border: Exploring the Impact of Shifting Study Abroad Online on Learners' Learning and Intercultural Competence Development During the COVID-19 Pandemic. *Online Learn* 25: 182–194. https://doi.org/10.24059/olj.v25i1.2471.

Mauro Toselli, "Sustainable Development Goals SketchNote Project (#SDGsketch)," Mauro Toselli Blog. Accessed June 17, 2024. https://www.maurotoselli.com/blog/sustainable-development-goals-sketchnote-project-sdgsketch.

Michel, Jessica Ostrow. 2020. Mapping out Students' Opportunity to Learn About Sustainability Across the Higher Education Curriculum. *Innovative Higher Education* 45: 355–371.

Mossman, Amy. 2018. Retrofitting the Ivory Tower: Engaging Global Sustainability Challenges Through Interdisciplinary Problem-Oriented Education, Research, and Partnership in U.S. Higher Education. *Journal of Higher Education Outreach and Engagement* 22(1): 35–60.

O'Dowd, R. 2018. From Telecollaboration to Virtual Exchange: State-of-the-Art and the Role of UNICollaboration in Moving Forward. *Journal of Virtual Exchange* 1: 1–23. https://doi.org/10.14705/rpnet.2018.jve.1.

Partnership for 21st Century Skills. 2011. Framework for 21st century learning. Partnership for 21st Century Skills.

Ramírez-Marín, F., Núñez-Figueroa, L. d. C., and Blair, N. 2021. "Collaborative online international learning: Language and cross-cultural experiences of university students." *Matices en Lenguas Extranjeras* 14 (1): 118–162. https://doi.org/10.15446/male.v14n1.92144.

Rubin, J., and S. Guth. 2015. *Collaborative Online International Learning: An Emerging Format for Internationalizing Curricula*, 15–27. Abingdon: Routledge.

Rubin, J. 2017. Embedding Collaborative Online International Learning (COIL) at Higher Education Institutions. *Internationalisation of Higher Education* 2: 27–44.

Scoones, Ian. "Chapter 14: Sustainability." In *Deconstructing Development Discourse Buzzwords and Fuzzwords*, ed. Andrea Cornwall and Deborah Eade. Rugby. Warwickshire: Practical Action Publishing.

Stevens Initiative. 2019. *Virtual Exchange Impact and Learning Report*. The Aspen Institute. Accessed June 18, 2024. https://www.stevensinitiative.org/wp-content/uploads/2019/11/Virtual-Exchange-Impact-and-Learning-Report.pdf.

United Nations. 2021, January 14. SDG knowledge: Sustainable Development Goals. https://sdgs.un.org/goals.

Vahed, A., and K. Rodriguez. 2021. Enriching Learners' Engaged Learning Experiences Through the Collaborative Online International Learning Project. *Innovations in Education and Teaching International* 58: 596–605. https://doi.org/10.1080/14703297.2020.1792331.

PART III

Enacting Principles of Human and Environmental Flourishing in Teaching

CHAPTER 20

An Afrocentric World View on Sustainability

Anthony S. Dandridge

Introduction

> Injustice anywhere is a threat to justice everywhere. We are caught in an inescapable network of mutuality, tied in a single garment of destiny.
> Rev. Martin Luther King Jr., *Letter from Birmingham Jail*

The dominant paradigm or fundamental structure of thought and belief systems driving contemporary reality cannot sustain our planet and people into the future. The idea that we are separate, isolated beings with no deeper connection to one another is the single greatest threat to our happiness and security. By embracing an Afrocentric World View on Sustainability (AWVS) that is fundamentally interdependent rather than individualistic, we can move toward greater environmental, social, and economic sustainability. An Afrocentric way of looking at the world provides the deeply meaningful and pragmatic historical example, ideological consistency, and scientific validity needed to promote equity and

A. S. Dandridge (✉)
Department of Black Studies, State University of New York, New Paltz, NY, USA
e-mail: anthonydandridge@gmail.com

S. M. Wyman and J. Korenblat (eds.), *Sustainability Learning for Action and Community Engagement*,
https://doi.org/10.1007/978-3-032-11698-7_20

inclusion of all human beings and non-human beings. We can honor African traditions and their empirically grounded holistic spiritualities for ecological global harmony and species survival as we come to understand how at all times individuals are shaped by and shape culture (Markus 1990). Be warned that this chapter is, at best, a broad outline of a new framework, grounded in marginalized traditions, and exposes many outdated extensions of radical individualism, which will be eradicated with equitable shifts toward ecological belonging, and a sustainable world view, grounded in interdependence. Sustainability is a worldview problem.

The Afrocentric World View

I am an Afrocentrist who studied at Temple University under the direction of Dr. Molefi Kete Asante, the scholar who forged and institutionalized the term *Afrocentricity*. The Temple School of thought centers the experiences of people of African descent while it prioritizes *culture* as the weighty lens by which one evaluates phenomena. While some people might talk about a topic and reference Albert Einstein, John F. Kennedy, or Donald Trump, an Afrocentrist may talk about the *same phenomenon* and intentionally reference W. E. B. Dubois, Marcus Garvey, M.L. King Jr., or Barack Obama, providing agency for the voices and experiences of people of African descent. This paradigmatic approach was originally offered not simply as a strategy to combat the oppressive epistemic conditions imposed by European colonization but also as an attempt to institutionalize orientations that emphasize pluralism and defy the hierarchy that has dominated our present conditions. At the same time, Afrocentricity provides a space for sanity in a world that overwhelmingly denies Africans the ability to participate boldly in the global marketplace of ideas and concepts. In recognition of how every culture operates within a wordview—a shared set of assumptions, values and relational understandings—through which reality, knowledge, and ethical responsibility are interpreted. Sustainability debates, therefore, are never culturally neutral but reflect the worldviews that shape how human-environment relationships are understood and governed; it is from this position that an Afrocentric World View on Sustainbility (AWVS) is introduced. An AWVS is a paradigmatic framewrok that understands sustainability as an ethical and relational condition grounded in interdependence, collective responsibility, and the insepрability of human, ecological, and spiritual realities.

An Afrocentric Critique of Knowledge

One significant but not exclusive way to legitimately gain knowledge is to go through the process of thinking about *where* your information comes from, *whose* perspectives are included, and *whether* this knowledge, justified as true belief, withstands the rigors of evidence and rationality. When speaking about a worldview, this is best understood as a structure that maintains and expresses the ideological integrity of ideas and concepts that inform one's beliefs and behaviors in the widest way, as a necessary component of explanation, identity, and meaning. An Afrocentric World View shares this structural meta-orientation with other worldviews, as it minimally includes and/or implies cosmology, ontology, epistemology, and ethics. Yet it is important to assert that not all expressions or worldviews are as neat or maximally consistent as one may want. These perspectives inform methodologies, beliefs, and behaviors in a dynamic, mutually constitutive way. Yes, there is an instability in our ways of *being* in the world that transfers into our ways of *understanding* the world. But that instability is at no time fully chaotic or entirely orderly. Africology, the Afrocentric study of African phenomena trans-generationally and transcontinentally, may afford some levels of incongruity, not unlike other fields of inquiry (Asante 2018). So, this should not come as a surprise: the quiet or loud coexistence of alternative interpretations within any field of study may foster growth in addition to decadence.

However, as a worldview that can support sustainability is spreading in a variety of spaces within academia and everyday living; we are engaging different people in ways we have not engaged them previously. So, as we move away from a hegemonic worldview of reality or of institutional ideas and practices that exclude Black, Brown, and non-white voices and visions, we welcome and amplify those voices that are implicitly and explicitly a reflection of how we literally see the world, and also how we see ourselves in the world. Ultimately, identity and worldview are mutually constitutive, each continually informing the other, not ulike the relationships between methods and methodologies (Mazama, 2003).

Now, as those committed to and/or invested in the dominant worldview seek solutions to climate change, environmental justice, and other branches of inquiry within the sustainability tree of possibilities, many are trying to understand the full consequences of this shift, while others are willing to teach those following the standard, (*status quo*) approach those consequences. Those *others* tend to be the ones from indigenous cultures that have always valued interdependence, communitarianism, ecological conservation, and spirituality. In this effort, we need to find and agree upon various categories to understand what is going on in this world from a AWVS view and how all perspectives can contribute to cross-cultural clarity and validity.

Interdependent Consciousness, Community, and Justice: Afrocentric Theory and Praxis

In African philosophy, the concept of Maat, often translated as Truth, embodies a holistic worldview that emphasizes an interconnected order of righteousness. This concept was not only a norm, but also a guiding principle institutionalized within African cultures. Scholars such as Obenga (1995), Kamalu (2000), and Karenga (2006) have elucidated Maat in contrast to the dominant Western individualistic worldview. They argue that Western science has not yet formulated a comprehensive conceptual framework capable of fostering deeply ethical imperatives as Maat does within African thought. Maat embodies not just a notion of Truth but it also a way of being and relating to the world that is polysemic as it provides an algorythmic foundation for ethical, poltical, religious, legal, social and natural dimensions. Its holistic nature suggests a paradigm that fully integrates scientific understanding with ethical considerations, offering a potential alternative to the fragmented worldview often associated with Western thought. In other words, as it can be proven that we are connected to both our natural and social environments, then neutrality is an illusion, and responsibilty is unavoidable and best conceptualized as an interdependence; connection with obligation.

The African philosophical conceptualization of interdependence emphasizes the communal idea that "I am because we are, and since we are, therefore I am," as articulated by John Mbiti. This worldview locates interdependence as a foundational principle that informs the ethical and social structures of reality. While not universally embraced as a primary principle in all philosophical or ethical systems, interdependence is regarded as a fundamental truth in certain holistic contexts. Within these frameworks, it serves as a cornerstone concept that shapes our perception and comprehension of the world, highlighting the intricate connections between all beings and the importance of collective existence over individual isolation. Ultimately, this becomes realized as a shfit away from an egocentric identity to an ecocentric identity, where the self is embedded in a living socio-natural web. The assertion of this evidence based identity and its corresponding way of life, then further compels one to think about one's identity as an expression of the same creative force that animates the universe.

With this in mind, the pinnacle of self-realization is the recognition of one's divinity. This concept posits that all reality is an expression of a singular entity or order of being. Black psychology has significantly contributed to this understanding of identity, emphasizing the interdependence of individuals within a larger spiritual and cosmic reality (Nobles 2006; Akbar 1998; Myers 1993). This perspective extends beyond individualistic notions of selfhood to embrace a broader understanding of identity rooted in a deep connection to the divine and the universe.

In pursuit of this objective, transcending the ego beyond the easily recognizable sociological identity and sustainability identity of the eco, to further embrace a *cosmic identity*, allows for a deeper understanding of the African worldview and its moral imperatives (Danon 2019). Others may refer to this cosmic identity as optimal psychology, extended self, or optimal consciousness. This shift in perspective moves beyond the limitations of individualistic identities, adding a profound recognition of, and identification with, the interconnectedness of all beings and the universe. This then is not a claim of dominance, but an ethical call to live in balance, justice, and reverance for all life. Subsequently, the self/other binary is transcended, as one idntifies with the creative force that animates all of existence. This cosmic identity is at the same time spiritual, and aligns with the core principles of African thought, highlighting the moral obligations that arise from this evidence-based interdependence. It underscores the importance of establishing an additional identity that more accurately recognizes one's place within the broader cosmic order that fosters a sense of responsibility and ethical conduct. This cosmic identity which extends beyond and reconciles other identites such as race, gender, class, sexuality and much more, by situating them within a larger unified field of existence, holds difference without hierarchy. Allows these identities and all things to co-exist as expressions of a deeper ontological oneness.

When thought and practice coincide, we increase the chances of achieving many goals, including the *Principles of Environmental Justice* (1991), laid out by delegates to the First National People of Color

Environmental Leadership Summit in Washington, DC. In the words of co-author Robert Bullard, this famous manifesto involves the safety of workers, the rights of Indigenous peoples, and the honoring of nature (Funes 2023). It thus calls for an understanding of both the inner (mental, consciousness) and outer (physical, material) dimensions of sustainability (Wamsler et. al. 2021) and the human element in all pedagogical undertakings (Landorf et. al 2008). Aspects of realistic humanity fundamental to African and other indigenous worldviews tend to be more holistic and non-dichotomous (Topa 2024; Obasi 2002; Nkrumah 1964; and others). In most cases, these conceptual and cultural frameworks are in categorical opposition to many European non-indigenous and/or atomistic worldviews, while at the same time being grounded in science.

On another level, as consciousness is of the world and in the world, thought shapes the world at all times, without exception, irrespective of its internal and/or opaque situatedness. In fact, the inner life of humanity is not hidden from the realities of existence. Your consciousness in this very moment is co-creating the interdependent world that we inhabit. This is a more holistic, non-binary way of understanding the world that connects the mental and physical as being mutually constitutive of one another. In this non-dichotomous theory of the human, there is an acknowledgment that humans are not simply composed of discrete parts or dualistic elements (such as mind and body or individual and environment), but that humans are complex, interconnected systems in an interdependent reality. This worldview emphasizes the interdependence of human existence as we identify the uniqeness of phenomena without negating reality's systemic continuity which has, physical, psychological, social, cultural, spiritual, and environmental dimensions.

Accordingly, if you think that you can separate your consciousness from our ontologically shared, interdependent reality, you are mistaken. Your presence and absence impact reality at all times. I like to remind my students: when you drove to campus or however you got here today, you did not do it by yourself. All it takes is a car to cross that center yellow line, and many sorts of tragedies may occur. A student came to class and raised their hand during the open-sharing period that starts my classes. They said, "Remember that story you told us about interdependency and how a car crossing the line, or not crossing it, impacts us? My friend did not get home last night because a car crossed the line." Our beliefs, behaviors and consciousness impact reality at all times.

We are *always, without exception*, dependent on things that are going on around us. Our being is defined by relationships and impacted by things that we do, and we do not do. In fact, your attention to this article right now is impacting your family members, loved ones, and the universe by your physical absence and psychological (conscious) inattentiveness to those spaces.

Systems, Culture, and Praxis: Afrocentric Foundations of Sustainability

Systems theory aligns with this perspective by viewing individuals as dynamic systems that influence and are influenced by their complex environments, at all times. According to systems theory, individuals are not isolated entities but are part of larger systems, such as families, communities, societies, and ecosystems, simultaneously. These systems are interconnected and interdependent, and changes in one part of the system will have measurable ripple effects throughout the entire system.

By embracing systems theory and a non-dichotomous theory of the human being as connected to nature and to others, my students recognize the intricate web of relationships and interactions that shape human experience. This encourages a more holistic understanding of individuals and emphasizes the importance of considering the whole person in their extended environment, rather than focusing solely on the individual as isolated. This personal ecology approach is valuable for all disciplinary fields including sustainability, psychology, sociology, anthropology, physics, chemistry, and biology as it provides a framework for exploring the complexities of human behavior, relationships, societies, and our natural world (Porter 2008). In the end, this is the foundation for a unified theory of science that is particularly illusive in cultures commited to dichotomous, fragmented, atomostic worldviews.

According to Frantz Fanon (2021), culture can be stated as the "… combination of motor and mental behavior patterns arising from the encounter of man with nature and with his fellow man." For humanity to meet twenty-first-century sustainability challenges, we as a human species will have to adapt and amplify our mindsets to an evidence-based, changing landscape for individuals, societies, and the natural environment. Some cultures are more aligned with this worldview than others are. Culture is at all times, social and natural. This will require a reexamination of our norms and values and a concerted exploration of existing wisdom traditions that have married thought and praxis in sustainable ways. As I tell my students, *thought without practice tends to be empty; practice without thought tends to be blind* (Nkrumah 1964). Reflections on meaningful change emphasize the importance of integrating theory and action as we seek to embrace the reality that we are, at all times, agents of change. Fanon's phrase underscores the need for a balanced approach that combines theoretical understanding with practical application in a way that links how one's beliefs (worldview) it impacts one's behavior.

The critical reflection on concepts and principles is essential for generating meaningful hypotheses and designing effective solutions. On the one hand, scientific inquiry—driven solely by theoretical speculation, without practical experimentation—risks being disconnected from empirical reality. This empty approach can lead to abstract and impractical ideas that do not contribute significantly to scientific progress. On the other hand, engaging in experimentation and action without thoughtful reflection can result in aimless activity that lacks direction or purpose. This approach may lead to inefficiency and wasted resources, hindering scientific progress.

Therefore, any solution must strive to effectively integrate both thought and practice in its work, ensuring that its theoretical insights are grounded in empirical evidence and that its practical efforts are guided by thoughtful analysis. By doing so, humanity can contribute meaningfully toward a culturally sustainable future and further positive change through research.

Yet, as one shifts into the maintenance and development of the conceptual framework of an Afrocentric Worldview, it is important to grasp how important it is to be consistent with this cultural orientation that both thought and practice shape the world at all times, without exception! In many ways, thought is practice. In an African World View, the universe itself is conscious, while more individualistic worldviews remove consciousness from the universe and place consciousness as awareness, reflection, and intentionality exclusively in the heads of human beings. This belief severs mind from matter, isolates the human from the ecological whole, and turns consciousness into a private posession rather a shared field of relational energy. This fragmented individualistic worldview also denies intelligence to ecosystems, the sentience of the land, air and water, in addition to the wisdom of African, indigenous and other ancestral memory, with a cultural closure to the measurable value of these very human traditions.

Epistemic Closure: The Effects of Racist Ideologies of Dehumanization

The term *epistemic closure* can describe the impact of the dehumanization that accompanies resistance to learning about and appreciating entire populations that, racially, are not white or dominant. Epistemic closure occurs at the moment of assumed complete knowledge of a phenomenon (or population) that prevents one from pursuing efforts of inquiry, of asking the questions that need to be asked, of being curious about others (Gordon 2016). Those in the US may have witnessed a recent rise in conversation and debate around Diversity, Equity, and Inclusion initiatives. Many have embraced the importance of supporting culturally diverse identities and interests. Such is the way of the Afrocentric World View. In the end, for a particular group to be centered is to prioritize cultural *agency* over cultural *othering* (Asante 1980), where cultural agency prioritizes one's own contributions to world history while not dehumanizing others in a multicultural landscape. Cultural othering erases the contributions of others as inferior and/or inhuman. Subsequently, multicultural approaches open the door to reducing the erasure of the humanity of the First Nations people of the world, whose ways of being, ethics, perspectives, and spirituality have been unjustifiably attacked (Asante and Dove 2021). As many marginalized cultures emphasize interdependency and communitarianism rather than individual autonomy and dominance, this erasure manifests itself in the domination and dehumanization of whole peoples and cultures (Freire 1970). In other words, there is a correlation between peoples from cultures that emphasize interdependence and communitarianism and systematic efforts to colonize, exploit, terrorize, and erase the culture and the very humanity of these populations by cultures with individualistic worldviews that deny their obligations to others. Ultimately, the core values of connection, interdependence, and interbeing are required if the planet and its creatures are to survive together into the future. This epistemic closure is not only dehumanizing, but it tends to proceed from a radically individualistc worldview about what it means to be human.

Defining Humanity

While African American philosophy has often argued *we are human too,* African diasporic philosophers have noted that there are certain culturally privileged notions of *humanity* that are in opposition to their own empirically-based traditions of thought. Moreover, the uncritical adoption of these definitions issued by your oppressor can deeply impact your identity in a way that reproduces similar kinds of harm that you are attempting to overcome. This internalization of destructive views can impact individual and group well-being and possible socio-natural futures. The current oppressive and individualistic idea of humanity is one that tends, in part, to contribute to the distribution of privileges racially and even anthropocentrically.

Historically, but not exclusively, Anglo-Europeans in positions of political power have deemed certain bodily features that are adapted to equatorial climates as inferior. Naturally occurring features, the result of natural selection for hot climates, have been used as racial markers and devalued. For example, advantageous aspects such as, increased melatonin to protect against strong sun rays (Yuen and Jablonski 2010), wider noses to improve cooling (Zaidi et al. 2018), and kinky hair to disperse heat away from the body (Lasisi et al. 2023) have been associated with deficiency and overwhelmingly, considered markers of inhumanity.

If people of African descent are not deemed human, then how can we value them and their contribution to the global marketplace of ideas and concepts that are available for the co-creation of sustainable futures? When selected phenotypical adaptations become markers of inferiority in the dominant worldview, then entire peoples tend to be disqualified and dismissed from the progress of humanity by not being categorized as human. By extension, a process of *othering* becomes culturally ingrained, a product of racist ideology, where the racialized (human) *self* is viewed as being firmly distinct and apart. This supports the hegemonic discourse of the powers that be.

In resistance, our planet has a rich legacy of folk who insisted on the humanity of people of African descent, often by refusing to ignore the point of insult. They do this by pointing to laws that institutionalize the perceived threat of the Black man as did Frederick Douglass (1852) or by simply insisting on the inner life and soul beauty of the race W. E. B. Du Bois (1903). While very bold and meaningful, saying that you are human too, can be dangerous. If you identify with a philosophical, anthropological ideal defined by the colonizer as the standard, without recognizing the role it has played in their dehumanization of others, you also may miss the point that this mindset is, additionally, a significant factor in widespread environmental and social destruction. Instead, a "cultural" framing of the socio-natural human condition, which assumes value and meaning within the traditions of the oppressed, opens up an evidence-based search within the traditional ways of being by these peoples that may yeild significant benefits that take you beyond the norms of the oppressor, that have the capacity to rescue both the oppressed and oppressor from destruction (Friere 1970).

This critical approach tends to have better outcomes for the community and planet in the past, present, and future: the natural world and empirical scientific evidence become centered and available in a way that transcends the current dominant and unjustified biases, which constitute significant foundations for exclusion and, at best, a meandering genocide. Moreover, the critical can lead to shift in worldview, and a shift in worldview can lead to a new critical disposition. In the end, critique destabilizes inherent assumptions while worldviews can provide the horizon from which those assumptions can be reinterpreted and possibly transformed. While there are many pathways to these alternatives in an conceptually interdependent world, dissonance in the current worldview; encountering an embodied alternative; the prevalence of racial injustice; and spiritual emptiness due to a search for meaning, are just a few factors that can become powerful catalysts for change. Such transformation is not confined to a single discipine; it can be sparked by insights from any field of study or an experience which disrupts the established ways of knowing.

That being said, as one takes the systemic nature of racism seriously, one must also understand that legitimacy itself is often occupied by white space, and we people of color may further legitimize it consciously and unconsciously as the vestiges of racism, as its foundations operate within a network of systems one may not fully recognize (Blue et al. 2016). As Black people can get dismissed as soon as they enter a room, walk down the street, apply for a job, or even submit to a journal for publication, we must take the complexities of a racist system seriously. In the North American context, whiteness, from names to hair styles and beyond, brings particular levels of legitimacy to these arguments and strategies about sustainability, and even humanity. All folk need to be careful not to internalize skewed and false notions while we also, at times, seek to use the realities of these privileges against harmful systems.

Holistic Indigenous World Views: A Way Forward

Wahinkpe Topa (Four Arrows), also known as Don Trent Jacobs (2024) of the Lakota Nation, North America, has contrasted *common dominant worldview manifestations* with *common indigenous worldview manifestations* in ways that always resonate with my students. The themes he lays out correspond with many philosophical approaches to social cohesion and sustainability across the great diversity of cultures on the African continent. But before we move any further, it is important to note that when the category of *Indigenous* is invoked in networks of sustainability and otherwise, rarely are Africans included in that universe. Adequately grasping the reasons for such closure is somewhat perplexing. But the short answer is that this is a function of the systemic and hierarchical nature of anti-Black racism. Moving on, while Four Arrows' worldview chart provides a useful comparative framework, my work advances the Afrocentric World View on sustainability by centering such Indigenous cultural logics as non-hierarchical thinking, respect for various gender roles and fluidity, embracing earth and all systems as living and loving, acknowledging words as sacred, emphasis on responsibility and obligation, humor as an essential tool for coping, justice as a return to a community context, and responsibilities to both victims and victimizers. This list is not exhaustive and not intended to homogenize the multiplicity of indigenous cultures across the globe. Nor is it meant to romanticize the lives or idealize the people of Africa or those from other nations. Yet these characteristics are prevalent in certain, often oppressed, and marginalized societies who have much to lose in the face of climate instability and increasing social inequity. We will look at a list of such features that would constitute An Afrocentric World View on sustainability.

Here is a set of proposed principles that aim to honor and integrate the Afrocentric Worldview on Sustainability

1. Interdependence: Recognize the interconnectedness of all human and non-human elements of reality, reflecting the African concept

of MAAT, which emphasizes community and mutual caring for the social and natural world.

2. Stewardship: Emphasize the role of humans as caretakers of the earth, responsible for preserving the natural world for future generations in line with traditional African stewardship practices.
3. Equity: Ensure fair access to resources and environmental benefits; aggressively address disparities, particularly those affecting African communities, both locally and globally.
4. Ancestral Wisdom: Incorporate traditional knowledge and practices in sustainable living that have been passed down through generations in African cultures.
5. Community Voice: Prioritize the voices and leadership of African communities in environmental decision-making processes.
6. Holistic Health: Promote environmental policies that support the physical, mental, spiritual, and social well-being of communities in harmony with holistic African health practices.
7. Resilience Building: Foster community resilience against environmental hazards through traditional and innovative adaptive strategies.
8. Cultural Integrity: Protect and nurture the cultural heritage and practices that connect African communities to their environment.
9. Land/Air/WaterRights: Advocate for the protection of land/air/water rights, recognizing the special relationship between African peoples and their existential reality.
10. Sustainable Practices: Encourage sustainable agricultural and environmental practices that align with traditional African methods of cultivation and natural resource management.
11. Youth Engagement: Involve young people in sustainability efforts, ensuring they receive the knowledge and skills to continue the legacy of environmental stewardship.
12. Gender Inclusivity: Address the unique impacts of environmental issues on women and ensure their equal participation in sustainability initiatives, respecting the matriarchal aspects of many African societies.
13. Economic Justice: Support economic policies and practices that provide equitable opportunities for all, reducing poverty and enhancing the quality of life without compromising the environment.

14. Spiritual Connectivity: Acknowledge and integrate the spiritual relationships that many African cultures maintain with the natural world in environmental policies and practices.
15. Biodiversity Conservation: Prioritize the conservation of biodiversity, valuing the vast variety of life forms that exist in African ecosystems.
16. Restorative Justice: Implement policies aimed at repairing and reversing environmental damages and their social implications, particularly in exploited and marginalized communities with the intent of minimizing harm on all communities.
17. Precautionary Principle: Adopt a cautious approach to environmental management, preventing harm before it occurs, in alignment with the foresight often embedded in African philosophy, spirituality, proverbs and storytelling.
18. Transparency and Accountability: Ensure that environmental policies and practices are transparent and that leaders are accountable to the communities they serve.
19. Celebration of Diversity: Recognize and celebrate the diverse cultures, languages, and traditions within African communities and their contributions to sustainability.
20. Global Solidarity: Foster solidarity and collaboration between African communities and other populations around the world facing similar environmental challenges, irrespective of whether they are considered victims or victimizers.

These principles are designed to reflect an Afrocentric perspective that emphasizes values, practices, and views inherent to African traditions and worldviews while aligning with global efforts toward environmental justice, climate justice, and other sustainable strategies.

My college students tend to respond to these ideas with a mix of curiosity, contemplation, and perhaps some resistance or skepticism, depending on their backgrounds and prior exposure to these concepts. Some students resonate deeply with the notions of interdependence, communal identity, and cosmic consciousness, finding them to be refreshing and insightful perspectives that offer a new lens through which to view the world. They appreciate the emphasis on ethical, environmental imperatives and moral obligations that these ideas logically entail, seeing them as guiding principles for a more harmonious and sustainable way of life.

On the other hand, some students struggle with these ideas and seek to validate more individualistic or Western-centric worldviews. Some have claimed that the obligatory weight feels too heavy. They find the concepts of transcending the ego and embracing a cosmic identity to be abstract or difficult to grasp as a liberatory project, particularly if they have not been exposed to these ideas before. One of the most radical things one can do within an individualistic worldview is to claim one is God. But again, this monistic ontology does not give humans any exclusive right to this universal claim. A slice of my students have been able to process these ideas by referencing such films as *Avatar*, *The Matrix*, and *Star Wars*. A few of them question the practicality or relevance of these concepts to their own lives and academic pursuits, wondering how these ideas can be applied in a concrete way. Teaching at a predominantly white institution, most of my students are white. I find that those with some interest in diversity grasp the ideas much easier. As their academic exposure tends to be a significant factor, my seniors tend to understand the impact of this worldview more than my first-year students. And my Black Studies, Women, Gender and Sexuality Studies, and Latin American Carribean and Latin X Studies students also tend to more easily embrace and acknowledge the relationships between worldviews and sustainability issues.

Overall, students are likely to engage with these philosophical concepts with a range of perspectives and reactions, leading to rich discussions and deeper explorations. As an educator, you can facilitate this process by creating a supportive and inclusive learning environment where students feel comfortable expressing their thoughts and challenging their assumptions. In teaching courses in Black Studies, it's imperative that students feel empowered not only to challenge other students and faculty but also, most importantly, to challenge themselves as they foster a deeper understanding and appreciation of these ideas.

Conclusion

The African philosophical concepts that support the African World View of interdependence, communal identity, and cosmic consciousness offer profound insights into our understanding of the earth and our place within it. These ideas challenge us to transcend individualistic and ego-centric perspectives, embracing a broader eco-centric and cosmic sense of identity and moral responsibility that extends beyond ourselves to encompass all of humanity and the universe. As college students engage with these ideas, they are invited to consider new ways of thinking and being that can enrich their lives and contribute to a more sustainable and harmonious future for all. Through open-minded exploration and dialogue, we can deepen our understanding of these concepts and their implications, paving the way for a more inclusive, interdependent, and sustainable world. In the end, any worldview that falls short of illuminating the divine dimensions of the self and thereby activating the full expression of one's humanity, must be understood as fundamentally insufficient for global sustainability (Obasi, 2002). While many cultures articulate parallel ethical and spiritual values, the Afrocentric expression of these values is distinct in its explicit commitment to confronting and repairing the historical erasure of African humanity, knowledge systems, and worldview orientations.

References

Asante, Molefi Kete. 1988. *Afrocentricity*. Trenton, NJ: Africa World Press.

Asante, Molefi K., and Maulana Karenga. 2006. *Handbook of Black Studies*. Thousand Oaks, CA: Sage Publications, Inc., Sage Knowledge, 2 Jul 2024, https://doi.org/10.4135/9781412982696.

Asante, Molefi Kete. 2018. The Relentless Pursuit of Discipline: An Africological March toward Knowledge Liberation. *Journal of Black Studies* 49 (6): 531–541. https://doi.org/10.1177/0021934718788646.

Asante, Molefi Kete. 1980. *Afrocentricity, the Theory of Social Change*. Chicago, IL, USA: Amulefi Publishing Company.

Asante, Molefi Kete., Dove, Nah. 2021. *Being Human Being: Transforming the Race Discourse*. New York, NY, USA: Universal Write Publications LLC.

Blue, Stanley, Elizabeth Shove, Chris Carmona, and Michael P. Kelly. 2016. Theories of Practice and Public Health: Understanding (Un)Healthy Practices (Online). ISSN 1469–3682 (Online) | Critical public health | The ISSN Portal. https://portal.issn.org/resource/ISSN/1469-3682.

Danon, Marcella. 2019. "From Ego to Eco": The contribution of ecopsychology to the management of the contemporary environmental crisis. Ecopsiché—Scuola di Ecopsicologia. https://doi.org/10.13135/2384-8677/3261

Douglass, Frederick. 1852. "What to the Slave Is the Fourth of July?" Presented at the Rochester Ladies' Anti-Slavery Society.

Du Bois, W.E.B. 1903. *The Souls of Black Folk*. Chicago: A. C. McClurg and Co.

Energy Justice Network. 1991. Principles of Environmental Justice. Ejnet.org. https://www.ejnet.org/ej/.

Freire, Paulo. 1968/1970. *Pedagogy of the Oppressed*. New York: Bloomsbury Academic. https://envs.ucsc.edu/internships/internship-readings/freire-pedagogy-of-the-oppressed.pdf.

Fanon, Frantz. 2021. Toward the African Revolution. *Power and Inequality*, ed. Levon Chorbajian. London: Routledge.

Funes, Yessenia. 2023. A New Chapter Begins. Atmos. https://atmos.earth/the-frontline-end-goodbye/.

Gordon, Lewis R. 2016. *Disciplinary decadence: Living thought in trying times*. London: Routledge.

Markus, Hazel. 1990. in *Discovering Psychology: Cultural Psychology. Cultural Psychology*. https://www.learner.org/series/discovering-psychology/cultural-psychology/. Los Angeles, CA: Annenberg Learner.

Kamalu, Chukwunyere. 2000. *Foundations of African Thought: A Worldview Grounded in the African Heritage of Religion, Philosophy, Science, and Art*. Red Sea Press.

Karenga, Maulana. 2006. *Maat, the moral ideal in ancient Egypt: A study in classical African ethics*. The University of Sankore Press.

Landorf, Hilary, Stephanie Doscher, and Tonette Rocco. 2008. Education for Sustainable Human Development: Towards a Definition. *Theory and Research in Education* 6 (2): 221–236. https://doi.org/10.1177/1477878508091114.

Lasisi, Tina, James W. Smallcombe, W. Larry Kenney, Mark D. Shriver, Benjamin Zydney, Nina G. Jablonski, and George Havenith. 2023. Human Scalp Hair as a Thermoregulatory Adaptation. *Proceedings of the National Academy of Sciences of the United States of America* 120 (24): e2301760120. https://doi.org/10.1073/pnas.2301760120.

Mazama, A. 2003. The Afrocentric Paradigm. Africa World Press.

Myers, Linda James. 1993. *Understanding an Afrocentric World View: Introduction to an Optimal Psychology*. Dubuque, IA: Kendall/Hunt Publisher.

Na'im Akbar. 1998. *Know Thy Self*. Tallahassee, Fl: Mind Productions & Associates.

Nkrumah, Kwame. 1964/2009. *Consciencism: Philosophy and Ideology for De-Colonization and Development with Particular Reference to the African Revolution. JSTOR*. New York University Press. http://www.jstor.org/stable/j.ctvwrm4jh.

Nobles, Wade W. 2006. *Seeking the Sakhu: Foundational Writings for an African Psychology*. Chicago, IL, USA: Third World Press.

Obasi, Ezemenari M. 2002. *Reconceptualizing the Notion of Self from the African Deep Structure*. EBooks, January, 52–74. Sage. https://doi.org/10.4135/9781452229119.n5.

Obenga, Théophile. 1995. *A Lost Tradition: African Philosophy in World History*. Princeton, NJ, USA: Source Editions.

Peng, Kaiping. 1990. in *Discovering Psychology: Cultural Psychology. Cultural Psychology*. Los Angeles, CA: Annenberg Learner. https://www.learner.org/series/discovering-psychology/cultural-psychology/.

Porter, Terry B. 2008. Managerial Applications of Corporate Social Responsibility and Systems Thinking for Achieving Sustainability Outcomes. *Systems Research and Behavioral Science* 25 (3) (May): 397–411. https://doi.org/10.1002/sres.902.

Topa, Wahinkpe. 2024. Restoring Sanity and Remembering Spirit in Psychology: Reclaiming Our Pre-Colonial Worldview. *Review of General Psychology* 28 (2). https://doi.org/10.1177/10892680231226387.

Wamsler, Christine, Gustav Osberg, Walter Osika, Heidi Herndersson, and Luis Mundaca. 2021 (November). "Linking Internal and External Transformation for Sustainability and Climate Action: Towards a New Research and Policy Agenda." *Global Environmental Change* 71. https://doi.org/10.1016/j.gloenvcha.2021.102373.

Yuen, A.W.C., and N.G. Jablonski. 2010 (January). Vitamin D: In the Evolution of Human Skin Colour. *Medical Hypotheses* 74 (1): 39–44. https://doi.org/10.1016/j.mehy.2009.08.007.

Zaidi, Arslan A., Brooke C. Mattern, Peter Claes, Brian McEcoy, Cris Hughes, and Mark D. Shriver. 2018. Investigating the Case of Human Nose Shape and Climate Adaptation, ed. Greg Gibson. *PLOS Genetics* 13 (3). https://doi.org/10.1371/journal.pgen.1006616.

CHAPTER 21

Illustrating the Geographies of Justice: Collaborative Storytelling About Sustainability Issues

Joshua Korenblat

Introduction

In 2010, Maria and Juan moved their family to tiny Kettleman City, California, seeking the stability of a rural farming community. Then, their children began to experience mysterious symptoms of chronic illness, nausea, and lethargy. And they weren't alone. Families in their San Joaquin Valley farm town showed troubling signs of similar health issues. These predominantly Hispanic residents lived near oak woodlands and river-fed habitats characteristic of this flat, hot, and dry landscape. But a large hazardous waste landfill also shadowed them, just three and a half miles from town. Industrial farms further contributed a plethora of environmental contaminants. This crisis could have remained another untold story of environmental injustice. Instead, it became a model for change when Juan and his neighbors formed *El Pueblo para el Aire y*

J. Korenblat (✉)
Department of Design, State University of New York, New Paltz, NY, USA
e-mail: korenblj@newpaltz.edu

S. M. Wyman and J. Korenblat (eds.), *Sustainability Learning for Action and Community Engagement*,
https://doi.org/10.1007/978-3-032-11698-7_21

Agua Limpia de Kettleman City to advocate for clean air and water. Despite initial dismissals from public officials, support from the advocacy group Green Action led to the discovery of benzene and arsenic contamination in the water and unsafe air quality. Together with the two grassroots groups, the community filed a civil complaint with the federal Environmental Protection Agency (EPA). Their coordinated environmental justice advocacy yielded an eventual victory: Their advocacy led to a civil rights settlement and commitments to monitoring, cleanup, and protecting the community.

Based on actual events, this story of the resilient people of Kettleman City was articulated and amplified by an illustrated podcast video created by our undergraduate students. Through simple drawings and heartfelt narration, students transformed conventional academic research about unnecessary harm into an accessible, vivid, shareable, emotionally resonant story that illuminated complex sustainability issues from a human perspective.

In 2019, as part of our college's Sustainability Faculty Fellows program, Professor Melissa Yang Rock (also a contributor to this book) and I brought together two typically siloed undergraduate disciplines: Geography and Graphic Design. Our semester-long project extended student-learning beyond classroom walls. Geography students began by writing research reports on social justice topics such as food justice, transgender rights, and reparations. They then partnered with Graphic Design students to transform these reports into illustrated podcast videos that paired cartoons with narration, music, and minimal animation. This approach transforms students' engagement with sustainability concepts from fact-based acquisition to interpersonal competency. Instead of confronting abstract systems or daunting statistics, they're drawn together in practice with the material through story and simple drawings. Students learn the classic benefits of cartooning for complex ideas. Cartoons can exemplify a story, facilitate compare-and-contrast scenarios, emphasize critical moments, and highlight ideas in ways that evoke emotion and interactivity through interpreting and reflecting upon the content (Murray 2021, 119).

This chapter examines how collaborative storytelling through visual media can transform traditional sustainability education. By bringing together undergraduate Geography and Graphic Design students, we created a learning environment where research and creativity converged, academic rigor responded to a call for emotional resonance, and students could explore complex social justice issues through the accessible medium of illustrated storytelling. With this interdisciplinary approach, students moved beyond buzzwords and jargon that can disavow lived realities to craft narratives that made sustainability concepts tangible and meaningful. Fig. 21.1 illustrates some of the work done in this collaboration.

Our hand-drawn, tactile, and multivocal approach to sustainability education embraces what West Indies poet and literary critic Édouard Glissant calls "a trembling way of thinking," or a "poetics of trembling" that "allows us to be in real contact with the world and with the peoples of the world" (Glissant, quoted in Obrist 2021), eschewing absolute answers and fixed categories. "Trembling together," our students had to be open to multiple perspectives and possibilities. Like the wavering lines in our students' hand-drawn illustrations, this trembling approach acknowledges the tenderness of human vulnerability, often hidden under the hard shell of political and media discourse about social justice topics. From his vantage point in the archipelago islands—a crossroads of imperialism,

Fig. 21.1 Kettleman City story frame excerpt by Project Clean Beans: Credit: Livi Carlen, et al. This short, illustrated podcast video tells the story of one California agricultural community and their grassroots activism for environmental justice

trade, and tourism—Glissant searches for art that expresses uncertainty, exploring new ideas without seeking perfection or complete comprehension. Quick sketches exemplify this mindset. They're fast, messy, and not always planned. When learners sketch ideas rapidly, they distill a scene, a moment, an emotion, an action, or an idea without worrying about making it perfect—just like Glissant's idea of letting currents of thought sweep, flow, and connect in surprising ways.

Rough cartoon drawings harness energy and emotion and convey process over product. This kind of sketching is concordant with Glissant's belief that visual communicators and storytellers can embrace mystery and complexity instead of immediately making ideas clear and straightforward. The trembling line becomes a metaphor for how we might approach sustainability education: with humility, openness, and a recognition of our shared humanity. The products of this work, however, can be clear and simple. Here, students shift from a mindset of dialog and reflection to one of communicating (Fig. 21.2).

At first glance, cartoons and serious social justice topics appear mismatched. Yet, as our Kettleman City project shows, cartoons offer students a tender, approachable way to grasp meaningful topics that have become clouded in everyday speech and political rhetoric. Children's book author A. A. Milne understood how simple cartoons can impart wise lessons in resilience, compassion, friendship, and acceptance with bare dialog and thoughtful vignettes. There's a reason his work has been called *The Tao of Pooh*. He imagined his cartoon bear character Winnie-the-Pooh sitting "at the foot of the tree, putting his head between his paws," thinking about a puzzling buzzing noise. *That buzzing noise*

Fig. 21.2 An exemplary rough thumbnail: part of a sequence of an *animatic* (simple animation) story about travel required for a student to leave a food desert. (*Credit* Artemis Folk)

means something. You don't get a buzzing noise like that, just buzzing and buzzing, without its meaning something (Milne 1926, chap. 1). Pooh discovers honey amidst the buzz. We can serve this up as a student-learning metaphor. Like Pooh, students in social justice-themed courses must go beyond buzzwords, pointed verbal critiques, and statistics. By making collaborative illustrated podcasts, our students practiced drawing as a cultural tool to facilitate conversations and discover and develop relatable stories of resilience within challenging topics. When put into hands-on practice through cartooning, social justice education becomes not divisive but collaborative, and it can even model what the Ancient Greeks called *paideia* and *eudaimonia*: self-development in education and human flourishing in a community.

Hand-drawn illustrations achieve this goal of relationality by what cartoonist Scott McCloud calls "amplification through simplification" (McCloud 1994, 30). By removing incidental detail, cartoons help viewers relate to characters' struggles and hopes regardless of their background differences and incidental details. This is why cartoons, from the Simpsons to Manga and Marvel Comics, travel so well across cultures. This also might be why Manga, with its repertoire of magnified facial features and expressions, works so well with readers with Autism Spectrum Disorder (ASD) who might otherwise have challenges reading subtler facial expressions (Korenblat). A photograph of Kettleman City might show us precisely what Juan looks like. However, a simple cartoon of his story allows viewers to see themselves in his struggle for environmental justice. At the same time, those trembling, sometimes awkward lines of student drawings leave a personal imprint that further conveys the artist's presence and a willingness to test fixed identities and categorical thinking.

Why Stories Matter in Social Justice Education

The Kettleman City story illustrates a simple yet oft-overlooked truth about sustainability: it's rooted in meaningful, caring relationships. This truth was beautifully evoked by Ami Dar, founder of idealist.org, during an informal conversation we held for this book. Speaking of his wife's home country, Ami observed: "It's lovely to walk around small Spanish towns and see older women having lunch in their 70s and having it the same way since they were sixteen years old…This is also sustainability."[1] His observation distills sustainable development: it's about creating contentment and belonging and maintaining practices that support the long-term well-being of people and the planet. But these relationships demand the right conditions to flourish.

Consider a community garden. Some community members will be impeded or excluded without social justice foundations—equal access to plots, distributed resources, inclusive decision-making processes, and multilingual signage. However, with these elements, the garden transforms into a community hub where bonds form across generations and cultures and sustainable practices take root in daily life. Like those Spanish friends sharing lunch across decades, the garden becomes a space where sustainability is lived—even unconsciously—rather than merely discussed. These stories exemplify how social justice creates fertile ground for sustainability work. The United Nations Global Goals also show this: ten of their seventeen sustainability goals pertain to social justice issues, from reducing inequality to achieving quality education and gender equality. Even the most well-intentioned environmental and economic initiatives might fail without addressing these social foundations.

Yet, teaching social justice through traditional academic approaches can feel abstract and disconnected from students' lives. Students learn critical thinking and communication skills about complex real-world problems when they work within structured yet flexible methods while discussing the source learning materials and engaging with their peers' viewpoints (Samuelsson and Lindström 2022, 211). Research shows that many students view *sustainability* as a vague idea about protecting the environment over time. They less often relate the concept of sustainability to immediate social issues, including environmental justice, food justice, ecofeminism, extinction threats, and environmental ethics (Kruger et al. 2020, 513). Vivid stories can help students negotiate normative sustainability principles—ideals of how the world ought to be—into real-world actions. This diminishes the vagueness of concepts and encourages learners to identify with agents of positive change, hope, and resilience no matter the distancing effects of research report deadlines or the details of lives, geographies, and circumstances they may perceive as different from their own.

[1] Ami Dar, conversation with the author and Sarah Mead Wyman, February 22, 2024.

As philosopher Martha Nussbaum argues, storytelling develops “the ability to think what it might be like to be in the shoes of a person different from oneself, to be an intelligent reader of that person’s story, and to understand the emotions and wishes and desires that someone so placed might have” (Nussbaum 2016, 95–96). Through stories, students engage reason and emotion to define and access their own values. Absent this capability of narrative imagining, we risk educating students who see other people as mere instruments and see other animals and the rest of nature as resources for us to exploit. Nussbaum notes that if we get to know others and value them intrinsically, we can better protect them from *injustice*, which she defines as impediments placed deliberately or by negligence that thwart an individual’s striving toward something of value to them (Nussbaum 2023, Introduction, para. 10).

The Learning Activity

From Theory to Practice: Our Process

Having briefly discussed why stories matter for social justice education, let’s sketch how we put these principles into practice. Our process transforms abstract concepts into tangible student work through eight carefully structured steps that move from individual reflection to public presentation. Our students learn how to apply a palette of verbal and visual techniques to make traditional research reports about social justice issues more accessible and engaging. They practice facilitating conversations and translating abstract concepts into concrete examples through storytelling and drawing. By making cartoons, students also learn ways to compare-and-contrast scenarios and emphasize critical beats, pivot points, and key ideas in a story. They create work that evokes emotion, helping themselves and their intended audiences access their values. Here, emotions are guides for reasoning about social justice.

For detailed lesson plans, please refer to Earthrise Commons.

A Journey from Private Ideation to Public Sharing

1. Generate themes: Students identify social justice topics that matter to them as individuals.
2. Research and write stories: Geography students create initial research reports about social justice topics featuring people who are advocates for change.

3. Form groups: Students from each course form collaborative teams around primarily generative themes.
4. Create sketches and storyboards: Teams visualize their stories through simple drawings on sticky notes, index cards, and storyboard templates.
5. Develop motion: Students refine their story arrangements, craft and polish the work, digitize it, and bring it into presentation software, where they can do basic transitions and animation moves such as changes in position, scale, rotation, and creation/destruction that give the feeling of live drawing and erasing. Students also create *Ken Burns effects* where appropriate: a technique built into Apple's presentation software based on the popular documentary filmmaker's technique of panning and zooming over still photographs.
6. Record narratives: Teams layer voice and sound into their illustrated stories.
7. Build social justice websites: Students create platforms that combine illustrated podcast videos, infographics, resources, and calls to action. Mission-driven organizations often pair engaging social media videos with detailed supporting materials, allowing visitors to quickly grasp main ideas and then go deep to explore the structural and systemic context. Designers build trust with this multi-layered approach to community engagement.
8. Share work publicly: At the end of the semester, students present their illustrated podcasts on a large screen with surround sound at a campus theater event.

We intentionally designed this progression from private brainstorming to public thinking and making. Writing educator Peter Elbow notes that sharing work is a form of "mini publishing" that promotes responsibility. And the timing of this sharing really matters. We began with individual study, discovery, and development of ideas. This private work allows students to develop their initial thoughts without as much fearful self-talk and concern about what others might think. Yet Elbow observes that students who wish to communicate ideas can't dwell in silence. Why? Silent work on paper can sometimes allow you to "withhold some piece of yourself, to keep your fingers crossed behind your back." The real transformation happens when students move their ideas into the public sphere: "Reading your words out loud" or sharing drawings "is a vivid outward act that amplifies your sensation of responsibility for your words." Elbow notes, "That's why oaths and promises must be spoken out loud to work best" (Elbow 1998, Chap. 3, para. 7). Consider, "I now pronounce you…." Here, private thoughts become a public commitment, like a marriage vow.

Creating Space for Projects, Peers, Passion, and Play

Even though social justice efforts aspire to the type of peace, harmony, and consistency Ami Dar observed in a small Spanish town, many of our students face another reality. As MIT Media Lab's Mitch Resnick writes:

> Today's young people will be confronted with new and unexpected situations throughout their lives. They must learn to deal creatively with uncertainty and change–not only in their work-lives, but also in their personal lives (how to develop and sustain friendships in an era of ever-changing social networks) and their civic lives (how to participate meaningfully in communities that have ever-shifting needs and boundaries). (Resnick and Robinson 2018, 4)

Resnick calls for a less sclerotic, more nimble approach to education, which he likens to lifelong kindergarten. Resnick and Robinson 2018 articulate a principle of *Four P's of Creative Learning:*

1. Projects. People learn best when they are actively working on meaningful projects—generating new ideas, designing prototypes, refining iteratively.
2. Peers. Learning flourishes as a social activity, with people sharing ideas, collaborating on projects, and building on one another's work.
3. Passion. When people work on projects they care about, they work longer and harder, persist in the face of challenges, and learn more in the process.
4. Play. Learning involves playful experimentation—trying new things, tinkering with materials, testing boundaries, taking risks, iterating again, and again (Resnick and Robinson 2018, 16)

When our Geography and Graphic Design students finally came together in small groups, they found occasions for joy and laughter amidst the serious content in their project-based learning. Working with vintage materials—paper, pens, sticky notes, crayons—they created peer-to-peer spaces where complex ideas could be explored through play. Rather than engaging each other directly, students debated through their drawings, making their thinking visible and collaborative. This approach embodied my earlier discussion of "trembling thinking" in these ambiguous times—students could explore complex topics while searching for the human story of connection, challenge, or creativity within (Heath and Heath 2013, 204).

We wondered how to spark passion, which seems like an intrinsic quality that can't be intentionally planned for in a learning experience. Let's examine generative themes as a potential way to create idea sparks.

Generative Themes

Following education reformer Paulo Freire and Ira Shor's approach, we encouraged three types of themes:

- Topical themes from around the world
- Academic themes from disciplinary knowledge
- Student-generated themes from their lived experience

At the start of our semester-long collaboration, students generated social justice themes of real-world significance to them. We provided structures and frameworks, yet we stepped back from overly orchestrating work and created space for students to envision what they wanted to make, riff on ideas, thrive in random discovery, and co-create. Here, storytelling is the product and also a process between students that helps them discover new ideas and negotiate interests.

Our learners' self-selected *generative themes* are one part of a tripartite liberatory education approach advocated by Freire and Shor, who argue that educators should introduce themes from around the world that students may not know (such as the lives of the one billion people living on US $2 or less per day). Shor writes, "The (first) topical theme is a social question of key importance locally, nationally, or globally that is not generated directly from the students' conversation. It is raised in class by the teacher." A second set of themes arises from the discipline and is typically introduced by the teacher with roots in "formal bodies of knowledge studied by specialists in a field" (Shor 55). In my class, for example, this might mean literature related to the cognitive and social benefits of visual thinking and simple drawing.

Generative themes turn everyday topics that students discuss informally outside of class into opportunities for critical analysis. For instance, students might talk about social media trends among each other; a generative theme would prompt them to examine how these platforms affect their relationships and mental health. Or they might discuss a genre of popular music; with a generative theme, they'd analyze how song lyrics reflect broader social issues. These themes connect students' familiar worlds to new academic concepts and perspectives. As Freire explains, generative themes capture the complexity of an era in miniature, bringing together "ideas, concepts, hopes, doubts, values, and challenges in dialectical interaction with their opposites striving towards their fulfillment"(Freire Institute). Think of how a class discussion about smartphone use might introduce themes of connection versus isolation, privacy versus transparency, and innovation versus dependency.

This approach holds value for creative projects, where students often struggle with open-ended assignments or doubt their creative abilities. By anchoring projects in themes that already matter to students—like sustainability, social justice, or digital culture—educators can help students find intrinsic motivation and momentum to navigate better a creative process that's often ambiguous and non-linear.

Real-World Models

We drew considerable inspiration from two visual journalism projects that emphasize education in social justice communication: *Historia de Zainab*, about a Syrian refugee girl, created by an interdisciplinary Spanish design and journalism team (Cabrera et al., n.d.), and StoryCorps. We sought to emulate these projects for their use of cartoons for serious topics, the vividness of their storytelling, and their appeal to our common humanity. Students paired their video podcast narratives with an infographic, a technique I have used in my professional work with mission-driven organizations to introduce a topic emotionally through a single story and then show how it fits into a bigger picture told through charts and maps. I shared the paired story and infographic methods of *Historia de Zainab,* presented in an accordion book form that can be folded and unfolded;

one side shows Zainab's experience in storyboards, the other in a hand-drawn infographic. Each element reinforces the other: the story side of the book provides emotion and details; the infographic: the big picture. Like painters examining details and stepping back to comprehend the whole, we need both approaches to create a whole visual communication that appeals to the heart and mind.

To use another analogy, telling a visual story using only charts and graphs would be like watching a movie with only wide, establishing shots. Film editors know that viewers need a mix of shots to be emotionally engaged, including frames or views that take us closer to the actors. As Charlie Chaplin is often attributed to have said, "Life is a tragedy in close-up up, but a comedy on the wide" (Williams 2023). In this context, the close shot is the storytelling side, while the wide shot is the info-graphic side: certainly not comedic, but distancing and objective enough that emotions become muted. By working with such close shots and literal, pictorial images rather than abstract ones, students learned how to mix these techniques to show a setting, characters, and challenges that resonate with the heart and not just the mind.

The illustrated podcast videos also drew inspiration from StoryCorps, a non-profit journalism team whose mission is "to help us believe in each other by illuminating the humanity and possibility in us all—one story at a time" (*About StoryCorps* 2024). StoryCorps combines podcast narratives with charming, hand-drawn illustrations about serious topics, told by people. StoryCorps is what researcher Cynthia Kurtz (2014) calls a "story bank"—a collection of community narratives where stories aren't just stored but contribute to community understanding and actionable projects. These stories can help us understand our present reality and potential future. Unlike commercial storytelling, which often provides fleeting, temporary entertainment through fictional narratives, story banks are rooted in a community. These authentic stories become seeds of collective knowledge we attend to and nurture for information and wisdom within communities (Kurtz 2014, 54–57).

Educational Outcomes

Our approach makes sustainability more tangible and vivid for college students in four interconnected ways:

1. Students study real people taking practical actions toward sustainable living.
2. Students gain hands-on experience through collaborative storytelling.
3. Students create intimate, handcrafted content that counteracts potentially dehumanizing social justice narratives.
4. Students apply their informal digital social media literacies—reading, making, and sharing stories that have pictures and text—toward more formal academic practices. By integrating everyday social media and academic literacies, students see the relevance of sustainability concepts in their daily lives while honing their ability to communicate these ideas using their existing knowledge and wisdom.

Though it's difficult to measure or calculate learning outcomes for this creative work, beyond valuable technical skills in design and execution, educators can create opportunities for students to self-reflect in a structured way as part of the assessment process. Through writing, students can explore the social and emotional part of their developing sustainability competencies (Brundiers et al. 2020). As one student wrote in a 2019 reflection (*used with permission*), "I learned something new and interesting about the course material and myself every time I was in the class. It's because of this class that I know where I want to be in the future. I found something I never knew I needed in that class.... no other class has made me feel so certain about what I want to be doing with my life and future studies."

Conclusion

Sustainability needs new definitions and stories. Over one semester, our Geography and Graphic Design students developed new verbal and visual communication skills to create stories about sustainability. They worked with methods of information gathering, vivid picturing, and coherent narrative imagining and arranging (Samuelsson and Lindström 2022, 211). They explored generative themes alongside academic and topical issues that forefront our world's pressing challenges. The practice of storytelling enabled students to understand the source material and each other better. A methods-based approach to critical thinking, communicating, and sharing complex real-world issues works best when students engage in a dialog with their material and the perspectives of their co-creators.

While research shows many students at first view sustainability in a narrow way, as long-term environmental protection, storytelling helped them connect sustainability to immediate social justice issues like environmental conservation and food justice (Kruger et al. 2020, 518). This need for communicating the social justice foundation of sustainability is supported by a 2019 study of the Intergovernmental Panel on Climate Change (IPCC) Reports, and their visual communications support our approach. The investigation analyzed how climate change impacts and adaptation were visually framed in IPCC assessment reports. The authors found that the visuals primarily framed climate change from a prevention-oriented perspective, emphasizing global and regional scales across past, present, and future timeframes. However, few visuals addressed effective adaptation and those that did lack clarity in their themes and scales. The study suggests that more thoughtful visual framing—a better balance between highlighting problems, strategies, and solutions—would support readers who want to prevent climate change impacts and learn about how people respond positively to this dire challenge (Wardekker and Lorenz 2019, 273).

Education for Sustainable Development (ESD) situates learning and growth within these human stories and ecological and social contexts. *Learning for a Sustainable Future* (n.d.) states, "ESD is about respecting and preserving our histories, valuing culture and community, caring for others and the environment, and taking action to create a fair, healthy, and safe world for all beings. ESD also supports flexibility, creativity, critical reflection, and fosters a sense of personal responsibility for the economy, society, and environment" (Learning for a Sustainable Future, n.d. as quoted by Landorf et al. 2008, 226). Building on this foundation, Hilary Landorf and colleagues propose reframing ESD as Education for Sustainable *Human* Development (ESHD), to define it as an "Educational practice that results in the enhancement of human well-being, conceived in terms of the expansion of individuals' agency, capabilities and participation in democratic dialogue, both for now and for future generations" (Landorf et al. 2008, 232). Sustainable development is not just about environmental, social, and economic concerns. It includes the spiritual dimension of *human* development. This aspect begins with learning: the unfolding of a learner's imaginative powers and interpersonal competencies uplifted by others.

Sustainability educator David Orr grounds us, however: “Education can teach us to emphasize the whole system and develop our cleverness,” he warns, but it can lead us to believe “we are smarter than we truly are” (Orr, quoted in Landorf et al. 2008, 226). Consider the contrast between cerebral and hands-on learning. Unlike fact-based knowledge transmission and acquisition, hands-on learning is *stochastic*: we can’t perfectly control and replicate results. As artists, bakers, and motorcycle mechanics can attest, the unpredictability of working with one’s hands and tangible materials teaches us humility. But it also brings the work to life; a perfectly controlled world is a dead one (Rosa and Wagner 2023, 2). This outlook on sustainability education concords with my view that learning works best through gathering facts, visualizing them, and turning them into stories. Rather than prescribing solutions or indoctrinating ideologies, this method allows students to explore issues of genuine concern within the ritual and rhythm of a workshop setting. Students show and say and learn anew at the speed of drawing. Here, sustainability knowledge, skills, and values grow, emerge, and find lively expression in community.

References

Brundiers, K., M. Barth, G. Cebrián, M. Cohen, L. Diaz, S. Doucette-Remington, W. Dripps, G. Habron, N. Harré, M. Jarchow, K. Losch, J. Michel, Y. Mochizuki, M. Rieckmann, R. Parnell, P. Walker, and M. Zint. 2020. “Key Competencies in Sustainability in Higher Education-Toward an Agreed-Upon Reference Framework.” *Sustainability Science*.

Cabrera, Elena, et al. n.d. “Historia de Zainab.“ Infotics.es. Accessed June 30, 2025. https://project.infotics.es/historia-de-zainab/#portada. CC BY NC License.

Elbow, Peter. 1998. *Writing with power: Techniques for mastering the writing process*. 2nd ed. New York: Oxford University Press. Kindle Edition.

Freire, Paulo. “Concepts Used by Paulo Freire.” The Paulo and Nita Freire International Project for Critical Pedagogy. https://freire.education.mcgill.ca/content/concepts-used-paulo-freire (accessed October 6, 2023).

Glissant, Édouard. 2020. *Introduction to a Poetics of Diversity*. Translated by Celia Britton. Liverpool: Liverpool University Press.

Heath, Chip, and Dan Heath. 2013. *Made to Stick: Why Some Ideas Survive and Others Die*. New York: Random House.

Korenblat, Joshua. "Comics Make Learning Easier for Children on the Autism Spectrum and Everyone." UNESCO MGIEP. https://mgiep.unesco.org/article/comics-make-learning-easier-for-children-on-the-autism-spectrum-and-everyone (accessed February 19, 2025).

Kruger, Tina M., Natalie McCreary, Brianna L. Verhoff, Virgil Sheets, James H. Speer, and Stephen P. Aldrich. 2020. College Students' Understanding of Social Justice as Sustainability. *International Journal of Sustainability in Higher Education* 21 (3): 513–530.

Kurtz, Cynthia. 2014. *Working with Stories in Your Community or Organization: Participatory Narrative Inquiry*, 3rd ed. N.p.: Kurtz-Fernhout Publishing.

Landorf, Hilary, Stephanie Doscher, and Tonette Rocco. 2008. Education for Sustainable Human Development: Towards a Definition. *Theory and Research in Education* 6 (2): 221–236.

McCloud, Scott. 1994. *Understanding Comics: Writing and Art*. New York: Harper Perennial.

Milne, A. A. 1926. *Winnie-the-Pooh*. London: Methuen.

Murray, Michael D. 2021. "Cartoon Contracts and the Proactive Visualization of Law." *University of Massachusetts Law Review* 16, no. 1: Article 2. https://scholarship.law.umassd.edu/umlr/vol16/iss1/2.

Nussbaum, Martha C. 2011. *Creating Capabilities: The Human Development Approach*. Kindle. Cambridge, MA: Harvard University Press.

Nussbaum, Martha C. 2016. *Not for Profit: Why Democracy Needs the Humanities - Updated Edition. Princeton*, NJ: Princeton University Press.

Nussbaum, Martha C. 2023. *Justice for Animals*. Kindle. New York: Simon & Schuster.

Obrist, Hans Ulrich. "'The Earth is Trembling': Édouard Glissant in Conversation with Hans Ulrich Obrist." 032c, December 20, 2021. https://032c.com/magazine/edouard-glissant-and-hans-ulrich-obrist.

Resnick, Mitchel, and Ken Robinson. 2018. *Lifelong Kindergarten: Cultivating Creativity Through Projects, Passion, Peers, and Play*. Cambridge, MA: The MIT Press.

Roam, Dan. 2011. *Blah Blah Blah: What to Do When Words Don't Work*. New York: Portfolio/Penguin.

Rosa, Hartmut, and James C. Wagner. 2023. *The Uncontrollability of the World*. Cambridge: Polity Press.

Samuelsson, Lars, and Niclas Lindström. 2022. "Ethics Teaching in Education for Sustainable Development." *Athens Journal of Education* 9.

Shor, Ira. 1992. *Empowering Education: Critical Teaching for Social Change*. Chicago: University of Chicago Press.

Stibbe, Arran. 2014. "Ecolinguistics and Erasure: Restoring the Natural World to Consciousness," in*Contemporary Critical Discourse Studies*, 113–132. London: Bloomsbury Academic.

StoryCorps. "About StoryCorps." https://storycorps.org/about/ (accessed May 28, 2024).

UNESCO. 2017. *Education for Sustainable Development Goals: Learning Objectives*. Paris: UNESCO.

Wardekker, Arjan, and Susanne Lorenz. 2019. The Visual Framing of Climate Change Impacts and Adaptation in the IPCC Assessment Reports. *Climatic Change* 156: 273–292.

Williams, Andrew. 2023. "Life Is a Tragedy in Close-Up, but a Comedy on the Wide." LinkedIn, August 22, 2023. https://www.linkedin.com/pulse/comedy-wide-tragedy-close-up-andrew-williams/.

CHAPTER 22

Using the University Campus to Teach Environmental Advocacy

Brian Obach

A vital aspect of sustainability education involves teaching students to advocate for sound environmental policy and practices. Programs may effectively educate students about the science of climate change and other ecological crises, but without the advocacy skills needed to press for policy changes to address these issues, we fail to optimize the use of this knowledge. University campuses provide an opportunity for students to develop advocacy skills while advancing achievable policy reform. As microcosms of the broader social-political-economic order, campuses involve a complex web of interests, hierarchy, power, norms, roles, conflicts, and every other facet of the broader society in which they are embedded. Thus, experience advocating for change at the campus level helps to develop important transferable skills that students can employ after graduation.

Faculty who utilize the campus as the focus for policy change, can play a vital role in imparting valuable social change skills to students through

B. Obach (✉)
Department of Sociology, State University of New York, New Paltz, NY, USA
e-mail: obachb@newpaltz.edu

S. M. Wyman and J. Korenblat (eds.), *Sustainability Learning for Action and Community Engagement*,
https://doi.org/10.1007/978-3-032-11698-7_22

internships, independent study, and volunteer opportunities. Among the skills that students can develop through such experience:

- Understanding the difference between individual and structural impacts
- Developing campaign strategy
- Data gathering and constructing evidence-based arguments
- Analyzing power and interests
- Implementing pressure tactics
- Building coalitions and enlisting allies
- Overcoming bureaucratic inertia
- Ensuring transparency, accountability, and sustained pro-environmental policy
- Negotiation and the acceptance of incomplete/ imperfect victories

Policy change does not come easily. While campus administrations and other stakeholders may readily adopt certain types of environmental reforms, as with policy at the municipal, state, or federal levels, most significant change requires persuasion, organizing, and pressure. Learning to carry out such reform campaigns should be a central part of sustainability education.

In this chapter, four adaptable case studies describe how students participated in institutional reform campaigns and the skills they learned in the process. The case studies describe ways in which these learning goals are achieved in projects involving improving recycling, divesting from fossil fuels, eliminating single-use plastics, and expanding sustainable food options.

Common Limitations of Student Environmental Activism: Individualization and Volunteerism

Through classes and extracurricular programming, there has been much progress in educating students about environmental challenges. Once students become aware of environmental issues, many of them seek ways to act on their concerns. In most cases, this initially comes in the form of lifestyle choices. Recycling, switching to reusable bottles, adopting a vegetarian diet and other practices are obvious ways individuals feel they can *do their part to make a difference*. This is to be expected in a highly individualistic Western culture, especially given that powerful institutional actors have an interest in steering action in the direction of personal responsibility as opposed to structural change. Unfortunately, modest lifestyle changes, especially when only adopted by a relatively small percentage of committed and conscientious individuals, are not sufficient to address the massive ecological crises that are now rapidly unfolding (IPCC 2023; World Wildlife Fund 2022). This "individualization" of the problem also obscures the powerful interests and structural forces that propel environmental degradation (Maniates 2001).

The more ambitious among our environmentally conscious students will join or create organizations that seek to achieve greater changes than can be accomplished individually. Yet, many of these more coordinated efforts focus almost exclusively on "consciousness raising.“ Educational efforts typically include environmental films or speakers that make other students aware of problems but often fail to move beyond awareness-raising to concerted action to bring structural change. At best, participants may be asked to sign a petition or join the organization to be involved in further consciousness-raising or individual lifestyle change efforts. It is perhaps a credit to our profession that many students see the root of positive change in education. But it is a mistake to believe that education in itself will yield that change. Learning about issues is an essential first step, but in most cases, the overall ecological footprint of those who are ecologically conscious is only marginally smaller than those who are ignorant of the issues or who make no concerted effort (Wackernagel and Beyers 2019). Only structural change will shift individual practices to those that are more ecologically sustainable.

Volunteer activities make up another common form of action among many new student environmentalists. This is also the type of environmental activity most encouraged by campus administrations. Such activities build community while avoiding difficult questions regarding structural causes or the interests served by ecologically unsound practices. Community clean-up days or tree-planting initiatives are typical examples of this type of voluntarism. The learning that occurs in the course of organizing such events is valuable, but again, volunteer campaigns, as important as they may be for building community and providing direct benefits, are insufficient to overcome the pace of ecological destruction carried out in the normal course of conventional societal functioning. It is evident that structural change is needed to truly address the ecologically unsustainable practices built into our major institutions (Hawken

2017). Yet relatively few campus activists engage in environmental action focused on structural reform, and formal education on how to carry out such campaigns within the institutional setting is sparse.

There have been, however, some important campus-based environmental movements geared toward structural change. The fossil fuel divestment movement is the most obvious example. Drawing on a tactic used in the 1980s to pressure South Africa to end apartheid, hundreds of higher education institutions around the world have divested endowment funds from fossil fuel holdings under pressure from student activists concerned about climate change. This campaign to starve the fossil fuel industry of needed funds was started in 2012 by scholar-activist Bill McKibben, the founder of the climate advocacy group 350.org (McKibben 2013). Since then, thousands of students worldwide have participated in the fossil fuel divestment campaign. A key point here is that student activists were provided with a tool kit for how to advance this change on their campuses, strategies that in most other cases, student activists improvise as they go along (gofossilfree.org). The fossil fuel divestment guide provides step-by-step instructions on how to build a team, develop a strategy, take action, and see a campaign through to policy change victory. While a practical guidebook is valuable, there is much that faculty can do, through internships and independent study, to provide direct support for advocacy skills development, and build this toolbox.

Other examples of strategically sophisticated campus-based movements, such as PIRGs (Public Interest Research Groups) operate on many North American college campuses (www.pirg.org). Originally founded by Ralph Nader, PIRGs employ campus-based professional staff members who impart social change skills to student volunteers and interns as they advocate for consumers, students, and the environment. Organized as statewide chapters with campus affiliates, most PIRG campaigns focus on state-level policy, not on individual campus practices. This provides valuable experience for student activists, but students will rarely see state-level campaigns through to completion during their brief time at university. In contrast, campus-level policy reform is very achievable within the time frame of an individual's college career.

Strategically informed, institutionally oriented campus change efforts are more the exception than the rule. There is a dire need for advocacy skills training for students who seek to be effective institutional change agents and to overcome the tendency toward the "individualization" of environmental problems in which personal lifestyle choices and voluntarism are viewed as the most important expressions of environmental concern. College campuses reflect all of the complexity of larger, more powerful governing units, such as states or nations, but they are smaller, more nimble, and more responsive. Faculty can play a vital role in imparting valuable social change skills to students through internships, independent study, and volunteer opportunities directed toward campus-level environmental policy reform.

Case Studies in Successful Structural Reform on Campus

With faculty guidance, students at my university, a public comprehensive college located in upstate New York, have carried out a number of institutional environmental reform campaigns on campus. Through this work, students learn about the importance of structural change and the skills needed to advance it. Most of the "lessons" described here were carried out in discussion sessions where faculty used critical pedagogical techniques to facilitate institutional analysis and movement strategy. Several guidebooks provide the basics of social movement strategy for faculty not well versed in activism (Bobo, Kendall and Max 2010; Hunter 2019; Moyers 2001). In the spirit of critical pedagogy, even less experienced faculty can learn along with students how to be effective institutional change agents.

Case Study #1: Learning Persuasion, Overcoming Institutional Inertia: Building an Effective Recycling System

In the early 2000s, a group of environmentally conscious students sought to increase recycling on the campus. As is common, they began with an effort to "raise consciousness" among the student body about the importance of recycling, the logic being that once individuals are aware of its importance, they will adopt the practice. This yielded few identifiable results. The main problem was the poor recycling infrastructure. Small, blue, unlabeled recycling bins were tucked haphazardly in various corners of the campus. Even those community members whose consciousnesses were fully raised had difficulty acting on their desire to recycle. Through discussion sessions with three students doing an internship for credit to address the recycling issue, the interns shifted their perspective away from promoting individual awareness and toward institutional reform, namely the need for improved recycling infrastructure.

Next, students were tasked with identifying the key decision-maker who had the power to improve that infrastructure. This required that they make some phone calls and examine the organizational structure of the institution. Ultimately, they concluded that the vice president for Facilities Management was the individual with the power to address the issue. This ability to recognize the need for structural reform, to analyze organizations, and to identify decision-makers is key to effective advocacy. Too often, young activists fail to recognize the need for structural reform, or when they do, unfocused general calls for change are made without consideration of the appropriate "target"—the individual with the power to implement the change they want. These insights can be facilitated using common critical pedagogical techniques (Freire 2017). Questions designed to help students evaluate the problem and its sources: *Why aren't more students recycling? What would make it easier to recycle? Why isn't appropriate disposal easier now? Who decided to create the system as it is?* Such questions can help facilitate the development of core social change skills and understandings.

Other lessons learned in this case involve institutional inertia and persuasion. Upon meeting with key decision-makers, students encountered resistance to changing existing practices. Facilities managers blamed students for their failure to use the recycling bins already in place. More recycling bins would require extra cost and labor, the need for retraining, and generally reforming long-established trash collection practices, none of which was viewed as desirable by decision-makers. Waste is collected at hundreds of places across campus and adding one or more recycling bins in each location would be a costly undertaking.

The students resolved to analyze the system and gather data to build their case. They persuaded Facilities Management to allow them to do an experiment in one academic building. They first conducted a trash audit. For a week, they measured the amount of recyclable material that was going into waste baskets in each classroom throughout the building. They then removed all of the individual waste baskets from the classrooms and placed well-marked recycling/trash stations in each hallway. A follow up audit revealed that recycling rates increased dramatically. The removal of individual waste baskets from each classroom saved considerable labor for the custodial staff, and, to the surprise of facilities managers, there was no significant increase in trash left in the classrooms. This simple change in the structure of the recycling system dramatically altered individual practices without the need for an extensive consciousness-raising campaign. The appropriate container for each type of material was clearly marked, and even the most ardent anti-environmentalist would have a hard time failing to recycle, given its ease under the new system.

The lessons here were multifold. Critical analysis allowed students to see that the failures of recycling were not individual but institutional. A structural change was needed to truly address the problem. Institutional inertia created resistance to change from decision-makers. Nevertheless, the students were able to use empirical data and evidence to make a persuasive case for recycling reform. This smoothed the way for more far-reaching changes to the recycling system throughout campus.

Case Study #2: Analyzing Interests and Deploying Pressure Tactics: Fossil Fuel Divestment

In the recycling case, students learned that it is possible to achieve structural change by analyzing institutions, identifying decision-makers, and using effective persuasion. But persuasion is not always sufficient in itself. Decision makers often have a vested interest in maintaining the status quo (otherwise, they would likely change it themselves). The call for fossil fuel divestment is an example of how those in power (university presidents, trustees etc.) may have a personal or institutional interest in resisting change. If fossil fuel investment brings funding to the campus, those managing the campus budget will be motivated to maintain that revenue flow. Thus, universities have been hesitant to divest (Tollefson 2015).

When persuasion fails, activists must turn to pressure tactics to alter the interest assessments of decision-makers. Analyzing all potential, influential actors is key to any effective campaign. The question activists must consider is, *What could we do that would alter the decision makers' interest calculus such that they would more favorably view what we ask?* Movement strategists have identified various types of pressure that can be brought to bear (Bobo, Kendall and Max 2010). Elected officials can be pressured through the threat of vote withholding. Business leaders can be pressured through boycotts or strikes that harm their financial interests. University presidents, usually the decision-makers when it comes to divestment (directly or indirectly), are not subject to direct electoral or financial pressure. They are not elected by students, and a targeted boycott (not enrolling in the university) is not a viable tactic. But in the grand scheme of things, university presidents do have a responsibility to maintain the reputation of the institution, and that is where student activists have brought pressure to bear. "Naming and shaming" can be very effective. With regard to fossil fuel investment, activists have many means at their disposal to regularly and publicly associate the president and the institution with a shortsighted investment strategy that purports to help students while condemning them to a future of ecological disaster. Letters to the editor, pickets, demonstrations, sit-ins are all tactics that can call attention to this failure and facilitate reconsideration by decision-makers about how best to protect the interests of the college. Returns on fossil fuel investment may help the university financially at the margins (or not), but this must now be weighed against the damage to the institution's reputation now that activists are bringing it to light. Effective pressure tactics compel targeted decision-makers to ask themselves, *How much is gained through fossil fuel revenue and how much is lost by the ongoing bad publicity?*

Student activists on my campus, working with an informal faculty advisor, made fossil fuel divestment a goal in 2016, a time when the movement was gaining momentum worldwide. There were no academic credits issued in association with this campaign. In the case of this college, resistance to divestment was rooted primarily in institutional inertia as opposed to significant financial interests. The state university operates primarily on public funds and tuition revenue; the college's endowment is insignificant compared to most private institutions and the portion invested in fossil fuels is marginal.

Students first appealed to the college president. It is always appropriate to make a formal, polite request of decision-makers before a pressure campaign of any kind is even considered. In this case, the president was initially non-committal and offered a common diversionary response: that he had no power to grant their wishes because the trustees make all such decisions independently. While technically true, in most cases, the president has significant influence over the trustees on such matters. Diversionary and delay tactics on the part of campus decision-makers are frequently used and often effective given the relatively rapid four-to-six-year turnover in the student population. Many campaigns fizzle simply because administrators occupy students with drawn-out "listening sessions" or bureaucratic delays until lead activists graduate and move on.

In this case, even knowing that the president had significant influence over the trustees, the students decided to target the head of the Board of Trustees directly. A strategy session identified a set of escalating tactics that could bring pressure to bear on the chair of the board, placing him increasingly in the public spotlight for investing college funds in climate change industries.

The best tactics place targeted decision-makers in a novel and challenging situation. The students decided to pressure the board chair, who was the president of a regional bank, with a postcard campaign directed at his workplace, something that he had likely never experienced before. Tabling in public locations on campus, the students gathered hundreds of signed postcards asking the chair (usually politely) to divest college funds from fossil fuels. These cards arrived in several waves at the bank where they certainly gained attention and made it clear that the students were organized and determined. This unusual experience undoubtedly caused the board chair to consider, *What might they do next?* In fact, there were student-run plans to follow up with a media-based strategy that would have more publicly linked the board president to climate change.

The strategy session also identified a potential ally, a member of the Board of Trustees, who was known to have pro-environmental sympathies. Direct outreach to him provided the students with an advocate on the board. Students also used their faculty allies to secure a resolution of support from the college faculty senate. Together, these pressure tactics yielded the desired outcome. However symbolic, given the small amount of revenue involved, the trustees agreed to end all direct investment in fossil fuels.

As with the recycling campaign, this effort was developed by students with active faculty guidance using critical pedagogy techniques to facilitate analysis of institutional interests and to develop a campaign strategy. Movement guidebooks often provide a format for discussing and assessing pressure tactics, as they are central to all politics. Rarely do decision-makers get to do exactly what they want. All decisions derive from an assessment of interests given competing constituencies of varying power and influence. Activists, who generally possess little institutional power, must strategically assess their capacity to affect the interests of those in decision-making positions. This can involve discomfort and risk. Decision-makers will often use personal appeals and "we're-all-on-the-same-team" claims to diffuse pressure. This can be effective in relatively small campus communities, especially when powerful campus officials make students feel special by giving them their time or even including them in a task force to "study the issue." This may feel sincere and may in fact be so in some cases, but good activists do not lose sight of their core goal, and they do what is necessary to advance it in a timely manner. Both students and faculty, especially those without tenure, need to be cognizant of the extent to which they are willing to sacrifice personally or professionally for a cause they believe in. Retaliation is common in politics and not unheard of on university campuses. These are all important issues that students must consider and learn from.

Case Study #3: Capitalizing on Institutional Power: Opposing Single Use Water Bottles

In the divestment case, students identified tactics designed to bring pressure to bear on decision-makers. Another important lesson related to this: it is best to *be* the decision-maker. Students are often unaware of the formal power they have within most higher education institutions. Various governance structures often have seats designated for students. This is in addition to student governance, which provides them with a powerful platform to voice concerns directly to the administration and, in many cases, independently controlled financial resources in the form of student fees that are formally designated for their use. An entire branch of social movement theory emphasizes the importance of resources in movement mobilization (Klandermans 1984; McCarthy and Zald 1977). Funds are needed for everything from fliers to buses to bullhorns. While expenditures may be restricted in certain ways, student activists should recognize the value of these independently controlled assets. In the next case study, students made use of their own institutional power within the university governance system.

The problem of plastics in our environment has gained increasing attention. In response, some campuses have undertaken efforts to reduce single-use plastics. At this college, a group of students working with faculty support sought to end the sale of single-use water bottles on campus. At many institutions, dining services, book sales, vending, and other auxiliary services are carried out by not-for-profit corporations created solely to serve a specific campus. Such is the case at this college. Like all corporations, the Campus Auxiliary Services (CAS) corporation has an executive director overseen by a board of directors. Anti-plastic activists used persuasion, including a formal presentation detailing the hazards posed by single-use plastics, to try to convince the CAS board to adopt their proposal. There was resistance to ending the sale of these products by the executive director and other campus officials who held seats on the CAS Board out of concern for lost revenue and potential consumer complaints. While bottled water was introduced for mass consumption relatively recently, beverage corporations have been particularly effective at normalizing this item of consumption, despite the reported health effects, pollution impacts, and easy availability of free drinking water (Barlow 2005).

Despite resistance from some on the board, students were able to capitalize on the fact that they held a number of seats on the CAS board. Student government leaders, residence hall representatives, and various at-large positions gave students over one-third of the voting power. Together with faculty on the board, they constituted a majority that eventually voted to end the sale of single-use plastic water bottles.

Effective advocacy is ultimately about analyzing power and interests. The key lesson for students, in this case, is that holding positions of institutional power allows them to directly shape policy to their liking without the need for persuasion or pressure or the expenditure of additional time, energy, and financial resources. While some eager activists may shun the drudgery of formal student governance, a careful assessment of power and interests will at times, reveal that it is best to advance goals by securing formal decision-making positions. Over the years, several local student activists recognized this principle and went on to run for public office after graduation.

Case Study #4: Institutionalizing Change: Sustainable Food

Movement victories can turn to defeat if policy achievements are later reversed or if the measures put in place fail to deliver expected outcomes. In another example involving the non-profit Campus Auxiliary Services, students learned the value of institutionalizing policy changes while also understanding that you don't always win a perfect victory. In the early 2000s, a group of students had a focused interest in sustainable agriculture. They were unhappy with the conventional food served on campus by an international food service provider that had won the CAS food service contract. Initially, as is typical, students sought an individual solution: to exit the mandatory meal plan so that they could purchase their own food from off-campus sources. But eventually, they adopted a more institutional perspective, and sought to change the campus's food purchasing practices.

In the course of this campaign, there were some partial victories. The student cohort persuaded CAS officials to require the food service contractor to utilize at least some locally grown goods, such as apples. Previously, the dining hall carried apples from as far away as Washington state and Chile, despite the fact that the college sits in the middle of one of the largest apple-growing regions in the country. But progress in implementing other reforms was slow and inconsistent. Some local or organic items would appear for a period, then be gone from the menu when attention waned.

Through an analysis of the situation, students came to recognize the fundamental challenge of their food service provider's ongoing interest to rely on their own national food distributors, not a hodgepodge of small local producers. Transformation would require constant vigilance and pressure to ensure the maintenance of their reforms and further progress. The desired practices needed to be institutionalized to be feasible and sustainable. Some national movement organizations came to a similar conclusion. The Real Food Challenge is a national organization born of this realization. The RFC created a standard measurement of how well colleges are providing healthy, sustainable food on campus. Such rating systems have proven to be an effective means of incentivizing environmentally sound practices. Various "green rating" measures for colleges have encouraged their presidents to vie for recognition as environmental leaders in hopes of attracting students and bolstering their reputation. The most well-recognized and comprehensive of these is the STARS rating system used by the Association for the Advancement of Sustainability in Higher Education (stars.aashe.org). The national environmental group, the Sierra Club, also issues a ranking of "green colleges" (Andrews 2023). The Real Food Challenge is designed to foster that pursuit specifically on the basis of campus food service. This objective was good in theory. The annual rating mechanism would give CAS and the food service provider an ongoing incentive to improve using a clear, measurable standard. But the problem of vigilance did not subside. The regular evaluation and documentation of food provided on campus was complex, tedious, and costly. While initially amenable to adopting the Real Food Challenge system, in time, CAS administrators allowed this practice to fall by the wayside. Without ongoing pressure, the food monitoring system was abandoned, and food service largely returned to prior practices.

Institutionalizing policies is essential to avoid backsliding when the movement goes into abeyance, as it inevitably will (Sawyers and Meyer 1999). Thus, policies must be designed to ensure simplicity, transparency, and accountability. Had advocates insisted that a permanent position be created to monitor food purchases or if sustainability performance were built into the job evaluation of the CAS executive director, perhaps more progress would have been made. In this case, the complexity of monitoring millions of dollars of food purchases, vagaries associated with measurement and commitments, and a lack of transparency made this level of assessment unviable.

There was no formal end to this campaign nor comprehensive debriefing, although that would have been useful to discuss the major lesson from this campaign: *you don't always win*. Most of the activists had dispersed years into the effort, either discouraged by the lack of progress or hopeful that the Real Food Challenge program would ensure continual progress. Contending with setbacks is essential for social change activists. Ideally, such defeats are never viewed as definitive failures, but rather delays in the ongoing struggle for a better world. There are always new tactics to try when others fail. Indeed, efforts to make food services more ecologically sustainable are ongoing at this campus. Much of this effort is being carried out as a result of earlier victories by which gains were institutionalized, the most significant being the creation of an Office of Campus Sustainability and the hiring of a full-time director. This office provides resources to advance sustainability work (including funds to pay student workers and interns) and a locus of accountability. The green rating systems cited earlier are the main means by which decision-makers are incentivized to make continuous, measurable environmental progress.

Conclusion

Achieving social change is hard. Those in positions of power tend to favor the status quo in which they are...in power. Those seeking change are generally less powerful. Thus, the deck is stacked against change advocates. It is only through careful strategic analysis that relatively less powerful people can devise ways to overcome those with relatively more. In the present context, many of those in positions of power have benefited greatly from practices that have yielded ecological devastation. For many it is *because* they were engaged in ecologically destructive practices that they achieved positions of power.

Young people today face far more costs than benefits from perpetuating the current system. Educating students about ecological threats and the need for change is essential. But it is equally important to provide them with the tools needed to achieve that change. The individual lifestyle changes that populate common "10-things-you-can-do-to-save-the-planet" lists are not adequate to meet the challenges we face. Acts of charity and volunteerism are also admirable but insufficient to address the current ecological crisis. Institutional changes are needed, and this requires collective political mobilization. But effective political action requires skills ranging from the ability to analyze power and interests to effective persuasion and devising pressure tactics. There are few venues through which these skills are taught. University campuses provide a training ground for developing the skills necessary to bring needed change. Faculty can play an important role in helping to guide students into effective campus activism. Reforming environmental policies on campus is a worthy goal in itself. But as with other aspects of student life, college is primarily designed to prepare students for what comes next. The advocacy skills that students learn while changing policies on campus can later be applied to change the world.

References

Andrews, Avital. 2023. *The Complete Ranking*. Sierra Club. https://www.sierraclub.org/top-ten-americas-greenest-universities.

Association for the Advancement of Sustainability in Higher Education. 2023. "The Sustainability Tracking, Assessment and Rating System." https://stars.aashe.org/

Barlow, Maude, and Tony Clarke. 2005. *Blue Gold: The Fight to Stop the Corporate Theft of the World's Water*. New York: The New Press.

Bobo, Kimberley, Jackie Kendall, and Steve Max. 2010. *Organizing for Social Change: Midwest Academy Manual for Activists*. Santa Clara, CA: Forum Press.

Freire, Paulo 2017. *Pedagogy of the Oppressed*. London, England: Penguin Classics. gofossilfree.org. 2023. "How to Run a Campus Divestment Campaign". Accessed September 5, 2023. https://gofossilfree.org/usa/divestment-guide/.

Hawken, Paul. 2017. *Drawdown*. New York, NY: Penguin.

Hunter, Daniel. 2019. *Climate Resistance Handbook: Or I was part of a climate action. Now what?* Lulu.com.

IPCC. 2023. *Climate Change 2022: Impacts, Adaptation, and Vulnerability.* Working Group II Contribution to the Sixth Assessment Report of the Intergovernmental Panel on Climate Change. Ed. Hans-Otto Pörtner, Debra C. Roberts, Melinda M. B. Tignor, Elvira Poloczanska, Katja Mintenbeck, Andrés Alegría, Marlies Craig, S. Langsdorf, et al. Cambridge, UK: Cambridge University Press. https://doi.org/10.1017/9781009325844.

Klandermans, Bert. 1984. Mobilization and Participation: Social-Psychological Expansions of Resource Mobilization Theory. *American Sociological Review* 49 (5): 583–600.

Maniates, Michael F. 2001. Individualization: Plant a Tree, Buy a Bike, Save the World? *Global Environmental Politics* 1 (3): 31–52. https://doi.org/10.1162/152638001316881395.

McCarthy, John D., and Mayer N. Zaid. 1977. Resource Mobilization and Social Movements: A Partial Theory. *American Journal of Sociology* 82 (6): 1212–1241.

McKibben, Bill. 2013 (February). The Case for Fossil Fuel Divestment. *Rolling Stone.*

Moyer, William. 2001. *Doing Democracy: The MAP Model for Organizing Social Movements.* Gabriola, British Columbia, Canada: New Society Publishers.

Sawyers, Traci M., and David S. Meyer. 1999. Missed Opportunities: Social Movement Abeyance and Public Policy. *Social Problems* 46 (2): 187–206. https://doi.org/10.2307/3097252.

Tollefson, Jeff. 2015. Fossil-Fuel Divestment Campaign Hits Resistance. *Nature* 521: 16–17. https://doi.org/10.1038/521016a.

Wackernagel, Mathis, and Bert Beyers. 2019. *Ecological Footprint: Managing Our Biocapacity Budget.* Gabriola, British Columbia, Canada: New Society Publishers.

World Wildlife Federation. 2022. *Living Planet Report 2022 – Building a nature-Positive Society.* Ed. Rosamunde Almond, Monique Grooten, Diego Juffe Bignoli, and Tanya Peterson. Gland, Switzerland: WWF

CHAPTER 23

Practicing Positive Masculinity: Navigating Cultural Constructs for Healing and Relationship

Joel Oppenheimer

Storytelling and Emotional Abundance

Why do we tell stories? Because we need to. Because stories let others know who we are, what we have been through, and how we have interacted with the world. Stories unite us, free us, open us, make sacred the mundane. There is a special moment when you're listening to a true story, a story with vulnerabilities, a story that is risky to tell, a story that will elicit judgment, a story with weight. Heavy. *Why do we equate emotions with weight? Is there a finite amount of emotional weight one can carry? On the shoulders? Heft you can hold?* This moment of telling transforms the story from burden to medicine. Its sacred nature, no matter how horrible, heals us through its relational power. It becomes us, defines us, unites us, earthing us closer to the true emotional abundance of our original instructions.

J. Oppenheimer (✉)
Psychological Counseling Center, SUNY, New Paltz, New Paltz, NY, USA
e-mail: oppenhej@newpaltz.edu

S. M. Wyman and J. Korenblat (eds.), *Sustainability Learning for Action and Community Engagement*,
https://doi.org/10.1007/978-3-032-11698-7_23

I am standing in front of 300 incoming first year students on campus for orientation. I ask them:

> If someone you know, or maybe someone you don't know, asks whether they can share something they're going through, would you listen to them? If so, raise your hand.

Everyone raises their hand. Next question.

> Raise your hand if, most of the time, you don't talk to a friend about something you're going through because you don't want to be a burden to them.

Almost everyone raises their hand. Do you see what just happened? We all have the emotional capacity to listen to someone and yet we think what we are going through is a burden to someone else. How did we determine that there is a limited amount of emotional capacity in others? Emotional scarcity. What if there were an endless amount of emotional space for you? Emotional abundance.

This feeling of emotional abundance, of sacred space, is what participants report during the monthly Positive Masculinity Workshop on our university campus. What does emotional abundance feel like? To know limitless capacities. When someone chooses to tell a real story; when our hearts are opened by someone's vulnerability; when they are sharing deeply: we enter a place where the instructions for being change. A different version of *you* rises to the surface to meet this energy. You have entered a sacred space. This story is healing, feeling, meaningful, and yet you have not experienced any part of it directly. There is something underneath this story, something vast and endless, something unexplainable that validates your existence.

The Positive Masculinity Workshop

The Positive Masculinity Workshop was formed to counter masculine individualism and support healthy interdependence, which prioritizes relationships and creates a space to form a new way of being outside of the oppressive constructs of patriarchy, competition, and perfectionism. Participants bring their stories and wisdom traditions to collaborate on ways of supporting and caring for each other and Earth. We ask the question, *what could life be like?* This chapter explores the myths of masculinity that have brought Western culture in the global north to a place of separation and unnecessary suffering due to values and behaviors that compromise our relationships to each other and the planet. The Positive Masculinity Workshop provides a template for transformation, an experience of interdependence, emotional abundance, and truth that activates our shared indigeneity.

In telling stories, we nurture ourselves and each other. One participant shared,

> My father was there to provide for me but really, he wasn't there for me emotionally. My father was strong and smart but wasn't present, probably didn't know how to be. I realize now that he was doing what he knew how to do. Using the tools he had. It must be hard trying to be a parent trying to figure stuff out, you know, when you haven't had the best parent yourself. I have held these angry feelings toward my father for a long time. But hearing these other stories I know I'm not the only one. I think I need to forgive my father. Not for him but for me. Well, maybe for him too.[1]

In 2019 a New Paltz community member shared a story about his relationship with his sister.

> I grew up not knowing any gay people, or not accepting them.... When my sister came out as a lesbian, I did not accept her. That's how I was back then. I was confused, it was hard to understand... I don't know if she will ever truly trust me and how my mistakes must have changed the way she sees the world.

Many of us can relate to the experience of finally seeing a person more clearly and perhaps the remorse of not seeing them this way earlier. He continued, *I see other people being homophobic now, and it turns my stomach because that was me.*[2]

When stories of vulnerability are shared, some part of our indigenous self is awakened. Some part that knows peace and honors all life rises within us. During the workshop, we stand in a circle with our imaginary

[1] Paraphrased to protect confidentiality and identity of the participant.

[2] Paraphrased to protect confidentiality and identity of the participant.

fire in the center, bringing together our hearts, minds, and the original instructions we all know. This circle includes all life, all our ancestors, and Earth (Tinker 2024). We name the power of this circle, and admit that we know where we are from; we have just gotten lost. This context, this acknowledgment, this story, is true. Now, on common ground, we trust, listen, support, and model equity. Participants tell stories they have never shared before. This is the goal of the Positive Masculinity Workshop: to experience the way we truly want to live in the world. Once we feel it, we know we can create it.

The Origins of Individualism and Domination in a Patriarchal Context

To understand individualism in the context of patriarchy and masculinity, we must look back to the Doctrine of Discovery. Named by Steven Newcomb (2008) as the *Doctrine of Domination*, this creed was a set of papal bulls declared by the Vatican in the 1400s that gave permission to European explorers to claim land and kill "savages" (Native People). This policy constituted the original authority, entitlement, and privilege by European colonizers to take land, which the US government later legalized through written laws, broken treaties, and assumed rights. This Domination led to the genocide of millions of native people whose connection to Earth was deemed unnatural by European colonizers (Tinker 2022).

Earth itself, regarded by the colonizer mind as wild and untamed, became the next target of domination. By establishing civilizations through the (dis)possession of land, the control of resources, and enslavement or killing of native people, colonizers effectively began to change both human ecology and the climate (Johnston 2022).

Native people had been sustainably managing the ecosystems for thousands of years, so their displacement and elimination immediately impacted the natural world. What native people identified in the colonizers as greed, taking more than what is needed, the colonizers saw as their God-given right to own, harvest, and extract what had never been owned (Johnson 2022). This concept of land ownership did not exist in Native cultures, and there was no word for ownership in most Native languages (Ghosthorse 2018). The process of dividing land for individual owners further embedded European individualism in what is now North America, and it remains cemented in our culture today, as well as in many other countries around the world (Tinker 2021). The ownership of land by men fueled modern manhood, then branched out forming vast interwoven hierarchies including racism, sexism, status, and capitalism, to name a few. People were then forced into stratified systems, habituated to consume and strive while blindly complicit in the cultural construct of individualism.

Patriarchy, in the global north, dictates that there is a man at the top who has the inherent knowledge, ability, and power to lead the family or community without help. Patriarchy puts men at the head of all hierarchies, including gender, sex, work, education, religion, government, military, criminal justice, and production of all material goods (Johnson 1997). Help or dependence indicates weakness and vulnerability. Competition for resources, scarcity, pits individuals against each other in a contest of competency and might. Feminism has been pointing to this oppressive structure for decades, suggesting, in some instances, that women must act "like men" in order to hold leadership positions, or simply survive in social systems dominated by men (Johnson 1997).

This patriarchal construct is deeply ingrained in higher education as well. We see this in the leadership structures at colleges and universities and in students' behaviors and cultures. Most blatant among students, yet the most hidden by administrations, is *rape culture*. This patriarchal entitlement to rape women is the same patriarchal entitlement used to rape (extract) the planet of resources and rape (control) populations of people through genocide, slavery, apartheid, caste systems, and denial of resources (poverty). Rape culture is difficult to eradicate because it is rooted in historical structures of domination that are driven by, and serve to maintain male power. Men are also harmed by these traps of entitlement and oppression, forced to objectify women, be strong and knowledgeable, not show emotions, lead without help, and always compete (Porter 2016).

Our students in higher education have tremendous levels of stress from the fallacy of competition. This manifests in fears of being vulnerable, of being an emotional burden to others, and ongoing existential crises of purpose, meaning, and belonging. College students have historically questioned their existence, as is appropriate for their developmental stage, but this may be the first generation to experience climate change as a threat to their existence. Our separation from the earth and our complete dependence on patriarchal leadership leaves many feeling anxious, depressed, and powerless, waiting for answers these white men don't have and certainly cannot solve alone. Meanwhile, Native people, who have suffered the most from rape culture, white supremacy, extraction, and control, *do* have answers that can move humanity toward healing and wholeness (Hawken 2017).

Patriarchal constructs also power perfectionism, a close relative of individualism. Perfectionism is perhaps the least discussed social construct, yet its forces are incredibly impactful. Perfectionism secretly formulates, often subconsciously, the ideals and images we strive for. It builds the pressures and threats that loom with potential failure. In fact, perfectionism *creates* the false duality of success and failure. Fueled by media images of wellness, health, accomplishment, and physical appearance, we create an unobtainable mental picture of who and what we want to be. Relationship with Earth is not in this picture. The pressure to meet these manufactured standards continues to be a significant factor in the mental health crisis of our time.

Capitalism provides the money, status, and the most important reward for individualism: ownership. Ownership of land and possession of material goods may be the single greatest contributor to the destruction of our natural environment. Individualism is so linked to the artificial rights of convenience, comfort, autonomy, and access that we can barely see past its illusion to imagine living collectively or in sustainable relationships with each other and Earth.

Vanessa Machado de Oliveria, in her book *Hospicing Modernity* (2021), explains that humans' sense of intrinsic value is based in our relationship with earth. Now, since we are separated from Earth in modern society, we strive to obtain our self-worth through wealth, status, accomplishments, ownership, comfort, convenience, and progress. This intrinsic value, now based on the individual, is exemplified by the masculine *hero*, a man with "good intentions" who believes in himself and destroys the villain, often represented as the savage *other*. Our strong male heroes protect this "civilized society" with their leadership, laws, and policing and force everyone to assimilate. With this illusion of progress, safety, and civilization deeply embedded in our belief systems, our disconnection from Earth remains. And our unsustainable consumption and comfort remain unquestioned according to the fallacy that we have earned the right to it.

Conclusion

In the Positive Masculinity Workshop, participants heal by gaining awareness of how individualism, patriarchal structures, and damaging concepts of masculinity have impacted them. They build trust with each other and experience the transformative power of co-creating possibilities of another way to be in the world. The workshop shifts our sense of intrinsic value to each other and our relationships. This is one step toward creating a culture of sustainability that recognizes the necessity of our collective to survive. Humans *are* nature and our survival depends on our bio-psycho-diversity as well as the full range of knowledge and experiences we share. In the workshop, we make space for this wisdom and lived experience to inform, heal, and inspire. We ask the questions,

> What would it be like to be fully seen, heard, and held, in all our identities? Can we create a space that simulates the village of our native lands? With our bodies and identities present and acknowledged, can we co-create relationships with each other and Earth, based on truth, vulnerability, and listening?

The cultural construct of individualism has trapped us in many ways. Just as colonization divided our lands, individualism separates us from each other and the true source of our intrinsic value: Earth. Masculinity further isolates us, restricting feelings and pushing us to achieve everything on our own in a stratified system that shames weakness and valorizes power over, not power with. We must look to Mother Earth for answers, and she will teach us to grieve, pray, heal, and question. We are not stuck. We have the capacity to transform, apologize, to ask Mother Earth, *please forgive us while we learn*. Being nice is not enough to save our planet; we've got to be deeply loving, radically creative, habitually uncomfortable, and enormously brave to become the world that will sustain all life.

References

Akwesasne Notes. 2005. *A Basic Call to Consciousness: The Haudenosaunee Address to the Western World*. Native Voices.

Hawken, Paul. 2017. *Drawdown: The Most Comprehensive Plan Ever Proposed to Reverse Global Warming*. Penguin.

Johnson, Allan G. 1997. *The Gender Knot: Unraveling Our Patriarchal Legacy*. Temple University Press.

Johnston, Lyla June. 2022. *Architects of Abundance: Indigenous Regenerative Food and Land Management Systems and the Excavation of Hidden History*. University of Alaska Fairbanks, Dissertation: Philosophy in Indigenous Studies.

Lilien, Jadina and Tiokasin Ghosthorse. 2018. *Butterfly Against the Wind*. Self-published.

Machado de Oliveira, Vanessa. 2021. *Hospicing Modernity: Facing Humanity's Wrongs and the Implications for Social Activism*. Berkeley: North Atlantic Books.

Newcomb, Steven. 2008. *Pagans in the Promised Land: Decoding the Doctrine of Christian Discovery*. Chicago Review Press.

Porter, Tony. 2016. *Breaking Out of the Man Box: The Next Generation of Manhood*. Skyhorse.

Tinker, Tink. 2021. "Relationship—Not Ownership: Indigenous Lands and Colonial Occupation." *Tribal Studies*.

Tinker, Tink. 2022. "How the Eurochristian Invasion of Turtle Island Created the Environmental Crises: Focus on an Early 'Immigration.'" *Displacement Climes: Shifting Climates, Shifting People*, edited by Miguel de la Torre, 19–30. Pilgrim Press.

Tinker, Tink. 2024. "The Land, Creation, Sovereignty, and Property." In *T&T Clark Handbook to the Doctrine of Creation*, edited by Jason Goroncy. Bloomsbury.

CHAPTER 24

Psithurism: What Musicians and Other Artists Bring to the Climate Justice Movement

Christiana Fortune-Reader

Standing in the glade I heard a quiet psithurism*, just straddling the line between music and noise.
from ψιθυρίζω (psithurízō, "I whisper")

The Arrow and the Song
I shot an arrow into the air,
It fell to earth, I knew not where;
For, so swiftly it flew, the sight
Could not follow it in its flight.

**Psithurism: Noun (obsolete):* a rustling or whispering sound, such as leaves in the wind; susurration

C. Fortune-Reader (✉)
Department of Music, SUNY New Paltz, New Paltz, NY, USA
e-mail: readerc@newpaltz.edu

S. M. Wyman and J. Korenblat (eds.), *Sustainability Learning for Action and Community Engagement*,
https://doi.org/10.1007/978-3-032-11698-7_24

> I breathed a song into the air,
> It fell to earth, I knew not where;
> For who has sight
> so keen and strong,
> That it can follow the flight of song?
> Long, long afterward, in an oak
> I found the arrow, still unbroke;
> And the song, from beginning to end,
> I found again
> in the heart of a friend.
>
> -- Henry Wadsworth Longfellow

Introduction

The connection between art and sustainability: an appeal to the imagination as a possible model for community engagement.

What can artists bring to the climate justice movement that scientists cannot? I asked a colleague in the Engineering department this question. As a musician and music educator in the United States, I was slightly embarrassed to be at a meeting about sustainability, dubious that musicians could contribute anything to something so science-centric. His response was that an intellectual understanding of the scientific facts has not moved people to act. If we want our students to move toward meaningful change, we need to engage the other side of their brains. We need to approach the problem with our imaginations. The people best suited to help us do this, he believes, are artists.

Increasingly, artists are using their work to cultivate a sense of curiosity and wonder about the natural world, using themes of endangerment and extinction to convey the unparalleled threat posed by humans to this planet's existence. Communication is a vector for social change in the sustainability movement. The arts provide affective components of environmental education: feelings, values, and motivations that can drive pro-environmental behavior. The arts evoke emotion, spark dialog, encourage innovative thinking, present diverse perspectives, cope with ambiguity and non-linearity, and influence the development of cultural norms. Iris Murdoch posited "A deep motive for making … art of any sort is the desire to defeat the formlessness of the world and cheer oneself up by constructing forms out of what might otherwise seem a mass of senseless rubble" (Murdoch 1999).

The arts handle the fuzziness (or multivalence) of action for sustainability, presenting multiple realities without necessarily having to resolve them with each other (Eernstman and Wals 2013) Scientists have valued the growing body of eco-art for its ability to reveal and explain a distinct overlap between the arts and science, as it makes data vivid and accessible. This work is fueled by reactions to climate change and global warming and is generally intended to overcome humans' hardwired tendency to overvalue personal experience and to disengage from data-based representations. Oceanographer Greg Johnson translated the 2000-page 2013 IPCC climate report's recommendations into 19 watercolor-illustrated haiku poems, delivering a compelling visual guide with clear, concise, and powerful talking points. Stripped of jargon and unfathomably large numbers, limitations and scales of confidence that confound and distract the layperson, art can be an arresting and informative entree into the science of climate change. Some visual artists collaborate with scientists and incorporate information into museum exhibitions, as in the work of Jill Pelto, featured on the cover of TIME Magazine's 2020 special issue on climate change entitled "One Last Chance" (Peleto 2020).

Broadly, many composers have written music that explores the intersection of music and climate. Some iconic programmatic works render an extramusical narrative, such as Antonio Vivaldi's *Four Seasons*. The music of the nineteenth century Romantics Ludwig van Beethoven, Franz Schubert, Johannes Brahms, and Gustav Mahler all birthed from experiences in nature. Beethoven's 6th Symphony "Pastoral" which he called "more the expression of feeling than painting," with its shepherd's pipe, birds singing, streams flowing, and a thunderstorm, is the pinnacle of programmatic ideas, as embodied in the first movement, "Erwachen heiterer Empfindungen bei der Ankunft auf dem Lande" (Awakening of cheerful feelings on arrival in the countryside). Claude Debussy's depiction of the sea in *La Mer* "avoids monotony by using a multitude of water figurations that could be classified as musical onomatopoeia: they evoke the sensation of swaying movement of waves and suggest the pitter-patter of falling droplets of spray" (Potter 2007).

Similarly, contemporary composers in Western music are focused on a more direct connection between visual art, music, and the natural world. A wide array of environmental concerns can be addressed through music: the natural world, family, community, interconnectedness, a sense of altruism to make the world better, economic development, climate change, imbalances of power/wealth, politics, consumerism/greed, abdicating responsibility, Indigenous peoples, and spirituality (Publicover et al. 2018). Szymon Weiss and Szymon Sutor wrote *Lost Seasons* (2018) for a documentary on climate change. They warped Vivaldi's iconic work, imaginatively adapted it to modern unstable weather conditions, and reflected the distorted future that scientists predict. The electro-classical piece premiered on stage at COP24 with a live orchestra against a dystopian backdrop: drone footage in monochrome of Poland's biggest coal plant, Belchatow.

Art Inspires Collective and Civic Action

Music engenders a sense of human connection through programming that builds individual and collective engagement.

The visual arts provide possibilities to reframe the climate crisis, expanding beyond a direct avenue for visual data delivery. Olafur Eliasson's work straddles sustainability, climate change, and collective activity. Many works result from research into complex geometry, motion patterns, and color theory. They explore Eliasson's deep engagement with society and the environment, revealing what an artist's perspective can bring to issues of climate change, energy, migration, and architecture.

Eliasson's *Ice Watch* coincided with the meeting of world leaders at the 2018 COP24. Created from a harvest of fallen glacial ice chunks in Greenland, *Ice Watch* contains twelve 1.5–6-ton blocks of ice placed in a circle resembling a watch face to echo time passing in relation to stopping or slowing down climate change. When viewers wandered through it, they could feel the ice, placing their ear to the frozen surface to hear tiny pings cracking and popping. Eliasson remarked that these bubbles trapped fresh air spanning millennia. Esthetically, the ice was remarkable in that it was very deep blue in places with occasional transparent or opaque spots. As it melted over several days, the shapes changed, becoming at first soft and then very angular. Like many of his installations, this sculpture had a defined expiration date as the material disappeared to nothing. Eliasson states (2016):

> I believe that one of the major responsibilities of artists ... is to help people not only get to know and understand something with their minds but also to feel it emotionally and physically. By doing this, art can mitigate the

numbing effect created by the glut of information we are faced with today and motivate people to turn thinking into doing.

John Luther Adams' *Inuksuit* (2009) invites listeners to consider the context of rising temperatures through an aural experience. Some musical events encourage communities to take stock of their surroundings, but very few fold so seamlessly into the environment itself that they become part of a community's memory and imagination. *Inuksuit,* a concert-length, large-scale work for 9–99 percussionists, is to be performed outside, where musicians are widely dispersed and move throughout a large, open area. Listeners may move around freely and discover their own listening points. Adams writes:

> "*Inuksuit* is inspired by the stone sentinels constructed over the centuries by the Inuit in the windswept expanses of the Arctic. The word "Inuksuit" translates literally: "to act in the capacity of the human." This work is haunted by the vision of the melting of the polar ice, the rising of the seas, and what may remain of humanity's presence after the waters recede." (Adams 2009).

This work is intended to expand our awareness of the never-ending music of the world in which we live, transforming seemingly empty space into a more fully experienced place.

This approach is rooted in the mid-twentieth century tradition of aleatoricism (from the Latin word *alea*, meaning "dice") in which some element of the composition is left to chance and/or some primary element of a composed work's realization is determined by its performers. Adams resists Western classical music's composition tradition of creating a master score. Instead, he calls the work a "folio" that contains "a collection of musical materials and possibilities for musicians to use in creating a unique realization of the work. Each performance is different, determined by the size of the ensemble, the specific instruments chosen, and by the topology and vegetation of the site — even by the songs of the local birds." Adams focuses on the location of the performance, the distribution of performers, the acoustical properties of the performance site, and how sound can define the listening space and the experience of it (Fonseca-Wollheim 2014).

Contemporary Activist Musicians

Students tend to resonate with contemporary musicians and composers who have assembled to create ecologically driven works. Pop groups such as Coldplay limit their tours' carbon emissions with eco-friendly choices like using bicycle-generated power and electricity-producing dancefloors, building sets from bamboo, and planting trees for tickets sold. Multi-platinum-selling indie pop group AJR invites arena audiences to do "the wave," a vast, coordinated ripple as the concertgoers throw their arms up instantly, uniting the room to harness massive, coordinated energy. The band collaborates with local non-profits in each city to inspire concertgoers to take local, policy-based action to help reduce the impacts of human-caused climate change right there in the arena. Landscape Music, a consortium of musicians and composers, deepens public appreciation of the natural world by providing a platform for contemporary composers and performers whose music engages with landscape, nature, and place. LM musician Ryan Suleiman questioned the role composers play when writing about his piece, *Burning* (2017), dedicated Californian firefighters, first responders, climate scientists, and wildfire victims:

> So what does it mean for our art when the natural balance of the environment is in a state of crisis? Should we continue writing pieces about the idyllic beauty of nature in its most pristine form in order to encourage people to value it? Should we write protest pieces criticizing the inaction of governments and the elaborate cover-ups of corporations? Should we write pieces about the environmental catastrophes we see in front of us? Should we put our pencils down and become activists? What is our responsibility?

Similarly, composers and performing musicians are collaborating actively with other artists and scientists. The Fry Street Quartet based in Utah developed the *Crossroads Project: Rising Tide and Emergence* together with physicist and educator Dr. Robert Davies to address pressing issues of global sustainability. Titled for a time and place where the scientific ability to identify unprecedented risk intersects with a societal inability to respond, this features an experimental and continuing collaboration between science and the arts.

These original multi-disciplinary performance pieces are meditations that integrate visual artists, filmmakers, actors, and composers. Libby Larsen's score assumes the role of one person's inner voice in *Emergence*, featuring video projections and a Shakespeare-inspired soliloquy by Davies to take audiences from understanding to action. *Rising Tide* combines music, information, and imagery along with original works by painter Rebecca Allan and photographer Garth Lenz. The music is written by composer Laura Kaminsky and sectioned in four movements: H2O, Bios, Forage, and Societas. Kaminsky says (2016),

> I was motivated to bring together artists and scientists to create a compelling work of art incorporating scientific research that would inspire deep reflection, and, ultimately, action, among those who experience this performance. In addition, I was greatly enriched by working with a quartet as artistically rigorous and intellectually vibrant as the Fry Street Quartet. My goal was to create a work with a persistent, and at times, subliminal pulse that carries through the four movements, making the quartet itself a symbol for the measure of life.

Where *Rising Tide* is about what we know, *Emergence* is about coming to terms with this knowledge on a deeply personal level. Continuing in this vein, the *Crossroads Project* invites audiences to engage with activism to make eco-friendly choices and then share them on social media via the #makeityoursxrds and #makeityoursusu hashtags. In this same vein, the Brooklyn Rider Quartet pursued a major new commissioning and programmatic venture in their 2022–2023 season, *The Four Elements*. It explored the four classical elements (earth, air, water, and fire) as a metaphor for the complex inner world of the string quartet and the current health of Earth.

Composer Gabriela Lena Frank developed a commissioning initiative in Boonville, CA. Its composers recognize that climate disruption is a bona fide civilizational emergency. Through the marriage of scientific knowledge and artistic imagination, composers become climate-aware, informing themselves and stirring their creative psyches. With a range of genres and instrumentations, Composing Earth achieves broad dissemination of these necessary stories of protecting the earth, avoiding cultural and socio-economic niches that limit exposure. Frank writes (2023):

> The very interdependent nature of the musical life and our longstanding custom of listening to and conveying human stories will be our community's strength in navigating the elusive space between alarm and aspiration as we face the environmental crisis. It is indisputable that change is upon us, but we can be the stewards of that change if we dare to imagine and

> propose alternative actions even if they eventually prove to be wrong, ineffective or outdated. The point is to move, to act, and be ready to discard what doesn't have a meaningful impact while demonstrating compassion towards one another.

Another group, Ocean Music Action presents multi-sensory concerts centered around a shared love of the ocean, collaborating with and commissioning new works from diverse artists. Concerts are paired with a volunteer waterway cleanup or other activity in partnership with a local conservation organization, giving audiences and musicians the opportunity to work hand in hand toward the common goal of cleaner oceans for all. The UK-based groups Julie's Bicycle and Music Declares Emergency provide a treasure trove of resources to create a community of support among artists, music industry professionals, and music fans alike for climate change advocacy on a policy level. Finally, San Francisco-based artist Stephan Crawford's group, Climate Music, combines the talents and expertise of world-class scientists, composers, musicians, artists, and tech pros to "harness this universal language [of music] to tell the urgent story of climate change to broad and diverse audiences in a way that resonates, educates, and motivates."

Discussion: Arts Integration in High Education

Institutions of higher education (HEIs) are well-positioned to provide a rich soil for seeding pro-environmental behavior. Music serves as a catalyst for behavior change and social activation, nudging a shift from an individualistic perspective to group harmony. The development of projects on Classical eco-friendly or contemporary activist musicians (see Earthrise Commons; earthrisecommons.org) can center outward-facing actions as simple as talking more often about sustainability, building a palpable sense of community, and resolving local challenges, especially when framed as collective projects.

Participation in a musical event centered on sustainability provides a model for meaningful community engagement that reduces the height of the academic ivory tower. According to Wals, there are signs of institutions developing and introducing new forms of transdisciplinary learning that can help students understand and engage in sustainable development (Wals 2014). Among these new forms of learning are many musical possibilities that require multi-stakeholder interaction, meaning-making, negotiation, dealing with competing claims, handling a diversity of perspectives, and the resolution of real issues as they emerge in everyday life at home in the university itself, or in the community.

Learning Experience: Music Appreciation Course

The transformational potential of sustainability education became clear in a standard general education music appreciation class, presenting Western music for non-music majors at our state university. Students received an experiential learning assignment to find a comfortable place outside to sit in silence for ten minutes, to notice their surroundings, and then listen to John Luther Adams' *Become Ocean*, a 40-minute orchestral mediation on the oceans of Alaska and the Pacific Northwest, inspired by John Cage. The composer states, "life on this earth first emerged from the sea. As the polar ice melts and sea level rises, we humans find ourselves facing the prospect that once again we may quite literally become ocean." The piece won the 2014 Pulitzer Prize as part of a trilogy of works, *Become River* (2010) and *Become Desert* (2017). The monumental orchestral palindrome depicts water in irresistible motion in the tradition of music by Wagner, Debussy, and Sibelius. However, Adams goes one step further than those nineteenth-century giants with an intention beyond mere description of the ocean. The piece submerges the listener in a swirling, churning wash of sound, presenting dynamic peaks of waves gradually rising and falling. At key moments, the motion of various groups in the orchestra synchronizes to create giant crests of sound. This inspires the audience to switch from an active listening mode to a more reflective mode, as it presents an opportunity to unearth a sense of agency in one's approach to living with the reality of climate change. Adams continues:

> I'm not a propagandist. I'm not trying in my music to teach a lesson or tell a story. I'm trying to discover a place and invite you into that landscape to find your own way and — if I'm lucky — get hopelessly lost in it. It is impossible for us to separate who we are from where we are.... As an artist, my primary responsibility must be to my art as art—and yet, it's impossible for me to regard my life as a composer as separate from my life as a thinking human being and a citizen of the earth....If my music can inspire people to listen more deeply to this miraculous world we inhabit,

> then I will have done what I can as a composer to help us navigate this perilous era of our own creation (Fonseca-Wollheim 2014).

Adams gets to the heart of music appreciation: why music matters and the importance of learning how to listen more carefully. Upon listening to the piece, students reflect on their experiences and question their role in their chosen field of study. One Engineering student wrote (*used with permission*):

> Adams attributes his success in music to the idea of listening. Throughout the piece, I noticed that the conventional aspects of harmonies and melodies are less distinct … They all seem to blend in along with the rhythm of the piece to create something entirely unique from standard forms of music. This allows the audience to listen more to the sound of the piece; the way it breathes and lives brings light to the important environmental warning it aims to deliver. Politics and social aspects often distract from the actual cause of the issue. The ability to drown out noises and find the true cause of a problem is the first step. Adams' job as an artist is the ability to "inspire people to listen more deeply to this miraculous world we inhabit." Similarly, my purpose as an engineer will be to listen to the world and, more specifically, society so that I can identify the problems within it. The ability to develop or improve technologies that better society, and the future of our planet is something that would give me a sense of purpose in this world.

Integrating Music and Sustainability Across the Higher Education Curriculum

One way to begin the process of incorporating music and sustainability in HEIs is to add a single module into a semester and gradually work towards an arts-integrated approach across the curriculum, joining arts and science to move towards action. By way of an introductory course, students could study the work of Susie Ibarra, who takes underwater microphones to the Himalayas to record the sound of glaciers breaking and melting and uses that recorded experience as the immersive backdrop to her compositions. The resulting album of field recordings, *Walking on Water* (2021), part of *Water Rhythms*, became an elegy to the climate crisis and an homage to human resilience in dire circumstances.

Water Rhythms is a story told by ice and water about the dualism of water, the universal connector of nature and humans, an intergenerational story of the rapid change and depletion of our glaciers and freshwater sources worldwide. Ibarra writes on her website:

> Humans are inextricably connected to the Earth's freshwater; the same rhythms of glacial meltwater that flow from the mountains to the sea flow through our bodies, our histories and our music. In recording the sounds of glacier melt, we discovered that these sounds were not simply ambient, but indeed very musical. They have precise rhythms, a precise tempo/bpm (beats per minute). These tempos match the sweet spot at which pieces of music from all over the world are created and played. They also match our heartbeats at birth. Our connection to water is hence so much more intimate than we think. A world losing its flowing freshwater is not only a world of increasing ecological precarity, it is a world losing its music, its culture, and its humanity.

Other works of Ibarra's include *Sky Islands*, a musical tribute to the earth's rich and fragile ecosystems inspired by the distinct rainforest habitats of Luzon, Philippines. This work integrates the Western European instrumentation of a string quartet together with traditional Filipino instruments and sound sculptures with the aim of connecting people to our fragile and beautiful ecosystems and drawing awareness to changing climate and global community practices.

Building Community Through Interactive Projects

Finally, one approach that incorporates the community into music-making is the work of Phyllis Chen, who invites the public to participate in the development of her work. In her compositions, audiences create paper music box strips that are incorporated into the harmonies of her compositions. Her piece, *Floating Verses* (2018), is a global community music-art piece that invites participants to create their own music box strip. At the premiere of one of her pieces *In Plain Air*, at the 2019 Philadelphia Fringe Festival, she hung the crowd-sourced music box compositions in bird houses in trees surrounding the concert venue, where audience members could play the music boxes before and after the concert. The music boxes produced a collection of corresponding pitches, determined by the holes punched in the strips, but also by the result of chance. The community members could not have exactly predicted nor imagined the final composition, much like our future on earth.

Chen's *Resonant Echoes* (2021) commissioned by the author, Christiana Fortune-Reader, invokes the spirit of Pauline Oliveros' *Sonic Meditation* (1971) "V: Native"which gives the instructions, "Take a walk at night. Walk so silently that the bottoms of your feet become ears." For *Resonant Echoes*, Chen invited community members to submit recordings of a walk outside on a snowy day when roads were closed, asking them to listen to the sound of snow. Chen then incorporated those recording fragments with written music. This approach does not require any prior knowledge of music and engenders a sense of community and togetherness.

Conclusion

I spent the COVID-19 pandemic at home, on the edge of a state park containing untouched old-growth forests, one of only five remaining virgin forests within several hundred miles of New York City. I felt lucky to experience the lockdown's isolation walking among the trees and fresh air as I slowly grew accustomed to wearing masks in indoor spaces. As a musician, I became increasingly starved for the experience of live music as all in-person concerts had been canceled. I started recording nature sounds on my phone as a way of focusing my listening, learning to distinguish simultaneous bird calls, comparing the sound of a rushing stream at the bottom of the dam after a heavy rainfall to its corresponding trickle on a dry day, laughing at the dogs barking to each other across the neighborhood, trying to dictate the rhythm of humming insects in concert with frogs, imitating the calls of migrating geese, looking for insistently loud and industrious woodpeckers and chatty squirrels in the trees, reveling in the sound of cracking ice and crunching snow beneath my feet. The sounds created their own concert hall under the tree canopy. It took my musically trained attention to experience it that way.

The one sound I could not capture was the sound of wind, one of my favorite elements reminiscent of my childhood on the Kansas prairie. Loud, blustery days powerfully affect my mental state, transporting me to a more grounded and calm frame of mind. In looking for advice on how to get a snippet of sound for my growing portfolio, I was struck by the metaphor that this venture provided. Most recorded wind sounds in television and film are artificially generated in recording studios using random objects in isolated chambers. Achieving the sound of wind in a natural environment without overloading mic capsules requires different senses, namely sight. Wind flows around objects and creates eddies, like water. One doesn't ever want to record wind itself: All one can do is record wind *in* or *through* something else. Finding these "wind eddies"aids the positioning of one's mics to get that sound without being buffeted directly by the wind itself. Apart from a wide-open desert, wind eddies/buffers are everywhere. What matters is their proximity to the desired sound to be recorded (Fig. 24.1).

At first glance, the connection of music with sustainability seems suspect when scientists and engineers can provide more technical solutions. But music provides a wind eddy for capturing elusive sounds and the human imagination as listeners engage with the natural world. Whether through singing, playing, or listening, participation in music provides a catalyst for the layperson to build a personal connection to ecological awareness. The differentiation between hearing and listening guides the music consumer to become an active participant.

Olafur Eliasson (2016) relates the arts to building momentum for sustainability solutions and mindsets:

> Engaging with art is not simply a solitary event. The arts and culture represent one of the few areas in our society where people can come together to share an experience even if they see the world in radically different ways. The important thing is not that we agree about the experience that we share, but that we consider it worthwhile sharing an experience at all.…I am convinced that by bringing us together to share and discuss, a work of art can make us more tolerant of difference and of one another. The encounter with art – and with others over art – can help us identify with one another, expand our notions of we, and show us that individual engagement in the world has actual consequences.

Participating in music wields vast influence on emotional identities. The ensemble experience provides insight into the microcosm of living in a community, highlighting the individual and the collective, considering the self as a part of the larger group. Such involvement reveals the nature of interdependence, the self as a node in a larger organism. The experience of a listener in a concert setting generates empathy and shared experience, creating a sense of community among increasingly isolated lives. Listening to and participating in music-making provides a sense of connectedness that our virtual world has eroded. The study of various

Fig. 24.1 *Forest* (2024) Andrea Kantrowitz

artists places the musician and music appreciator in the artist's role as a citizen in a democratic society.

What matters is our situational awareness, something that music can help reveal. This approach demonstrates how persuasive messaging can inspire individuals to honor their natural surroundings, creating spillover and eddying effects to persuade others. Borrowing from climate activist and writer Bill McKibben (2005), we are reminded that each individual contributes to the existence of our species as a whole, most engaged by our imagination:

> We are all actors in this drama, more of us at every moment. [...] It may well be that because no one stands outside the scene, no one has the distance to make art from it. But we've got to try. Art, like religion, is one of the ways we digest what is happening to us, make the sense out of it that proceeds to action. Otherwise, the only role left to us — noble, but also enraging in its impotence — is simply to pay witness. The world is never going to be, in human time, more intact than it is at this moment. Therefore it falls to those of us alive now to watch and record its flora, its fauna, its rains, its snow, its ice, its peoples. To document the buzzing, glorious, cruel, mysterious planet we were born onto, before in our carelessness we leave it far less sweet.
>
> Time rushes on, in ways that humans have never before contemplated. That famous Earthrise picture of our planet from outer space that Apollo beamed back in the late 1960s — already that's not the world we inhabit; its poles are melting, its oceans rising. We can register what is happening with satellites and scientific instruments, but can we register it in our imaginations, the most sensitive of all our devices? (McKibben 2005).

References

Adams, John Luther. Programme note on *Inuksuit* (2009). Wise Music Classical. https://www.wisemusicclassical.com/work/57056/Inuksuit--John-Luther-Adams/. Accessed 3 May 2026.

Eernstman, Natalia and Arjen Wals. 2013. "Locative Meaning-making: An Arts-based Approach to Learning for Sustainable Development." *Sustainability 2013* 5(4): 1645–1660. https://doi.org/10.3390/su5041645

Eliasson, Olafur. 2016. Why art has the power to change the world. *World Economic Forum*. https://www.weforum.org/agenda/2016/01/why-art-has-the-power-to-change-the-world/.

Fonseca-Wollheim, Corinna da. 2014. A Composer Attuned to the Earth's Swirling Motion: A New York Premiere for John Luther Adams's Become Ocean. *The New York Times* 5 May 2014 https://www.nytimes.com/2014/05/06/arts/music/a-new-york-premiere-for-john-luther-adamss-become-ocean.html. Accessed 3 May 2026.

Frank, Gabriela. 2023. Climate commitment. *Gabriela Lena Frank Creative Academy of Music*.. https://www.glfcam.com/climate-commitment.

Ibarra, Susie. n.d.. "Water rhythms." Susie Ibarra. Accessed 14 May 2023. https://www.susieibarra.com/about-3-1.

Kaminsky, L. 2016. "Crossroads project." Fry String Quartet. Accessed 14 May 2023. https://www.thecrossroadsproject.org/risingtide.

McKibben, Bill. 2005. What the warming world needs now is art, sweet art. *The Grist*. https://grist.org/article/mckibben-imagine/.

Murdoch, Iris and Peter Conradi. 1999. *Existentialists and mystics: Writings on philosophy and literature*. New ed. New York: Penguin.

Peleto, Jill. 2020. "One Last Chance." TIME magazine.

Potter, Caroline. 2007. "Debussy and nature". *The Cambridge companion to Debussy*. Simon Trezise (ed). Cambridge: Cambridge UP. ISBN 978–0–521-65243-8.

Publicover, Jennifer L., Tarah S. Wright, Steven Baur, and Peter N. Duinker. 2018. "Music as a Tool for Environmental Education and Advocacy: Artistic Perspectives from Musicians of the Playlist for the Planet." *Environmental Education Research* 24(7): 925–936.

Wals, Arjen. 2014. "Sustainability in higher education in the context of the UN DESD: A review of learning and institutionalization processes," *Journal of Cleaner Production* 62: 8–15, ISSN 0959-6526, https://doi.org/10.1016/j.jclepro.2013.06.007

CHAPTER 25

Food Justice as Social Justice: Making Hidden Socioeconomic Realities Visible Through Arts and Science Education

Laia Solé Coromina and Arnau Amat

This is a letter written by four hands. They belong to Laia Solé Coromina, artist and art educator and to Arnau Amat, biologist and professor in science education. This reflective piece explores threads of thoughts, projects, and actions through diverse ways of thinking, doing, and telling—that have been developing since 2020. We maintain an interdisciplinary approach in various projects on sustainability in diverse settings: from higher education with student teachers at Universitat de Vic-Universitat Central de Catalunya, Spain, in cultural centers with young children, or assessing and educating teachers in local networks of schools for sustainability. We are committed to the embodied way of knowing

L. S. Coromina (✉) · A. Amat
Departament de Didàctica de Les Arts I Les Ciències (FETEP), Universitat de Vic-Universitat Central de Catalunya, Vic, Spain
e-mail: laia.sole@uvic.cat

A. Amat
e-mail: arnau.amat@uvic.cat

S. M. Wyman and J. Korenblat (eds.), *Sustainability Learning for Action and Community Engagement*,
https://doi.org/10.1007/978-3-032-11698-7_25

necessary in the arts, allowing for the resonance and *amplification* of an experience of reality within the human body (Wagensberg 2015). Meanwhile, science allows us to approach socio-environmental issues by looking at connections and the possibility of generating conceptual models with which we and our students can make informed decisions. This holistic approach embraces the hands, heart, and mind and the world outside, promoting an open and collaborative engagement with phenomena, where senses, experience, and imagination are key to building knowledge (Korenblat & Coromina, 2023). This approach can foster innovative responses to familiar and unfamiliar issues and simultaneously open spaces for new possibilities.

We believe that there is no sustainability without social justice, which includes food justice. This work requires an approach that integrates methods from the arts and sciences and attention to individual experiences of common problems.

Interdisciplinarity is not just thinking conceptually about a problem in different ways. It is also about combining real practices and methods from art and science, such as creative explorations and fact-finding methods. Both epistemologies allow us to make visible things that are usually invisible or difficult to express through words alone (Klee 1961). As Nicholas Mirzoeff (2016) suggested, our way of visually experiencing reality can be restrictive; our customary practices may develop an authoritarian aspect that simultaneously diminishes or negates the possibility of seeing otherwise. Opposed to this limited perceptual experience, there is the *right* to see (metaphorically): a right to gain real knowledge, comprehension, and the ability to take action to positively transform our reality.

Our way of living as humans, based on the exploitation of natural resources, has contributed to the disruption of natural cycles—*metabolic rift* as Marx named it. Capitalism addresses this disruption by turning a blind eye, and, thus, leaves the consequences for future generations. Or, as suggested by Kohei Saito (2017), capitalism solves issues by creating other issues. Works in sustainable living have inspired us, such as *The Mushroom at the End of the World: On the Possibility of Life in Capitalist Ruins* by anthropologist Anna Tsing (2015). Tsing proposes an alliance between the natural sciences and what she calls *the arts of noticing* as a way of doing and being. This mode of living has the potential to allow the discovery of alternative possibilities to the current reality. We are also inspired by Agnès Varda and her documentary *The Gleaners and I* (2000), devoted to the art of gleaning: the practice of collecting what has been left

Fig. 25.1 Fragment of a drawing that describes a sausage's life cycle, where we can see an idealized depiction of cows grazing, a bloody depiction of a slaughterhouse representing food transformation, followed by truck transportation to a supermarket

and discarded after the harvest. This documentary speaks of food waste, of what gleaning may be in our times, and also about the role of artists as gleaners of realities, which often remain obscured and invisible.

From art education and scientific education, we draw upon the capacities to glean, to restore, and to validate alternative ways to "see" and relate to our issues and contexts. We understand that our task is to render visible the blind spots generated by capitalism, such as the exploitative or deadening means and relationships of production and excessive consumption. Indeed, we use the metaphor of *blind spots* to explain the socio-environmental injustices created by our way of living, which consciously or unconsciously, remain obscured. Recently, we analyzed drawings made by children on the life cycle of food (Fig. 25.1). This research has allowed us to identify a series of often overlooked areas within the food system: the absence or sugarcoating of pressing issues—namely, the impact of the food industry on the landscape, the road system expansion for truck transportation, the tensions experienced by farmers in the face of increasingly energy-intensive models of food production, and the pollution of the soil and water.

We are interested in inhabiting these problematic areas instead of escaping from them. Or, as philosopher Donna Haraway (2019) proposes, we want to *live with the trouble*: see it, create relationships, and build *possibilities*. Simply put, we want to integrate alternative ways of being and knowing into the language of critical thinking around sustainability topics.

Case Study: Interdisciplinary Project on Sustainable Food

Three years ago, at our university, we developed an interdisciplinary project between three courses, Art Education, Science Education, and Social Science Education, that addresses socio-environmental issues related to food sustainability. This is a project of great significance for the context in which we live: an important agricultural and livestock farming region in Catalonia in Northeast Spain, where the production and distribution of vegetal and animal food has shaped and continues to shape the economy and culture. Our regional infrastructure represents many food system problems and challenges that communities worldwide face.

The project starts with our students investigating and writing a critical essay on an issue related to sustainable food practices: agricultural production models, livestock farming, fossil fuel-powered transportation, water pollution, and food access. We use scientific facts to understand the consequences of damaging production and transportation practices. Then, we explore how artists have addressed these problems as an opportunity to know alternate ways of relating to complex phenomena.

Artists' models of doing and thinking about an issue assist our students in the next step: designing and developing an educational activity with children from nearby elementary schools. Some of these projects have taken the shape of a curricular kit, reminiscent of Fluxus kits or Duchamp's *Boîte en Valise*. This consists of a container with images, objects, and materials that allow children to explore and build knowledge about food systems and socio-environmental issues. Some versions of this educational activity have included walking trails in the countryside. This year, inspired by artist Antoni Miralda, who delves into *gastronomic ethnology* with his performances, activities and projects such as *Foodcultura* (Gray Center, nd), we look forward to designing menus and cooking meals, which will be conceptualized and presented as performances. Thus, this project requires *inhabiting the problem*. Our students are positioned as consumers, cooks, visual artists and educators, making decisions about what to buy, where to buy it, how to cook it, and how to present it according to the issue they are working with.

Reflections on the project (Sabido-Codina et al., 2023) showed that 60% of students understood this learning experience was useful for their professional development. Moreover, 89% of the students changed their perception about what constitutes pedagogical material and saw the curricular kit as a good resource for meaningful learning. It is critical to choose images that allow children to interact and reflect upon socio-environmental topics and to design good questions to help them explore these issues.

Conclusion

As science educator and teacher educator Derek Hodson (2010) says: "A politicized ethic of care entails active involvement in a local manifestation of a particular problem or issue, exploration of the complex sociopolitical contexts in which the problem/issue is located, and attempts to resolve conflicts of interest" (201). The arts have the potential to disrupt everyday life, to shake the appearance of continuity, or as the philosopher of education Maxine Greene (1986) said, to remove what is *taken-for-granted* in our reality. We understand this as becoming conscious again of our interdependency with respect to other beings, of our finitude, and our capacity to build alternatives. Arts-informed scientific education can empower students to act in ways that transform our world. When learners engage in artistic practices, they have the potential to question the limits between what is possible and what is not. These integrated, interrogative activities invite us to reflect and to imagine possible worlds and to inspire actions that can transform our contexts. Our students perceived the making of a curricular kit as a potent resource to explore food justice topics; it allowed children to engage with socio-environmental issues that otherwise would remain beyond their imaginations.

References

Foodcultura. n.d. Retrieved from https://www.foodcultura.org/en/

Gray Center. n.d. Retrieved from https://graycenter.uchicago.edu/people/antoni-miralda

Greene, Maxine. 1986. The Art of Being Present: Educating for Aesthetic Encounters. *Journal of Education*, *166* (2): 213–135.

Haraway, D. 2019. *Seguir con el problema: Generar parentesco en el Chthuluceno.* Consonni.

Haraway, D. 1995. *Ciencia, cyborgs y mujeres: la reinveción de la naturaleza.* Ediciones Cátedra.

Hodson, D. 2010. Science Education as a Call to Action. *Canadian Journal of Science, Mathematics and Technology Education* 10 (3): 197–206.

Klee, Paul. 1961. *The Thinking Eye. The Notebooks*, Vol. 1. Ed. Jurg Spiller. G. Wittenborn.

Korenblat, Joshua and Laia Solé Coromina. 2023. *Undesigned world: what educating through walking means in the Anthropocene*. In Pau Alsina y Andrés Burbano (coords.). "Posibles". Artnodes, no. 32. UOC. https://doi.org/10.7238/artnodes.v0i32.411197

Mirzoeff, Nicholas. 2016. *The right to look*. https://doi.org/10.12795/IC.2016.i01.01

Sabido-Codina, J., L. Solé, and A. Amat. 2023. Projectes interdisciplinaris centrats en l'alimentació sostenible mitjançant maletes didàctiques: Canvis perceptius de l'estudiantat. *Didacticae* 2 (14): 197–213. https://doi.org/10.1344/did.2023.14.197-213.

Saito, Kohei. 2017. *Karl Marx's Ecosocialism: Capial, Nature, and the Unfished Critique of Political Economy*. New York: New York University Press.

Tsing, Anna. 2015. *The Mushroom at the End of the World: On the Possibility of Life in Capitalist Ruins*. Princeton, NJ: Princeton University Press.

Wagensberg, J. 2015. *El pensador intruso: El espíritu interdisciplinario en el mapa del conocimiento*. Booket.

PART IV

Building a Sustainability Culture in Institutions

CHAPTER 26

Build It and They Will Come: The Faculty Learning Community Approach to Infusing the Curriculum with Sustainability Content

Will Hong

Introduction

Perhaps the biggest question that we face as teachers is how best to prepare our students to hit the ground running once they leave campus and begin their professional careers. With respect to sustainability issues, this means instilling in them not only a wide range of skills with which to begin collaboration with others in solving problems (quantitative and qualitative, local and global), but also providing a communicative starting point, a common dialect and concrete objectives using that dialect with which they can begin their work with professionals in other fields. The United Nations (UN) Sustainable Development Goals (SDGs) serve as an ideal starting point for conversation and action that inch us closer to a sustainable global living. The SDGs provide a widely accepted framework of collective action, one that has been put to use by actors both in the

W. Hong (✉)
SUNY New Paltz, New Paltz, NY, USA
e-mail: hongw@newpaltz.edu

S. M. Wyman and J. Korenblat (eds.), *Sustainability Learning for Action and Community Engagement*,
https://doi.org/10.1007/978-3-032-11698-7_26

public and private sector. It is therefore crucial that, regardless of major or field of study, university students have classroom experience that affords them a solid grasp of what the SDGs are, how they provide a framework for action, and where they point us as a civilization.

One problem, then, is that systems-thinking approaches (upon which the SDGs foundationally rest) and the SDGs themselves are rarely explicitly taught in the classroom. Ratified by the UN in 2015, the goals are only a few years old. Although they were rolled out to great fanfare within professional development circles, scant attention has been paid to the goals beyond that space, even within the academy. The societal shifts that sustainability research tends to put forth as necessary often fall squarely within the bounds of the politically or economically infeasible. For instance, the Green New Deal proposed by the progressive wing of the US Democratic Party, marries climate change action with social justice progress, a systems-thinking perspective that squares with the SDGs, yet it has been roundly criticized by conservatives for being wholly impractical, with even some centrist-leaning Democrats calling it purely "aspirational" (Axelrod 2019; King 2019). Over three years since their creation, the SDGs have yet to make their way into the classroom substantially, and a crucial opportunity for this generation of students goes by the wayside.

How to change this, then, especially given the growing number of budgetary shortfalls that befall many higher education institutions, especially those public universities that do not benefit from generous endowments? The sustainability-focused faculty learning community (SFLC) model is one that has been effectively deployed for decades. The SFLC can be created without incurring prohibitive financial cost to the institution. This program leverages an already present interest in sustainability on the part of students and instructors, it expands the learning space to the community of stakeholders beyond the walls of the campus and the natural environment in which the university lies. The SFLC can be tailored to meet the needs and limitations of any higher education institution. In this chapter, I will describe the model and its history in the United States, how colleagues and I at my home campus of SUNY New Paltz created an SFLC and have seen substantive, increasing success with the program. The chapter will also outline the steps one might take in building an SFLC at one's home campus even if there are challenges concerning limited time and resources. I hope to persuade readers that this is something concerned campus citizens can create on their campus. The SFLC value to the community far outweighs its costs and offers a straightforward blueprint with which to start the process of creating one's own version of the program on other campuses.

A Brief History of the Sustainability-Focused Faculty Learning Community

The Faculty Learning Community (FLC) model has proven to be effective in promoting professional development in higher education. Defined loosely as a "community of practice" (Engin and Atkinson 2015), the FLC model addresses a number of management problems pervasive in higher education institutions. These problems include the compartmentalization of faculty within specific disciplines, lack of opportunity to openly network and share research with colleagues across departments and schools, and structural disincentives to even temporarily relieve the "expert" label burden in the pursuit of learning something entirely new. Faculty at all levels, particularly junior faculty and adjunct instructors, routinely report time constraints and work creep as they are continually asked to accept more administrative responsibilities, while not seeing a commensurate rise in compensation. Add in the inevitable repetition of teaching a set repertoire of classes, and it is no surprise that faculty burnout and restlessness are common. All this can and often does lead to diminished morale and a sense of collective stagnation.

The FLC model, based on the social theory of learning, brings together interested faculty from different disciplines under the directive of a particular theme or subject matter and can help to address these issues. Four characteristics of professional development make the FLC model generally appropriate in a higher education context with faculty. The FLCs are collective, collaborative, contextualized in a meaningful way, and problem-based (Elgin and Atkinson 2015).

Summarizing previous research on FLCs, Elgin and Atkinson expanded on these four characteristics noting that in faculty learning communities, like-minded participants (here, university faculty and staff) work towards a common goal (Eckert 2006). They develop over the course of the FLC activities a sense of collegiality that includes sharing ideas and advising one another as appropriate (Weller 2009, Wenger 1998). There is also a focus on real-world, timely issues (Eckert 2006, Wenger 1998). The FLC also finds efficacy in being problem-based, with participants collectively working towards concrete solutions (Elton 2009, Kelnowski, Askew, and Carnell 2006). Banasik and Dean (2015) have further observed that the

FLCs that include non-tenure track faculty contribute to those members experiencing an increased sense of community and belonging at the institution as well as exposure to new pedagogy and teaching techniques. One might conclude that the very act of bringing faculty together to share teaching and research ideas under the aegis of investigating a subject of common interest has proven an effective way to promote campus community and professional growth.

It seems a logical step, then, for the FLC model to be used within a sustainability framework, and indeed, there is a meaningful history of this being the case. In 1992, Tufts University introduced the Tufts Environmental Literacy Institute (TELI) program (Cortese 1992), a week-long environmentalism-oriented faculty training program, which ran for five years and saw a lasting impact at the school even after the program itself ended. Twelve years after the training program, Tufts had over 70 TELI-trained faculty members who were still actively engaged in environmental issues, and the school had grown to offer 21 different environmental degrees (Barlett and Rappaport 2009; Natkin and Kolbe 2016).

Building off the example set by TELI, Geoffrey Chase and Paul Rowland began the Ponderosa Project at Northern Arizona University in 1995 (Bartlet and Eisen 2002). Taking cues from Tufts and keeping an eye on sustainability-infused curriculum development as the ultimate goal, they established a two-day program during which faculty would be trained in the fundamentals of sustainability on the first day, and also be offered time to reflect and work on a specific syllabus. Either a revision of an existing class or an entirely new course, during the second day of the workshop would be done. This, in turn, led to Peggy Barlett bringing the Ponderosa model to her institution, Emory University in Atlanta, Georgia, and in 2001, with the guidance of the Northern Arizona University leaders, she created the Piedmont Project, now the gold standard for this sort of sustainability faculty professional development program (Bartlett and Eisen 2002, Eisen and Barlett 2006). Moreover, Barlett and Chase have partnered and team-taught workshops in conjunction with the Association for the Advancement of Sustainability in Higher Education (AASHE) helping faculty and sustainability staff from institutions across the United States to design and promote similar programs at their home institutions. By 2012, hundreds of participants in these workshops, representing over 275 institutions had completed the training and brought the SFLC model back to their home campuses (Barlett and Chase 2012).

Given that different institutions have different needs and thus require specific tailoring of any program in order to best serve their faculty and students, the SFLCs born of the Barlett and Chase workshops vary in scope and duration, among other parameters. However, there are some common elements put forth as goals by the two pioneers (Barlett and Chase 2012):

1. Foster creativity.
2. Emphasize faculty expertise.
3. Build faculty community.
4. Encourage an interdisciplinary approach.
5. Welcome diverse pedagogical approaches.
6. Make connections.
7. Focus on learning outcomes.

Note should be taken that these goals are oriented around encouraging faculty, typically expected to be authorities in any professional setting, to momentarily let go of that encumbrance and once again position themselves as learners, beginners with experience. This taps into what one imagines as the fuel that drives any person to head for a career in teaching and academics in the first place, namely a persistent thirst for knowledge and understanding and the joy that comes with being in a position to satisfy that thirst. Good teachers were first good students, and creating the conditions for faculty to return, if only for a couple of days, to that state, is much of where the value of the experience lies. The SFLC design asks its participants to use the connections, community, and resources provided by the workshop to draw individual conclusions regarding how best to develop their future class plans and syllabi to better align with current sustainability content.

The importance of this dynamic cannot be emphasized enough. Most faculty who begin to sense that class content is being dictated to them will likely and rightfully turn away, protective of their charge. To that end, the Piedmont/Ponderosa model focuses on providing sustainability material and simply creating the conditions for discovery and collegial sharing of techniques and approaches in order to inspire each workshop participant to choose, from this newly presented content, what might fit best in their classrooms. The personal connections made with other faculty who they may not have otherwise met and, in many cases, community partners who wish to engage students in practical projects beyond the walls of the university, is a particularly eye-opening element. As such, participants across many of the now established SFLCs have reported the SFLC as being valuable and applicable to course revision.

The SUNY New Paltz SFLC: A Case Study

In June 2015, two New Paltz faculty members, one from the School of Business and another from the Department of Digital Media and Journalism, attended a week-long program in sustainability leadership at the University of Vermont in Burlington, Vermont (UVM). The faculty did not have extensive backgrounds in sustainability studies. UVM's Professional Certificate in Campus Sustainability Leadership (PCCSL) program trains higher education faculty and staff in the fundamentals of sustainability while showcasing the substantive progress that UVM has made as a campus. Such showcasing includes the overhauling of UVM's energy system to leave a smaller carbon footprint, initiatives like a campus bicycle-sharing system to help ease transportation pressures, and the construction of platinum LEED-certified buildings, the cynosure of their campus infrastructure. The program also emphasizes UVM's connection to the community and local institutions like Shelburne Farms, a school and estate outside of Burlington with a long-standing sustainability mission. The week serves as much tangible case study as an information-based learning experience for those coming from universities that have yet to take similar progressive steps.

Most impressive at UVM is that they have not only been successful in promoting sustainability in the curriculum, but that they have gone so far as to make sustainability coursework a general education requirement for all undergraduate students. This is largely due to the momentum created about sustainability content in the classroom. This momentum has been further fueled by the UVM Sustainability Faculty Fellows Program, their SFLC modeled after the Piedmont/Ponderosa initiatives. Spearheaded by two UVM faculty members who attended one of Barlett and Chase's AASHE workshops in 2009 (Natkin and Kolbe 2015), the UVM program, as of 2015, boasted 103 trained faculty from 35 different academic disciplines, across the schools on campus and including ranks from Lecturer to full Professor. Over the life of the program, trained fellows produced 113 sustainability content-rich courses for the UVM's student population of roughly 7500 students (Natkin and Kolbe 2015).

At New Paltz, a regional public university in upstate New York with a student body of approximately 7500, the two UVM-trained faculty members returned in the fall of 2015 determined to create a similar SFLC-style workshop at home. After presenting the idea to the university Sustainability Committee and receiving approval to develop a formal proposal to be presented to the administration, a sub-committee consisting of three faculty members (including the two who had gone to UVM and the campus Sustainability Coordinator) was formed.

The first order of business was to create the goals for the New Paltz program. The goals are as follows:

1. Create a community of faculty/staff who are committed to infusing interdisciplinary approaches to environmental sustainability into the curriculum.
2. Starting with the UN SDGs and a systems-thinking approach, enhance the understanding of environmental sustainability concepts among faculty and students, particularly those not trained in environmental fields.
3. Explore teaching and course design strategies that will engage students in sustainable development from a multidisciplinary approach.
4. Encourage and fund curricular development without content prescription from above.

Recall that in 2015, the UN released the updated Millenium Development Goals (MDGs) as the now 17 UN SDGs, and as we created the program at New Paltz, it made sense to use these SDGs as the conceptual foundation for the program's training. The SDGs were preferred for their comprehensiveness and global application and their immediate usefulness and relevance to the world beyond the university. In this context, one of the great assets of the SDGs is their breadth—almost any teacher can come to the SDGs fresh and find their work and field of study located somewhere within the seventeen SDGs. This means that training such that an SFLC offers can be structured as widely inclusive from the start, and the training can take advantage of the many resources that the

UN and United Nations Educational, Scientific, and Cultural Organization (UNESCO) have already published with respect to the SDGs. This provides a framework of action that is systems-based from the start, and in the realm of curricular development, UNESCO's document on learning objectives is particularly useful (Tang 2017).

We also chose to make explicit the founding ethos of the training, which is to avoid content prescription and not require participating faculty to teach any particular content or to teach using any specific pedagogical styles. We would ask for concrete curricular change but structure the training to allow each participant to decide what to change and how to include the SDG content in their work. We set expectations for the faculty participants in the following way:

- Attend four program workshops/breakfasts across the academic year, two each semester.
- Participate in a two-day training institute during the winter session in January.
- Redesign or develop new course material to integrate the concepts of sustainability.
- Evaluate the redesigned course to gauge student attitudes and knowledge regarding sustainability.
- Demonstrate enthusiasm for developing interdisciplinary partnerships in sustainability teaching and scholarship.

Our program, based on UVM's adaptation of the Piedmont/Ponderosa model, consists of two workshops each semester (consecutive fall and spring) that sandwich the centerpiece of the program, a two-day January Institute during which the participants, or "sustainability faculty fellows," receive the lion's share of their training in sustainability content and resources. Faculty participants at New Paltz have had varying levels of experience with sustainability in general, so the two meetings in the fall introduce the fellows to one another and are geared towards laying out basic sustainability concepts. The two meetings in the spring reinforce the work done in January and prepare the fellows to present their revised syllabi at the end of the school year.

The January Institute is divided into one day of intensive learning led by a sustainability expert during which the cohort of program fellows, as well as community partners are invited to participate in the workshop agenda and a second day located off-campus. During the second day, fellows are given the opportunity to reflect on the previous day's content and begin to think through what they want to do with their course revision or class development in adding sustainability content.

To help incentivize participation from already time-and energy-taxed faculty and staff, fellows receive a $250 stipend upon completion of the training and the submission of their revised syllabus. Fellows also have access to additional funding for new interdisciplinary team-taught classes to come out of the training with a sustainability focus. Offering financial compensation is a common element of SFLCs based on Piedmont/ Ponderosa model, though the amount offered to participants can differ quite a bit. The Piedmont workshop at Emory offers participants $1000, whereas Santa Clara University, where their Penstemon Project follows the Piedmont example, offers $500 to each member of each cohort (Barlett and Chase 2012). At New Paltz, given a strict budgetary climate and a decision on the part of the organizing sub-committee to place less emphasis on the financial compensation and more on the professional development opportunity, we set our stipend a bit lower than other programs have. Initially, we considered having no stipend at all. However, we decided in the end that even a nominal amount of money provided a good-faith gesture to applicants that was likely to be worth the outlay.

This does touch on the question of why faculty would voluntarily take on additional training such as that put forward by an SFLC. The short answer for us is that there was a latent interest among faculty and staff on campus in becoming more active concerning sustainability, and that interest exists as a great untapped reservoir of positive change and action. Many faculty members have a deep personal interest in sustainability, and once introduced to the SDGs were eager to dive deeply into that content. However, many were sure how to bring that interest to their research or the classroom. That the program has steady interest and seen continual positive results speaks to the value of providing the opportunity for instructors to connect personal values with professional development. In other words, the monetary benefit helps, but was never seen as the primary appeal of the program. Had it been, surely the program and the others that had inspired it would have failed long ago.

With the initial proposal of two years of funding for up to 15 faculty and staff participants per cohort per yearly, the sub-committee presented an official proposal to the Provost's Office, who after a round of minor revisions approved the project, including funding for fellow stipend remuneration and funding for course development. In the end, the New Paltz SFLC is an endeavor supported by the Office of Sustainability, the Sustainability Committee, and the Office of the Provost.

To gain the Provost's support was crucial not only for the budgetary help, but also for the imprimatur of the administration in the eyes of those who would become fellows. Some of the hesitance to branch out individually and move beyond silos for faculty seems born of a feeling that to do so sees little professional reward. When asked about barriers to progress in developing sustainability-related coursework and working beyond departmental needs, fellows have mentioned "institutional inertia/resistance," "bureaucracy," and "no institutional framework or support for community collaboration at the university level." To have the stamp of approval from the Provost announced to the faculty at New Paltz that not only was it safe to spend time branching out and learning something completely new but that it was officially encouraged. This opened the doors for many to apply to the program and ultimately receive the training.

After a successful two years of the program at New Paltz, the Provost has reinstated funding for a third year, including supplemental funding for the development of interdisciplinary team-taught classes.

Assessment of the Program and Results

As the January Institute has served as the focal point of the training, we surveyed fellows after each retreat in order to assess programming and ultimate pedagogical and material value for the participants. The results show positive responses from the majority of the fellows primarily along the vectors of increased knowledge of sustainability concepts, interpersonal connection with other faculty and staff, and introductions to potential community partners (Table 26.1). Respondents were asked to reply to each statement with a value of 1 = strongly disagree, 2 = disagree, 3 = neither agree nor disagree, 4 = agree, 5 = strongly agree.

Additionally, the survey solicited comments from the fellows asking both what they found most valuable about the training and what they thought might be improved with the next iteration of the program. Some representative comments regarding the positive value of the retreat are shown in Box 26.1.

Table 26.1 Participants' response to the Sustainability Faculty Learning Community

Question	*2017 Retreat (n = 16)*	*2018 Retreat (n = 14)*	*2019 Retreat (n = 13)*	*Total Average*
The retreat aided in my understanding of sustainability in general and the specific sustainability issues relevant to my teaching	4.56	4.00	4.27	4.28
The retreat aided me in shaping the course I intend to develop/revise with sustainability content	4.63	3.67	4.00	4.10
The retreat exposed me to new information and resources	4.94	4.67	4.82	4.81
I met colleagues whom I might otherwise not have met	4.94	4.44	4.64	4.67
I am interested in learning more about sustainability education	4.56	4.56	4.50	4.54
I am interested in learning more about sustainability initiatives and trends on campus outside of the academic division	4.75	4.60	4.60	4.65
The retreat met my expectations	4.81	4.30	4.80	4.64

Box 26.1: Respondents assessment on the value of the training and what could be improved

"An excellent experience that transformed how I'll motivate my students in content-rich information design courses. "

"Thank you for hosting this! It was great to connect more with the academic faculty because so many times we do not have the opportunity to share what's going on outside of the classroom!"

"...I was able to establish internal sustainability-based collaborations, which I never had the occasion to initiate in over two years at New Paltz. Being in the same room with the right people fueled discussions which will likely lead to joint work on sustainability problems. ...It raised my spirits and allowed me to redirect my frustrations."

"I enjoyed learning more about community partners and about how to integrate sustainability into my courses."

"I liked how there was a larger number of people from the community and other institutions than I expected."

"The opportunity to share ideas and hear what other faculty are thinking about doing with respect to integrating sustainability content in their classes."

"The interaction with my colleagues from all over campus. I learned a lot beyond the intent of the workshop."

"Was very inspiring and increased my enthusiasm for teaching economics. In the context of the SDGs, economics takes on more meaning."

"I loved hearing the stories of the community members as well as the professors who are trying to bring the issues of sustainability into their classrooms."

"This is a wonderful initiative. I hope it continues to grow with leaps and bounds."

"I'm honored to be part of a campus that values sustainability to such a high degree and have been so impressed with the work that the Sustainability Office is doing."

Common themes to emerge from these comments are that the program and the two-day January retreat, in particular, succeed in boosting faculty morale. This comes out through assembling faculty bonded by a common interest in sustainability and by providing the learning context within which to share ideas, pedagogy, and resources. Access to community partners with whom student projects and events can be designed and used in teaching and learning is another cited positive as it presented opportunities for practical application of the SDGs and sustainability principles in general. These results correspond with those found by Peggy Bartlett and Ann Rappaport in their comprehensive assessment of the TELI and Piedmont projects where they found both programs changed participants' teaching and knowledge of environmental and sustainability issues substantially (Barlett and Rappaport, 2009). Critical survey comments tended to focus on the availability of resources and support beyond the two-day retreat and logistical concerns regarding the organization of the two days. Further responses from the participants are presented below.

> Introductions to the community people were rushed, and it would have been good to have more time to hear from everyone. It would have helped make better connections.

> The two days were overloaded with content...sometimes a bit too much time sitting. It would be refreshing to break out into 'think-pair-share' small groups with specific assignments for discussion or something specifically directed.
>
> Since a lot of the ideas presented were new to me, I needed more time to comprehend them.
>
> There was a bit too much information given a bit too quickly.
>
> I would have preferred getting information sooner (prior to the morning of) so I had more time to prepare....

These comments have helped us think about scheduling and small programmatic changes with each new iteration of the program. The observations further assist us on how to strike the right balance between a broader fundamental approach for those new to sustainability and more targeted delivery of resources for those already well-versed in sustainability issues. During 2017–2018, our second year, the cohort included some faculty members who already work steadily in sustainability, and their response to the training was more muted, that while the opportunity to network with others still led to a positive overall experience, the information and resources provided felt too basic for their needs. In the other years' cohorts, the training seemed to be closer to the aggregate sustainability experience level of the fellows.

Finally, we asked participants about future intentions having come through the January Institute training and received many positive responses indicating concrete plans. The responses are shown in Box 26.2. We saw enthusiasm and ambition for changes in the classroom and in research agendas given the SDGs as a guiding framework.

Box 26.2: Required future intentions after experiencing the January Institute training

"...I plan to push my department to introduce a concentration in sustainability for elementary ed majors."

"I would love to travel to some more sustainability training that has specific links with psychology and education."

"In one of my modified classes, I plan to ask students to work in small groups to take action on a local sustainability issue. My intention is to have them report on their success or hand over the project to a group in the subsequent semester. I think this could be a really good study about what teachers do to help students see that their actions make a difference."

"I would like to investigate the PFOA/S water problem in [neighboring] Hoosick Falls and Newburgh."

"If we can help my students identify a local business/organization that has a product or service for sustainability, and could potentially be marketed overseas, it'll be great help for my MBA class project."

"I am arranging a meeting with the local watershed alliances to initiate new projects."

"Work on aspects of communicating climate change information."

"Continue to pursue sustainable natural materials for industrial application."

"...working on community currencies and how systems can generate positive feedback loops on income generation."

Regarding curriculum results from the two completed years of our SFLC, the program has produced 27 revised or newly developed courses (as verified by fellows submitting syllabi to a common SFLC online repository). The submitted syllabi come from across the university's schools (LA&S, Engineering, Education, Fine Arts) and disciplines (including Mathematics, Digital Media and Journalism, Geography, Economics). Additionally, six SFLC fellows have been funded to develop entirely new sustainability-infused courses, with the first team-taught class to come out of the program being offered to students during the fall of 2018. An additional 13 revised or new sustainability-infused courses as well as proposals for developing new interdisciplinary team-taught courses were expected in the next cohort.

Proposing and Establishing an SFLC at Your Institution

Given our experience at New Paltz in developing the proposal for a Piedmont/Ponderosa-style SFLC on our campus, some thoughts on how to approach developing a similar program at your home institution including suggestions as to how to position a proposal when speaking with potential funding sources or administrative heads can be offered.

First, it is worthwhile to at least informally assess the local conditions to see if they are conducive to being receptive to this sort of program. Meaning, it is important to get a sense that there is some of the aforementioned latent interest, from faculty and administration, in increasing sustainability content in the classroom. The SDGs framework remains key to such endeavors. In the majority of circumstances, you will be able to find such interest simply by talking with faculty or, as we did in our case, speaking with our Sustainability Coordinator who already had a good sense of the campus dynamics and temperature. This is an important step because you will likely have to find a group of committed faculty or staff facilitators who are willing to the legwork to get the program off the ground, both in the proposal stage and the active programming stage. For our facilitators, this has meant service in addition to our normal workloads as faculty and sustainability staff, so it is good to know going in that such will be required and that program leaders will have to take that into account given their other responsibilities.

Once you have a group ready to develop a proposal, and depending on whose approval you seek, you need to tailor your argument accordingly. Given that New Paltz does not have the financial resources that many private institutions have (e.g., a large private endowment, high student tuition to help fill the coffers), we knew that for an SFLC to be approved we would have to demonstrate substantive returned value for a relatively low capital investment as well as programmatic relevance to the entire university, not simply a select campus group. To that end, having the SDGs to point to as the framework for the training was of great help because, again, the UN goals are comprehensive by design and allow any faculty member/potential SLFC participant to locate themselves somewhere within their parameters. To reiterate this key point: a strong rationale for centering the SDGs within an SFLC is that by their very design the SDGs invite inclusion in action and projects already underway in both private and public sectors, and they serve as a very accessible *communicative* starting point, one that promotes *ipso facto* a systems-thinking approach and easy cross-disciplinary exchange of ideas and collaboration. That the SDGs have been accepted globally as universal objectives connects even those new to sustainability with much larger global context providing real-world immediacy for faculty and, ultimately, for their students.

Related to this universality point, we also argued the inevitability of higher education moving towards sustainability-related pedagogy if only because the world is quickly approaching a tipping point with climate change (Masson-Delmotte, V., P. Zhai, H.-O. Pörtner, D. Roberts, J. Skea, P.R. Shukla, A. Pirani, W. Moufouma-Okia, C. Péan, R. Pidcock, S. Connors, J.B.R. Matthews, Y. Chen, X. Zhou, M.I. Gomis, E. Lonnoy,

Maycock, M. Tignor, and T. Waterfield (eds.) 2018, Watts 2018, Wu 2018) and societal systems across the board will have to be reconsidered in the context of a warming planet. Though the academy is slow to embrace institutional change, the issue will be forced by a world that is evolving faster than the pace of academic adaptation. A concurrent shift in curricula that will allow schools to deliver the practical, interdisciplinary skills that students will need as they join a global community faced with previously unaddressed culturally embedded problems is unavoidable. Thus, ways to begin to push that sort of curricular change will need to be found, and efforts like these, put forth and tried. The time to foster creative change is now, in other words, and additionally, the pressure will most certainly come from the bottom up.

Student demand for courses and majors that focus on sustainability topics is rising, and again given the SDGs as a starting-point framework, this includes not only environmental and ecological disciplines, but also related sustainable economic, political, and social justice areas. Generational research on the attitudes of Millennials (ages 22 to 37) and Generation Z (under the age of 27) shows an increasing belief in government and public policy to solve problems, higher levels of belief in man-made climate change than previous generations, and an increasing acceptance and expectation of diversity, inclusion, and social equity as societal norms (Parker, Graf, and Igielnik 2019). These attitudes will translate into prospective students looking for schools that offer curricula that align with that expectation and understanding. Institutions that want to continue to attract the best and brightest will do well to develop corresponding coursework if it is not already in place.

The potential penalty for not getting ahead of that curve will be to see applications and enrollments drop. The benefit of actively developing the corresponding course work, however, is the potential to position the institution as a regional sustainability leader, or in coarser terms, to boost the brand of the school. Some universities have already taken the lead in this respect, partnering with AASHE to showcase their respective sustainability-focused faculty training programs as AASHE Centers for Sustainability Across the Curriculum (CSAC). Achieving a reputation as a school that has ample offerings in sustainability has the potential to be a key marketing point for any higher education institution, the long-term recruitment value of which would far outweigh the nominal short-term capital investment in an SFLC program.

Tied to student demand is satisfying the demand on the part of the faculty to introduce sustainability in their teaching. We found that especially junior faculty expressed interest in the training offered by the SFLC, which again, is not that surprising given the research expressing the preferences of Millennials as noted above. This is not to discount or exclude senior faculty but more to reiterate the larger point that in creating an SFLC, one creates the conditions for faculty with a personal interest in sustainability and the SDGs to bring that interest to their professional work and collaboration in an institution-supported context. What we have discovered, also confirmed by Barlett and Rappaport, is that much of what seems to be missing from business as usual in the academy is an opportunity for interdisciplinary conversation, events that capitalize on the wealth of expertise and experience of any school's greatest human resource, its teachers. In the context of the SDGs and sustainability in general, the coming together of the experts on campus has a multiplier effect of shared knowledge, affirmation of values, and genuine collegiality and cross-fertilization of ideas with a deeper connection to the practices going on globally.

Finally, we presented the results as reported by other institutions, beginning with the UVM SFF program (since this was the one we most directly modeled our program on), but we also pointed to the history of success nationally as outlined above. Again, since we needed to demonstrate a likely high return on investment, having so many success stories to point to was invaluable in validating our proposal. To this end, we owe much to Cortese, Barlett, Chase, et. al., our pioneer predecessors, in leading the way and showing that simply creating the conditions for faculty to move confidently in a particular direction can have a hugely positive and lasting impact on an institution, even if the program that fostered the sustainability or environmental learning itself eventually disappears (Barlett and Rappaport 2009).

Conclusion

The future of the sustainability-based faculty learning community is one that will hopefully see continued growth in higher education on campuses globally. As discussed in this chapter, in 2017, AASHE partnered with a group of universities in the United States (with now also a center in Hong Kong) to serve as centers for sustainability-based learning for college instructors. At these centers, faculty from neighboring schools can receive training to bring back to their institutions, furthering progress in delivering sustainability content to university classrooms.

Moving forward, there is call for further research into a standardized means of assessing SFLCs as they continue to proliferate in higher education (Natkin and Kolbe 2016). With more institutions creating similar programs, a common means of evaluating success would be greatly valuable, and as of this writing, AASHE and some of their CSAC leaders are gathering information from institutions in an effort to understand commonalities across programs.

However, perhaps Dickinson College in Carlisle, PA, one of the AASHE centers, reflects the next stage of advancement where they have no discrete major or program in sustainability but instead have reached a level where sustainability is so infused throughout the school's curriculum that there is no need for a separate academic course identifier or study requirement—sustainability is simply considered to be ingrained in all fields of undergraduate study. A wonderful example to help us see what is achievable with strong leadership, vision, and a bit of persistence, an aspirational goal for all of us.

The sign of a good idea is that it grows legs, and the SFLC model, has been up and running for some time now. I encourage you to review the literature for additional data and results, but the bottom line is that the reason that you'll see such consistently positive reports from schools that have adopted the program is that it's simply a good idea for which, again, the credit must go to the pioneers who first gave it a shot. It is now vetted, time-tested, and up to you to bring the idea to your campus. After all, sustainability requires not only research, but the application of that research towards progress, concrete action. All it takes is for a few people to step up and decide to do it. As former chess world champion and political activist Garry Kasparov likes to say, "If not you, then who?"

References

Axelrod, T. 2019, February 12. Klobuchar on Green New Deal: 'I see it as aspirational.' The Hill, website. https://thehill.com/homenews/campaign/429719-klobuchar-on-green-new-deal-i-see-it-as-aspirational. Accessed 24 April 2019.

Banasik, M. D., and J. Dean. 2016. Non-Tenure Track Faculty and Learning Communities: Bridging the Divide to Enhance Teaching Quality. *Innovative Higher Education* 41: 333–342.

Barlett, P. F. 2008. Reason and Reenchantment in Cultural Change: Sustainability in Higher Education. *Current Anthropology* 49, 1077–1098 (October 5, 2018). https://doi.org/10.1086/392425

Barlett, P. F., and A. Rappaport. 2009. Long-Term Impacts of Faculty Development Programs: The Experience of TELI and Piedmont. *College Teaching* 57 (2): 73–82.

Barlett, P. F., and G. W. Chase. 2012. Curricular Innovation for Sustainability: The Piedmont/Ponderosa Model of Faculty Development. *Liberal Education* 98(4). https://www.aacu.org/publications-research/periodicals/curricular-innovation-sustainabilitythe-piedmont-ponderosa-model Accessed 27 September 2018.

Barlett, P. F., and G. W. Chase. 2013. *Sustainability in Higher Education: Stories and Strategies for Transformation*. Cambridge, MA: MIT Press.

Barlett, P. F., ed. 2005. *Urban Place: Reconnecting with the Natural World*. Boston: MIT Press.

Dich, L., K. M. Brown, J. H. Kuznekoff, T. Conover, J. P. Forren, and J. Marshall. 2017. Growing Lemon Trees from Lemons: Lessons Reaped from a SoTL Faculty Learning Community's Research 'Failures.' *Journal of the Scholarship of Teaching and Learning* 17 (4): 1–16.

Dmochowski, J. E., D. Garofalo, S. Fisher, A. Greene, and D. Gambogi. 2016. Integrating sustainability across the university curriculum. *International Journal of Sustainability in Higher Education* 17 (5): 652–670.

Eisen, A., and P. Barlett. 2006. The Piedmont Project: Fostering Faculty Development toward Sustainability. *Journal of Environmental Education* 38 (1): 25–38.

Engin, M., and F. Atkinson. 2015. Faculty Learning Communities: A Model for Supporting Curriculum Changes in Higher Education. *International Journal of Teaching and Learning in Higher Education* 27 (2): 164–174.

Hurney, C. A., C. Nash, C. B. Hartman, and E. J. Brantmeier. 2016. Incorporating sustainability content and pedagogy through faculty development. *International Journal of Sustainability in Higher Education* 17 (5): 582–600.

Kaza, S., L. W. Natkin, and T. Rowse. 2016. Developing sustainability leadership through faculty professional development. *Journal of Environmental Studies and Sciences* 6 (2): 437–444.

King, L. 2019, March 14. Pelosi call Green New Deal 'a list of aspirations,' and her Democratic party isn't sold on it either. *USA Today*, website. https://www.usatoday.com/story/news/politics/2019/03/14/green-new-deal-democrats-split-strategy-address-climate-change/3164072002/. Accessed 24 April 2019.

Masson-Delmotte, V., P. Zhai, H.-O. Pörtner, D. Roberts, J. Skea, P. R. Shukla, A. Pirani, W. Moufouma-Okia, C. Péan, R. Pidcock, S. Connors, J. B. R. Matthews, Y. Chen, X. Zhou, M. I. Gomis, E. Lonnoy, Maycock, M. Tignor, and T. Waterfield (eds.). 2018. IPCC, 2018: Summary for Policymakers. In: Global Warming of 1.5°C. An IPCC Special Report on the impacts of global warming of 1.5°C above pre-industrial levels and related global greenhouse gas emission pathways, in the context of strengthening the global response to the threat of climate change, sustainable development, and efforts to eradicate poverty. Resource document. World Meteorological Organization, Geneva, Switzerland, 32 pp. https://www.ipcc.ch/sr15/chapter/summary-for-policy-makers/. Accessed 24 April 2019.

Natkin, L. W., and T. Kolbe. 2016. Enhancing sustainability curricula through faculty learning communities. *International Journal of Sustainability in Higher Education* 17 (4): 540–558.

Parker, K., Graf, N., Igielnik, R. 2019. Generation Z Looks a Lot Like Millennials on Key Social and Political Issues. Resource document. Pew Research Center. http://www.pewsocialtrends.org/2019/01/17/generation-z-looks-a-lot-like-millennials-on-key-social-and-political-issues/ Accessed 17 January 2019.

Rands, M. L., H. Bender, M. T. Gillette, and L. Orgler. 2017. The Role of Faculty Learning Communities in Supporting Team-Based Learning. *The Journal of Faculty Development* 31 (3): 61–78.

Rusinko, C. A. 2010. Integrating sustainability in higher education: A generic matrix. *International Journal of Sustainability in Higher Education* 11 (3): 250–259.

Tang, Q. 2017. Education for Sustainable Development Goals: learning objectives. Resource document. UNESCO Digital Library. https://unesdoc.unesco.org/ark:/48223/pf0000247444. Accessed 24 April 2019.

Vogel, J. R., and A. T. Rogers. 2017. Bridging the Gap with Faculty Learning Communities: Creating Community between Onsite and Offsite Faculty. *Journal of Higher Education Theory and Practice* 17 (9): 81–91.

Watts, J. 2018, October 8. We have 12 years to limit climate change catastrophe, warns UN. *The Guardian*, website. https://www.theguardian.com/environment/2018/oct/08/global-warming-must-not-exceed-15c-warns-landmark-un-report. Accessed 24 April 2019.

Wu, K. 2018, October 8. The Worlsd Was Just Issued 12-Year Ultimatum On Climate Change. *Smithsonian.com*. https://www.smithsonianmag.com/smart-news/world-was-just-issued-12-year-ultimatum-climate-change-180970489/. Accessed 24 April 2019>

CHAPTER 27

Focus Groups for Evaluating Student Perspectives on Sustainability Education

Jessica Kate Crowell and Kate Bellody

Introduction

While the path to achieving sustainability may not be easy or well-defined, scholars and practitioners agree that higher educational institutions (HEIs) can—and should—play a role in this process (Bergstrom and Lovejoy 2023; Caiero and Azeiteiro 2020; Cortese 2003; Stephens and Graham 2010). The potential impact of HEIs is significant given their vast size: according to UNESCO (2023), HEIs worldwide currently serve over 235 million, a number which is only expected to grow in the coming years (UNESCO n.d.). Yet, while higher educational institutions increasingly recognize the importance of building a sustainability-focused

J. K. Crowell
Department of Digital Media & Journalism, SUNY New Paltz, New Paltz, USA

e-mail: crowellj@newpaltz.edu

K. Bellody (✉)
Sojourner Truth Library, SUNY New Paltz, New Paltz, USA
e-mail: bellodyk@newpaltz.edu

S. M. Wyman and J. Korenblat (eds.), *Sustainability Learning for Action and Community Engagement*,
https://doi.org/10.1007/978-3-032-11698-7_27

curriculum, previous research has demonstrated that there are persistent gaps in students' knowledge about sustainability and that students have very different definitions of sustainability as a concept (Décamps et al. 2017; Kruger et al. 2020; Owens and Legere, 2015; Stough et al. 2018). Given these gaps and challenges, researchers have advocated for varying approaches to assess the impact of sustainability education on student experiences and learning, such as creating sustainability "competencies" or developing assessment tools such as the "ASK" (Assessing Sustainability Knowledge) model which tests knowledge on issues like climate change (Brundiers et al. 2021; Viegas et al. 2016; Zwickle et al. 2014). Research also indicates that informal learning opportunities play a key role in shaping student knowledge of sustainability (Alm et al. 2022; Gramatakos and Lavau 2019; Kruger et al. 2020). Therefore, a challenge of assessing the impact of sustainability education is that sustainability itself cuts across disciplines and that students' experiences with sustainability do not fit neatly into existing campus borders or boundaries.

Given the varied and intersecting dimensions of sustainability education, this chapter advocates for a qualitative intervention in sustainability education research to help us better capture the full impact of sustainability educational experiences. We argue qualitative research can help generate more detailed and nuanced information regarding what our students are learning—and what they value—regarding sustainability. Our project asserts that the focus group methodology aligns well with a student-centered research perspective to help us understand and evaluate the impact of sustainability educational opportunities on learning. Positioned within the social information world students inhabit, focus groups can be especially useful in the evaluation of shared "knowledge or attitudes about curriculum" or the "effectiveness" of sustainability education (Hennink 2014; Palmer, Boardman, and Bauchner 1996; Pugsley 2018; Williams and Katz, 4). Moreover, as the focus group methodology mimics a classroom discussion format, it can facilitate deeper student reflection and engagement with sustainability content. We also emphasize that guided focus group activities like concept mapping—which have been used in previous sustainability education research such as Lourdel et al. 2007—can be productive reflection tools to help students map the complex dimensions of sustainability. This chapter also argues that qualitative research is well-positioned to collect data on how sustainability may support other institutional values and priorities within higher education, such as efforts to advance diversity, equity, and inclusion (DEI).

While this chapter calls for a qualitative approach to understanding the impact of sustainability education on students, it also intends to offer a practical roadmap for researchers who are interested in using focus groups to understand student perspectives on sustainability education at their own institutions. Accordingly, we begin with an overview of the sustainability education offerings at our own US university and outline how our institutional needs guided the research framework, research questions, and research design of our focus group project. Second, we discuss focus group methodology, including its history as a research tool, offer concrete advice on how to run a focus group, and reflect upon ethical considerations. Third, we review concept mapping as a tool for assessing student learning within the focus group setting, including how it has been used effectively in previous education research to capture complex sustainability relationships. Finally, we present two of the major findings from our research study, specifically evidence that showcases the unique strengths of a qualitative lens and the focus group methodology. We then offer concluding thoughts, including a brief discussion of some of the limitations of the focus group approach that should also be considered when designing a sustainability education research study. Having reviewed the key points of this chapter, we turn to a discussion of the research site and design which guided our own sustainability education research project.

Research Site and Design

Sustainability is a valued principle within our home university, SUNY New Paltz, and a distinguishing feature of our campus curriculum. In addition to offering Environmental Studies as a major and minor academic program, our campus boasts a vibrant and robust campus Sustainability Faculty Learning Community. The learning community supports a related Sustainability Fellows Program, composed of over 90 fellows, that supports faculty and staff in integrating sustainability into their courses or programs. Recently, our university pioneered a new sustainability course designation to identify sustainability-related and sustainability-focused courses within our undergraduate course catalog (Hong 2026). Notably, our university has a dedicated campus Sustainability Coordinator who works to advance sustainability within our campus and an undergraduate Sustainability Ambassadors program to engage students in sustainability efforts. The campus hosts a range of relevant speakers and events, such as the annual Earth Day celebration. We currently hold an institutional membership with the leading global association working towards advancing sustainability within higher education, the Association for the Advancement of Sustainability in Higher Education (AASHE), and our faculty and staff are a visible presence within the organization. Indeed, our commitment to sustainability is crystallized in our campus mission statement which affirms that the "knowledge and ability to engage others in progress toward a more sustainable world" is an institutional learning outcome for all SUNY New Paltz students (SUNY New Paltz n.d.). In short, our university has a unique investment in sustainability education which it upholds as a core value.

Yet, despite our university's deep engagement with sustainability and rich course offerings on sustainability topics, we had limited research insight into the impact of these programs on students. We wondered: *do students value our sustainability efforts and educational experiences? Are students able to make connections between sustainability and our institutional values, specifically that of diversity, equity, and inclusion (DEI)?* To answer these questions, we designed a qualitative research project, "Evaluating Sustainability Education at SUNY New Paltz," with several key goals:

- Create a student-centered definition of *sustainability* and better understand what students value about campus and community sustainability efforts.
- Conceptualize how sustainability was integrated into the undergraduate educational experience from the student perspective.
- Examine the value and impact of the sustainability educational experience to students and/or the campus community.
- Map student connections between different campus sustainability efforts, for example, taking a general education sustainability-related course or attending a sustainability campus event.
- Learn whether students drew connections between sustainability and our campus's broader efforts to promote DEI.

The research questions for this project were:

1. RQ1: What is "sustainability," and why is it important to students?

2. RQ2: How has sustainability been addressed and/or integrated into the student educational experience (such as course curriculum or campus events, etc.)?
3. RQ3: What is the impact of sustainability education on the SUNY New Paltz student experience and/or students' future goals?
4. RQ4: How do sustainability concerns connect with DEI from the student perspective? How do sustainability frameworks illustrate the potential for equal access to rights, opportunities, and protections on the local *and* global levels?

To answer our research questions, we utilized a qualitative focus group methodology, selected due to its unique ability to harness group interactions to generate shared social understandings of terms and concepts, such as producing a student definition of *sustainability*. For our recruitment strategy, we were interested in assessing the wide impact of sustainability educational experiences. We wanted the composition of our focus groups to represent the diversity of sustainability education among our undergraduate student population. Accordingly, recruitment targeted students who participated in sustainability-designated classes, sustainability-related events, and other relevant campus groups.

Ultimately, we were successful in our recruitment efforts, with a total of 33 participants involved in the study. We held four focus groups in February 2024. Our participants had diverse sustainability educational experiences. We had participation from students in all five of our university's major academic schools: Business, Education, Fine and Performing Arts, Liberal Arts and Sciences, and Science and Engineering.

To speak to our ethical considerations, this project was approved by the university's Human Research Ethics Board. All participants received a twenty-five dollar gift card. It is important to note that there are additional important ethical considerations with the focus group methodology, which will be discussed in the following section. Having sketched out the structure and approach to this project, we will turn to a discussion of the utility of the focus group methodology in studying the impact of sustainability education on students.

Focus Groups: History, Utility, and Ethical Considerations

What are focus groups and how can they be used to evaluate the impact of sustainability education? Focus groups are generally composed of 6–8 members who are selected based on some similar characteristic, capacity, or knowledge (Hennink 2014). The focus group first emerged as a methodological tool in social science media effects research of the 1920s and 1930s, most famously utilized in the foundational work of mass communication scholars such as Lazarsfeld and Merton (1948). However, decades later in the 1950s, the methodology became highly popularized as a tool for marketing and consumer-focused research.

Today, focus groups are an established tool used in the social sciences across a range of disciplines (Barbour and Kitzinger 1998). One strength of this methodology is that it can be "highly efficient" allowing researchers to gather perspectives from multiple group members in a relatively short period of time (Patton 1990, 335). Focus groups can also capture nuances of the human experience and shared knowledge that cannot be easily quantified, such as, in our case, student definitions of sustainability (Berg 2001; Williams and Katz 2001). Focus groups are "particularly useful for exploring people's knowledge and experience and can be used to examine not only what people think but how they think and why they think that way" (Kitzinger 1995, 299). Therefore, for our research project, utilizing focus groups would not only help us create a student-centered definition of sustainability but also understand *how* and *why* students hold those definitions. Madriz (2003) succinctly argues that focus groups are a "collectivist rather than an individualistic" methodology, for example, based in the social world (836).

Focus groups can also be useful research tools for studying the impact of sustainability education on student learning and experiences. Focus groups are well suited to evaluating a "service, program or intervention" because they center group experiences, allowing researchers the opportunity to observe social processes (Hennink 2014, 16). Moreover, scholars have argued that the focus group methodology is especially well tailored to evaluate student views on curriculum and its perceived effectiveness (Palmer, Boardman, and Bauchner 1996; Pugsley 2018). Similarly, Williams and Katz (2001) argue that focus groups may even be the *best* methodology for evaluating educational curriculum and related programs, stating:

> Focus groups may also generate rich data that can facilitate decision-making and provide useful information for the development, evaluations, and modification of curriculum, learning tools, and programs – information that might not be accessible from other research methods. (7)

Simply put, focus groups "generate rich data" on student attitudes and unique perspectives, often the information most useful to curriculum planning and evaluation.

As they mimic a classroom discussion format, focus groups can also be a more comfortable way for students to share their thoughts on educational services, a particular curriculum, or other academic programs because they place less pressure on an individual student to respond compared to one-on-one interviews. Another significant benefit is that focus groups can be empowering or validating, as they allow research participants to feel that their ideas are being valued and held as important (Byron 1995). In short, students who participate in sustainability education focus groups not only generate valuable information, but they may also learn, connect, and grow as change agents through their participation in the research process.

Having sketched out the utility of focus groups for an evaluation of sustainability education, it is important to provide practical information on how to approach, organize, and facilitate a focus group. According to Merton, Fiske, and Kendall (1990), a focus group moderator should try to uncover a range of perspectives with depth and specificity. At all times, the researcher should not be centered in the conversation, as the goal of a focus group is to capture the interactions between participants (Johnson 1996; Kitzinger 1994; Parker and Tritter 2006). Therefore, the goal of a moderator is to harness the power of the group dynamics in such a way as to generate "a kind of momentum" among the participants, promoting a deeper exploration of themes and ideas (Parker and Tritter 2006, 26; Kitzinger 1995). Even disagreements between participants can help other group members clarify their own opinions and perspectives; however, in these cases, it is incumbent upon the moderator to help the group maintain a positive, constructive tone (Kitzinger 1995, 301). Whenever possible, a moderator should also try to understand the personal context that frames the participant's response, which may sometimes require asking probing or follow-up questions (Merton, Fiske and Kendall 1990).

Despite the core strengths of the focus group methodology, there are ethical challenges unique to focus groups that researchers must consider before selecting this approach for their own sustainability education research (Sim and Waterfield 2019). Because participants are interviewed in groups, it is extremely difficult for researchers to ensure full confidentiality and anonymity, as other participants could easily communicate responses that were shared during the focus group. One potential way that researchers can mitigate this risk is to remind participants of the need for privacy of responses not only at the start of the focus group but at relevant points throughout the group session as well (Sim and Waterfield 2019). As another ethical point, even though focus group participants are not intended to reflect a randomized sample, to ensure the integrity of the research process, researchers must ensure that recruitment is intentional (Krueger and Casey 2009; Parker and Tritter 2006). Researchers should clearly articulate what types of participants are needed and why they are needed at the outset of the focus group. Ultimately, the research questions and scope of the project should govern the recruitment strategy and participant selection process.

The focus group format is useful in evaluating the impact of sustainability education on the student experience because it gives participants a social space to reflect on ideas and concepts. Yet, as instructors also know, hands-on exercises and small breakout group discussions can promote deeper engagement with complex material in group settings. Therefore, building on previous sustainability education research conducted by Lourdel et al. (2007), we also engaged our focus group participants in a concept mapping exercise where we asked them to chart any potential connections between sustainability and DEI. Overall, this was an effective strategy, as it painted a richer portrait of students' understanding of sustainability that may not have been apparent through discussion alone. Furthermore, previous research suggests that students may not draw clear connections between sustainability and social justice in quantitative approaches like surveys (Kruger et al. 2020). While we did provide participants with a sample concept map on an unrelated topic to illustrate the structure of this approach, we elected to not explain our choice to use them with the participants to ensure that their approaches to the map were more organic. Having addressed the utility of the focus group methodology as a whole, we will turn to a discussion of concept mapping, including how it has been applied in similar research on sustainability education.

Concept Mapping: Capturing Complex Relationships

Focus groups can harness the power of group dynamics in the same way as the seminar format when applied in a classroom setting. Given that asking students to trace connections between diverse sustainability concepts and DEI values is a complex task, we decided to incorporate a concept mapping exercise within the focus group to allow participants more space to organize and process their thoughts. This exercise captured a more nuanced student understanding of the relationship between sustainability and values included in our campus mission, such as inclusion.

Concept mapping and related activities have been used effectively in previous research on sustainability education. For example, Lourdel et al. (2007) used cognitive maps as a tool to assess student understanding of sustainable development before and after sustainability education training. Notably, they found that prior to training, students primarily focused on the environmental dimensions of sustainability within their maps, but after coursework, students articulated more complex connections, such as mapping the social and cultural aspects of sustainable development. Given this synergy with our own research questions, we decided to build on this research by using a related concept mapping exercise.

To provide additional background, concept maps are visual organizations of knowledge. They are ordered by an underlying "focus question" and utilize a hierarchical structure (Novak 1995). The maps often incorporate shapes, such as boxes or circles, to denote important components, whereas lines or "crosslinks" between the shapes denote relationships. Although the terms "mind mapping" and "argument mapping" are commonly used interchangeably, there are some key differences in these mapping approaches (Davies 2011, 280). Concept mapping is more organized since it intends to "outline relationships between ideas" and can be a useful approach when you want students to connect new concepts or ideas to their preexisting knowledge (Davies 2011, 282; Hay 2007; Machado and Carvalho 2020). Because a main goal of this research project was to assess how well students connected the values of DEI to their preexisting knowledge on sustainability, concept mapping was selected as the most appropriate exercise.

Beyond providing helpful visualizations, concept mapping was also implemented because it can cut across disciplinary boundaries. The exercise gave students the flexibility to draw on any existing concepts they had learned (within class or without) to make their maps. Due to the transdisciplinary and interdisciplinary nature of sustainability education itself, we hoped the concept map exercise would help students refine and capture complex and interrelated concepts. Above all, the mapping activities supported the social construction of knowledge, making this a tool well aligned with our research questions and the focus group methodology more broadly (Machado and Carvalho 2020; Wu, Chen, and Hou 2016). While there were some challenges during the exercise, such as disagreements within the small groups on how to approach the process, the exercise was effective in helping students demonstrate a more sophisticated understanding of sustainability than would have been possible through group discussion alone.

Findings: Capturing Nuance Through the Social

The focus group format provided nuanced and detailed social insights on the impact of our university's sustainability education on undergraduate learning and experience. To speak more concretely to our findings: first, the focus group format and concept mapping exercises showed that students with sustainability educational experiences had a more nuanced understanding of sustainability than they may have articulated at the beginning of the discussion. Through their concept maps, students were able to draw multiple links between sustainability concepts they had learned and our institutional values of DEI, though this was not frequently mentioned by students at the outset of the session. Second, we gained detailed insights into why sustainability was important to students, learning that a student's social networks and K-12 educational experiences played a definitive role in shaping their understanding of sustainability prior to arriving at university. This was a result that we had not previously considered, thus, it would not have been captured at all if we had designed our study in another format such as a student survey.

To elaborate on the first major finding, when asked to provide individual definitions of *sustainability* at the outset of the focus group, students primarily focused on environmental dimensions, remarking that "sustainability is environmentalism" or "an ecological perspective." Yet, when asked to draw concept maps connecting sustainability to DEI, students demonstrated a more sophisticated understanding. For instance, some used their maps to outline the ways sustainability supports social justice movements, while others traced linkages between sustainability and global and local issues. One group drew on the theoretical framework of ecofeminism they had learned in a Women and Gender Studies class to argue that women in the global south may be more heavily impacted by climate change as they often source their family's food and water. Accordingly, a student shared the following thoughts on ecofeminism and global warming while describing their concept map:

> …we also talked about how it also negatively impacts women, more especially women in like, the Global South, and whatnot, because they are the ones who are primarily, like sourcing food and making sure that there's food and water and electricity.

Here is a depiction of this group's related concept map:

As the complex relationships mapped between sustainability and DEI contained in Fig. 27.1 illustrate, by pairing focus group discussion with an experiential activity, students were able to demonstrate a complex understanding of sustainability that would not have been well captured through quantitative approaches alone. Notably, this finding also supports Lourdel et al. (2007), who argued that students were able to better map the social and cultural aspects of sustainable development after completing sustainability coursework and training.

Focus groups also permitted us to gain deeper insight into student attitudes towards sustainability and how those attitudes were shaped by prior social experiences. To return to an earlier point, Kitzinger (1995) argued that focus groups can help us understand "not only what people think but how they think and why they think that way" (Kitzinger 1995, 299). Indeed, all students in our focus groups agreed that sustainability is important, and many argued that sustainability education had inspired them to pursue different careers and change personal habits and practices, such as reducing their consumption patterns. Most importantly, when asked *why* sustainability was important to them, most cited K-12 educational experiences and/or their social networks, such as parents or friends, rather than their university experiences. As an illustration, this student shares their view on why sustainability is important to them:

> ….my first real introduction to sustainability was like, from my mom, actually, because I feel like the way I grew up we always composted. All of my neighbors, they use pesticides on their lawn. We never did that because my mom was like, "That's terrible for the environment." ….so, I feel like

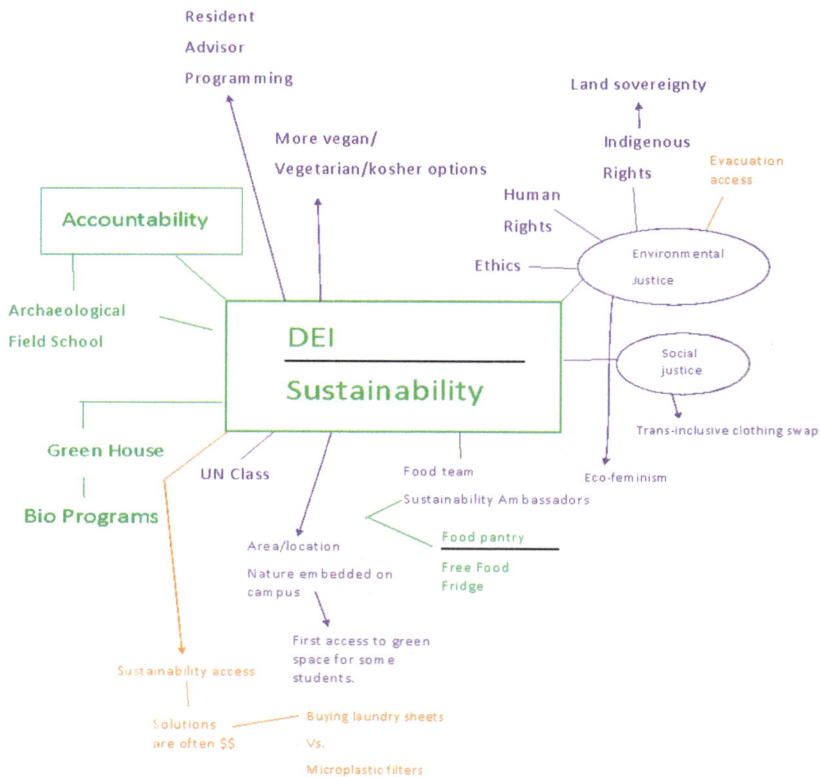

Fig. 27.1 Student concept map of DEI and sustainability links

> that was sort of ingrained in me from a young age. And then I got to high school. And it was always in the back of my mind....and I ended up joining my....my high school's environmental club. And we would... run like a district-wide recycling drive.

To touch on a few key points from the above quote, it is important to recognize that the current generation of US university students typically arrives on campus with significant preexisting knowledge and K-12 engagement with sustainability. As another illustration, several students mentioned their middle or high school participation in the student walkouts to protest climate change organized by activist Greta Thunberg as a formative experience. This experience may vary greatly, however, depending on North American or global region. Furthermore, many US students have learned about sustainability in their social lives, such as through their family upbringing. Thus, in many cases, college-level educational experiences may simply reinforce prior educational and personal experiences. This suggests that we should approach our curriculum with an eye to extending, rather than beginning, university students' understanding of sustainability and related concepts. While we gained other insights through our focus group research, we have included two key findings here that illustrate the potential of qualitative methodology to extend our understanding of the impact of sustainability education on student learning and experiences. Having reviewed some sample findings, we will turn to concluding thoughts and address some limitations of our study.

Conclusion

The effective integration of sustainability coursework within higher education will clearly require a transformation of pedagogical practices and working across disciplinary boundaries and borders. This work is vital not only for maintaining higher education's relevance in our rapidly changing world but also for preparing our students for an uncertain yet hopeful future. Our research suggests our approach should remain optimistic and action-oriented because sustainability education that only provides "depressing facts about the environment" sends the unintended message that students are "doomed no matter what," and so learning about sustainability is, in short, futile. How do we respond to such significant educational challenges in a way that is hopeful, fresh, radical, imaginative, and transformative? How do we adapt our curriculum to an ever-shifting social and environmental landscape? In brief: where do we go from here?

Notably, this chapter argues that our educational approach must center on our students, considering their own thoughts and beliefs as we move forward in our sustainability curriculum (Seatter and Ceulemans 2017). Though for some institutions, the work of integrating sustainability education into the curriculum may only be beginning, for those already engaged in this work, we suggest that it is important to pause and take inventory to check-in with our students. Why? Because how can we know where to go if we do not know where our students see themselves in relation to sustainability challenges, use prior experience and knowledge, and see their own potential to make an impact? With this goal in mind, it is our intention that this chapter can provide a roadmap for other sustainability educators who are interested in learning more about the impact of sustainability education on the student experience at their own institutions. Given their social situatedness, we assert that a qualitative focus group methodology can provide unique insights into student definitions, understandings, and experiences of sustainability. Our focus groups revealed that our students had a far greater background in sustainability education from their K-12 experiences and social experiences than we expected. While students may define "sustainability" in ways that focus on environmental elements, when asked to map connections between sustainability and concepts like *equity* in small groups and the way these dynamics are applicable when scaled up to larger social systems, they applied what they learned in class—such as ecofeminist frameworks—to articulate an extremely nuanced understanding of sustainability.

Though focus groups are useful tools for assessing the impact of sustainability education on student learning and experience, it is worth noting that focus group research is most effective when used in concert with other types of data, such as aforementioned student surveys (Parker and Tritter 2006). Whereas focus groups can tell us why participants think what they do, this methodology does not capture larger trends or attitudes at a population-level (Kitzinger 1995, 299). For instance, our research was focused on a smaller group of students within our own institution, for example, those with sustainability educational experiences. Therefore, it does not offer insight into how sustainability education is perceived or understood by the average university student. While this chapter advocates for a qualitative approach, there is also value in a quantitative methods approach. For example, survey data on student attitudes towards sustainability curriculum at our own institution would clearly develop our understanding of the focus group data we have already collected. Future research may explore the most common touchpoints for sustainability education throughout our student body, such as sustainability courses, participation in sustainability-related events, demographics, etc. That said, this chapter has provided a practical look at the utility of the focus group approach in sustainability education research. Thus, it is our sincere hope that it can offer a roadmap for others to assess the impact of this critical work at their own institutions.

Disclosures

This research project was funded by a SUNY PRODiG (Promoting Recruitment, Opportunity, Diversity, Inclusion, and Growth) grant, awarded for 2022–2024.

Acknowledgements Thank you to Lisa Mitten, Campus Sustainability Coordinator; the Co-Coordinators of the Sustainability Faculty Learning Community; and our students for their support of this project.

References

Alm, Karin, Thomas H. Beery, David Eiblmeier, and Tarek Fahmy. 2022. Students' Learning Sustainability-Implicit, Explicit or Non-Existent: A Case Study Approach on Students' Key Competencies Addressing the SDGs in HEI Program. *International Journal of Sustainability in Higher Education* 23 (8): 60–84.

Barbour, Rosaline, and Jenny Kitzinger. 1998. *Developing Focus Group Research: Politics, Theory and Practice*. Sage.

Berg, Bruce L. 2001. Qualitative Research Methods for the Social Sciences.

Bergstrom, Apryl, and Thomas E Lovejoy. 2023. *The Future of Sustainability Education at North American Universities*. University of Alberta.

Brundiers, Katja, Matthias Barth, Gisela Cebrián, Matthew Cohen, Liliana Diaz, Sonya Doucette-Remington, Weston Dripps, Geoffrey Habron, Niki Harré, and Meghann Jarchow. 2021. Key Competencies in Sustainability in Higher Education—toward an Agreed-upon Reference Framework. *Sustainability Science* 16: 13–29.

Byron, Suzanne. 1995. *Computing and Other Instructional Technologies: Faculty Perceptions of Current Practices and Views of Future Challenges; A Focus Group Study Conducted for the Information Resources Council and the Office of the Provost*. Washington, D.C.: Institute of Education Sciences. https://eric.ed.gov/?id=ED390381.

Caeiro, Sandra, Leyla Angélica Sandoval Hamón, Rute Martins, and Cecilia Elizabeth Bayas Aldaz. 2020. Sustainability Assessment and Benchmarking in Higher Education Institutions—A Critical Reflection. *Sustainability* 12 (2): 543

Chattopadhyay, Sutapa. 2019. Infiltrating the Academy through (Anarcha-) Ecofeminist Pedagogies. *Capitalism Nature Socialism* 30 (1): 31–49. https://doi.org/10.1080/10455752.2019.1574846.

Cortese, Anthony D. 2003. The Critical Role of Higher Education in Creating a Sustainable Future. *Planning for Higher Education* 31 (3): 15–22.

Davies, Martin. 2011. Concept Mapping, Mind Mapping and Argument Mapping: What Are the Differences and Do They Matter? *Higher Education* 62 (3): 279–301. https://doi.org/10.1007/s10734-010-9387-6.

Décamps, Aurélien, Guillaume Barbat, Jean-Christophe. Carteron, Victoria Hands, and Carole Parkes. 2017. Sulitest: A Collaborative Initiative to Support and Assess Sustainability Literacy in Higher Education. *The International Journal of Management Education* 15 (2): 138–152.

Gardner, Charlie J., Aaron Thierry, William Rowlandson, and Julia K. Steinberger. 2021. From Publications to Public Actions: The Role of Universities in Facilitating Academic Advocacy and Activism in the Climate and Ecological Emergency. *Frontiers in Sustainability* 2: 42.

Gramatakos, Anastasia Luise, and Stephanie Lavau. 2019. Informal Learning for Sustainability in Higher Education Institutions. *International Journal of Sustainability in Higher Education* 20 (2): 378–392.

Hay, David B. 2007. Using Concept Maps to Measure Deep, Surface and Non-learning Outcomes. *Studies in Higher Education* 32 (1): 39–57. https://doi.org/10.1080/03075070601099432.

Hennink, Monique M. 2014. *Focus Group Discussions*. Oxford University Press.

Hong, Will. 2026. The Sustainability Course Designation as Key Stakeholder Communication. In *Sustainability Learning for Action and Community Engagement*, Chapter 29. Palgrave/Springer Nature.

Johnson, Alan. 1996. 'It's Good to Talk': The Focus Group and the Sociological Imagination. *The Sociological Review* 44 (3): 517–538. https://doi.org/10.1111/j.1467-954X.1996.tb00435.x.

Kitzinger, Jenny. 1994. The Methodology of Focus Groups: The Importance of Interaction between Research Participants. *Sociology of Health & Illness* 16 (1): 103–121. https://doi.org/10.1111/1467-9566.ep11347023.

Kitzinger, Jenny. 1995. Qualitative Research: Introducing Focus Groups. *BMJ (Clinical Research Ed.)* 311 (August): 299–302. https://doi.org/10.1136/bmj.311.7000.299.

Krueger, Richard A., and Mary Anne Casey. 2009. *Focus Groups: A Practical Guide for Applied Research*. 4th ed. Sage Publications.

Kruger, Tina M., Nicholas McCreary, Brandon L. Verhoff, Virgil Sheets, James H. Speer, and Stephen P. Aldrich. 2020. College Students' Understanding of Social Justice as Sustainability. *International Journal of Sustainability in Higher Education* 21 (3): 513–530.

Lazarsfeld, Paul F., and Robert K. Merton. 1948. Mass Communication, Popular Taste and Organized Social Action. *Media Studies: A Reader* 2: 18–30.

Lourdel, Nathalie, Natacha Gondran, Valérie Laforest, Bruno Debray, and Christian Brodhag. 2007. Sustainable Development Cognitive Map: A New Method of Evaluating Student Understanding. *International Journal of Sustainability in Higher Education* 8 (2): 170–182.

Machado, Cristiane Tolentino, and Ana Amélia Carvalho. 2020. Concept Mapping: Benefits and Challenges in Higher Education. *The Journal of Continuing Higher Education* 68 (1): 38–53. https://doi.org/10.1080/07377363.2020.1712579

Madriz, Esther. 2003. "Applied Ethnography." In *Collecting and Interpreting Qualitative Materials*, ed. Norman K. Denzin and Yvonna A. Lincoln, 2nd ed., 835–50. Sage Publications.

Merton, Robert K., Marjorie Fiske, and Patricia L. Kendall. 1990. *The Focus Interview: A Manual of Problems and Procedures*, 2nd ed. New York: Free Press.

Novak, Joseph, D. 1995. Concept Mapping: A Strategy for Organizing Knowledge. In *Learning Science in the Schools: Research Reforming Practice.*, 229–45. Hillsdale, NJ, US: Lawrence Erlbaum Associates, Inc.

Owens, Katharine A., and Sasha Legere. 2015. What Do We Say When We Talk about Sustainability? Analyzing Faculty, Staff and Student Definitions of Sustainability at One American University. *International Journal of Sustainability in Higher Education* 16 (3): 367–384.

Palmer, Diana A., Barbara Boardman, and Howard Bauchner. 1996. Sixth and Eighth Graders and Acquired Immunodeficiency Syndrome: The Results of Focus Group Analysis. *Journal of Adolescent Health* 19 (4): 297–302.

Parker, Andrew, and Jonathan Tritter. 2006. Focus Group Method and Methodology: Current Practice and Recent Debate. *International Journal of Research & Method in Education* 29 (1): 23–37. https://doi.org/10.1080/01406720500537304.

Patton, M. Q. 1990. *Qualitative Evaluation and Research Methods*. SAGE Publications, inc.

Pugsley, Lesley. 2018. "Focus Groups, Young People and Sex Education. In *Gender and Qualitative Research (1996)*, 114–30. Routledge.

Seatter, Carol Scarff, and Kim Ceulemans. 2017. Teaching Sustainability in Higher Education: Pedagogical Styles That Make a Difference. *Canadian Journal of Higher Education* 47 (2): 47–70

Sim, Julius, and Jackie Waterfield. 2019. Focus Group Methodology: Some Ethical Challenges. *Quality & Quantity* 53 (6): 3003–3022. https://doi.org/10.1007/s11135-019-00914-5.

Stephens, Jennie C., and Amanda C. Graham. 2010. Toward an Empirical Research Agenda for Sustainability in Higher Education: Exploring the Transition Management Framework. *Journal of Cleaner Production* 18 (7): 611–618.

Stough, Talia, Kim Ceulemans, Wim Lambrechts, and Valérie Cappuyns. 2018. Assessing Sustainability in Higher Education Curricula: A Critical Reflection on Validity Issues. *Journal of Cleaner Production* 172: 4456–4466.

SUNY New Paltz. n.d. "Mission Statement." *About New Paltz*. Accessed June 4, 2024. https://www.newpaltz.edu/about/mission.html.

Trencher, Gregory, Xuemei Bai, James Evans, Kes McCormick, and Masaru Yarime. 2014. University Partnerships for Co-Designing and Co-Producing Urban Sustainability. *Global Environmental Change* 28: 153–165.

UNESCO. n.d. "Higher Education." *UNESCO*. Accessed June 4, 2024. https://www.unesco.org/en/higher-education#:~:text=The%20number%20of%20students%20in,along%20with%20international%20student%20mobility.

Viegas, Cláudia. V., Alan J. Bond, Caroline R. Vaz, Miriam Borchardt, Giancarlo Medeiros Pereira, Paulo M. Selig, and Gregório Varvakis. 2016. Critical Attributes of Sustainability in Higher Education: A Categorisation from Literature Review. *Journal of Cleaner Production* 126: 260–276.

Williams, Amanda, and Larry Katz. 2001. The Use of Focus Group Methodology in Education: Some Theoretical and Practical Considerations, 5 (3). *IEJLL: International Electronic Journal for Leadership in Learning*.

Wu, Sheng-Yi, Sherry Y. Chen, and Huei-Tse Hou. 2016. Exploring the Interactive Patterns of Concept Map-Based Online Discussion: A Sequential Analysis of Users' Operations, Cognitive Processing, and Knowledge Construction. *Interactive Learning Environments* 24 (8): 1778–94. https://doi.org/10.1080/10494820.2015.1057740

Zwickle, Adam, Tomas M. Koontz, Kristina M. Slagle, and Jeremy T. Bruskotter. 2014. Assessing Sustainability Knowledge of a Student Population: Developing a Tool to Measure Knowledge in the Environmental, Economic and Social Domains. *International Journal of Sustainability in Higher Education* 15 (4): 375–389.

CHAPTER 28

Growing a Sustainability-Focused Campus and Fostering Faculty Development

Lisa Mitten

When I first arrived on campus as the SUNY New Paltz Campus Sustainability Coordinator in 2013, I remember posting sticky notes on my cubicle with thematic sustainability operational areas: materials/waste, energy, water, and food. I was thrilled to launch my dream role in protecting our planet and the natural world. But as those first years passed, I realized that ours was not strictly an environmentally focused university. Then in 2015, the United Nations members agreed upon the seventeen Global Goals for Sustainable Development, visually and powerfully articulating the three environmental, social, and economic dimensions of sustainability. It's not *just* about the earth. It's about care for people *and* planet. Listening to our campus, I realized though ours may not be at its core an *environmental* campus like SUNY Environmental Science and Forestry, we have a campus culture that is committed to environmental, social, and economic justice, human thriving, and

L. Mitten (✉)
SUNY New Paltz, New Paltz, NY, USA
e-mail: mittenl@newpaltz.edu

S. M. Wyman and J. Korenblat (eds.), *Sustainability Learning for Action and Community Engagement*,
https://doi.org/10.1007/978-3-032-11698-7_28

protecting the earth for future generations. We are *genuinely* a three-dimensional sustainability campus. Add wellness as a fourth dimension, as Erika Bailey-Johnson has done at the Bemidji State University in Minnesota, and you approach the heart of SUNY New Paltz's values and culture.

In the first year of my job, our university was experiencing a positive budgetary environment, and I received strong encouragement from my supervisor, Assistant Vice President of Facilities Management John Shupe, to learn and grow through professional development. In my thirteenth month, I found myself at a week-long Campus Sustainability Leadership Training at the University of Vermont. I was inspired and moved by Debra Rowe, one of the workshop leaders and President of the US Partnership for Education for Sustainable Development. I remember asking her, "What do I do when I feel uncomfortable being the only woman in the room surrounded by a bunch of white male Facilities Managers and consultants?" "Get over it and speak your mind!" Debra replied. I appreciated her directness, her passion, and her sense of urgency in the midst of the climate crisis.

Soon after I returned from the Vermont sustainability leadership training, there was a call from campus administration to submit ideas for one-time funding. I thought to pass forward what my boss had so generously granted me: a commitment to professional development. I requested $50,000 from campus administrative leadership to support the professional development of campus employees to learn about sustainability. I also included funding for materials and supplies for faculty and student sustainability research and money to hire students to support this work.

This one-time professional development funding led to a chain of events around learning and community engagement for students, faculty, staff, and regional community members from 2015 to the present. This campus's commitment to teaching, scholarship, student success, creative activity, and community wellbeing continues to grow. Our Sustainability Faculty Learning Community and the commitment of the Student Sustainability Ambassadors energized our campus and led us to compile this book. Without this coordinated and collaborative beginning with its focus on professional development, we never would have arrived where we are today with justice and sustainability woven into our University Strategic Plan vision statement and sustainability embedded across our

Strategic Plan strategy core areas including educational innovation for sustainability and student career development for sustainability.

With $50,000 allocated, I sent a call to the university's faculty/staff listserv inviting my colleagues to attend the 2015 University of Vermont Campus Sustainability Leadership Training program. I asked interested persons to reflect briefly on why they wanted to attend and what they intended to get out of the training. Two professors responded to the call with thoughtful reflections: Will Hong of Digital Media & Journalism and Michael Sheridan of Business Management. I sent them both at considerable expense for the in-person week-long program.

Hong and Sheridan returned with a mission: Will brought back a drive to replicate the University of Vermont's Sustainability Fellows program and house it as a project of the Sustainability Committee. Michael Sheridan and the Sustainability Committee Co-Chair Rosemary Milham of the School of Education joined us. We approached the Office of the Provost for financial support and our institution's interim Provost Stella Deen suggested that we call the program a *Sustainability Faculty Learning Community* to bring together academic and professional faculty to become Sustainability Faculty Fellows. Stella and I emphasized the *community* elements of the program, as we understood the potential of this program to establish connections, boost morale, and foster interdisciplinary collaboration. Many institutions have academic faculty professional development programs in sustainability, but the ongoing sense of *community* surrounding ours sets the SUNY New Paltz version apart.

We completed our ninth year of the Sustainability Faculty Learning Community in 2025. 101 academics and professionals (from all academic divisions) have participated in the program as currently employed or emeritus Sustainability Faculty Fellows. We have mentored 126 undergraduate student Sustainability Ambassadors. Our institution has 344 full-time faculty, 323 part-time faculty, and 6197 undergraduate and 1036 graduate students.

Looking back over at over a decade of public service for sustainability at SUNY New Paltz, I offer my academic and professional colleagues around the world some principles that have guided my journey.

Top 10 Sustainability Leadership Strategies

1. Education and Action for Sustainability is about the *collaborative* mission, the work, the actions, and the solutions. It's not about you as an individual. It's about the community. Embrace *we, not me; us, not I*.
2. Leadership is about building relationships. Recognize the power of a shared cup of coffee, getting to know one another over conversation.
3. Deeply listen from your heart whenever you can access it.
4. Don't direct anything. Guide, support, and facilitate.
5. Recognize and value individual approaches. Learn what makes each colleague tick and what their passions are. Then, you know how to engage them in contributing to the betterment of the program and the campus. Their gifts and wisdom can be held by the community.
6. Pass it forward. Be generous. For me, this means investing in the sustainability professional development of my colleagues.
7. Experiment! Try new things. Fail. Evolve the program. Let those failures chart a pathway for the next success.
8. Let the urgency of the climate crisis and social inequality drive students and faculty towards positive change and action and away from doom and gloom.
9. Focus on solutions and change agency. Strive to transform what is in your area of influence. Envision the future we wish to see.
10. There is no hierarchy (a tip from Debra Rowe). I began my journey with grassroots campus community connections and the leadership of students and faculty. Support from the top is helpful, but a top-down directive may not shift culture.

Engaging Academic Faculty in Sustainability Education and Action

In partnership with tenured faculty, pre-tenure faculty, lecturers, and adjuncts, we've fostered a grassroots movement and achieved a critical tipping point for our campus community and culture. I have facilitated all this from outside the Academic Division, as I'm housed within Facilities Management. According to the Association for Sustainability in Higher

Education (AASHE) 2023 Sustainability Staffing Survey, 42% of sustainability professional roles are housed within Facilities Management. The next most common place for directors of campus sustainability is in the CFO/Administration and Finance division, far from the academic core of the university.

Sometimes, working with academic faculty on how to infuse sustainability concepts and materials into their existing curriculum challenged me to define my own role in the institution. Earlier in my career, not having a PhD preoccupied me, but over the years, I've learned that plenty of tenured faculty and full professors without PhDs hold terminal Master's degrees. I may not have such a degree, but I do have two Master's credentials: in Public Policy and Education, as well as training in business, entrepreneurship, government, and politics. I also have broad educational and work experiences in the dramatic arts and my family's small business that connect with all five of our academic schools: Fine & Performing Arts, Education, Liberal Arts and Sciences, Business, and Science and Engineering.

So, how did I successfully partner with a rotating steering committee of deeply committed academic faculty and a growing cohort of fellows devoted to campus-wide service for sustainability?

Here are my guiding principles for partnering with faculty:

1. Stay open to interdisciplinary areas for growth. All disciplines have skills, strategies, and knowledge to contribute toward advancing a just and sustainable world. Furthermore, interdisciplinary research can be refreshing and generative for mid-to-late career faculty.
2. Learn, grow, and evolve. Academics may have grown up in one discipline with a narrow area of focus outside an explicit, intentional sustainability orientation. As they develop and advance in their careers, many academics seek to tie their disciplinary teaching and research to advancing approaches and solutions to tackle the most urgent challenges of our time.
3. Find a space for yourself within sustainability. Nobody and no department *owns* sustainability as a discipline. This unleashes creativity and allows leadership to flourish.
4. Don't worry about the money. Faculty participation in sustainability professional development and research does not necessarily depend on money or funding. Our program offers a $250 stipend

for a year-long participation in the Fellows program. After taxes, that doesn't amount to much.

5. Let personal passions and global urgency drive student and faculty motivation for education and action for sustainability.
6. Celebrate community. Academic faculty come to our Fellows Program for content and stay for community, connection, purposeful work, and intellectual fulfillment. Being a part of an interdisciplinary community is intellectually stimulating and rewarding. Also, academic faculty welcome the opportunity to make professional and collegial connections beyond their departmental homes. They might enjoy less drama and departmental tensions. Our Sustainability Faculty Learning Community is an impactful employee retention tool. Not everybody stays. But most do.
7. Promote consensus-based, non-hierarchical decision-making and shared leadership, all the time in both faculty and student-run contexts. Meet in circles. Begin your meeting by going around the circle with a check-in that is not about business at hand.
8. Develop a robust program over time. Incremental change, one cohort per year, requires patience. On an individual faculty level, understand and embrace the vast dimensions of sustainability pedagogy and action that require years of thinking, processing, reading, reflecting, sabbaticals, iteration and learning from students.
9. Offer multiple sustainability frameworks (https://www.newpaltz.edu/sustainability/for-faculty/sustainability-faculty-learning-community/sustainability-frameworks/) for an inclusive, broad, transdisciplinary approach with many options and many starting points for academic faculty to connect with and explore. Our list of frameworks grows each year, and our website has become a resource for others in the field. Will Hong started us off with the United Nations Global Goals for Sustainable Development. Based on faculty feedback, we've broadened this to offer over a dozen sustainability frameworks.
10. Encourage academic and professional faculty to articulate how they can approach this multidimensional sustainability work starting from a place of strength: this often means a focus on one dimension of environmental/social/economic sustainability.
11. Reveal the work already underway. Academics attracted to a faculty sustainability community are likely already doing the work and may

not conceptualize this with sustainability language. Becoming a sustainability-informed academic faculty member may be simply a matter of reframing existing curricula within an overarching sustainability framework that links to a particular discipline's approach to sustainability.

12. Use the evolving Sustainability Competencies (Brundiers et al. 2021; Wiek et al. 2016) to contribute to defining the emerging interdisciplinary practice of Education and Action for Sustainability. These competencies help shape learning outcomes and assess student learning.
13. Institutionalize the campus commitment to sustainability by creating a sustainability course designation in the course catalog (Hong 2025). Offer additional support structures like cohort-based learning events and a yearly institute or summer retreat.
14. Leverage up and share out beyond your individual course, class, and/or research (Rowe 2019). Leveraging up is our group approach to writing this book as an open access resource.

Conclusion

When I was hired as SUNY New Paltz's first Campus Sustainability Coordinator, the hiring manager shared that one thing that made me stand out from the other sixty-five candidates was my emphasis on *listening* in my multi-stage interview process. When the search committee posed the question: "What's one word that sums you up? I immediately replied, "Community." Listening and community building are two core approaches woven throughout my decade-plus career in sustainability in higher education and have allowed me to thrive in this position. I have grown into this job in collaboration and coordination with the faculty, students, administrators, and Facilities staff who have helped shape our campus community's commitment to sustainability pedagogies and practices. This work has been celebrated in the Hechinger Report, "How Colleges Can Become 'Living Labs' for Combating Climate Change" (Preston 2024). In this way, we prepare our students to become leaders in social, environmental, and economic sustainability, serve the Earth and forward the potential for its people and creatures to survive and thrive.

References

Association for the Advancement of Sustainability in Higher Education (AASHE). 2023. Salaries and Status of Sustainability Professionals in Higher Education. https://www.aashe.org/wp-content/uploads/2023/10/2023-Staffing-Survey-Report-1.pdf

Brundiers, K., et al. 2021. "Key Competencies in Sustainability Higher Education—Toward an Agreed-Upon Reference Framework." *Sustainability Science* 16: 13–29.

Hong, Will. 2020. *Build it and they Will Come: The Faculty Learning Community Approach to Infusing the Curriculum with Sustainability Content*. Springer.

Hong, Will. 2025. "The Sustainability Course Designation as Key Stakeholder Communication." In *Sustainability Learning for Action and Community Engagement*. Palgrave.

Preston, Caroline. 2024. "How Colleges Can Become "Living Labs" for Combating Climate Change." *The Hechinger Report*. July 30. https://hechingerreport.org/how-colleges-can-become-living-labs-for-combating-climate-change/

Rowe, Debra. 2019. Keynote Address to SUNY New Paltz Sustainability Faculty Fellows. January.

Wiek, A., et al. 2016. "Operationalizing Competencies in Higher Education for Sustainable Development." In *Handbook of Higher Education for Sustainable Development*, ed. M. Barth, et al., 241–260. London: Routledge.

CHAPTER 29

The Sustainability Course Designation as Key Stakeholder Communication

Will Hong

As our SUNY New Paltz campus community deepened its commitment to sustainability-informed education and operations, we considered ways to reflect these changes in our curriculum. Our goals were to encourage innovative teaching and curriculum development as we provided relevant and rigorous course content to our students. By the fall of 2020, we at New Paltz had already made serious headway towards infusing our curriculum with up-to-date sustainability content. Not only had our traditionally aligned departments (e.g., Geography, Environmental Studies) continued with their sustainability-related course offerings, but with the ongoing success of our Sustainability Faculty Learning Community (SFLC) program (Hong 2020), sixty recently trained Sustainability Faculty Fellows had added new sustainability content to their courses, in some instances for the first time. They did so across departments and schools on campus. By 2025, the number of fellows reached 101.

W. Hong (✉)
Department of Digital Media & Journalism, SUNY New Paltz, New Paltz, NY, USA
e-mail: hongw@newpaltz.edu

S. M. Wyman and J. Korenblat (eds.), *Sustainability Learning for Action and Community Engagement*,
https://doi.org/10.1007/978-3-032-11698-7_29

With this new critical mass of sustainability-aligned course material in the university curriculum, what was the next step toward a sustainable campus writ even larger? And how could we make the changes that were afoot visible to those interested stakeholders?

While the curricular changes were the first phase in providing students with sustainability content, it appeared that the next logical structural move would be to ensure interested students could *find* this new content when they were considering classes during their registration period and perhaps when looking over the entire curricular offerings in choosing a major or general direction of study. After all, course titles and even more explicit course descriptions often do not provide enough detail to inform students of sustainability content that would be covered in the class. Furthermore, as many different sections of a single course were taught, often by different professors with slightly different approaches, students had no easy way to see whether a particular instructor and class section had sustainability content. Especially in disciplines not traditionally aligned with sustainability (such as my own, Digital Media and Journalism), the newly created content, as valuable as it might be to the engaged student, would likely remain hidden.

An official **sustainability course designation** seemed an important structural piece to create and implement something that students could easily see in the online Schedule of Classes next to the other existing designations/class attributes (for example, Writing-Intensive, Liberal Arts-qualifying, etc.) as they chose core classes and electives for their upcoming semesters.

But where should we put such a designation in place? Several questions immediately presented themselves as we created this designation:

1. What criteria should the designation require for approval, and how might we design it so that it would be rigorous and inclusive enough to both invite applicants to apply and still be academically meaningful?
2. Presuming this would be an ongoing designation, what would the approval procedure look like, and which campus body would take the review responsibility in the future?
3. Through which administrative procedure would a designation be approved and included in the live Schedule of Classes? What information would need to be provided to whom?

4. What records would be kept to ensure any assessment or audit of the procedure continued to meet university administrative standards?
5. Finally, would such an effort even be worth it? What ultimately would be gained and by whom?

I'll take these questions one by one and describe how we, the members of the university's Sustainability Committee (SC), approached each of them.

As the Sustainability Committee considered developing a sustainability course designation, the first obvious question was, "How would a course qualify for the designation?" The SC at New Paltz comprises faculty members and non-academic staff who wouldn't necessarily need to chime in on the question (since it was strictly curriculum-related). In the fall of 2020, a sub-committee was formed of four faculty members from the Sustainability Committee and Lisa Mitten, the campus Sustainability Coordinator and co-chair of the SC, to research and explore possibilities for presentation and recommendation to the full committee.

In defining the subcommittee's goals, chief among them was to propose set criteria such that they could strike a balance between academic rigor and inclusivity. The sub-committee had a clear sense of the current courses offered that might be ready for a sustainability designation, given that the SFLC had been successfully running for four years. The faculty fellows who had completed that training program had (as a matter of the program's requirements) each produced changes to their courses that brought more sustainability content to their respective classrooms. Some of these curricular changes were minor and some decidedly major, so with that variance of content infusion in mind, the sub-committee sought designation criteria that could be a means for students to both find courses with sustainability content and give them a sense of *how much* sustainability material might be covered given that course's disciplinary context. Striking exactly the right criteria-related balance was a bit of a Goldilocks proposition where we wanted to make sure we set the threshold just right—not too high or too low.

Again, the designation would need to be academically valid and stand up to scrutiny in that regard, and yet not so difficult that faculty just entering the space would be discouraged from applying. Additionally, one of our longer-term goals with the designation was to make visible *a critical mass* of courses offering sustainability content so that the larger argument might eventually be made for a cross-disciplinary Sustainability major/minor or a GE requirement.

The sub-committee began to look for reference models from campuses that had already created a designation for sustainability courses to gain a sense of what had been successful elsewhere. Of the examples considered, the course designation at Dickinson College (a private liberal arts college in Carlisle, PA) seemed to the sub-committee members the most applicable to our situation, one that would address the Goldilocks problem and which proved easy enough for faculty and students to understand, while straightforward enough for us to manage administratively.

Dickinson College has committed wholeheartedly and top-down to becoming a sustainable campus, and Lindsey Lyons, Director of Sustainability Learning at Dickinson's Center for Sustainability Education, generously provided the sub-committee with information regarding their sustainability course designation and the process surrounding its administration. Just as the University of Vermont's sustainability faculty fellows program served as a useful model for our SFLC, so, too, Dickinson's model of sustainability in their curricular requirements served as a useful model. Dickinson chose to create *two* levels of sustainability content in their designations, a Sustainability Connections (SCON) course and a more rigorous Sustainability Investigations (SINV) course, to make a distinction as to how much sustainability content was covered. Both designations denote a course that qualifies for Dickinson's graduation requirement that all students take a qualifying sustainability course and to demonstrate the ability to "think critically about a sustainability question, problem, and/or potential solution and articulate connections between a field of study and sustainability."

While we don't yet have a corresponding sustainability general education requirement, the dual-level designation made sense to us given the varying levels of content in the courses at New Paltz that would likely qualify for designation. Again, some faculty were relatively new to sustainability frameworks and concepts while others had been in the intellectual/scientific space for the entirety of their careers, so parsing that level of content for students would likely be helpful as they chose classes to take (given *their* varying levels of commitment to sustainability as well).

The sub-committee then proposed the following to the Sustainability Committee as a whole (giving explicit credit to Dickinson):

Two Available Sustainability Course Designations:

A **Sustainability-Related (SUSR) course** must:

1. Introduce students to one or more definitions of sustainability as related to the content of the course;
2. Present sustainability as having environmental, social, and economic dimensions; and
3. Include student learning outcomes on the course syllabus that indicate that students will be able to think critically about sustainability and articulate connections between the course's discipline and sustainability.

A **Sustainability-Focused (SUSF) course** must fulfill the above sustainability-related requirements **AND.**

1. Include a student learning outcome indicating that students will be able to apply disciplinary or interdisciplinary methods to analyze and address a sustainability problem;
2. Devote a substantial portion of the course material to sustainability content; and
3. Give significant weight to sustainability content in evaluating student performance in the course.

Related to the curricular training that faculty fellows experienced in the SFLC, a handful of sustainability conceptual frameworks would also be offered to help faculty meet the requirement of inclusion of a definition of sustainability in their course material. This list includes the 2015 U.N. Global Goals, likely the most widely used framework beyond academia. The SUSR designation would cater to faculty and students who were interested in bringing *some* content to the classroom, while the SUSF designation built on the foundational level of work while adding an *application* element to the coursework, putting conceptual knowledge to practical work in addressing a problem in the physical world.

Interested faculty would choose a level designation and then submit their syllabus for the specific course and fill out a short questionnaire that would help the Sustainability Committee evaluate their application. The goal was to keep the application process as streamlined and frictionless as possible and make the prospect of applying inviting to faculty. Applicants were also asked to consider the U.N. Global Goals with respect to their course and to indicate which goals they thought most relevant to the course, information that could help the Sustainability Committee with future programming and initiatives. In the Spring of 2021, the entire Sustainability Committee reviewed the subcommittee's proposal and approved the dual-level designation, readying the stage to solicit applications from the faculty.

Since the fall of 2021 (when the official sustainability designation application was first sent to faculty), 70 successful applications have been received, with the majority of submissions receiving the SUSR designation and roughly one-third of the submissions receiving the SUSF designation. The designations themselves went live on the university's Schedule of Classes for the Fall 2022 semester and have been updated each subsequent semester.

As has been implied, the administrative task of soliciting applications for the designation and performing the evaluations of the applications rests with the Sustainability Committee (professional staff members notwithstanding) and so presented additional work for the committee to do each semester, a concern that was raised early on in the development process. Certainly, the SC was the logical and most appropriate campus body to evaluate these submissions, but the concern was that application evaluation would demand too much time of the Committee to be manageable given its other obligations across any given semester or academic year.

As it turned out, however, the demands on the Committee's time were managed by having appropriate academic members of the committee assigned to pre-screen applications from their particular school/field (LA&S, Business, Visual Arts, Physical Sciences, etc.) before the SC meetings proper and then simply summarize and make recommendations to the Committee who would then discuss and vote for or against a designation. This divided the duties reasonably and fairly and proved an efficient means for a proper look at each submission. Moreover, with the initial introduction of the sustainability designation to the larger campus community, the bulk of the submissions arrived with the first wave (the attribute being a new one on the books), and with each subsequent semester, only new courses or courses taught by professors for the first time or newly developed sustainability courses have been submitted for the designation, many fewer in number than during the initial rush. So, the additional workload for the Sustainability Committee over time in evaluating courses for the designation has proven more manageable than initially anticipated.

That said, some additional administrative work has come with the formation of the designation, primarily around record-keeping and continued communication with stakeholders, including campus departments to make sure that the approved designations appear correctly in the Schedule of Classes and in a timely manner.

First, record-keeping. The Sustainability Committee's goal here is to imagine that, were the processes surrounding the sustainability course designation to be audited by an outside party at any point, the auditor would find detailed records regarding each approved (and denied) course regarding the application itself and the SC's deliberations. So, all applications have been recorded, collected, and stored in Microsoft's online Office365 ecosystem (adopted by the college universally). Detailed voting records of each submission's SC discussion have also been recorded and saved. All records are accessible by Sustainability Committee members at any time and are open to applicants and administration members should they want to take a look. Maintaining such transparency and meticulous record-keeping is important to the integrity of the process and for the sustainability of the designation itself in the long run.

As for the additional communication responsibilities related to the designation process, this has proven to be a bigger lift, as there are many moving parts, all of which are important.

The central goal of having a designation at this stage is to let students know about the classes with sustainability content so that they can make informed choices during their class registration period or in considering overall academic direction (such as choosing a major). Of course, this means having up-to-date information available to students before they register for classes, which typically happens during the middle of the semester before a class runs, which thus requires the entire designation submission and evaluation process to be completed roughly a semester before any course is offered. Adapting to this anticipatory timeline took a bit of adjustment. Producing the application and informational webpage and then soliciting applications from faculty with enough time for us to make proper evaluations and decisions felt a bit rushed during the first rounds of submissions. Initially, Records & Registration (who create and maintain the Schedule of Classes) suggested deadlines for our designations that felt too early in each semester for us to accommodate. Asking for submissions during periods when faculty are either on vacation or otherwise preoccupied with non-course-related research (summer and winter sessions) felt less than optimal time-wise.

However, the Sustainability Committee is fortunate to have a member who is a staff member of Records and Registration, and upon seeing the time crunch that we struggled with, she suggested that we simply work directly with her to have new designations implemented through the system which, cutting a bit of standard red tape, did buy us additional time at the beginning of each semester. She could provide a much quicker turnaround in assigning designations or fixing errors than were we to go through normal channels. This has proven very helpful, not only clarifying the schedule that the Sustainability Committee sets for itself to follow but also providing an understanding of the administrative demands in play and how our process might reasonably fit in with the larger workings of the university and class registration.

An additional concern, initially, was that the Provost's Office would require justification for each decision and round of designations as we went along. As labyrinthine a bureaucracy as even a small university campus is (New Paltz is a medium-sized comprehensive college of 6000 undergraduates and 2000 graduates), one might presume that trying to get anything that influences the entire student body passed would require many steps and almost by definition face friction if not outright resistance along the way. Often, good reasons exist—consistency of form, forcing seriousness of purpose, and maintaining reasonably defined academic standards—but sometimes, the presumption of institutional resistance can lead to hesitation, inaction, or avoidance.

Thankfully, however, it quickly became clear that I was wrong to have expected this pushback. As mentioned, working with Records & Registration proved easier than expected once we began assigning designations and said committee member helped us through the administrative stumbling blocks, but were she not on the committee, continued communication to R&R would have ensured that placing this information on the Schedule of Classes was less of an administrative hurdle than it initially seemed. Since going live in the fall of 2022, there have been a few instances where the wrong designation was applied to a course or where a designation was given to the wrong course taught by an applicant, but making corrections as soon as we spotted them (or they were brought to the committee's attention) has been simply a matter of asking R&R to make the change ASAP, which they've been happy to do. We did, again for the sake of transparency, reach out to the Provost's Office in the spring of 2021 to describe the sustainability designation and our intentions with it, and we found no resistance to the idea or its implementation so long as the Sustainability Committee could find a method to get things done in a consistent and fair way.

Regarding communication with faculty applicants, as might be expected, they are typically inundated by myriad demands, and timing aside, it's a challenge to get their attention for a moment to have them consider a new initiative, especially if it asks that they complete a detailed form that doesn't immediately reward them for time spent. As simplified as the application process is, it does take a bit of time and thought, thinking through how an existing (or proposed) course might fit specific criteria, and faculty members might not be blamed for thinking that they needn't bother given more pressing matters at hand. Happily, we did have our SFLC Sustainability Faculty Fellows to constitute an instructor demographic primed and, for the most part, eager to take a bit of time to get the ball rolling with the designation going live. The hope is that with more and more designations visible on the Schedule of Classes when faculty advisors and students alike discuss class planning, momentum and curiosity will fuel continued and growing interest in the designation and encourage other instructors to follow suit with their own course innovations when applicable.

For the most part, the submissions received have proven to be complete and ready for either the SUSR or SUSF designation, but in the few instances where the Committee felt that changes were necessary before a syllabus designation could be awarded, a committee member (usually the faculty/SC member who prescreened the application) reaches out directly to the applicant to explain what further action is required. Usually, this means the inclusion of a specific sustainability-related student learning outcome on the syllabus or to more explicit language regarding which sustainability framework the instructor planned to discuss with the class (see Earthrise Commons; earthrisecommons.org), changes that were simple to make. Again, the purpose of the Committee in these instances was not to be rigidly prescriptive in dictating class content but rather to help suggest language in a syllabus to clearly communicate to students what sustainability content would be covered by the instructor, in which assignments or discussions, and in what context.

One communication matter that the committee could still improve upon at this point is in letting applicants know when and how the Sustainability Committee has evaluated their submissions. As mentioned, in those few cases, when the committee feels the need for an applicant to revise their materials to qualify for a designation, a member of the committee reaches out directly and passes along relevant information. However, with submissions that meet approval easily, the applicant has not always been made aware of that decision. Going forward, we have made sure to have a simple congratulatory email sent out to each successful faculty member as soon as the committee's decision to award a designation has been made official. This personalized communication element is one to always keep squarely in mind with such an administrative process to maintain goodwill in the community.

Furthermore, communication to students that there is sustainability content available for them to experience should they be looking for it, which research suggests they are (Crowell and Bellody 2026) is, again, at the heart of this whole endeavor. Many students arrive on our campus with little exposure to the concepts of environmental/social/economic sustainability as articulated in formal academic disciplinary terms. The designation also communicates to them that the overall commitment to sustainability by the faculty and institution is real and potentially *growing* (as more designations appear on the books) and that we respect *their* interests and desire to learn this material. Theirs is the future that sustainability content purports to take most seriously, so acknowledging and amplifying this growing priority on campus is systemic communication of an essential kind. This speaks to prospective students as well as current ones. For students to be able to see right there on the live Schedule of Classes that there is sustainability content across departments and disciplines and not simply relegated to, say, one Physical Sciences department also allows them to absorb the cross-disciplinary nature of sustainability and to internalize the interconnectedness of the knowledge systems and techniques that they garner during their college careers. If systems thinking is an epistemological philosophy of sorts, the designation helps to make the basic tenet of that philosophy tangible and real: that responsibly solving complicated problems effectively requires embracing the complexity of the systems involved.

Additionally, part of the original idea of supporting the development of more courses with sustainability content was to work towards a sustainability major once enough courses at all levels existed. Subsequent to the creation of the SFLC in 2016, an interdisciplinary Environmental Studies major was created at New Paltz (2021), however, that box seems now to be well-checked, but perhaps with additional designated courses, more and more qualifying electives in that major could be offered and accepted by the body that oversees the new official major.

In retrospect, is all this worth the ongoing effort? In addition to the near-term benefits, it's important to think of the designation as a step in a longer evolution of the curriculum. Once again, looking at the Dickinson College curricular requirements as a model, they currently require all students to take at least one course that qualifies as a Sustainability course. This seems like a worthy goal for us to adopt at New Paltz, and as we have recently seen the SUNY system embrace courses that qualify under the Diversity (DEISJ) umbrella as now required of all students. It's reasonable to imagine Sustainability following a similar path to wider acceptance and promotion. Whether falling under general education requirements or as a standalone category, reaching the point where students at New Paltz are required to take a sustainability course (either SUSR or SUSF) before graduating seems like the logical next step and for the argument to be made that such is both reasonable and feasible, we must demonstrate that there are enough courses across a breadth of departments to support such a requirement. The sustainability designation can now track those numbers accurately and when a critical mass of courses earns the designation, perhaps this next curricular step can be discussed as a possibility.

Finally, though the sustainability course designation currently feels a vital step in reaching stakeholders with these new developments and information, curiously, its ultimate success may well be defined by its elimination. Once sustainability content is commonplace enough to be accepted as part of the bargain regardless of a student's intellectual path through the college, one can imagine the situation as one that obviates the need for any sustainability designation at all because the content would be saturated in the curriculum enough so that one need not go searching for it—it would be implicitly part of the deal. Larger top-down strategic initiatives with respect to sustainability (as seen on campuses like Dickinson's) might well have such an outcome as a long-term goal and direct resources towards achieving such, but the goal might also be achieved from the bottom up with more and more courses having students engage with sustainability content as a matter of greater emphasis on the interconnectedness of knowledge systems and a commitment to embracing complexity at all levels. Let us hope that we, across the larger institution of higher education, have the gumption and wherewithal to see this through.

References

Barlett, P. F., and G. W. Chase. 2013. *Sustainability in Higher Education: Stories and Strategies for Transformation.* Cambridge, MA: MIT Press.

Barth, Matthias, Gerd Michelsen, Marco Rieckmann, and Ian Thomas. 2015. *Routledge Handbook of Higher Education for Sustainable Development.* London: Routledge.

Crowell and Bellody. 2026. Focus Groups for Evaluating Student Perspectives on Sustainability Education. *Sustainability Learning for Action and Community Engagement*, Palgrave.

Hong, Will. 2020. *Build it and they Will Come: The Faculty Learning Community Approach to Infusing the Curriculum with Sustainability Content.* Springer.

Kaza, S., L. W. Natkin, and T. Rowse. 2016. Developing Sustainability Leadership Through Faculty Professional Development. *Journal of Environmental Studies and Sciences* 6 (2): 437–444.

Rowe, Debra. Sustainability Improves Student Learning (SISL). Accessed February 24, 2024. https://serc.carleton.edu/sisl/index.html

CHAPTER 30

Foundation Art Program Branches Out, Makes Connections, Builds Community, Finds Common Roots

Suzanne Stokes

Introduction: Building Community & Belonging Through Outdoor Arts Education

The introductory Foundation program in the Art Department at SUNY New Paltz, US, recognizes the need for building community with first-year and transfer students new to the college campus and often away from home for the first time. Our students are diverse culturally, racially, economically, and regarding their art experience. Many of our learners are also first-generation college students who arrive without the context for higher education academic culture that others may have. The community and lasting relationships that form amongst students during the Foundation year often continue throughout their undergraduate experience and

S. Stokes (✉)
Art Department, Foundations Coordinator, SUNY New Paltz , New Paltz, NY, USA
e-mail: stokess@newpaltz.edu

S. M. Wyman and J. Korenblat (eds.), *Sustainability Learning for Action and Community Engagement*,
https://doi.org/10.1007/978-3-032-11698-7_30

beyond graduation into critique groups, job market opportunities, and more.

We believe it is vital to catch students early on their journey and set them up for success in college, their future careers, and their lives. Over the years, we learned that outdoor education provides opportunities for embodied learning and holistic pedagogies (Clarke and Mcphie 2014). Students must view their existence as part of nature and resist Western culture's anthropocentric emphasis on separation and dominance (Rose and Cachelin 2018). Together, faculty and students explore the metaphors of individual trees and the community of actual trees by entering forests and studying and responding to a series of videos and podcasts. We ask students to write about what most surprises them in their research on tree communities. During each of the workshops at the *Ashokan Kick-Off* and *Chalk It Up!* events, we discuss this new information and its personal relevance to each participant, as well as to our collective humanity. Students are open and eager to share their thoughts about the community of trees, outdoor education, creating artwork from natural materials, and the ethereal, sometimes fleeting, quality of that work. Through these activities, they construct new knowledge they can own and share further.

Our curriculum also includes students working collaboratively in a variety of interdisciplinary ways. We provide experiences where learners are immersed in making artwork in and with the natural environment as much as possible. By rethinking *what an art object is* and how it can *make a mark on the world*, students achieve a new orientation to the planet and self that emphasizes process and connection rather than consumption and production.

The Foundation faculty has determined that focusing on the environmental, social, and economic dimensions of sustainability from the start is important for incoming students and helps connect their skill-building and art practices to current realities. As educators, we center the following questions:

- *How do we create community amongst beginning Art students, both first-years and transfers?*
- *How can we best assist students in learning about environmental and social sustainability so they genuinely embody this information as life-long learners/educators?*

- ***What can learners create individually and collaboratively with the materials in their surroundings?***

We decided, therefore, to think about ways to welcome these students and foster connection and a sense of belonging that would increase their ability to commit to college and their education. This chapter presents two successful models for environmental education, one wilderness-based exploration and one more adaptable to urban environments. These holistic learning activities introduce students to basic sustainability concepts and allow them to enter a campus-wide conversation on planetary and human thriving.

The Foundation Program

The Foundation Year at SUNY New Paltz is an entire experience, more than a series of classes. Students are challenged to ask critical questions in art, craft, and design and interrogate their roles as thinkers, makers, and participants in visual culture. Assumptions are challenged, risk-taking is encouraged, new concepts are explored, techniques are developed, and traditional and digital studio methods and media are introduced. Foundation emphasizes community, collaboration, and issues around sustainability through program-wide activities and events. The Foundation Year experience prepares first-year students for creative and critical engagement and provides a basis for continued study in all areas of art and design. Our Foundation courses are Drawing: Visual Thinking I & II; Design: Color, Design: Form, and Art Seminar.

The Foundation Year Student Learning Outcomes are as follows:

Art/Design History, Theory, and Critical Thinking

Students will:

- Gain awareness of the social responsibilities of the artist and the designer.
- Learn to engage in discourse, analyze works of art and design perceptively, and evaluate them critically—both written and oral.

- Develop an understanding of the common principles, elements, and vocabulary of art and design, appreciate their interaction, and be able to use this knowledge in analysis.
- Synthesize critical reading in context with both individual work and class dialog.
- Acquire the ability to place works of art and design in historical, cultural, political, and interdisciplinary contexts.

Studio Methodology

Students will:

- Achieve competence in principles of visual organization, including the ability to work with visual elements in two- and three-dimensional, time-based media and emerging technologies.
- Learn methods of generating ideas and subject matter through expansive research.
- Understand the importance of developing standards of craft and material appreciation.
- Become familiar with the historical achievements, contemporary issues, processes, and directions in a variety of disciplines in art and design.
- Learn about Sustainability through materials and cultural context.
- Be afforded opportunities to exhibit their work and to experience and participate in critiques and discussions of their work and the work of others.
- Present work that demonstrates openness to varied methods of production, conceptual development and awareness, and expansion of technical facility in specific areas of interest in art and design.
- Critique, justify, and present choices in the process of analyzing, selecting, curating, and presenting artwork for informal exhibitions in the fall and the formal Foundation Student Exhibition in the spring.

Current Sustainability Learning Outcomes

- Define and research social/ environmental/ economic dimensions of Sustainability.
- Learn about Sustainability through materials and cultural context. (Methodology)
- Gain awareness of the social responsibilities of the artist and the designer. (Critical Thinking)

Pedagogical Frameworks

As Art Department faculty, we help our incoming students learn that one can make art from any material, and we focus on the process more than the product. We also explore issues around preciousness and perfectionism while making art in nature, as perfectionism can stop students in their tracks and stifle artistic progress. The process of working with unfamiliar materials and focusing on making and process rather than the product is freeing for students and often allows for more creative experiences and use of materials than they would have ever imagined. Aligned with inclusive Universal Design for Learning methodologies, faculty talk with students about what we are trying to do pedagogically and why (Wiggins and McTighe 2005). Setting up a democratic, student-centered, and student-driven learning environment facilitates a transformative experience and encourages learners to let their guard down a bit and be open to the possibilities of creating in a variety of ways and seeing themselves as creative beings.

In many of our Foundation courses, students research the United Nations Sustainable Development Goals, preparing them for the *Foundation Kick-Off Event* at Ashokan Reservoir in the Catskill Mountains and for *Chalk It Up!* activities.

Sustainability Competencies

We also engage several sustainability competencies through these events to prepare learners to be effective advocates and critical thinkers (Brundiers et al. 2021). The Interpersonal Competency is at the forefront of our introductory activities, which involve a series of icebreakers: large and small group collaborations that require working together to achieve a goal. The Systems-thinking Competency was addressed through learning about trees as a connected and interdependent community, which students identify as a metaphor for our human society. For the Values-thinking Competency, we addressed the value of *nature*, the myriad of information points we can learn from it, and its importance to us as humans and learners. Students articulate the need to sustain our environments and the people within them. Strategic thinking and Problem-solving Competencies are inherent in making art, as artists solve problems through their creative practices and negotiate how to represent, express, and use materials in a variety of ways. When researching, thinking, and writing responses to questions such as *What will our trees, world and society look like in the future?* and *How can we plan for that?*, we address issues relevant to the Futures-thinking competency.

Learning Experience #1 Foundation Kick-Off Event: Wilderness Field Trip

We have the great fortune of the Ashokan Center in Olivebridge, about thirty minutes from our campus. We are in our ninth year of taking Foundation students to this sustainability-focused learning center by the Ashokan Reservoir to immerse them in nature. This enables students to look closer at their surroundings, learn to be comfortable in natural settings, and make art onsite. Many of our students have little exposure to natural settings as they come from the five boroughs of New York City or they may not have made experiencing nature a priority. We also want to introduce them to the history and culture of the Hudson Valley and the surrounding region.

The most recent *Foundation Kick-Off Event*, attended by 220 students, was at the Ashokan Center three weeks into the fall semester. In August, as the Foundation Coordinator, I send out a letter to the incoming students about this event so they can mentally prepare and bring the proper clothing to campus. In the letter, I let them know that this will be a memorable day, co-existing with a group of artists/designers/craftspeople/visionaries-in-the-making. There will be workshops and the opportunity to enjoy the Hudson Valley landscape. We ask students to bring their sketchbooks, camera/phone, comfortable walking shoes, rain gear, insect repellent, sunscreen, towel, and swimsuit. We also ask them to bring their curiosity and good humor. This event is required for all students (majors and minors) and is funded by a portion of the course fees and a small grant. Students are excused from their other courses for the day.

Students leave early to enjoy a lovely breakfast at the Ashokan Center, where they meet staff, new peers, student mentors, and additional visitors from our campus. This is a critical part of the day, as students meet all the Foundation faculty and several former Foundation students who act as mentors/assistants. This is also a great time to put names and faces together and meet the Dean of Fine & Performing Arts and the Chair of the Art Department. Students will be working with this team of people for their entire time at SUNY New Paltz; it is extremely important to the incoming students, and they begin to feel at ease asking questions, talking with all of us, and sharing details about their lives.

The day unfolds into a series of workshops. During free time, students are encouraged to go canoeing, swim near an extraordinary waterfall, hike the grounds and trails, and make wreaths out of found materials to wear on their heads. They can also nap under a tree, work in their sketchbooks, take photos, or simply hang out with friends. Each year, we hear from students how much they enjoy these open-ended activities (Fig. 30.1).

Fig. 30.1 *Sticks and Stones*, Ashokan Center, (2023) James Fossett

Workshop Overview

The team-taught workshops, focused on Environmental and Social Sustainability, include a mix of options from the Ashokan Educational team and from the SUNY New Paltz Foundation Art faculty. We invite other Art Department faculty to join us and teams teach from a menu of workshops:

Sticks & Stones where student groups work collaboratively to create a series of sculptural installations on the Ashokan land. They collect sticks, twigs, stones, and a wide variety of organic materials (without tearing up the forest) to make large organic sculptures. Their focus is on structure, patterns in nature, motion, and fun!

Nature's Palette, where students explore how to make traditional and inventive drawing materials and tools by foraging for natural resources on site. This workshop also involves making and drawing with black walnut ink, a new experience for most students.

Sounds Like Nature is a workshop where students collect resources they cannot touch to make art they cannot see. By relying on the experiential progression from hearing to sound, they listen deeply to the natural world and form a collective, sound-based reply to the environment and each other.

Drawing in Nature has participants draw a natural form in its setting. Close observation of the subject helps one to create a detailed foreground. In the background, the landscape will be loosely sketched for context. This workshop focuses on an eye/hand connection to the subject and a heightened awareness of the unique place.

Blind & Double-Blind Contour Drawings where students walk silently and calm their minds as they move through the trees. When they feel settled, they stop and look around for what captures their attention. Working in their sketchbooks, they begin with blind contour drawings without looking at the page. They walk a bit more and move to double-blind contour drawings. They are asked to look straight ahead and to slowly follow the contours of a tree-trunk, a branch, a rock with their hand, moving their pencil over the paper at the same slow speed.

Charcoal in the Expanded Field is a workshop where students learn to make their own charcoal. They forage for a variety of woody plants and carbonize them in a large portable retort/furnace fueled by fire. They test the resulting char as drawing medium and ground pigment, and also explore the role of carbon as a sculpture material.

Invasive Vine Collaborative Sculpture is a workshop that considers the ecosystem impact of invasive plants and uses Ashokan-sourced vines to create a kinetic sculpture that transforms linear elements into a dynamic three-dimensional form.

One double-blind drawing student enthused:

> I really enjoyed the tranquil vibe of the first workshop I selected, the blind and double-blind contour drawings in the woods nearby the Ashokan Center. The activity had us focus on our surroundings and really ground ourselves ... getting us intimate with the subjects we chose to draw. The activity where we would be basing our contour on the feel of the woodland objects without looking at them was particularly interesting as it had me drawing based on a sense besides visuals.

Creating and using charcoal, one student dreamed of its eco-art potential:

> My friend and I worked together in the charcoal workshop to collect dried twigs, nuts, leaves and other materials to be carbonized....I thought it was remarkable that charcoal could be made from nearly anything. Before this year I had not made artwork from foraged/natural materials, and this was very inspiring. I hope that I can build a self-sustaining and environmentally conscious art practice for myself using similar methods.

Sculpting with vines, one student observed:

> Working with the invasive vines came easier to me than expected. I found that the temperamental nature of the vines forced us to experiment in understanding how to use them in a sculpture. The abundance of these materials allows for trial and error in said experimentation, which prevented the activity from being frustrating.

The students' feedback about this day in the woods is consistently positive and *used with permission*. It can be challenging for some who do not usually spend time outside or engage in physical exercise, but those are often the ones who have the most memorable experiences and gain the knowledge to continue these types of activities later on their own. Several years ago, for example, I was walking up a moderately steep trail with five female students who were having a hard time keeping up with me. I asked why they thought that was, and they all agreed that they just weren't used to walking or hiking much and wished they were in better physical shape to be able to enjoy nature in these ways. Within a month after going to Ashokan, these young women had formed a hiking club and invited other interested students to join. They explored numerous trails near campus throughout the year and came to tell me all about it with proud smiles on their faces. I was pleased to know that their experience at Ashokan set an idea into action that could quite possibly change the course of their lives.

Cultivating a Sense of Place Leads to Appreciation of Human Impacts

As a breathtaking landscape transformed by human technological intervention, the Ashokan Reservoir provides a history lesson in environmental and social sustainability challenges. Prior to visiting the Ashokan Center, students learn about the decisions made and the methods used to create the principal water source for New York City, 100 miles to the south. The reservoir, located 14 miles west of the City of Kingston, was formed in 1915 by the construction of 5 ½ miles of dams and dikes within the Esopus Creek drainage. This 8,315-acre reservoir is a major component of the Catskill System of the NYC water supply (dec.ny.gov). Students learn about how European farmers took up residence along Esopus Creek in a series of settlements destined to be consumed by the waters of the reservoir (Powell 2024). After the construction of the reservoir, some towns, including Brown's Station, Olive City, Broadhead's Bridge, and Ashton, ceased to exist. Other communities such as West Hurley, Shokan, and Boiceville, were relocated, a highly controversial decision with regard to environmental, social and economic sustainability. (http://publicwater.com/story-of-nyc-water/ashokan-reservoir/) When students walk through the Ashokan gorges and sit by the water with this deeper understanding of the history of a place and the environmental impact of creating this beautiful pseudo-wilderness, they appreciate how complex the environmental systems are and the implications of such human impacts.

Students tend to react strongly to the story of the creation of the Ashokan Reservoir as their critical thinking leads them to the complexity of this transformation of place: (*comments used with permission*).

> 2,000 families, 500 homes, 35 stores, 10 churches, 11 schools and 5 railroads had to either relocate or be destroyed. Graves even had to be dug

up and relocated. For me, it was the beginning facts that really got my attention.
The people couldn't believe they had to lose their homes to support New York City. Ashokan is called the "last of the homemade dams."
The people being moved from Ashokan were upset because they lost their land and memories, but they obviously didn't realize that the water system would provide an incredible benefit for the quality of life for people in NYC, which was/is also composed of immigrants, lower income, people of color, little to no education, etc.

Assessing Student Learning and Sense of Belonging

The Foundation professors have a discussion with each class to better understand the student experience at the Ashokan Center and how learning happens. We asked them about the itinerary and timing of the day in general, and whether they were comfortable at Ashokan and met new friends. We revisit the materials they read before going on the trip.

We are interested in the specific experiences students had in their two workshops. We invite them to characterize the collaborations that took place in many of the workshops and how they felt about what happened. They reflect on making art with the materials at hand and whether they learned any new skills or went more in-depth with skills they had coming into this day-long event. We asked them whether they would come to Ashokan again and what they would tell incoming students about their experiences. Rather than using a finished artwork to assess student learning, we probe student understanding and appreciation of the creative processes they initiated and experienced throughout the day.

Since we have been going to the Ashokan Center for nine years, we have collected significant data from the students and adjusted our plan each year according to survey comments from the years before. We are now able to prioritize accommodations for special needs students and help them to choose workshops that will best suit their abilities. We find that if there is a pattern of dining, workshop, free time (and then again) that timing works the best for all involved.

Students are generally very pleased to have spent the day at the Ashokan Center, immersed in nature in various ways, working collaboratively with other students without any judgment or expectation of an amazing final product. When we remove this expectation, students thrive in ways we do not usually see in the classroom. We now bring this process-oriented thinking and creating into the classroom, even more than before, with great success.

Learning Experience #2 *Chalk It Up!*

In the Fall of 2021, due to the Covid pandemic, we knew it was not wise to put students in enclosed spaces or buses to the Ashokan Center, so we collectively planned an Art Foundation-wide learning experience closer to campus that would meet our goals of social connection and environmental stewardship. We maintained our Foundation commitment to helping our students understand sustainability concepts and what they can do in their lives to help change the trajectory of issues specific to environmental, social, and economic thriving.

On an afternoon in mid-September, students, faculty, and staff of the Art Department transformed an empty parking lot into a space for artistic exploration and collaboration. The event was called *Chalk It Up! Foundation Community Event / Branch Out—Make Connections—Build Community—Find Common Roots.*

In keeping with our stated Foundation program goals, we used the same set of objectives for the *Chalk It Up!* event. We asked our first-year and transfer students to consider how trees form a community from the roots up and how these trees communicate as individuals in a forest. Students were also asked to consider how much social interaction relates to their growth as individuals and how their community contributes to that experience.

On that very warm (90° + F) day, 170 Foundation Art students met with Art faculty and staff at a reclaimed parking lot site to create their own forest together. This massive collaboration had students working in smaller teams to depict a large variety of tree species, which, throughout the course of the day, turned into a captivating forest full of known and fantastical trees, undergrowth, critters who exist within this world, and more for viewers to contemplate and enjoy (Fig. 30.2).

One goal for *Chalk It Up!* was for students to understand the science and supportive nature of how trees coexist as a community and how this relationship is a metaphor for learners themselves in terms of community building, branching out, making connections, and finding common roots. Another important goal was for students to work together to collaboratively create an ephemeral work of art and to have an unforgettable,

Fig. 30.2 *Chalk It Up!* (2021) Suzanne Stokes

positive experience. They created their forest by breaking into self-chosen groups of 10–15 students, often inviting in students they didn't know. The groups then organically negotiated the space they would work within and where there would be overlapping imagery. The chalk drawings from various groups merged together in surprising and unexpected ways, which included a variety of real and imagined trees, intricate root systems, undergrowth, woodland animals, birds, and insects that live both above and below the earth.

In preparation for *Chalk It Up!* Art students watched three videos and wrote about each video, focusing on the information that was the most surprising to them and brainstorming how they could be part of environmental change. They watched Real Science's "The Secret Language of Trees," Suzanne Simard's TED talk "How Trees Talk to Each Other," and Peter Wohlleben's "Talking to Trees." A Radiolab podcast, "From Tree to Shining Tree" (McEwan and Farrell 2016) provides an introduction to systems thinking:

> Trees require more than just carbon to grow. The nutrients fungi supply, including nitrogen, phosphorus and calcium all help trees with the essential mineral nutrients they need for survival. Fungi tubes that are wrapped around a tree's roots extract these minerals from soil in order to provide for the tree. Other methods of nutrient extraction such as carcass decay, soil dwelling insects, and sugar networks are all beneficial for the tree and fungi interconnection. These massive scale networks operate through large forests, making fungi an optimal source of nutrient exchanges for our environment.

After completing the research and writing exercises, students chose three trees that were of particular interest to them and carefully drew each one of them in their sketchbooks or on large pieces of drawing paper. Two of the trees they drew from observation—we asked them to go outside, locate their trees, get comfortable, and start drawing! They were to observe the trees very closely, paying attention to their specific forms; the way they move through space; color palettes that exist within them; texture of the bark; type and shape of leaf, seeds, and configuration; type of branches; their root systems (both seen and unseen); creatures they may attract; and settings where they grow well.

The third tree, in contrast, was drawn from students' imaginations. We provided examples of trees in films, literature, and other cultural references, including the Truffula Trees in Dr. Seuss's *The Lorax*, the Ents in J.R.R. Tolkien's *The Lord of the Rings*, the Groot in Marvel Comics' *Guardians of the Galaxy*, the Whomping Willow in J.K. Rowling's *Harry Potter*, the Mother Tree in James Cameron's *Avatar*, the apple trees in L. Frank Baum's *The Wizard of Oz*, Garden of Eden, the Tree of Life, Shel Silverstein's *The Giving Tree*, the Tumtum Tree in Lewis Carroll's *Jabberwocky*, and countless others.

Many of the students became very interested in the subject matter and decided to research further, e.g. Peter Wohlleben's *The Hidden Life of Trees*, Suzanne Simard's *Finding the Mother Tree: Discovering the Wisdom of the Forest*, Cesare Leonardi's *The Architecture of Trees*, Robin Wall Kimmerer's *Braiding Sweetgrass*, and a film by Louie Schwartzberg called *Fantastic Fungi*.

Discussion

When Foundation professors discussed the *Chalk it Up!* mural event with their classes afterward, they focused on the tree research, the process itself, and the outcome of the students' extraordinary efforts. We received extremely positive feedback from the Foundation Art students who were excited to learn about the complicated and intricate systems and communities of trees, mycelium, and other growth that supports or destroys this network. Many students wondered why they did not know this information before and were pleased to start digging into issues related to sustainability. They were also delighted to spend time with their classmates making work and strengthening their own social systems of sustainability within a public space where campus community members spoke to them and asked questions as they walked by. Many of the Art students felt empowered to disseminate the information they researched about the systems of trees and the surprising way the forests' ecosystems work together for the betterment of all.

During the debrief as a teaching group, we discussed the successes and challenges of this event. The primary challenge was getting permission from the college administration to work in the parking lot. There was a lot of paperwork involved, and much explanation was required. Before the students arrived to make the mural that day, the faculty went to the space and marked-out areas where imagery would go; as the parking lot was irregularly shaped, and we wanted to maximize the space. The students were eager when they arrived, got into their groups, and began brainstorming their approach. As faculty, we learned that we cannot predict final outcomes. For example, a few groups worked in a different orientation to the picture plane than the other groups despite our marked-out areas, which provided a good learning moment for us and the students.

Conclusion

The Foundation faculty choreographs learning experiences in physical spaces where students can be creative, where they can solve problems using art, where they can build relationships so that our challenges are shared, and where each can make a mark as a symbol of their impact on the world. Whether one is in a forest or a parking lot, we can see the impacts of climate change and we can see the impacts when people fail to care for their locations or each other. We feel that it is vitally important to set the stage for our students to enjoy their college experience and feel connected. We know that outdoor education is paramount. Our students explore their own creativity, working in any space with any materials, come to their own conclusions about the situation they are in, and investigate their potential to be change agents in our ever-changing world.

Through these events, students begin to understand that they are caretakers of the forests, waterways, and earth itself and rally for the environment. They value spending time in these important spaces, waiting, being quiet, and experiencing their surroundings organically, without imposed expectations or a need to produce. These activities are appropriate for many levels of learning, including K-12, undergraduate and graduate college students, elder communities, and non-profit arts, science, and sustainability organizations. This material can be altered for any situation or location with limited supplies. Students desire to learn more about environmental, social, and economic sustainability in our classes and to share their knowledge with others.

References

Brundiers, K., et al. 2021. Key Competencies in Sustainability Higher Education—Toward and Agreed-Upon Reference Framework. *Sustainability Science* 16: 13–29.

Clarke, D., & Mcphie, J. 2014. Becoming Animate in Education: Immanent Materiality and Outdoor Learning for Sustainability. *Journal of Adventure Education and Outdoor Learning*. Institute for Outdoor Learning.

Lima, Manuel. 2014. *The Book of Trees: Visualizing Branches of Knowledge*. Princeton Architectural Press.

McEwan, Annie and Brenna Farrell. 2016. From Tree to Shining Tree. Radiolab (July 30) https://radiolab.org/podcast/from-tree-to-shining-tree.

New York Department of Environmental Conservation—Ashokan Reservoir, https://dec.ny.gov/places/ashokanreservoir#:~:text=Ashokan%20Reservoir%2C%20located%2014%20miles,New%20York%20City%20Water%20supply.

Powell, Eric. 2024. Ghost Towns of the Ashokan Reservoir. *Archaeology* (May/June). Accessed 8 June 2024, https://www.archaeology.org/issues/554-2405/letter-from/12312-catskills-ashokan-reservoir.

Rose, J., and A. Cachelin. 2018. Critical Sustainability: Incorporating Critical Theories into Contested Sustainabilities. *Journal of Environmental Studies and Science* 8: 518–525.

Wiggins, G. and J. McTighe. 2005. *Understanding by Design*, 2nd ed. Alexandria VA Association for Supervision and Curriculum Development ASCD, https://www.researchgate.net/publication/318021095.

CHAPTER 31

Making Global Engagement for Sustainability Accessible to All

Ş. İlgü Özler

Introduction

The United Nations Sustainable Development Goals (SDGs) present a vision of a world where "no one is left behind." The Agenda 2030 of the SDGs envisions a holistic approach to development where people live in dignity in a healthy environment. Higher education institutions are at the center of carrying out these aspirations (UNESCO 2014). SDG 4 calls for "Ensuring inclusive and equitable quality education" and "Promoting lifelong learning opportunities for all." Public higher education institutions can provide quality education and upward mobility through programs that are intentionally designed to give students access to careers. Yet, some careers are not easily accessible to all students. This is especially the case for those who seek to engage in global policy making, advocacy

Ş. İ. Özler (✉)
Department of Political Science & International Relations, SUNY New Paltz, New Paltz, NY, USA
e-mail: ozleri@newpaltz.edu

Director SUNY Global Engagement Program, New York, NY, USA

S. M. Wyman and J. Korenblat (eds.), *Sustainability Learning for Action and Community Engagement*,
https://doi.org/10.1007/978-3-032-11698-7_31

and humanitarian work as a part of international organizations. Careers in these fields require access to networks that are often beyond the reach of working- and middle-class students at public universities. In this spirit, the United Nations Semester and the Global Engagement Program foster sustainable education in substance and program design by giving students an avenue to pursue global engagement at home in New York State, US.

The substantive educational components of these two academic programs include broad global issues. Students focus on ethical, normative, and legal practices around the three pillars of the United Nations: peace and security, human rights, and sustainable development. The courses reflect on the impact of current global practices and how they intersect with global justice, encouraging discourse around sustainability as a complex and contested issue (Scarff and Ceulemans 2017). Students discuss potential solutions and responses to climate change and environmental degradation, immigration and refugees, health policy, inequality, mass atrocities, and conflict. They question, define, and critique what it means to be a global citizen as they learn and practice in these programs (Croucher 2018).

In the course of their studies, students from across the disciplines directly engage with Agenda 2030 of the SDGs. The SDGs are assumed to be an essential component of the UN's ideals for a more peaceful world. The interdisciplinary and cross-campus engagement allows for increased diversity of ideas and a more holistic, substantive approach to sustainability (Cachelin et al 2019; Mossman 2018). Learning expands beyond the classroom as students interact with campus and broader community members as well as global actors and develop work, leadership, and advocacy skills (Chang et al. 2021). They find pathways into global governance institutions as advocates for sustainable change or as humanitarians who mend the wounds of global failures. Those who eventually join the private sector carry the notions of sustainability into positions in the for-profit world.

Public university students from rural and urban New York alike typically do not have access to international organizations that offer positions directly engaged with global sustainability. The programs discussed in this chapter lift barriers to students' opening up to the field of global affairs and related job opportunities by developing transferable twenty-first-century career skills, including critical thinking and ethical reflection on major global issues. Public universities can empower students from all backgrounds to broaden their perspectives and access globally engaged careers in sustainability-related fields virtually or in person. The *UN Semester* and the *SUNY GEP in NYC* improve inclusion, diversity, accessibility, and equity for students in international and multilateral sustainability fields. Creating pathways for students from low-income, rural, and immigrant backgrounds into positions of global engagement at the United Nations also allows for underrepresented views to enter into intergovernmental organizational settings, thus providing crucial perspectives. In addition, the students have the opportunity to make ethical choices in global action as they learn about the UN SDGs and the critiques associated with them. Thus, the pedagogical approach fits within the framework of key competencies put forth by Wiek et al. (2011, 2016) (Brundiers et al. 2021). These key competencies—systems thinking, normative/value thinking, strategic thinking, anticipatory/futures thinking, and interpersonal and collaborative competencies—are discussed in the context of the program descriptions below.

United Nations Semester Course

The Department of Political Science and International Relations has offered the United Nations Semester course since 1978. This course takes advantage of the United Nations' location, about one-and-a-half hours' drive from campus. While New Paltz has the geographic advantage, the virtual programming through the UN Speaker's Bureau of the Department of Global Communications makes all campuses proximate to the United Nations. Students can take virtual tours of the UN and hear directly from UN officials. The six-credit course historically was designed to have three credits on campus like a regular international governance course exploring the United Nations within the global governance literature of Political Science. For an additional three credits, the students hear from UN officials, representatives of country missions to the UN, and non-governmental organizations that advocate at the United Nations. The course is now in hybrid form, where students experience part of the briefings in person at the United Nations and the rest virtually. However, the course could be delivered entirely online.

The UN Semester course covers the three interconnected pillars of the United Nations: peace and security, sustainable development, and human rights. The course brings theory and practices together to expose students to the work of the United Nations in real-time through briefings presented by UN officials and others directly engaged in global efforts to advance the SDGs and related goals. Students engage in normative/values-thinking as they utilize academic literature to reflect on and analyze these communications. They discuss UN sustainability practices with those who directly engage in the work and are encouraged to engage in systems thinking through an analysis of the interconnections between environmental sustainability and economic development at a global scale. They study issues around climate change, immigration and refugee crises, health policy, mass atrocities, and interconnections between these issues, and their intersection with peace and conflict.

Students are also presented with real-life conflict areas each year and hear perspectives from various sides of international conflicts, including UN officials working on the issues as international civil servants. They hear from UN representatives who have recently returned from the field and are working to broker peace in conflict zones or aid in humanitarian crises. Representatives of many NGOs also present to students about their advocacy work at the UN.

Yet, most of the UN's work is on sustainable development and human rights. Thus, students mostly hear from those supporting the SDGs' review and implementation. Since 2015, much of the UN's work has been shaped by Agenda 2030, and nearly all briefings touch on the SDGs in some way. Students learn about how these 17 goals and 169 targets are "specified, applied, reconciled, and negotiated." Students have opportunities to question and discuss the practices' viability and value with those carrying out these goals in the field.

The UN Semester course offers students the unique opportunity to engage with practitioners, but not necessarily to break into careers in global governance institutions like the UN or its surrounding organizations. The SUNY Global Engagement Program in NYC was developed to bridge this gap and allow them to become globally engaged agents of change.

SUNY Global Engagement Program in NYC

Globally engaged intergovernmental and non-governmental organizations positions are highly competitive, and experience and network connections are required to enter the field. Students can learn a great deal about international governance and the efforts of a wide range of actors and organizations to bring the SDGs to fruition through the UN Semester Course, but this alone is often insufficient to successfully launch a career in areas where students can have an impact on sustainability. By merging theory with practice, the SUNY New Paltz Global Engagement Program in New York City (NYC) provides a practical stepping stone into this field. Students spend a semester in NYC, earning twelve to fifteen credits. They do six to nine-credit internships ranging from twenty to forty hours per week and take two courses at SUNY Global Center in the heart of New York City. The credits are issued through the Study Abroad program at SUNY New Paltz. There are three components to the program: 1) an internship with an organization engaged in global issues, 2) a seminar through which interns come together to share experiences and to process and learn about the complex challenges and contradictions associated with this work, and 3) a research colloquium that allows students to engage deeply with issues of their sponsor organization. The seminar and research colloquium are venues where students process their learning, situate their work in the broader context of social change processes, and analyze complex global issues from an ethics and justice perspective. The hands-on experience students gain during their internships provides the raw material for this learning. Below, I first describe the types of internships that are part of the GEP and the practice of building partnerships with globally engaged organizations. Then, I turn to how these experiences are contextualized and analyzed in the academic classroom components of the program.

From Practice to Theory and Back Again: Internships

Internships serve as the foundation of the GEP, and establishing good relations with internship providers is vital to any such program. In line with the sustainability education ideals, the internships are cultivated from organizations emphasizing public service, including the United Nations, country missions to the UN, international non-governmental organizations, and community organizations with a global focus. The program nurtures a strong relationship with these organizations by providing reliable, vetted interns upon whom organizations can rely to help carry out their work. In turn, supervisors appreciate the educational component of these internships, especially given that they bring in talented students from diverse backgrounds. Each student is interviewed to identify their interest areas and potential placements in an internship that is globally engaged. These students develop skills in presenting themselves to the outside world for the specific field of study. Supervisors consistently mention how refreshing it is to have interns from public higher education institutions who are hungry to learn, eager to perform well, and who care about the impact of their work.

The collaborative spirit of the internship program is not limited to partnerships with the internship site organizations; mutual support is facilitated among the students. During the internship placement process, students are encouraged to cooperate, share internship information, and offer other forms of support. They are informed that they increase their chances of finding the right internship by collaborating and sharing information about placement opportunities. The students are encouraged to develop specific skills through self-education over the summer preceding their fall placements if they do not possess these capabilities. For example, suppose an organization is looking for someone able to support a social media campaign, and the student is not familiar with the platforms used in this setting. In that case, they are encouraged to develop these skills using publicly available online tools. The internship search and work experience and sharing these experiences in a classroom setting develop interpersonal/collaborative competency skills that "motivate, enable, and facilitate collaborative and participatory sustainability research and problem-solving," as emphasized by Wiek et al. (2011, 211).

To facilitate reflection on their internship experience, students write a weekly essay describing their work and their assessment of their contributions. Students are encouraged to derive maximum benefit from their experience. Internships sometimes may involve mundane work that is helpful to the organizations. Thus, even if interns perform simple tasks like document filing, they are encouraged to learn what is in those documents and how they are relevant to the organization's work. This allows them to understand the difference between a career in sustainability contemplating substantive issues versus an administrative job. Attaching meaning to every contribution enables the students to understand the value of their work for these organizations. As students develop a sense of efficacy and appreciation for their labor, they become agents of change rather than remain idle receptors.

The practices in this internship program allow students to understand their micro impact in their organizations and what that may mean for a sustainable future. They carry this into a more substantive assessment during class discussions in seminar and colloquium settings.

Internship experience and the work that sponsor organizations carry out is contextualized in the seminar portion of the program. Students need to process their internship work and fit it into the larger effort to advance global goals. The seminar's main purpose is to grasp this bigger picture; to understand global problems and the student/individual role in finding potential solutions. The readings assigned for this class are social scientific, connecting theory and policy (see Earthrise Commons; earthrisecommons.org). Students are encouraged to practice critical thinking in questioning the shortcomings of theory and policy solutions. They situate their micro experiences as advocates or humanitarians on the ground within the macro impact of their organizations through the academic research they read. Ultimately, the class encourages the students to reflect on their global citizenship roles.

The GEP Seminar: Ethical Dilemmas and Contested Justice in Global Citizenship

Before sustainability intentions can become practice, learners need practical knowledge of what is ethically responsible. The course starts with the theoretical foundations of international relations and the most important global standards and agreements. Students learn about the UN Charter, Universal Declaration of Human Rights, other international human rights agreements, the SDGs, the Geneva Conventions, the Rome Statute, and the International Criminal Court. These norms and international agreements provide a base from which to understand complex aspects of human rights, humanitarian law, and sustainability.

Students build a solid understanding of current practices for peace and security, human rights, and sustainable development and then offer their critiques of these ideals using various critical perspectives. They identify and analyze the missing parts and contradictions in these documents and the challenges to fully implementing these aspirational norms by the world community.

The course material introduces global problems using a social scientific lens yet acknowledges that these measurements have certain biases. One example is based on a discussion of poverty and inequality (UN SDG Goal #1 No Poverty and Goal #10 Reduced Inequalities). Students are provided with multiple definitions of *poverty* from different theoretical perspectives. Ultimately, definitions of *poverty* and *inequality* have implications for how they are measured and the policy solutions offered to combat these problems. The students grapple with whether defining extreme poverty as income below $3.00 per day is a useful measure.

Students debate whether overcoming inequalities should be done through redistribution rather than the generation of further growth tied to the issues of climate change, consumption, and environmental destruction (Goals 12, 13, 14, 15). In another setting, students discuss the dilemmas of whether international NGOs should continue providing lifesaving aid in conflict zones even if some of the aid is used to feed combatants who are prolonging the war. These types of dilemmas are present in other settings as well. Students are presented with the debate among advocates on whether working with governments to improve the conditions of undocumented immigrant detainees is tantamount to legitimizing the criminalization of undocumented immigration. They discuss whether climate discussions should be prioritized around adaptation or mitigation and how organizations prioritize issues. While the international community has made much progress in the development of international law and norms to respond to complex global problems, solutions are not perfect. Students are encouraged to engage in futures- and strategic thinking regarding where and why there is progress— "what works"—to identify areas of failure, and learn from both.

Micro Impact with Global Implications

The seminar setting becomes a place for reflecting on students' substantive experiences at their internships tying theory and practice in a global context. When asked to reflect on the question: *how does one's micro impact fit within the larger global structures?* they develop an understanding of the distinction between advocacy vs. humanitarian work and policy vs. charity. Those who engage in direct humanitarian work can see their impact immediately, such as helping a refugee family integrate into New York City, supporting an undocumented immigrant as they apply for permanent residency, helping a local center run after-school programs for children in underserved neighborhoods, or empowering kids without housing as they gain agency to express their needs through theater and arts programs. Yet, students understand that these are not finite solutions; they are temporary stopgaps or limited measures to help those who suffer from global structural shortcomings and failures. In contrast to the immediate impact orientation of humanitarian work, those working in policy advocacy fields rarely see the concrete results of their efforts. Students working in non-profit organizations that advocate for such causes as the rights of persons with disabilities, gender rights, or climate measures, for example, understand that policy on the global scale moves very slowly, and it may take years to see tangible change.

Being in a seminar where students can situate their internship experience in the larger context of globally engaged organizations allows them to develop a more informed and nuanced position on the direction they want to take in their careers. They understand that there is no immediate visible progress for the policy advocate and there are never enough resources for the humanitarian. Thus, connecting students' day-to-day micro impacts to the larger context becomes key to motivating them to become effective advocates for positive change.

Global Citizenship as Ethical Practice

In addition to understanding potential impact, the seminar facilitates student engagement in values-thinking, reflecting on right or wrong, morally good or bad, and how we should live our lives to advance these values. In this context, examining what is acceptable and deemed ethical in relation to justice is imperative. We use the classroom as a democratic space to understand our global universal values from multiple perspectives. The ethical practices considered good for one group of people may come at a cost for others. For instance, an investment in a large hydroelectric dam project may bring clean energy to millions of people in a region while destroying Indigenous peoples' land, and impacting the biodiversity and livelihood of downstream communities. Globally, ethical practices include understanding our current values and their impact in an intercultural context and, most importantly, careful reflection on how ethics intersect with justice.

In the GEP Seminar, students grapple with the complex relationship between the ethical values of a cosmopolitan global citizen and "the citizen" (Croucher 2018). For example, practitioners of international relations, as well as scholars, consider global citizenship to be a positive virtue. Yet, some groups see the construction of global citizenship as an affront to many local traditions and values, while others welcome this connection. How do we reconcile the global and the local without compromising universal values? Global citizenship exists in the context of global capital, the neo-liberal agenda associated with colonial and neo-colonial legacies. Despite being linked to social democratic values, global citizenship, and cosmopolitan agendas, they still live in the context of the sovereign state. Therefore, the benefits offered through social democratic systems with high living standards are limited to citizens of rich Western countries. This citizenship is exclusive despite cosmopolitan aspirations. Students discuss who is "left behind" in the context of sovereignty in the neoliberal global era.

Research Colloquium: Building a Career Focused on Sustainability

The third component of the GEP is the research colloquium. The internship provides the immediate experience, and the seminar allows for reflection and contextualization of that experience. By requiring that students develop a research topic relevant to the organization for which they are working, the colloquium facilitates anticipatory/futures thinking. Students "analyze, evaluate, and craft" their papers on significant topics that have urgency and relevance to sustainability. Sustainability has been central to all global issues in their globally engaged internships and often centers student colloquium papers. Students are encouraged to write papers that are relevant to policy, where they can evaluate the current policy and offer potential solutions to global problems by engaging in strategic thinking. As in the seminar discussions, students engage in future thinking as they develop a response to their policy questions based on justice, fairness, and equity. They provide feedback to each other during the research presentations and develop a better systems-thinking understanding of the interconnectedness between different issue areas (Meadows 2008).

The colloquium project furthers the collaborative goals of the program. Some projects respond directly to the needs and mission of the organizations where students are interning. In addition to the reflection papers for their internship experience, the seminar paper is another way students understand the difference between building a career in these specific fields and having an administrative position in an organization. The program emphasizes that students can contribute to meaningful change by becoming experts in a specific issue area, an important component of building a career that contributes to sustainable change. This is part of the social capital that students attain from the GEP.

Engaging with the Community Beyond the Classroom

Another way both programs, *the UN Semester and the GEP*, have engaged outside of the classroom is at the community level. The students practice anticipatory/futures thinking and strategic thinking as they develop their interpersonal skills in exercises in classes open to other campus and extended community members.

In the Spring and Fall of 2014, the UN Semester class and the Global Engagement Program students respectively led roundtable discussions and invited community members other university and high school students. At this time, the UN SDGs were being developed, and the UN Secretary-General Ban Ki-Moon wanted input through a strong involvement by global civil society. In the United States, the United Nations Association-USA was chosen to lead the civil society feedback as the SDGs were developed. UNA partnered with stakeholder groups around the country, including the *UN Semester* and *the GEP*. The class space was used to organize round tables in the mid-Hudson Valley and New York City. These conversations focused on health, poverty, gender inequality, quality education, food security and nutrition, peace and justice, and climate change. The outcomes from the exchanges were submitted to the UNA to be incorporated into the final civil society report from the United States to the UN Secretary-General. The development of the SDGs thus included the voices of civil society.

The two program courses create a space of engagement beyond the traditional academic setting that benefits all involved because it allows SUNY New Paltz to deliver underrepresented views and voices to the discussion. In 2018, SUNY New Paltz ranked in the top three percent of all US colleges in social mobility and 101 on the Social Mobility Index out of 1,204 institutions as of February 2025. 27% of students are "low income," and the median early career salary is $45,000 (SMI). At least 35% of the students identify as students of color (https://www.newpaltz.edu/about/glance.html).

After adopting Agenda 2030, during the five-year review in 2020, the UN semester class once again took part in organizing roundtable discussions, asking: *How do we engage with the SDGs at the local community and campus levels*? These exercises educated campus and local community participants about the goals and invited them to reflect on how they envision sustainability locally. The goals were used to understand areas needing improvement and take stock of what we do best on campus and in the Hudson Valley. Using this model, the UN semester and the GEP continually engage the students in anticipatory futures— and strategic thinking as they connect the global goals to their communities and imagine potential for improvements. The workshops and roundtables in the Hudson Valley allow the views of rural, urban, and suburban communities to reach the global stage, including views of students from underrepresented demographics.

Conclusion

The courses discussed in this chapter practice sustainability through program design, content, and implementation. The students define their roles in achieving a sustainable future and the meaning of being a global citizen. They do this with an understanding of the SDGs' intention to "leave no one behind." The contested definitions and practices of *sustainability* are taken up in seminar settings, discussed with current practitioners directly through internships and research projects, and brought into the community for discussion. The students question what is ethical just as they study the most pressing issues of the day and contemplate the policy responses to these issues. About 72% of the program alumni are in public sector or non-profit jobs; this is much higher than the national average of these two sectors, representing 24% of the total workforce in the United States. The marriage of theory and practice makes the students gravitate collaboratively toward public service positions that seek to contribute to advocacy and humanitarianism.

References

Brundiers, K., et al. 2021. Key Competencies in Sustainability in Higher Education—Toward an Agreed-Upon Reference Framework. *Sustainability Science* 16: 12–29.

Cachelin, A., B. Clark, A. Darling, and K. Hackford-Peer. 2019. Weaving Equity and Sustainability into the Fabric of Higher Education: The University of Utah Experience. *Sustainability (United States)* 12 (6): 304–309.

Chang, C. H., Kidman, G., & Wi, A. (Eds.). 2021. *Issues in Teaching and Learning of Education for Sustainability: Theory Into Practice*. Routledge.

Croucher, S. L. 2018. *Globalization and Belonging: The Politics of Identity in a Changing World*. Rowman & Littlefield.

Meadows, Donella. 2008. *Thinking in Systems: A Primer*. Chelsea Green.

Mossman, A. P. 2018. Retrofitting the Ivory Tower: Engaging Global Sustainability Challenges Through Interdisciplinary Problem-Oriented Education, Research, and Partnerships in US Higher Education. *Journal of Higher Education Outreach and Engagement* 22 (1): 35–60.

Scarff, Seatter, and K. Ceulemans. 2017. Teaching Sustainability in Higher Education: Pedagogical Styles that Make a Difference. *Canadian Journal of Higher Education* 47 (2): 47–70.

Social Mobility Index. https://www.socialmobilityindex.org/. Accessed 19 February 2025.

UNESCO. 2014. *UNESCO Roadmap for Implementing the Global Action Programme on Education for Sustainable Development*. Paris: France, UNESCO.

Wiek, A., L. Withycombe, and C. L. Redman. 2011. Key Competencies in Sustainability: A Reference Framework for Academic Program Development. *Sustainability Science* 6 (2): 203–218.

CHAPTER 32

Wicked Questions and Writing: Program-Level Transformations

Matthew Newcomb

When educating citizens about sustainability, one significant challenge can be the piecemeal nature of the effort. The number of environmentally related programs has grown in the humanities, and frequently, an individual instructor or a particular course may emphasize sustainability topics. However, institutions may want to reach a broader spectrum of learners. At the same time, sustainability-related work requires complex and nuanced thinking and communication. One programmatic answer for reaching most learners while maintaining instructor flexibility around sustainability topics is the *wicked questions* approach, developed in a social policy planning context by Rittel and Webber (1973), who note that wicked problems have "no definitive formulation;" they can be set up many ways (161).

Wicked questions, problem-based learning, or topical issues have long been considered an effective way to think about the many perspectives required for education for sustainability (Singer-Brodowski 2016; Wals

M. Newcomb (✉)
Department of English, SUNY New Paltz, New Paltz, NY, USA
e-mail: newcombm@newpaltz.edu

S. M. Wyman and J. Korenblat (eds.), *Sustainability Learning for Action and Community Engagement*,
https://doi.org/10.1007/978-3-032-11698-7_32

et al. 2016). Such approaches work from a baseline in the pedagogical history of using questions. This ranges from Socratic methods to push students to work out answers themselves to contemporary question strategies as in *Understanding by Design* by Wiggins and McTighe (2005), which emphasizes finding questions to pursue or search out rather than simply answer. Wicked questions have no simple answers, are highly complex problems, cannot be fully addressed by any person or single discipline, and tend to combine both political and material factors. Questioning as a process fits well with continually seeking sustainability solutions-based approaches.

In English studies, for example, Beach, Share, and Webb's 2017 work on teaching climate change prioritizes student-led "critical inquiry," where initial readings prompt student exploration; students research their questions and present results to each other, and the interactions lead to small solutions and online publications around community action (29–31). The idea of wicked problems "having multiple problem definitions" can promote interdisciplinary experimentation that is often appropriate for entry-level composition courses (Gough et al. 2016, 121). In my field of Rhetoric and Composition Studies, this angle engages with Richard Marback's (2009) work in Composition on "wicked problems."

The term "wicked" can have negative connotations. Warren Berger has shifted to "beautiful questions" to emphasize how these big questions can create new solutions, innovations, and creative thinking (2016). Similarly, the "How-Might-We" questions approach (2024) from Stanford's d. school tries to "provoke meaningful and relevant ideas" in a solution-oriented sense. These are key perspectives to keep in mind, but I am using the term *wicked* because our writing program has used this terminology and as a reminder that these questions have no complete fixes. Climate instability is a clear example; it has no singular solution or agreement about who should work on it and how. The interdisciplinarity of this phenomenon makes the variety of educational activities and learners in Writing Studies an appropriate match for the topic.

The wicked questions General Education writing courses we offer at SUNY New Paltz are a preliminary step in our departmental commitment to engaging students and rendering course material relevant to their lives. The learners are usually newly in college and ready to offer a variety of perspectives on a common cultural issue/wicked problem. In fact, the transdisciplinary nature of such complex problems implemented at the preliminary program level, often through collaborative activities, can encourage learners to think about their majors or connections between GE courses and the perspective that another course or discipline might have on the issue.

Wicked Questions and Design Thinking

This chapter describes the institutional context for developing a wicked questions-based writing program and introduces that approach in more detail. A *wicked question*, derived from the idea of a *wicked problem*, is based on work in Design Studies "to draw attention to the complexities and challenges of addressing planning and social policy problems" ("Wicked Problem" 2023). This learning model will suggest potential advantages for learners in nuanced and long-term understanding of rhetorical communication and sustainability, for instructors who share flexibility with guidance and common purpose, and for the institution in terms of number of learners reached and setting a baseline for further sustainability-oriented education. The chapter also considers opportunities and practices around Teaching Assistant and instructor training in ways that do not place extra strain on them. In the context of thinking through sustainability, climate change is a prototypical wicked question whose players include the non-human. The *wickedness* of the question is based on the need to consider historical and geological time together as well as human/nature/cultural separations (Bonneuil and Fressozs 2016). This variety of species, disciplines, and approaches fits well for a General Education course with learners from any major. It can emphasize the collaborative nature and variety of expertise needed to improve local or global sustainability outcomes. Introductory, interdisciplinary courses and programs have been identified as High Impact Practices that further the retention of students, transfer of learning, and concept connections across disciplines, depending on implementation ("High Impact" n.d; Keup and Young 2017). A writing course section based on climate instability as a sustainability topic will be briefly presented as an example of how a specific wicked question course could function within the larger programmatic framework.

Beyond the traditional educational setting, a wicked, questions-informed method brings a more holistic perspective to the work and goals of various business, workforce development, and non-profit settings where there is potential to empower participants by developing similar skills in analyzing complex problems. Writing, design, and sustainability education interweave as a space of invention for more general work in environmental humanities, first-year seminar programs, and interdisciplinary education programs. These areas intersect through efforts to re-think the teaching of writing and institutional communication practices. This allows participants to move away from a mode of writing that emphasizes critique and toward one that prioritizes long-term audience engagement, felt opportunity, and innovation.

Our instructors (including graduate students) also engage with Design Thinking, another way of discussing revising communication practices around long-term problems and audience involvement in this context. As Lauren Garskie argues,

> The key to enacting meaningful learning via wicked problems is to adhere to the culture that supports design thinking, namely the courage to experiment with unknowns or ambiguity, and willingness to embrace failing. For learners, this means taking up more complex design issues, tinkering with new tools or technologies, trying unattempted approaches, and venturing into new terrains. (2022: 157)

In this sense, wicked questions and design thinking processes work together and push sustainability work less in the direction of preservation or conservation and more in the direction of experimentation and innovation. In terms of writing about sustainability issues, the emotional aspects are often central, whether environmental, social, or economic. Garskie calls for a balance between "creative tinkering" and the size of the problems addressed so "implementable outcomes" can demonstrate results rather than incite further frustration or even withdrawal from sustainability issues (158).

As an example of how a wicked problem matches with sustainability-informed approaches, consider the remediation of PFAS contaminants in Newburgh, New York. Not only does the process of initiating remediation require scientific studies on contaminants, environmental analyses, and technological tools, it also requires public communication, pressure on policymakers, legal knowledge, and community engagement (Lawson and Wyman 2026). Beyond the variety of fields of knowledge involved, the problem is a *wicked* one because there is no perfect or pure result. Contaminants may be removed from the Newburgh water supply, but many particulates are then sent to Indiana and burned—creating a different environmentally toxic impact on another community. This is not just greenwashing, where the goal is to look like environmental action is being done regardless of the actual effect. There are real positives and negatives to balance in a case like Newburgh's, including the impacts of forever chemicals on different communities. Bringing in questions such as jobs gained and lost through the remediation work, long-term health impacts, and what might replace the use of PFAS chemicals adds further complexity. Sustainability questions like this tend to be *wicked* because, for every "good" answer, there are "bad" or damaging results to be considered. The contexts for problems like this keep expanding until a conceptual wicked problem truly represents a socio-environmental issue in the broadest sense.

Put another way, Nicole Seymour uses the term "irony" in *Bad Environmentalism* to discuss the hard-to-disentangle outcomes of environmental action. Seymour uses the example of increased wolf populations near Chernobyl to note the positive effects of humans being absent in the area while still having the apparent dangers of major radiation there (2018: 4). As a wicked problem, with wildlife habitat preservation and biodiversity maintenance, the seeming good and bad arrive together and no pure or right solution exists. There are always more effects (and affects) that require new adjustments and experiments to respond to the dynamic dilemma.

Course Content in Context

First-year writing courses tend to have a long-standing debate about what their content should be. Writing skills are part of the course, but learners must write about something. Should it be about language, about writing itself, about current issues, about a theme the instructor finds compelling, about the varied essays in a textbook reader? While the history of answers to this question is too long to go into here (Berlin 1987), the wicked questions approach is also an attempt to provide another possible answer to the question of course content and interdisciplinary appeal. Taking action on a sustainability issue or wicked question also requires sophisticated rhetorical analysis and communication in a variety of contexts. This creates natural and practical relevance or connection between the topic, the writing activity, and the learner's own lived experience on an unsustainable planet.

Within Writing Studies and first-year composition programs, the subject of *ecocomposition* defines problem-solving approaches to writing and problem-based learning by centering rhetorical approaches and ecological or environmental themes (Dobrin and Weisser 2002). Given the history of interest in social and cultural issues in writing studies, ecocomposition becomes an area ready for further connections to sustainability education as it includes attention to human/nature intersections (sometimes termed *coalescence*) and the environment (which may be natural, urban, or imaginary). Like the wicked questions approach, Mary Wright and Florence Monsour base ecocomposition lessons on “essential questions” and “an interconnected systems thinking approach to curriculum writing” (2020: 204). As Wright and Monsour note, “Siloed disciplines housed within colleges may hinder a holistic view of sustainability, but the sustainability dilemma belongs to all and can bridge the disciplinary divide” (202). All writing work is interdisciplinary in this context, and the key questions or problems, rather than the discipline itself, guide the inquiry.

In the Composition Studies classic *Rhetoric and Reality: Writing Instruction in American Colleges, 1900–1985*, James Berlin covers a variety of ways public or social issues have served as central material for writing classes. These include the *Ideas approach* in the early part of the twentieth century, more political aspects of social rhetoric before WWII, up and epistemic versions of social/political rhetoric in the later years of that history (1987: 51, 81, 165). In this context, wicked questions are not a brand-new strategy; there is a history of composition studies addressing the rhetorical needs around public issues. New pedagogies, like a wicked questions approach, are appropriate because the historical context changes. In a time of ecological concern some call the Anthropocene, specific elements of complexity, global impact, and extended exigence are vital to fully investigate the impact of humans on environmental, social, and economic systems.

Institutional Transformation at the Program Level

Our institution’s Writing and Rhetoric program developed a wicked questions approach to its main required General Education course in response to several factors. The primary impetus for the change was simply a university-wide revision of the General Education program. The need to reconsider the work in all GE areas made for a natural transition moment. It helped bring some openness and flexibility since a change would happen anyway. Related to that, GE initiatives included a programmatic desire to move away from literary themes for GE writing classes into topics that would be more easily multidisciplinary, given that learners from all fields would be taking the courses. Finally, a partial institutional move toward a sustainability focus, including creating a Sustainability Faculty Learning Community (Hong 2020), gave a sense of potential support for course adaptations that might participate in education and action for sustainability.

A key early move for our program was to determine the structural mechanism that would continue to meet our writing requirements, allow freedom for faculty and others involved, and prioritize sustainability education in a new way. Working through the course theme or topic emerged as a primary way to alter the program and maintain faculty autonomy. Our compromise was to set the parameter of a wicked question as a course theme while leaving a great deal open to the instructor. Early wicked questions to drive the research and content of writing courses included "How can we live more sustainably?"; "What are children's rights?"; "How can the humanities address climate change?"; "What should we eat?"; and "How should we think about disability?" As readers can discern, some of these phrasings prioritized an environmental issue while others prioritized a social issue. The question of *what to eat* readily brings in environmental, economic, and social factors simultaneously. While sample questions were provided for instructors, and many shared the same wicked questions, one element we might do differently if given a chance would be to limit the number of wicked questions to a finite list. This would create space for more collaborative work among instructors, consequent refinement, and less reinventing the wheel.

A 2024 writing and rhetoric section from one of our instructors that utilizes a wicked question around environmental issues is a good example of the variety of sustainability topics that can arise. Students wrote papers on topics including the "aging electric grid in the US," "online vs. in-store shopping pollution," "food waste from bakeries," "fast fashion," "electric vs. gas cars," "landscapes' effects on culture," "digital immortality," "consumerism and Congo slavery," and "damage to American aquifers" among others (Christy-Robinson 2024). That course section includes action-oriented research for the assignments that might inform different publics. Taken together, the writing topics demonstrate a variety of sustainability possibilities and show how a course with an environmental question moves almost automatically to tie in social and economic factors.

In the introduction to *Prioritizing Sustainability Education: A Comprehensive Approach*, Joan Armon summarizes different names and approaches to sustainability education and focuses on "comprehensive sustainability education" for its willingness to join "sensory, imaginative, aesthetic, intuitive, and spiritual facets" to the "cognitive" (Armon 2020: 2). While the wicked questions approach appears to centralize cognitive efforts, the rhetorical aspect of considering possibilities for change and influence within the specifics of a situation absolutely calls for the inclusion of more holistic and varied factors in communication situations. Someone's affective experience or aesthetic sense can matter a great deal when addressing the complexities of a wicked problem as they define it. Armon's focus on comprehensive sustainability education emphasizes three factors: "cultivating an ethic of care," "fostering resilience and regeneration," and "advocating for life's flourishing" (4). This pedagogy's ethical and advocacy aspects serve as behind-the-scenes assumptions for the wicked questions approach.

Faculty Training and Co-Creation of Curriculum

The implementation of our wicked questions approach relied primarily on training instructors. Numerous Teaching Assistants staff our program, so there is high turnover, but a population ready to receive new ideas and approaches regularly. The rest of our program is staffed by part-time instructors, many of whom have significant experience in teaching writing but may not be as open to changing their curriculum honed over time and in a context where change often means extra work that may not be financially compensated.

The fact that the change in our GE program required a revision, no matter what, helped with the transition. A key second factor was voluntarily including all experienced part-time instructors in training sessions with Teaching Assistants so the experienced instructors could have a sense of investment and gain ideas for adjusting their previous versions of courses rather than designing entirely new courses.

For faculty, the United Nations' global goals for sustainable development provide the main sustainability framework used in early training, primarily as a starting point for defining and discovering wicked questions for instructors to use. Still, there is no philosophical dedication to that approach. The seventeen categories help push toward specific issues, wicked problems, and wicked questions, while some other frameworks tend toward the strictly conceptual (see Earthrise Commons; earthrise commons.org). Student familiarity with the UN, the eye-catching visuals, and the targeted curriculum enhance the framework's ease of introduction and use. Methodologically, the wicked questions approach fits with the Education for Sustainability discourse, where systems thinking is central to the class and shifts in mindset are part of the overall goals.

After a one-semester trial run in 2019 with a few wicked questions and a whole-staff retreat, the wicked questions approach became the basis for all sections of the required GE writing class. This course design initially encouraged earlier and more varied research by students when working on major papers, pushing our program to emphasize critical reading. An increased focus on audiences with varied values and considering cognitive biases also emerged from the complexities of the wicked question issues. It reflected faculty involvement and commitment to shaping the revised curriculum.

The learning outcomes for the writing courses that use the wicked questions approach are standard for contemporary composition classes. They include writing in different modes and rhetorical situations, ethically using research, and providing audience-appropriate evidence-based arguments. There are two additional learning outcomes (currently optional for instructors to include) related directly to sustainability that are revised by the United Nations Education, Scientific, and Cultural Organization (UNESCO):1. The learner is able to strategically communicate in ways that influence decision processes toward sustainable choices; and 2. The learner is able to question norms, practices, and opinions; to reflect on their own values, perceptions, and actions; and to take positions and actions in the sustainability discourse. Ultimately, we intend to make these learning outcomes more central to the variety of wicked questions-oriented writing courses.

Emerging Sustainability Competencies for Future Accreditation

The "Key Competencies in Sustainability in Higher Education" (2021) project seeks to find common competencies across institutions in higher education as the new discipline of education and action for sustainability emerges. Many of these competencies also work for citizen education outside of higher education. Katja Brundiers' et al. (2021) group of seventeen scholars have noted particular competencies that are directly part of a wicked questions approach: the "Integrated Problem-Solving Competency" and a "Systems-thinking Competency." While these kinds of thinking are central to addressing wicked questions and serve as implied goals across our writing and rhetoric program, the "Interpersonal Competency" and "Values-thinking Competency" are two named elements historically part of many writing and communication classes via peer review and self-reflection exercises. These key competencies can be explicitly made more central as twenty fist-century career skills, particularly in communicating across differences or seeking common values with others.

Evaluation

The wicked questions approach has served as a learning model to share across campus with others working on Education and Action for Sustainability. We have presented the approach to multiple groups on campus and seek ways to share it beyond our institution. One of the primary benefits has been the experimentation with a sustainability approach that others can revise in their particular contexts. Because Composition I and II are required courses for incoming first-year students, we provide a foundational understanding of basic sustainability concerns that will help them find coherence among their university courses as they complete their degrees. While we consider the approach generally successful thus far, we need a more formal assessment of sustainability outcomes and highlighting difficulties that have arisen.

There are a variety of barriers to furthering this approach to sustainability education. The learners do not necessarily identify the sustainability components of Composition in any direct way. Many sections of the course using the wicked questions approach do not use the word *sustainability* on their syllabi or introductory materials. This is perhaps both a strength and a weakness. Calling attention to an institutionally supported commitment to changes in perceptions, mindsets, and behaviors provides a chance to guide learners in connecting their writing and sustainability work or relevant content in other courses. However, learners with concerns about sustainability as an overused buzzword or who associate negative connotations with the arguably politicized word do not have to deal with those implications. Instead, they can focus on the wicked question—usually a new term to the learners. Some communication practices provide models for working with climate change skeptics (Hayhoe 2024; Roderick 2024). Classes may do their own definitional work around what makes a wicked question, and the approach lends itself well to learners from a variety of disciplinary or pre-disciplinary, and ideological perspectives.

The wicked questions approach has not significantly altered retention or pass/fail rates to the best of our understanding, but it has brought more coherence among the many sections of our first-year writing courses while still giving instructors a significant level of freedom and flexibility. An instructor does not even have to be especially active or interested in sustainability to deliver a wicked questions course, although this can help in prioritizing sustainability learning outcomes. From an organizational perspective, many other requirements pull on the Writing and Rhetoric courses, including General Education expectations, assessments of multiple types, institutional priorities around diversity and accessibility, and shifts to and from online course delivery. While each of these elements is important, perhaps vital, and all can be understood within a sustainability framework, administrators at multiple levels (from the Writing Program Administrator on up) must be interested in working with that approach. Regular changes in administrative personnel can shift priorities, and, in our case, the development of the sustainability focus with wicked questions may slow after initial implementation.

Present and Future

Several years into the wicked questions approach, faculty in our Writing and Rhetoric program have adopted the shift away from more literary or personal themes. The move to wicked questions has helped re-energize the thematic aspect of writing courses and encouraged complex writing situations to take center stage in courses. Regular faculty retreats involve discussions of different wicked questions and approaches to teaching the class. At the same time, the pressures to make sure the course is still a "real writing class" are significant and can de-prioritize sustainability topics. However, since communicating through complex sustainability issues and thinking in terms of systems are more significant goals for the course than specific sustainability-oriented content, the priorities may still be well-aligned. One next step is to further assess sustainability or systems thinking goals separate from the regular writing, analysis, argument, and other general thinking and communication goals already evaluated for the program. This approach does not push Education and Action for Sustainability toward becoming a more accepted or widespread academic discipline. Instead, it works to make sustainability principles and thinking part of already existing disciplines and structures. The approach is about changing goals and priorities within an institution (academic or otherwise) and culture rather than adding a sustainability section, topic, discipline, or other overlay. Readers who might want to try a version of a wicked questions approach could consider starting with a few or just a single wicked question and work from there, as these questions tend to be so big that other issues link to them rapidly.

References

Armon, J. 2020. In *Prioritizing sustainability education: A comprehensive approach (1–9)*, ed. J. Armon, S. Scoffham, and C. Armon. New York: Routledge.

Beach, R., J. Share, and A. Webb. 2017. *Teaching climate change to adolescents: Reading, writing, and making a difference*. New York: Routledge and the National Council of Teachers of English.

Berger, W. 2016. *A more beautiful question: The power of inquiry to spark breakthrough ideas*. New York: Bloomsbury.

Berlin, J. A. 1987. *Rhetoric and reality: Writing instruction in American colleges, 1900–1985*. Carbondale, IL: Southern Illinois University Press.

Bonneuil, C. & Fressoz, J. 2016. The shock of the Anthropocene: The earth, history and us. Translated by David Fernbach. New York: Verso.

Brundiers, K., et al. 2021. Key competencies in sustainability in higher education—Toward an agreed-upon reference framework. *Sustainability Science* 4 (1): 213. https://doi.org/10.1007/s11625-020-00838-2.

Christy-Robinson, H. Email. March 1, 2024.

Dobrin, S. I., and C. R. Weisser. 2002. Breaking ground in Ecocomposition: Exploring relationships between discourse and environment. *College English* 64 (5): 566–589. https://doi.org/10.2307/3250754.

Garskie, L. 2022. Wicked problems. In *Keywords in design thinking: A lexical primer for Technical Communicators & Designers*, ed. J. C. K. Tham, 155–159. The WAC Clearinghouse, Boulder: University Press of Colorado. Wac.colostate.edu/books/tpc/design/. https://doi.org/10.37514/TPC-B.2022.1725.2.29.

Gough, S., M. Mor, A. Sowter, and P. Vare. 2016. Ongoing and future directions of research on higher education for sustainable development. In *Routledge handbook of higher education for sustainable development*, ed. M. Barth et al., 114–125. New York: Routledge. https://doi.org/10.4324/9781315852249.

Hayhoe, K. 2024. How to talk about climate change so people will listen. *Chatelaine*. https://chatelaine.com/living/how-to-talk-about-climate-change/.

Hong, W. 2020. Build it and they will come: The faculty learning community approach to infusing the curriculum with sustainability content. In *Sustainable development goals and Institutes of Higher Education*, ed. G. Nhamo and V. Mjimba, 49–60. New York: Springer. https://doi.org/10.1007/978-3-030-26157-3_4.

"'How-Might-We' Questions." 2024. Hasso Plattner Institute of Design at Stanford University. https://dschool.stanford.edu/resources/how-might-we-questions.

Keup, J. R., and D. G. Young. 2017. Investigating the first-year seminar as a high-impact practice. In *The first year of college: Research, theory and practice on improving the student experience and increasing retention*, ed. R. Feldman, 93–125. Cambridge: Cambridge University Press. https://doi.org/10.1017/9781316811764.005.

Lawson, E., and S Wyman. 2026. STREAM: A theory of community-led engagement for health, wellness and healing. In *Sustainability Learning for Action and Community Engagement*, eds. S. Wyman and J. Korenblatt. New York:Palgrave Macmillan.

Marback, R. 2009. Embracing wicked problems: The turn to Design in Composition Studies. *College Composition and Communication* 61 (2): 397–419.

Rittel, H., and M. Webber. 1973. Dilemmas in a general theory of planning. *Policy Sciences* 4 (2): 155–169.

Roderick, L. 2024. "What I'm doing now." https://libbyroderick.com/now/.

Seymour, N. 2018. *Bad Environmentalism: Irony and Irreverence in the Ecological Age*. Minneapolis: University of Minnesota Press. https://doi.org/10.5749/j.ctv65sz3q.

Singer-Brodowski, M. 2016. Students' competency development in the context of self-organised and project-oriented sustainability seminars: Research at the Interface between self-description and real-life action. In *Routledge Handbook of Higher Education for Sustainable Development*, ed. M. Barth et al., 411–420. New York: Routledge.

Wals, A. E. J., V. C. Tassone, G. P. Hampson, and J. Reams. 2016. Learning for walking the change: Eco-social innovation through sustainability-oriented higher education. In *Routledge Handbook of Higher Education for Sustainable Development*, ed. M. Barth et al., 25–39. New York: Routledge.

"Wicked Problem." 2023. Stony Brook University. www.stonybrook.edu/commcms/wicked-problem/about/What-is-a-wicked-problem. Accessed 23 January 2023.

Wiggins, G., and J. McTighe. 2005. *Understanding by Design*. 2nd ed. Association for Supervision and Curriculum Development.

Wright, M., and F. A. Monsour. 2020. Developing a curriculum for sustainability education: Lesson planning for change. In *Prioritizing sustainability education: A comprehensive approach*, ed. J. Armon et al., 202–217. New York: Routledge.

CHAPTER 33

Sustainability Leadership at Universities: Why it Matters

Naif Alghamdi

A team of practitioners presented a "roadmap" proposal designed to promote sustainability practices to the top management team of one of the oldest and most prestigious public universities in the Kingdom of Saudi Arabia (KSA). The meeting was chaired by His Excellency, the President of the university. The Vice President and several top consultants to the President were among the attendees. The author of this chapter was also in attendance; he introduced himself as an expert in the subject by providing his educational and professional experience to make the case for reimagining sustainable practices in the higher education sector in the areas of academics, facilities management, and institutional commitment.

This meeting was important because it epitomized the context of enhancing sustainable practices in the region's largest economy and populace. In this light, this chapter will use the exchanges in this meeting to

N. Alghamdi (✉)
Department of Architecture and Building Sciences, King Saud University, Riyadh, Saudi Arabia
e-mail: naag@ksu.edu.sa

S. M. Wyman and J. Korenblat (eds.), *Sustainability Learning for Action and Community Engagement*,
https://doi.org/10.1007/978-3-032-11698-7_33

explore what is needed to lay the site-specific foundations for sustainability in the higher education sector.

To demonstrate that the university lagged behind international standards, the author presented PowerPoint slides that juxtaposed the current status of sustainability at the university with examples of what other higher education institutions, locally and globally, have accomplished to advance sustainability efforts on their campuses. This point was buttressed by showing how other higher education institutions align their approach with the World Universities Sustainability Rankings, such as the Green Metric University Sustainability Ranking (UI's GreenMetric), Sustainability Tracking, Assessment and Rating System (STARS), International Sustainable Campus Network (ISCN), and World University Ranking (QS). Additionally, the author reminded the attendees of their obligations by highlighting the national plan, by which the government of Saudi Arabia, through its Vision 2030 policy, pledged to cut its carbon emissions to net zero by 2060.

To meet this climate commitment, several national initiatives and programs have been presented, such as Saudi Green Initiative, Middle East Green Initiative, National Biotechnology Strategy, Green Riyadh, AlKhafji Desalination Plant, King Salman Park, King Salman Energy Park, Sakaka Solar Power Plant, Dumat Al Jandal Wind Farm, Low Power Research Reactor, to name but a few (VISION 2030). During the President's cabinet meeting, the author grounded these national obligations by presenting a 'sustainability roadmap' proposal, including the suggestion of establishing an 'Office of Sustainability' to assist in managing these efforts holistically on the university campus through a six-step approach (commitment, evaluation, planning, implementation, tracking, and reviewing). The 6-step approach is based on comprehensive research on sustainability at university campuses in Saudi Arabia and elsewhere (Alghamdi et al. 2017; Alghamdi 2018a, 2020). Using the roadmap as a foundation, the author pressed upon the university to take the lead in not only participating in the abovementioned national sustainability initiatives and programs but also advancing them to another level, given the existing capabilities and capacities of the university. The presentation finished with the potential outcomes of applying new sustainability practices and operations and possible advantageous scenarios. This includes local and global impacts on the campus community and beyond. The meeting ended with a compliment, acknowledging the proposal and promising to consider it, with a clear commitment to action.

The meeting's outcome laid bare the need for bold leadership in sustainable development. The main question to be raised then is: *What can be done if there is little interest and support from senior leadership?* A desk study approach was utilized to answer this question: the team reviewed and analyzed case studies and relevant literature.

We must first define leadership in the context of higher education. Leadership is about developing and articulating a vision, establishing direction, looking outwards, empowering colleagues, doing the right things, sharing inspiration, influencing and acting decisively, and making potentially revolutionary change (Yukl 1998). More specifically, university sustainability leadership entails the processes by which leaders, policymakers, and academics implement sustainable development policies and other initiatives within their organizations (Leal Filho et al. 2020). Leadership is a key ingredient in successfully planning and implementing sustainability on the university campus. Martin and Samels (2012) argue that leadership plays a crucial role in the success of sustainability implementation. Such projects have moved from purely operational advances to guidelines for much of the strategic thinking on campus, including curricular commitments. These definitions provide a broad concept of leadership, especially from the tertiary education perspective.

Leading sustainable practices appear to be axiomatic. On the one hand, it is known that the university campus is the best place for a sustainability revolution since it is considered a 'living lab' (König 2013). This means that universities should mobilize their campuses as dynamic incubators; education and research should emphasize sustainability topics, and daily lives in lecture halls, laboratories, and dormitories should exemplify sustainability principles. On the other hand, there are ever-growing significant roles and responsibilities in higher education institutions. Cortese (2003) raises the question, "If higher education does not lead the sustainability effort in society, who will?" Thomashow (2014) likewise underlines that "university leadership is our last best hope for addressing the global climate challenge, and campus sustainability initiatives are the foundation of that leadership." Bartlett and Chase (2004) confirm this by highlighting that support from above is critical.

Several barriers have negatively impacted sustainability leadership at universities, which can be highlighted here as follows:

First, there is a lack of commitment to the sustainability issue. This matter can be considered a significant influence that dramatically impacts the whole institution's movement toward sustainability practices and curricular commitments. To date, not a single university in the country has set a target to be climate neutral, even though the Kingdom has promised to cut its carbon emissions to net zero by 2060. No commitment from the university leadership means that they are under no obligation to become more sustainable in terms of energy use, green building practices, emissions impact or other considerations. Resembling many universities around the world, this old and prestigious university, as far as is known, has not signed any international declarations, charters, or partnerships in the form of written agreements to (a) inculcate environmental, social, economic, and educational sustainability in colleges and universities and (b) advance all aspects of sustainability in higher education institutions, including physical plant, purchasing policies, materials management, or consumption and emissions standards (Alghamdi 2018b).

Second, there is a lack of knowledge and awareness of sustainability and what values such initiatives can add to the institution. The sustainability leadership role in university campuses needs to "lead by example" by introducing a clear vision, mission, and goals to promote sustainable development and implement pro-planet and pro-socio environmental practices and operations on the campus (Alghamdi 2018a). Sometimes, the institution may consider "educating the educators" to increase sustainability knowledge and awareness for the students and faculty, staff, and other employees. Sustainability education, especially at universities, is key since students can be considered the primary *agents of change*. As Tolley (1996, 214) puts it, "They will progress to occupy influential roles in government, companies, or other organisations." Faculty and staff members need to be involved in the sustainability movement on campus, given that they are key agents in achieving lasting progress toward sustainability (Brinkhurst et al. 2011).

Third, financial support is a common issue worldwide, and it is no different in the Kingdom. There are several common ways to raise funds for sustainability initiatives at universities, including but not limited to national or international government or agency grants, subsidies, blended finance packages, innovative repayment schedules, sustainability bonds and contingent interest instruments, impact investors, and crowdfunding.

Diagnosing the problem is one step in the right direction. An emphasis on sustainability leadership as the new bottom line is also imperative for this and other universities to advance their efforts toward sustainable development: good governance requires wise leadership (Thomashow

2014). The highest administration level must demonstrate leadership in institutionalizing sustainability on the university campus at all levels (top-down approach). Although bottom-up grassroots initiatives have proven to be of great value, reinforcement from the top, supported by actions and budget, is far more effective and saves time, money, and considerable effort. It has been observed that green campus initiatives have a much higher success rate when the campus community knows that the university's top administration is committed, involved, and supportive of the sustainability programs (Simpson 1996). Therefore, this short essay highlights sustainability leadership at higher education institutions and demonstrates its roles and responsibilities on campus and beyond. Many universities worldwide with advanced leadership in sustainability, climate change mitigation, and resilience currently use their campuses as living laboratories. This means universities integrate teaching, learning, research, campus planning and operation, and outreach into active research and implementation workshops for sustainability initiatives (König 2013). University campuses should exemplify how sustainable practices and operations can be implemented, illustrating that "actions speak louder than words." Approaching sustainability holistically can be far more effective and influential when institutionalized. This requires bold leadership, commitment, vision, plans, targets, and tools to measure and report progress periodically.

References

Alghamdi, N., A. Den Heijer, and H. De Jonge. 2017. "Assessment Tools' Indicators for Sustainability in Universities: An Analytical Overview". *International Journal of Sustainability in Higher Education* 18 (01): 84–115.

Alghamdi, N. 2018a. Knowledge and Awareness of Sustainability in Saudi Arabian Public Universities. In *Handbook of Sustainability Science and Research*, ed. W. L. Filho, 103–127. Switzerland: World Sustainability Series, Springer International Publishing AG, Cham.

Alghamdi, N. 2018b. University Campuses in Saudi Arabia: Sustainability Challenges and Potential Solutions. https://doi.org/10.4233/uuid:7e494678-0bef-44e6-ab22-f6332eb9a525.

Alghamdi, N. 2020. "Sustainability Reporting in Higher Education Institutions: What, Why, and How". In *Filho and de Brito*, ed. International Business, 975–989. Trade and Institutional Sustainability: World Sustainability Series, Springer International Publishing AG, Cham, Switzerland.

Bartlett, P., and G. Chase. 2004. *Sustainability on Campus: Stories and Strategies for Change*. Cambridge, US: The MIT Press.

Brinkhurst, M., P. Rose, G. Maurice, and J. D. Ackerman. 2011. "Achieving Campus Sustainability: Top-down, Bottom-up, or Neither?" *International Journal of Sustainability in Higher Education* 12 (4): 338–354.

Cortese, A. 2003. "The Critical Role of Higher Education in Creating a Sustainable Future". *Planning for Higher Education* 31 (3): 15–22.

König, A. 2013. *Regenerative Sustainable Development of Universities and Cities: The Role of Living Laboratories*. Cheltenham, UK: Edward Elgar Publishing Limited.

Filho, Leal, João Henrique. Walter, Paulino Pires Eustachio, Adriana Cristina Ferreira. Caldana, Markus Will, Amanda Lange Salvia, Izabela S. Rampasso, Rosley Anholon, Johannes Platje, and Marina Kovaleva. 2020. "Sustainability Leadership in Higher Education Institutions: An Overview of Challenges". *Sustainability* 12 (9): 3761. https://doi.org/10.3390/su12093761.

Martin, J., and J. Samels. 2012. *The Sustainable University: Green Goals and New Challenges for Higher Education Leaders*. Baltimore, US: Johns Hopkins University Press.

Simpson, W. 1996. "Environmental Stewardship and the Green Campus". *Facility Managers Journal* 39–45. Available at: https://www.appa.org/membershipawards/documents/1997.pdf (Accessed 20 May 2017).

VISION 2030. 2024. Saudi Arabia's Vision 2030. Retrieved from, https://www.vision2030.gov.sa/en/ Accessed 15 Mar 2024.

Thomashow, M. 2014. *The Nine Elements of a Sustainable Campus*. Cambridge, US: The MIT Press.

Tolley, R. 1996. "Green Campuses: Cutting the Environmental Cost of Commuting". *Journal of Transport Geography* 4 (3): 213–217.

Yukl, G. A. 1998. Leadership in Organizations. Prentice Hall: Upper Saddle River, NJ, USA.

PART V

Engaging Communities: Campus and Region as Living Laboratories

CHAPTER 34

Collaborative Creation for the Campus Ecosystem

Ripley Butterfield

In the fall of my sophomore year of college, I spoke and read poetry at an open mic for the first time in my life. The event was *Sound Your Truth: Courage* on the SUNY New Paltz campus, US, and I was in front of a crowd of other students, professors, and neighbors. It was not something I had ever considered doing before, but the opportunity came at a dire moment.

My time at New Paltz had, up to that point, been extremely solitary. I couldn't seem to make friends or connect with classmates. It felt as though when I stepped out of the classroom or away from the desk at my work-study job, I became invisible. At times, the loneliness was overwhelming, but I dismissed this feeling; it was "just school" after all. My physical, day-to-day existence during the semester felt unreal, or at least unimportant, save for my academic efforts. Then, my tether to "real life" was cut. My partner back home ended our relationship due to religious

R. Butterfield (✉)
State University of New York, New Paltz, NY, USA
e-mail: butterfr1@newpaltz.edu

S. M. Wyman and J. Korenblat (eds.), *Sustainability Learning for Action and Community Engagement*,
https://doi.org/10.1007/978-3-032-11698-7_34

guilt and homophobia. The detachment I had from the campus environment suddenly became terrifying. Grief-stricken and confused, I badly needed a support system but came up dry. The cold, institutional climate had hindered me from finding community in the place I lived.

Luckily, I encountered *Sound Your Truth* the very next week. Students in one of my classes were prompted to collaborate on a site-specific sculpture for the event, which I had never heard of before. The whole gathering was built from various contributions of participatory art, performance, and active listening designed to amplify underrepresented voices on our campus. That evening, an otherwise neglected corner of our campus was transformed by those who showed up, constructing a space of care. When I recognized the chance to share my story, I didn't question the impulse to do so. The entire crowd was attentive and quiet as I spoke into the microphone. Former classmates and professors, along with complete strangers, approached me later to express sympathy and gratitude for my vulnerability that night. The compassion that manifested around me was unbelievable. I made lasting, vital relationships with several people who were present. I also developed trust for my voice as a creative writer. Participating in *Sound Your Truth* revealed pathways to interpersonal connection and community membership that were otherwise invisible to me in the daily performance of student life.

I was invited by members of the collective behind *Sound Your Truth*, Eddy at New Paltz, to share more about my university experience. Speaking with this group, I felt valued as a person rather than a nameless degree-seeker. I joined Eddy soon after. My intention today as a student collaborator is to create more possibilities for empowerment, communication, and mutual care to emerge in our campus ecosystem. Speaking from my own experience and observations over the past several months, public art engagements can break through the structures that keep us from making meaningful connections.

In one recent Eddy project, the broader campus community was invited to share knowledge, ideas, and art on ecological topics in the form of *zines*. Zines are folded, stapled, or sewn paper booklets that are independently created and easily reproduced. After announcing an open call for submissions, I took the lead in holding a series of workshops at our school library to provide zine-making instruction and materials to anyone interested. Participants made collages with recycled materials, wrote about environmentalist history, composed poetry, did botanical illustrations, and more. Holding a space for hands-on creation brought surprising results. Students who claimed they "don't draw" or "don't write" made invaluable contributions and enjoyed the process. One professor brought his class of student teachers to a workshop to demonstrate alternative forms of teaching literacy. He later invited me to make zines with seventh graders at the middle school where he teaches.

After the zine submissions were collected, I scanned and made copies of each original work. Together, these reproductions formed the Earth Day Zine Library. An outdoor, mobile sculpture was constructed out of repurposed materials to make the knowledge and artwork publicly available. The library made its debut at a campus-wide Earth Day event, where dozens of people stopped by to browse the beehive-shaped bookshelf. Zine contributors were proud to see their work displayed and archived, while those unaware of the project flipped through the booklets with curiosity.

I cannot say whether the content of the zines has made any impact on the global environment, but the process of their creation made ripples in our campus ecosystem. Conversations between workshop participants have evolved into connections, and the simple paper folding method I demonstrated has become a practice for some. Most importantly, students and educators across the siloed departments of the university were empowered to make something together. I am still an inexperienced organizer and artist, but in the future, I hope to facilitate projects that radically alter our perception of "campus life." Every person living, working, educating, and studying in our shared space should have access to the sustaining force of belonging (Fig. 34.1).

Fig. 34.1 Zine library, *Photo credit* Andrea Frank

CHAPTER 35

Building Accessible Bridges of Knowledge Through Disability Studies: The Value of a Social Framework in Academic Spaces and Beyond

Kimberly Sanford and April B. Coughlin

Building sustainable cities and communities across the globe depends on a radical commitment to the inclusion of all people and an ethic of equity. Disabled students in the United States today are still subject to implicit segregation in their learning environments at all levels, despite landmark legislation that affirms separate facilities for people of a particular (perceived minority) population can never be considered equal. The language used to justify the ongoing segregation of Disabled students is deeply embedded in the medical model of Disability, which places the responsibility of "overcoming" so-called ailments on the individual to

K. Sanford (✉) · A. B. Coughlin
School of Education, SUNY New Paltz, New Paltz, NY, USA
e-mail: sanfordk1@newpaltz.edu

A. B. Coughlin
e-mail: coughlia@newpaltz.edu

S. M. Wyman and J. Korenblat (eds.), *Sustainability Learning for Action and Community Engagement*,
https://doi.org/10.1007/978-3-032-11698-7_35

better assimilate into the non-Disabled population. This language is still widely utilized across educational institutions and within teacher preparation programs for both "general" and "special" educators (Valle and Connor 2019). Coded discriminatory language or repackaged segregation within a "general" learning environment continuously reinforces long-held ableist notions of separation and isolation that are reportedly in the "best interests" of Disabled students. Permitting the integration of Disabled and non-Disabled students within one school building but forcibly separating them once instruction begins is not the meaningful inclusion that Disability Rights advocates and forward-thinking educators have fought for (Heumann 2020; LeBrecht and Newnham 2020; von Tippelskirch and Linton 2014). Our work and advocacy are focused and based on grassroots efforts particular to the Disability Rights Movement and its implications in the US education system as it exists. These frameworks may be useful for others working in this area around the globe.

Like other fields of study focused on the history and culture of a marginalized group, Disability Studies (DS) did not start as a discipline conceived at the highest tiers of an ivory tower for the sole fulfillment of scholars or academics. Grassroots organizers, Disability Rights activists, and martyrs built DS into the blossoming field that it is today, and we continue to expand the subject from the roots of their honorable, action-focused work. These efforts are inherently multifaceted in nature. At the collegiate level, we have an opportunity to develop programs and establish DS resource repositories for incoming generations of students to use as they pursue any field in their major and minor path. We work toward this pursuit by engaging students in the principles of DS, using our campus as a living lab. Some of these include dismantling the strict medical/clinical model of Disability and prioritizing a social/humanistic model; (re)vising the way Disability is conceptualized as a natural form of human diversity; creating avenues for building physically and culturally accessible spaces; and centering the perspectives of Disabled people in dialogue about Disability Rights, justice, and acceptance (Society for Disability Studies 2021).

Disability Studies in the US has a strong record of using policy change and education at all levels to shift attitudes and behaviors that can further our potential to survive as a species and as a planet where the *environment* means more than green trees. If the physical environment, be it natural, urban, or institutional, cannot adequately accommodate the bodies and minds that pass through it, the community cannot be considered sustainable (SDG 11). As a recent example, the COVID-19 pandemic affirms this outlook as many undermine the risk the virus continues to pose to Disabled individuals who have, for too long, been considered unimportant casualties during periods of crisis (Andrews et al. 2021; Sabatello et al. 2020).

Attention to Discourse

Our collective energy as Disability-conscious and socially responsive educators is employed in constructing truly inclusive teaching ideologies that determine the role that Disabled students have in our classrooms and our greater community. This depends upon a meaningful (re)visioning of the language used formally and informally when discussing Disability (Dillard 2021). Part of this (re)visioning demands an acknowledgment of how many Disabled people in our general society are treated as second-class citizens and, disturbingly, as a population whose vulnerability should not concern non-Disabled people. Changing the way Disabled people are regarded and treated starts with this acknowledgment but does not begin to transform until the language utilized in everyday life is altered to place value and meaning on the lives of those whose existence is an act of resistance against ableist ideologies. Impairment rhetoric invades aspects of educational, medical, social, and cultural spaces, which is rooted in ableist ideology. Combating this subconscious undercurrent of ableist rhetoric enables learners to critically evaluate *all* the ways that our social landscape actively reproduces and reinforces deficit-based thinking about Disabled people.

Institutional and Programmatic Commitment—The Teacher Preparation Course

We begin this pursuit in teacher preparation programs at SUNY New Paltz, US, with the charge of enabling teachers of emerging generations to be critical thinkers and social justice advocates who inspire students. In alignment with the United Nations Global Goals, the further integration of Disability Studies (DS) into the teaching practices of future educators is an active form of transformational work in support of SDG 10 *Reducing Inequalities* in school settings and, by extension, their surrounding communities. Providing students with the language required to engage in discussions about the inequities they experience as members of underrepresented groups gives voice and, therefore, power to the injustices their communities face inside and outside of school. Expanding access to DS resources for students of all disciplines advances the work of Disability Rights advocates who have long fought for equal and equitable access to education. Our work to further integrate DS across our teacher preparation programs is an exercise in challenging notions of traditional education practices that often segregate Disabled students from their non-Disabled peers. The establishment of our annual Disability Studies Series highlights prominent scholars and activists and their ongoing work for Disability Rights and representation. Past speakers include Dr. Simi Linton, Disability rights activist, author, and film producer; Dior Vargas, Disabled woman and mental health activist; Nicole Ineese-Nash, Indigenous researcher, educator, and community worker; Dr. Sami Schalk, author of *Black Disability Politics;* and Judy Heumann, *"Mother of Disability Rights."* This series marks the emergence of DS as an important, meaningful, and urgent area of attention at our institution and maps progress over time in the growing academic field.

In alignment with our School of Education's pillars, Disability Studies in Education (DSE) "works to contextualize Disability within political and social spheres; privilege the interests, agendas, and voices of people labeled with a Disability/Disabled people; promote social justice, [and] equitable and inclusive educational opportunities" (Connor et al. 2008, 448). This effort is usually reserved for programs that focus on serving Disabled students as indicated by the Special Education Program's philosophy which is "founded on the belief that learners of all levels of ability can develop their intellectual, social, emotional and physical strengths with the help of teachers who are knowledgeable and passionate advocates for their growth." It is our belief, in contrast, that *all* teacher preparation programs should be founded on the principle that Disabled students have been and always will be an irreplaceable and immeasurably valuable presence in our classrooms, schools, and communities. Providing teacher candidates with the language and tools they need to emerge as advocates for and with Disabled students safeguards the future for further integration and an expansion of school support services.

Our effort to create greater access to DS initiatives at the collegiate level has a dual purpose; the first intention is to further validate DS as a legitimate academic discipline and place issues of access and inclusion at the forefront of our collective consciousness as we envision the future of higher education. Secondly, we hope to enable generations of teacher-leaders and, by extension, their students to continue to advocate for greater access and integration for Disabled people in all environments, including those beyond the bounds of academic spaces (Coughlin and Sanford 2022). Institutions of public education have long been considered the "great equalizers" with the intention to provide equal opportunity for all students to succeed academically, develop socially, and emerge into higher education or the workforce ready for the challenges to come. Unfortunately, the de jure and de facto segregation of marginalized students of past and present makes this democratic *ideal* of equality just that: a vision we still aspire to fulfill but have yet to achieve.

Disability Studies in Education Framework

Students in our Introduction to Disability Studies course come from various majors and programs across campus, as the class fulfills their GE diversity elective requirement. As a result, they develop an increased understanding of Disability as identity, culture, and pride and learn to resist ableist language and practices that further marginalize Disabled people. They also explore ways to engage in Disability advocacy and activism in their own social circles, current work environments, and future professions through inclusive pedagogical models.

Students in certain teacher preparation courses taught through a DSE framework are introduced to how DS principles and perspectives have transformed the way Disability is conceptualized as a social and political identity that can be acquired at any point in time. They learn how to engage differentiated instruction (DI) practices in their lesson plans while implementing the principles of Universal Design for Learning (UDL), which prepares them to plan for and meet the needs of *all* students. The classic example of how accommodations benefit humanity in general is curb cuts in sidewalks, which still do not exist in many countries. Curb cuts were first installed in 1945 to accommodate Disabled veterans using wheelchairs after World War II but improved access for other wheeled devices, including strollers, carts, and additional mobility aids. Teacher candidates share how a humanistic DS perspective revises the way they once conceptualized Disability as a strict biological reality that demands isolation or "fixing." This leads them to dismiss the general medical/clinical view of Disability in favor of the concept of Disability as a flexible state of being, a source of pride and identity, and a valued element of our diverse student body.

Integrating the Education System

This mission also directly aligns with UN SDG #4, *quality education*. A DS perspective allows teachers, students, and related community stakeholders to understand "quality" education from multiple perspectives. A segregated education system is automatically disqualified from this characterization and does an active disservice to students across a spectrum of ability and Disability. Disabled students do not receive educational experiences equal to their non-Disabled peers and many do not benefit from adequate educational opportunities. Moreover, Disabled students are still significantly overrepresented in school disciplinary records data (Annamma et al. 2014; Forman and Markson 2022; Miller and Meyers 2015). This statistic does not detail the instructional time lost because of disciplinary infractions and the social isolation that comes with being physically or academically separated from mainstream groups.

On the other side of this issue, non-Disabled students do not become closely acquainted with those who are Disabled since they are often physically divided from one another. Non-Disabled students then lack a fundamental understanding of Disability as a social identity and, as a result, are more vulnerable to the oppressive ideology of ableism. This mindset maintains the segregated status quo inside and outside of schools by privileging non-Disabled people. Without this preliminary understanding, students who emerge into higher education and the workforce often move through their entire educational journeys without putting much thought to Disability unless (1) they are Disabled themselves or (2) someone they know is Disabled. Of course, it becomes nearly impossible to conceptualize and then address issues of prejudice and oppression without specific language to recognize, identify and combat it.

Rolling Forward

Our aspirations for expanding DS go beyond teacher preparation programs and college campuses alone. We hope that working in the spaces we presently occupy will allow us to build more onramps for individuals everywhere to further explore contemporary issues Disabled communities face in every facet of our global society. In doing this work, we can build communication bridges with untapped demographics who can aid in the fight for high-quality education and eliminating inequalities.

Some additional steps that have already been taken at SUNY New Paltz to expand DS in higher education include a DS-focused professional development workshop at the Multicultural Education Conference, the establishment of the annual Disability Studies Series, and the construction of a DS library collection, all of which have been important contributions to making DS accessible to our students and to the regional community. We have further expanded this work by formally proposing a Disability Studies minor that will reside within the School of Education and take an interdisciplinary approach to this growing academic field. We also have a space to offer professional development workshops to current faculty, enriching the minds not only of those in pursuit of their educational or career goals but also individuals who are already in-service instructors. These are the leaders of our educational institutions who work with students daily to further color their world and expand their intellectual horizons. These are the people who shape the campus culture and set the priorities of the institutions they serve through the daily emotional and intellectual labor of leading a classroom. Top-down initiatives will have little influence if the individuals doing work on the ground do not commit to transformational pedagogy. Consequently, professional development cannot be viewed as a singular occurrence or administrative task. Teachers must be students, too; they must be willing to use workshops as a launchpad for their own pursuits of growing as educational leaders.

This is how we frame our work in developing programs for university students and faculty members: we lay the groundwork for further exploration beyond the institution's scope and look to provide participants with the vocabulary to embody the principles of DS in their daily lives. Through this integration, individuals can share what they have learned with others inside and outside the classroom and carry Disability-inclusive perspectives into their communities and workplaces. Our efforts are also focused on the generative work of creating new understandings, new perspectives, and adding new language to our collective lexicon as we traverse ever-changing sociopolitical, economic, and environmental landscapes over time. This is why the collaboration between educational leaders, students, and community members is essential to the meaningful success of DS as a discipline and sustainable way of living. We cannot compartmentalize the field of DS to only exist within academic institutions. To do so would defy the functionality of the discipline, which is fundamentally rooted in the betterment of society for *all* people.

References

Andrews, E. E., K. B. Ayers, K. S. Brown, D. S. Dunn, and C. R. Pilarski. 2021. No body is expendable: Medical rationing and disability justice during the COVID-19 pandemic. *American Psychologist* 76 (3): 451–461. https://doi.org/10.1037/amp0000709.

Annamma, S., D. Morrison, and D. Jackson. 2014. Disproportionality fills in the gaps: Connections between achievement, discipline and special education in the school-to-prison pipeline. *Berkeley Review of Education* 5 (1): 53–87.

Connor, D. J., S. L. Gabel, D. J. Gallagher, and M. Morton. 2008. Disability studies and inclusive education: Implications for theory, research, and practice. *International Journal of Inclusive Education* 12 (5–6): 441–457.

Coughlin, A. B., and K. D. Sanford. 2022. Expanding access to disability studies: Reflections from both sides of the classroom. *Journal of Teaching Disability Studies.*

Dillard, C. B. 2021. *The Spirit of our work: Black women teachers (re)member.* United States: Beacon Press.

Forman, K. & Markson, C. 2022. Students with disabilities and suspension rates: A cautionary tale for school districts. Journal for Leadership and Instruction, v 21 n 1 p 20-24 Spr 2022.

Heumann, J. E. 2020. *Being Heumann: An unrepentant memoir of a disability rights activist.* Beacon Press.

LeBrecht, J., and N. Newnham. 2020. *Crip Camp.* Higher Ground Productions.

Miller, C., and S. Meyers. 2015. Disparities in school discipline practices for students with emotional and learning disabilities and autism. *Journal of Education and Human Development.* 4. https://doi.org/10.15640/jehd.v4n1a23.

Sabatello, M., T. Blankmeyer-Burke, K. E. McDonald, and P. S. Appelbaum. 2020. Disability, ethics, and health care in the COVID-19 pandemic. *American Journal of Public Health* 110:1523_1527. https://doi.org/10.2105/AJPH.2020.305837.

Society for Disability Studies. 2021. SDS Principles–Society for Disability Studies. Society for Disability Studies. https://disstudies.org/index.php/sds-principles/.

Valle, J. W., and D. J. Connor. 2019. *Rethinking disability: A disability studies approach to inclusive practices.* 2nd ed. Routledge.

von Tippelskirch, C. (Producer & Director), & Linton, S. (Producer). 2014. *Invitation to dance* [Motion picture]. USA: Metuffer Films.

CHAPTER 36

Songbirds and Sustainability: How Birds Can Teach Us to Design Sustainable Communities on Campus and Beyond

Kara Loeb Belinsky

Urbanization is a rapidly growing threat to wildlife and global biodiversity (Gaston 2010, Grimm et al. 2008, Pickett et al. 2001, Seto et al. 2012). Biodiversity is the sum of all life on our planet that collectively maintains the ecosystems we live in, and biodiversity is in steep decline across the globe (Brondizio et al. 2019). Habitat loss, which includes loss due to urbanization, is one of the key drivers of this broadscale biodiversity loss. The effects of urbanization on local biodiversity are easily studied on university campuses with students as collaborators. Urbanization can be defined as replacing natural habitats (vegetation, landforms, and waterways) with human-centered habitats, such as those that feature ornamental landscaping, pavement, and buildings. This issue is central to both United Nations Sustainable Development Goals "Sustainable Cities and Communities" (11) and "Life on Land" (15). Most students and

K. L. Belinsky (✉)
Department of Biology, SUNY New Paltz, New Paltz, NY, USA
e-mail: belinskk@newpaltz.edu

S. M. Wyman and J. Korenblat (eds.), *Sustainability Learning for Action and Community Engagement*,
https://doi.org/10.1007/978-3-032-11698-7_36

most people in general have lived their entire lives in exurban, suburban or urban environments without realizing the effects of urbanization on biodiversity. Many are unaware of the natural habitats or wildlife species that are endemic to their home regions. Similar patterns of urbanization worldwide have created biological homogenization. Homogenization is the opposite of biodiversity; unique native species in each part of the world are depleted or lost as human-associated species that thrive in urbanized environments proliferate (McKinney 2002).

Within the field of conservation biology, urbanization is a widely recognized problem but it also provides an opportunity for us to learn how to design or restore urbanized habitats that are better for wildlife and humans. As humans, we control and create these habitats—we can learn to do this in less detrimental ways if we so choose. The largest land area currently undergoing urbanization is not the increasing size of city centers. It is rather the expanding rings of suburban development around cities and the isolated clusters of exurban development that hopscotch through rural areas (Hanson et al. 2005). These areas of lower-intensity urbanization result in huge areas of land-use change. Mid-sized developments, such as university, municipal, industrial, and commercial campuses, may be useful to regional conservation efforts because single decision-makers control these large parcels. If we are to act on the opportunity presented by urbanization to improve the way we build and maintain these growing suburban and exurban habitats, then we need to learn more about how the details of suburban design affect wildlife.

Birds are useful indicators of the effects of urbanization because they are plentiful in urban, suburban, and rural habitats and they respond quickly to habitat changes. Birds are also a prominent example of biodiversity loss because they are in steep decline. For example, recent research indicates we have lost over three billion birds since 1970 in North America alone (Rosenberg et al. 2019). Despite the decline, supplementary bird feeding is an extraordinarily common activity that is increasing throughout residential neighborhoods across the developed world (Amrhein 2014; Jones and Reynolds 2008; Robb et al. 2008). Feeders draw birds in and concentrate them in specific areas for easy identification and counting, which makes them a potential tool for assessing species diversity across habitats. Bird feeders are easy to install and maintain, and they may be welcomed on campuses simply because people enjoy seeing the birds. Students can watch and count birds at the feeders and learn to identify the species that use them with very little equipment (binoculars and video cameras can be useful but aren't required). The use of bird feeders for this type of assessment is limited in that feeders only attract a subset of species (seedeaters), but bird diversity and abundance observed at feeders likely measure the extremes of how far into a particular habitat those individual birds (or species) will travel to forage. As such, a network of bird feeders may be easily used to measure habitat suitability for birds that visit feeders across variable habitats, such as those in urbanizing landscapes, including university campuses.

Case Study: Campus as a Living Lab

My students and I used a network of bird feeders to compare bird diversity, abundance, and community composition with specific features of urbanization across our suburban university campus. This work could easily be replicated on campuses or public spaces with varying levels of urbanization and landscaping at locations across the world. Prior work at our institution (Belinsky et al. 2019a) using bird point count observations to compare bird communities on campus with those in nearby forest fragments revealed that our campus bird communities have lower species diversity as compared to the forest communities. Our forest sites alone support neotropical migrant bird species. These species are declining rapidly, are vulnerable to extinction, and are also some of the most unique species to our region. In addition, the bird communities on campus are dominated by invasive European starlings and house sparrows, the same species pervasive in many urban locations across the globe. We used our bird feeder network to examine how the specific features of suburban design across our campus affect the birds by comparing the bird communities with land cover measurements around each feeder station (Belinsky et al. 2019b). In this study, we asked what specific components of suburban design affect bird communities across campus, intending to make recommendations for improving our institutional facilities as a habitat for native birds. We predicted increasing bird diversity at feeders in areas with more retained natural habitats, such as native forests and landscaping that includes trees and plantings. Additionally, we predicted that invasive bird species would dominate feeders surrounded by more buildings and perhaps manicured shrubs.

Methods

Our Campus

We conducted our study on the suburban campus of the State University of New York at New Paltz. The Village of New Paltz has a population of 7070 people and is surrounded by smaller residential towns, forest fragments, and farmland (apple orchards and vegetable farms, along with some hayfields), with the closest city, Poughkeepsie, 15 km to the East, across the Hudson River. New York City is 137 km to the South, positioning New Paltz on the edge of one of the most intense urban expanses in the United States. However, several large, protected forests (including the 9000 ha Minnewaska State Park and 3000 ha Mohonk Preserve), are located on the Western border of New Paltz. The SUNY New Paltz campus serves 7628 students and consists of 87 ha of almost entirely developed land spanning a range of urbanization intensities. Like many campuses, ours includes a central core of high-traffic academic and administrative buildings, suburban residential areas where students live, groomed turf fields for athletics, and remnant forest fragments along the Southern and Western edges of campus.

Bird Feeder Network

We installed a network of 16 feeder stations across the campus, with four stations in each of four urbanization categories that we referred to as Forest Edge, Residential Campus, Turf, and Central Campus. Forest Edge feeder stations were placed in close proximity to natural forest fragments (closed tree canopy and natural leaf-litter ground cover) located along the edges of campus in areas with low amounts of pedestrian traffic and little development (pavement and buildings). Residential campus feeder stations were placed near dormitories in areas with few pedestrians. Turf feeder stations were located near athletic fields and parking lots, with large open expanses of lawn and low levels of pedestrian traffic. Central campus feeder stations were placed among academic and administrative buildings with the highest levels of pedestrian traffic.

Each bird feeder station consisted of an iron double-armed shepherd's hook style pole, with each hook holding one high-capacity tube bird feeder with six feeding perches/ports, enclosed in caging to exclude squirrels. The squirrel-proof caging was necessary but also excluded larger bird species, such as the invasive European starling. Each feeder was filled with sunflower seed hearts (to reduce the mess of dropped shells) during the study periods, and all feeders were emptied, washed, and sterilized with a bleach solution at the end of the spring, summer, and fall seasons (May, August, and December). Students watched each bird feeder, counting the highest number of individual birds of each species seen at any one time during a 10-minute observation each week in the fall and spring semesters and during summer sessions. We also experimented with using video cameras to observe the feeders for one-hour sessions at each feeder weekly. The advantage of the cameras is that students can watch the video at high speed to skip times with low feeder activity. The drawbacks are the costs of cameras, batteries, and computer hard drive space for large video files.

Summer Mist-Netting and Banding

To more accurately determine the population size of each bird species, the students and I used mist nets to capture, band, and release birds at the feeders during the summer. Banding involves closing a small aluminum band, like a bracelet, around the lower leg of a bird. Bird banding requires an experienced faculty or community member and Federal and State bird banding permits. The bird band allows us to identify each individual and make accurate population estimates through counts of individual birds, but this is not necessary for all studies. On our campus, banding took place on two dates at each feeder station between 20 May and 22 July 2016, with each session using 2 standard 12-meter mist nets for 3 hours between 7 am and 12 pm. Captured birds were identified by species, age, and sex, measured, banded, and then released.

GIS Land Cover Measurements

We mapped the bird feeder network and measured land cover area at two scales around each feeder station using GIS with help from faculty and students in our Geography department (Fig. 36.1). Using GIS allows for finer accuracy in measuring land covers but estimates simply using publicly available aerial photos and free software such as Google Earth would be sufficient. In our study, we used a Garmin Montana 650t GPS handheld device to record the exact location of each feeder station and added these points to a base map of orthophotos of the campus in ArcGIS software. We downloaded the aerial photos taken in 2013 from the New

York State Clearinghouse GIS database. Once the base map and GPS locations were combined, we created 10 m and 50 m buffers around each nest box. We then used the aerial photos to delineate the borders of each distinct land cover type: 1. natural forest (with closed canopy and leaf-litter ground cover), 2. lawn, 3. herbaceous plants (flower beds and ground cover bushes lacking woody stems), 4. woody shrubs (primarily pruned evergreen bushes), 5. trees growing over lawn, 6. trees growing over pavement, 7. water (stream/pond), 8. pavement (paved walkways, roads, and parking lots), or 9. buildings. We then used ArcGIS to calculate the total areas of each land cover type in the 10 m and 50 m areas around each nest box pair appearing on our map. Universities in different biomes may have different land covers, but all universities have buildings, pavement, and some kinds of landscaped or natural land covers to measure and compare. Conducting this environmental inventory can help position campus members to better understand the impacts of past construction choices and to use futures and strategic thinking to plan for the next steps.

Results and Discussion

In our study, we detected significant differences in bird diversity across our campus using our network of bird feeders. Overall, larger areas of natural forest within 50 m of feeder stations increased bird diversity and larger areas of pavement decreased bird diversity. Throughout the year, we observed a total of 21 species of birds. House sparrows were the most common species observed (we detected at least one house sparrow at 206 of the 256 ten-minute observations we completed). During the summer, we mist-netted and banded a total of 667 birds of 17 different species. Only one species, the house sparrow, was captured at every feeder station, and 225 of 667 birds we banded on campus were house sparrows. House sparrow dominance (% of birds that were house sparrows) was predicted statistically by larger areas of buildings and to a lesser degree, larger areas of herbaceous plantings within 50 m of feeder stations (Fig. 36.2).

The results support our hypothesis that the area of natural forest with a closed canopy and leaf litter covering the ground is the strongest predictor of suburban bird diversity. The forest is the native habitat at our study location, so while this result is not surprising, it does confirm that natural forested habitat is still important to species that are fairly well adapted to urbanized landscapes, as seed-eating birds that visit feeders generally are. Many other researchers have reached similar conclusions about the value

Fig. 36.1 GIS map of our campus with feeder station locations and 10 m and 50 m circles around each one depicting land covers within these circles. Created by C. Scully'17 (*used with permission*)

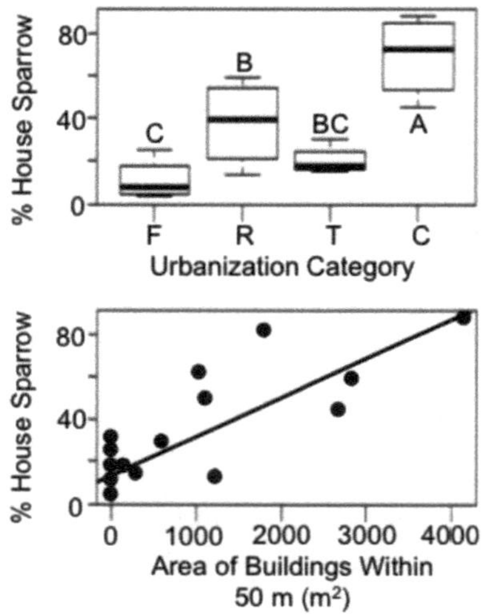

Fig. 36.2 (a) Top panel. Box plot showing the percentage of house sparrows observed at feeder stations in each urbanization category: F is Forest Edge, R is Residential Campus, T is Turf, and C is Central Campus. Letters indicate statistical significance of differences, indicating that house sparrows are most dominant at Central feeders, then Residential and Turf, then Forest Edge categories. (b) Bottom panel. Positive relationship between the percentage of house sparrows observed and the area of buildings within the 50 m circle around the feeder stations

of preserving native habitats in urbanizing locations (Gagne et al. 2015; Marzluff and Rodewald 2008).

Forests support bird diversity because trees and snags (dead trees) provide food, foraging or caching substrates (bark, foliage), and branches or cavities for nesting, while the closed canopy provides continuous cover from predators. The leaf litter on the ground also provides a foraging substrate and nesting location that is replaced with pavement or lawn in most anthropogenic habitats (Ignatieva et al. 2015, Nowak and Walton 2005, Sharpe et al. 1986). However, the bird feeder stations with less forest and more pavement in our study are also more altered in ways aside from land cover, such as having more pedestrian and vehicular traffic and more noise and nighttime lighting. These aspects of urbanization are known to affect birds, as well (Fernández-Juricic 2000, Patón et al. 2012, Platt and Lill 2006). Interestingly, when we removed the four feeder station sites containing natural forest from our analysis, the area of trees growing over lawn then predicted increased bird species diversity, although other attributes of habitat complexity, such as the area of shrubs or herbaceous plants, still had no measurable effect. This result has useful management implications for landscapes where forest restoration is not preferred or feasible and indicates that planting more trees, even over manicured lawns, may increase bird diversity.

House sparrows are well-known urban invaders in cities and towns across the planet; they are omnivorous generalists that nest in cavities, often on or in human-made structures. In our study, we found that house sparrow dominance was associated with the area of buildings. Campus buildings provide nesting habitat (we have observed active nests in the Central campus, and we banded many young fledglings there during the study), as well as food in the form of scraps, since the areas of campus with the largest areas of buildings also contain the most litter and trash receptacles. Buildings are also associated with house sparrows in several other studies (MacGregor-Fors et al. 2011, Nath et al. 2015, Wilkinson 2006).

In conclusion, our study provides empirical data suggesting changes in suburban landscaping and development practices to better support bird diversity. Our results confirm the importance of retaining remnants of natural habitat (Aronson et al. 2017; Gagne et al. 2015; Marzluff and Rodewald 2008) and point to the possibility of restoring similar habitats on campus. Bouma et al. (2013) reported encouraging results of a reforestation project at a small US suburban college (Calvin College, Grand Rapids, Michigan), where habitat patches reforested with native trees and shrubs had greater biodiversity (plants, insects, birds, and mammals) than lawns, trees over lawns and even forest remnants just four years later. Our results also indicate that replacing lawns or pavement with vegetation that best replicates native structures (trees planted over lawn, in this case), may be a feasible alternative where conservation or restoration of natural habitat is not possible or desirable.

On our campus, forest restoration patches would likely be most effective (and perhaps most palatable) if located in the Southern end of campus between athletic fields and bordering the walking paths around the artificial waterway and some of the Southern-most dormitories because these areas experience less human disturbance and are more connected to remnant forest fragments. Adding more trees along streets and paved walking paths on our campus could provide additional connectivity between restored habitats. Many new trees, including large native species, have recently been planted in areas of our campus, notably around the construction and renovation of several large buildings in the highly urbanized Northern end of the campus and along a new concourse (Mohonk Walk), which is an appropriate location for trees over lawn.

I was able to present the results of this study to our facilities team on campus and to an architectural team planning the renovation of a dormitory. I learned that adding trees to campus is costly and complicates lawn mowing and the movement of construction vehicles. I was unable to convince the planning team to add groves of trees with intact leaf litter around the renovated dormitory, but I was able to convince them to keep several large trees already on site that they had planned to remove. These experiences confirmed to me and my students that most landscapers, planners, and even landscape architects do not recognize the importance of native habitat features such as leaf-litter ground cover or limiting the area of lawns; in fact, most people think of lawns in a positive light, as the default "natural" green space (Ignatieva et al. 2015). In contrast, areas of native habitat, such as forests with leaf-litter ground cover, un-mowed meadows, shrub/scrub edges, and swamps, are often considered undesirably untidy (Nassaer 1995). This study and my new experience working with facilities and planning teams will help me more effectively advocate for more biodiversity-supporting campus landscaping designs in the future. This experience and advocacy could easily be replicated at any university campus or in any community interested in understanding their own biodiversity metrics and creating more vibrant wildlife habitats.

Involving students in research on the effects of suburban design on wildlife not only allowed me to begin affecting landscaping design on my campus, but it also changed my students' perceptions of the landscapes around them, regardless of whether or not they continued on to careers in environmental biology. Troy Ellick'17, a student who gathered a significant portion of the final data and co-authored one of the resulting papers, moved on to work in Information Technology after graduation. However, what he learned from our work together remains with him today. Troy wrote that before this project,

> I don't think I gave any more thought to what birds were around me than what type of car was driving by. Even years later after working on this project, I still find myself reflexively logging the species we share our homes

> with. The other day as I was walking through my apartment complex, I stopped in my tracks because I heard an odd song, a Blue-winged warbler! I interpreted its buzzy song as an encouraging sign that the complex had been developed in an environmentally conscious manner, else I would only hear House sparrows. Had either of those species been singing prior to my avian research experiences, I doubt I would have even registered their presence and I certainly would not have considered them as potential indicators of habitat quality. This "pause and listen" reaction has become very common in my daily life now; once you start to learn bird's names, niches, and songs you'll notice many new feathery neighbors you didn't see before. (*Used with permission*)

Laura Stark'17, a student who worked on the bird-banding portion of this project, sampled a variety of different jobs in environmental biology, from wildlife rehabilitation to sustainable agriculture, before going to SUNY College of Environmental Science and Forestry, studying invasive species and their effects on forest ecology. Laura wrote,

> Working with Dr. Belinsky on her campus bird feeder project was my first experience with ecological research. Prior to this, I was unaware that the field of ecology was a potential career path. I not only fell in love with the incredible bird species in our area, but I learned crucial skills in the scientific research process that I use to this day. I was surprised to learn the diversity of birds that visit campus feeders, which offered insight into the potential value of urban settings for wildlife and other organisms. (*Used with permission*)

Stephanie Hale'15, a student who monitored the initial set of bird feeders, went on to work in environmental biology and has come full circle to become a Senior Avian Specialist for an ornithology lab at Columbia University. Stephanie wrote,

> Working on the bird feeder network project changed the course of my life. I wouldn't be where I am without it. Through this project, I gained the skills to help me move on to different research projects across the US; from helping conserve endangered species in Texas to keeping the waters of NYC clean.... Learning about birds can really make you pay attention. Birds are everywhere, you can always see or hear them no matter where you are; they keep you company. You can lose yourself in observing and admiring their wonderful and weird behaviors. I am constantly looking at and for birds. Though I had numerous steps along the way, the feeder

> project directly translates to my current position, where part of my job is to maintain a network of bird feeders in NY from Suffolk County, LI up to the Catskills. (*Used with permission*)

In addition to effectively increasing suburban biodiversity, the conservation and restoration of native habitats and bird diversity may also improve local people's lives. Fuller et al. (2007) find that urban citizens in the United Kingdom report gains in psychological benefits (cognitive restoration, positive emotional bonds, and sense of identity) from urban green spaces with higher species richness of plants and birds. In a laboratory study, Alvarsson et al. (2010) reveal that students in Sweden recover faster from completing stressful mental arithmetic tasks when recordings of birdsong are played, as opposed to recordings of traffic or ambient noise. Furthermore, Methorst (2024) compared cross-sectional health data from the German National Cohort Study with citizen science bird diversity data in their home postal codes and found that higher bird diversity was positively correlated with better mental health outcomes. Interestingly, Methhorst (2024) found that this relationship was strongest among people with lower socioeconomic status, indicating that they benefited most from increased bird diversity around them. In conclusion, habitat conservation, restoration, and remediation are worthwhile for wildlife and humans alike in the near term, in addition to stalling and potentially reversing global biodiversity decline. I hope this work will continue to inspire my students and others to re-imagine their own future yards, campuses, towns, and cities for the common good.

References

Alvarsson, J. J., S. Wiens, and M. E. Nilsson. 2010. Stress Recovery during Exposure to Nature Sound and Environmental Noise. *International Journal of Environmental Research and Public Health* 7 (3): 1036–1046.

Amrhein, V. 2014. Wild Bird Feeding (probably) Affects Avian Urban Ecology. In *Avian Urban Ecology: Behavioural and Physiological Adaptations*, ed. D. Gil and H. Brumm, 29–37. Oxford, UK: Oxford University Press.

Aronson, M. F., C. A. Lepczyk, K. L. Evans, M. A. Goddard, S. B. Lerman, J. S. MacIvor, C. H. Nilon, and T. Vargo. 2017. Biodiversity in the City: Key Challenges for Urban Green Space Management. *Frontiers in Ecology and the Environment* 15 (1): 189–196.

Belinsky K. L., E. Keeling, and D. Snyder. 2019a. Bird Communities at the Suburban-rural Interface: The Role of Low-intensity, Small-scale Urbanization. *Urban Naturalist* 28.

Belinsky, K. L., T. Ellick, and S. LaDeau. 2019b. Using a Birdfeeder Network to Explore the Effects of Suburban Design on Invasive and Native Birds. *Avian Ecology and Conservation* 14 (2): 2.

Bouma, C., E. Huizenga, and D. Warners. 2013. Assessing a Reconciliation Ecology Approach to Suburban Landscaping: Biodiversity on a College Campus. *The Michigan Botanist* 52: 93–104.

Brondizio et al. 2019. Global Assessment Report on Biodiversity and Ecosystem Services of the Intergovernmental Science-Policy Platform on Biodiversity and Ecosystem Services. In *IPBES secretariat*, ed. E. S. Brondizio, J. Settele, S. Díaz, and H. T. Ngo. Bonn, Germany.

Fernández-Juric, E. 2000. Local and Regional Effects of Pedestrians on Forest Birds in a Fragmented Landscape. *Condor* 102: 247–255.

Fuller, R. A., K. N. Irvine, P. Devine-Wright, P. H. Warren, and K. J. Gaston. 2007. Psychological Benefits of Greenspace Increase with Biodiversity. *Biology Letters* 3: 390–394.

Gagne, S. A., F. Eigenbrod, D. G. Bert, G. M. Cunnington, L. T. Olson, A. C. Smith, and L. Fahrig. 2015. A Simple Landscape Design Framework for Biodiversity Conservation. *Landscape and Urban Planning* 136: 13–27.

Gaston, K. J. 2010. Urbanization. In *Urban Ecology*, ed. K. Gaston, 10–34. New York, US: Cambridge University Press.

Grimm, N. B., S. H. Faeth, N. E. Golubiewski, C. L. Redman, J. Wu, X. Bai, and J. M. Briggs. 2008. Global Change and the Ecology of Cities. *Science* 319 (5864): 756–760.

Hanson, A. J., R. L. Knight, J. M. Marzluff, S. Powell, K. Brown, P. H. Gude, and K. Jones. 2005. Effects of Exurban Development on Biodiversity: Patterns, Mechanisms, and Research Needs. *Ecological Applications* 15 (6): 1893–1905.

Ignatieva, M., K. Ahrné, J. Wissman, T. Eriksson, P. Tidåker, M. Hedblom, T. Kätterer, H. Marstorp, P. Berg, T. Eriksson, and J. Bengtsson. 2015. Lawn as a Cultural and Ecological Phenomenon: A Conceptual Framework for Interdisciplinary Research. *Urban Forestry and Greening* 14: 383–387.

Jones, D. N., and S. T. Reynolds. 2008. Feeding Birds in Our Towns and Cities: A Global Research Opportunity. *Journal of Avian Biology* 39 (3): 265–271.

Macgregor-Fors, I., L. Morales-Pérez, J. Quesada, and J. E. Schondube. 2011. Relationship between the Presence of House Sparrows (Passer domesticus) and Neotropical Bird Community Structure and Diversity. *Biological Invasions* 12: 87–96.

Marzluff, J. M., and A. D. Rodewald. 2008. Conserving Biodiversity in Urbanizing Areas: Nontraditional Views from a Bird's Perspective. *Cities and the Environment* 1 (2): 6.

Methorst, J. 2024. Positive Relationship between Bird Diversity and Human Mental Health: An Analysis of Repeated Cross-Sectional Data. *The Lancet Planetary Health* 8 (5): e285–e296.

McKinney, M. L. 2002. Urbanization, Biodiversity, and Conservation. *BioScience* 52 (10): 883–890.

Nassaer, J. I. 1995. Messy Ecosystems, Orderly Frames. *Landscape Journal* 14 (2): 161–170.

Nowak, D. J., and J. T. Walton. 2005. Projected Urban Growth (2000–2050) and Its Estimated Impact on the US Forest Resource. *Journal of Forestry* 103: 383–389.

Nath, A., H. Singha, P. Deb, A. K. Das, and B. P. Lahkar. 2015. Nesting in a Crowd: Response of House Sparrow Towards Proximity to Spatial Cues in Commercial Zones of Guwahati City. *Proceedings of the Zoological Society* 69: 249–254.

Patón, D., F. Romero, J. Cuenca, and J. C. Escudero. 2012. Tolerance to Noise in 91 Bird Species from 27 Urban Gardens of Iberian Peninsula. *Landscape and Urban Planning* 104: 1–8.

Pickett, S. T. A., M. L. Cadenasso, J. M. Grove, C. G. Boone, P. M. Groffman, E. Irwin, S. S. Kaushal, V. Marshall, B. P. McGrath, C. H. Nilon, R. V. Pouyat, K. Szlavecz, A. Troy, and P. Warren. 2001. Urban Ecological Systems: Scientific Foundations and a Decade of Progress. *Journal of Environmental Management* 92: 331–362.

Platt, A., and A. Lill. 2006. Composition and Conservation Value of Bird Assemblages of Urban 'Habitat Islands': Do Pedestrian Traffic and Landscape Variables Exert An Influence? *Urban Ecosystems* 9: 83–97.

Robb, G. N., R. A. McDonald, D. E. Chamberlain, and S. Bearhop. 2008. Food for Thought: Supplementary Feeding as a Driver of Ecological Change in Avian Populations. *Frontiers in Ecology and the Environment* 6 (9): 576–484.

Rosenberg, K., A. Dokter, P. Blancher, J. Sauer, A. Smith, P. Smith, J. Stanton, A. Panjabi, L. Helft, M. Parr and P. Marra. 2019. Decline of the North American avifauna. *Science* 366 (6461): 120–124.

Seto, K. C., B. Güneralp, and L. R. Hutyra. 2012. Global Forecasts of Urban Expansion to 2030 and Direct Impacts on Biodiversity and Carbon Pools. *Proceedings of the National Academy of Sciences of the United States of America* 109 (40): 16083–16088.

Sharpe, D. M, F. Stearns, L.A. Leitner, and J.R. Dorney. 1986. Fate of Natural Vegetation during Urban Development of Rural Landscapes in Southeastern Wisconsin. *Urban Ecology* 9:267–287.

Wilkinson, N. 2006. Factors Influencing the Small-Scale Distribution of House Sparrows Passer Domesticus in a Suburban Environment. *Bird Study* 53: 39–46.

CHAPTER 37

Role-Playing Case Study on Wind Farm Acoustics: From Technology to Environmental Justice

Heather L. Lai and Anne C. Balant

Introduction

This chapter addresses integrating sustainability competencies in STEM curriculum design through a summative case study with a role-playing component to help students develop sustainability-related twenty first-century career skills. We discuss students' academic experiences in mainstream STEM fields as opposed to interdisciplinary programs in

H. L. Lai (✉) · A. C. Balant
Division of Engineering Programs, State University of New York, New Paltz, New York, USA
e-mail: laih@newpaltz.edu

A. C. Balant
e-mail: balanta@newpaltz.edu

A. C. Balant
Department of Communication Disorders, State University of New York, New Paltz, New York, USA

S. M. Wyman and J. Korenblat (eds.), *Sustainability Learning for Action and Community Engagement*,
https://doi.org/10.1007/978-3-032-11698-7_37

Sustainability Science and examine the extent to which sustainability competencies are addressed in accreditation standards and/or curricular recommendations in STEM disciplines. This is followed by a description of the course's development, the design and implementation of the case study, and student reactions to the course's first offering. We close with a discussion of the potential application of this approach to STEM courses with different content as a model for preparing future generations of scientists and engineers to collaborate effectively with government agencies, environmental justice organizations, and industry to build a sustainable future.

As issues of sustainability and environmental justice have received increasing attention globally and at home, there is "a growing consensus on the need to increase higher education's contributions to developing more effective and equitable responses to a diverse array of sustainability challenges" (Mochizuki and Yarime 2016, 21). Institutions of higher education are responding to this call. According to the Association for the Advancement of Sustainability in Higher Education (AASHE n.d.), there are nearly 700 academic programs in sustainability in the US alone. Students can access everything from certificate programs to doctoral programs in myriad facets of sustainability. Nevertheless, it is unclear whether the graduates of sustainability programs have the specialized expertise necessary to address technological challenges and/or whether scientists and engineers graduating from more mainstream STEM programs achieve the required sustainability competencies. For example, Macias et al. (2022) report that although there is increasing interest on the part of engineers and scientists, their collaborations with environmental justice organizations are not always effective. Through structured interviews, the authors identified these organizations' technology and analysis needs and how they intersect with their strategies for effecting change. One of their key conclusions is that when scientists and engineers move from their more traditional contexts and collaborate with these organizations, they gain sustainability competencies through experience, increasing the social relevance of their technical work. However, these collaborations would likely be more successful if graduates of mainstream STEM programs already possessed a well-developed set of sustainability competencies, along with their technical skills and expertise.

At SUNY New Paltz, US, for example, there is a strong focus on sustainability. The university has a full-time Sustainability Coordinator and an active sustainability committee of faculty, professional staff, and students. The UN Sustainable Development Goals (SDGs) have been adopted as the framework for strategic planning for the campus. Currently, students may choose from among over thirty-eight sustainability-focused or related courses. At this writing, there are 23 students majoring in environmental science and 64 majoring in environmental studies. The campus frequently hosts speakers on climate and sustainability issues, and many students participate in environmental organizations such as the NY Public Interest Research Group. As with the environmental justice organizations discussed by Macias et al. (2022), although many of these students are passionate about environmental causes and climate justice, they lack the technical background to make realistic assessments of environmental problems and potential solutions and often seem unaware of the practical constraints that limit the efficacy of the strategies for which they advocate so fervently.

Meanwhile, the students in mainstream STEM majors who would be best qualified to have meaningful discussions about environmental justice issues appear to be least engaged with these opportunities. Although they may be aware of the issues, their rigorous academic programs leave them little time for extracurricular activities. Their curricula focus primarily on methods that will allow them to arrive at solutions to scientific and technical problems, with little consideration of the environmental or societal impacts of those solutions. They also typically do not observe their faculty instructors and mentors engage with the environmental community and activities on campus. This sets the stage to perpetuate the disconnect between environmental justice organizations and the engineering and scientific community, as Macias et al. (2022) described.

Sustainability Learning Outcomes and STEM Programs

The disconnection described above may be understood by considering the fundamental differences between mainstream STEM disciplines and Sustainability Science, the scientific/technical aspect of Education for Sustainable Development (ESD). Mochizuki and Yarime (2016) describe Sustainability Science as an interdisciplinary field that is "defined by the problems it addresses, rather than by the disciplines it employs" (18). They describe knowledge in relevant disciplines as "multidisciplinarity," the first stage in Higher Education for Sustainable Development (HESD). The next stage is "interdisciplinarity," the integration of knowledge from these domains to understand the interactions between humans and the environment to solve problems. The third and final stage is "trans-disciplinarity," wherein those solving the problems collaborate with and learn from members of the sector of society being affected by the problems and solutions (18–20). Typically, all three stages would be integrated within a single course, or the first two would be included, with a separate course or academic experience that introduces the third. Moving through these stages allows students to acquire a number of competencies in the area of sustainability.

Although different courses in sustainability will, of necessity, have diverse student learning outcomes (SLOs), specific competencies can be identified as common threads. When students explore a problem from multiple perspectives (transdisciplinary stage), they develop what Wiek et al. (2016) refer to as the "systems thinking" competency. The "strategic thinking" competency develops in the second stage as they integrate knowledge from these disciplines and begin formulating solutions. Because these solutions must take into account ethics, equity, and social justice, students necessarily begin to develop the "values-thinking" competency. As students move into the third stage, trans-disciplinarity, wherein they must collaborate and interact with affected members of society, they hone the "values-thinking" competency in addition to developing interpersonal skills. Ideally, as a summation of these experiences, they will develop the "futures-thinking" competency, giving them the tools to be change agents in their professional careers (242). However, mainstream major programs in STEM areas do not necessarily offer students the learning experiences that will help them develop these sustainability competencies.

Most departments in mainstream STEM disciplines offer general education courses that address topical issues to promote scientific literacy and help non-science majors understand the relevance of STEM to their lives and society. The programs offered for majors are far less likely to include courses that address societal issues, let alone develop sustainability competencies. Students must first acquire the specific knowledge and skills in their discipline. In some technologically intensive programs, students barely have time to complete their requirements and may only have exposure to topical issues if they participate in an undergraduate research program or external internship. Others may provide more such opportunities within the curriculum, especially if there is a professional association that dictates accreditation requirements or offers voluntary curriculum guidelines. For example, a recent draft of the American Chemical Society's guidelines for undergraduate programs includes learning outcomes in "green chemistry and sustainability" development of "expertise at the interface of chemistry to help ...solve problems that span scientific disciplines" (American Chemical Society 2023) and outcomes related to "systems thinking." The current guidelines of the Accreditation Board of Engineering Technology, ABET, have two out of the seven learning outcomes that relate to sustainability competencies:

> Outcome 2. an ability to apply engineering design to produce solutions that meet specified needs with consideration of public health, safety, and welfare, as well as global, cultural, social, environmental, and economic factors.
>
> Outcome 4. an ability to recognize ethical and professional responsibilities in engineering situations and make informed judgments, which must consider the impact of engineering solutions in global, economic, environmental, and societal contexts.

The first outcome requires students to integrate the "values-thinking" competency by analyzing engineering problems that have societal and/or environmental impacts (systems thinking) and proposing solutions (strategic thinking) that address the stated extrinsic factors. The second learning outcome places additional emphasis on values thinking and seems to lead toward the futures thinking competency. Any engineering program that is ABET-accredited will be able to pinpoint components of one or more courses that are intended to address these outcomes, but without going through the three stages of HESD described above, the degree to which sustainability competencies will be achieved is unclear. It can be challenging for educators to find opportunities for their students to develop sustainability competencies due to the demands of teaching the extensive technical material associated with the curriculum. For engineering students to meet these learning outcomes, their curricula must require them to grapple with technological challenges in sustainability, integrate multidisciplinary knowledge, reflect on human and societal values as they weigh possible strategies, and connect more holistically with the end-users and others affected by their decisions. Students typically have little or no experience grappling with these issues in their everyday engineering training. A well-designed case study involving multidisciplinary learning and role-playing can provide these opportunities.

Role-Play Case Study

Case studies involving sustainability and environmental issues provide a unique opportunity for STEM students to tackle technical and ethical aspects simultaneously and understand the importance of connecting them in their future professional responsibilities (Bartz and Deaton 1996; McConville et al. 2017). Including a role-playing component in a case study presentation provides additional opportunities for students to develop awareness of environmental justice issues (Martin et al. 2019; Walther et al. 2012).

A wind energy workforce development and engineering education grant allowed our campus to create a series of short courses related to wind energy, including one on wind farm acoustics. Wind farm noise can negatively impact residents living nearby and frequently is a barrier to community acceptance. Therefore, instruction in wind farm acoustics should consider human and environmental impacts, providing a rich opportunity for STEM students to develop sustainability competencies and meet ABET Student Outcomes 2 and 4 by considering real-world case studies of wind farm noise and community impacts. In this specific example, the interdisciplinary course includes the following subject areas: wind turbine design (engineering), physical acoustics (physics), sound perception (psychoacoustics), and environmental impacts (ecology).

Students develop the "systems thinking" competency by integrating the information from multiple disciplines. By exploring their perceptions of sounds, including wind farm noise, and "hearing" the voices of stakeholders through case study materials and sources, students develop the "values-thinking" competency. As they weigh possible solutions and consider how decisions they make can affect human and non-human species, they are developing the "strategic thinking" competency. Finally, they move into the "transdisciplinary" phase as they present, discuss, and debate their solutions in the role-play activity, enhancing their interpersonal skills and working toward the "futures-thinking" competency (Fig. 37.1).

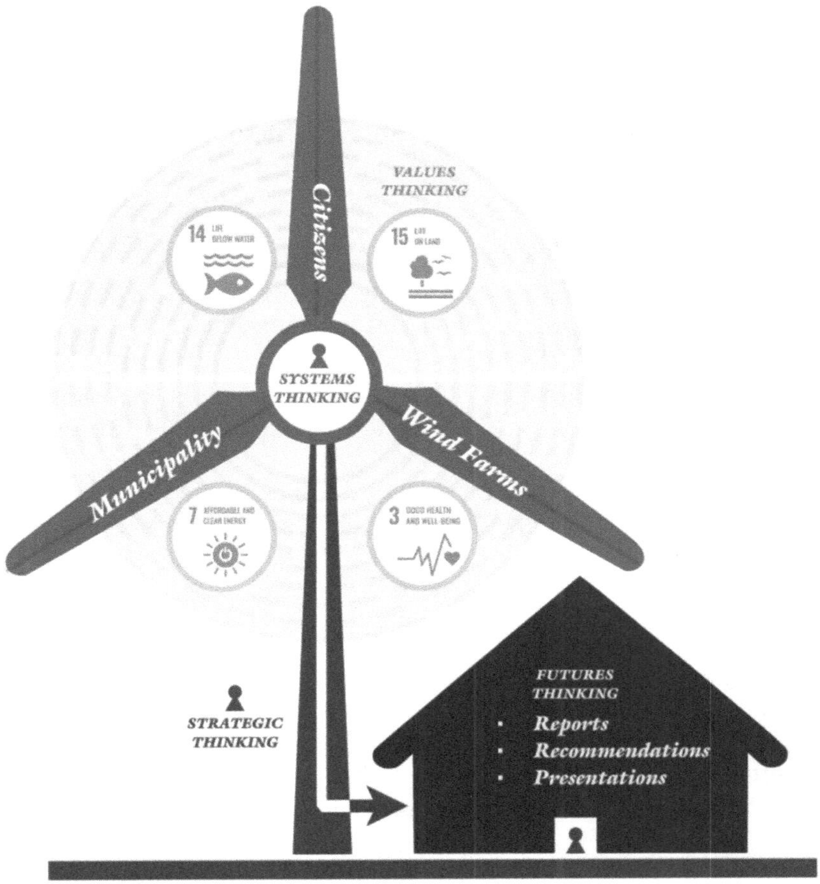

Fig. 37.1 This graphic illustrates the relationship between the UN Sustainable Development Goals and the Higher Education Sustainability competencies & ABET SLOs in the context of a role-playing case study. The UN SDGs that pertain to wind energy are the background of the turbine and frame the values-thinking of the students. The turbine blades represent the roles the students assumed in the case study as they use systems thinking to analyze the noise problem caused by the wind turbines from the perspective of their role. The turbine tower represents the integration of the proposed solutions using strategic thinking. As the students interact to work out solutions, they apply values thinking and hone their interpersonal skills. The result, illustrated through the house being powered by the wind turbine, represents the use of futures thinking in engineering problem-solving.

Wind Farm Acoustics Course Description

The goal of this course is to encourage science and engineering students to engage in analyzing systems; devise strategies to reduce human impacts of wind farm noise; and promote acceptance of this form of green energy within the context of global, societal and economic factors. The course is open to any student who has completed a course in introductory physics, including students with a major or minor in environmental science.

The UN SDGs are the preferred framework for this course because they are widely accepted and potentially useful to our students in their future careers. Specifically, the applicable global goals are:

SDG 3 Health and Well-Being—Impacts of wind farm noise on humans SDG 7 Affordable and Clean Energy—Wind Farm engineering design
SDG 14 Life Below Water—Impact of wind farms on marine species
SDG 15 Life on Land—Impact of wind farms on terrestrial species

Course Structure

The five-week course was designed as a "flipped classroom" hybrid, with 75-minute weekly in-person classes reserved for discussion, demonstrations, lab activities, and homework support. The first four weeks consisted of online readings, video lectures, quizzes, assignments and in-class activities. These components of the course have multiple goals (SLOs):

- Give students the technical background they would need to interpret the case materials and reports
- Introduce students to the perceptual aspects of wind farm noise through listening activities to help them understand the impacts of wind farm noise on nearby residents
- Introduce principles of noise control engineering so they can propose realistic technical solutions

The students prepared and presented their case study project during the fifth week.

Development of Role-Playing Based Case Study

The wind farm selected for the first iteration of this course was Nā Pua Makani Wind Farm in Kahuku, Honolulu, Hawaii. The siting, construction, and operation of this wind farm have been very controversial due to the proximity of the turbines to community areas such as schools, and the adverse effects experienced by local residents. Along with environmental and noise assessment reports, documents describing the residents' concerns are readily accessible online. Links to the primary source resources provided to students can be found on the Earthrise Commons (earthrisecommons.org). These include governance regulations, government and corporate information sites, required technical reports, data sites, scientific journal articles, news reports, documents, ethnographic research, public comments, environmental justice advocacy groups, direct contact with community stakeholders.

Identification of Information Sources

To develop a role-playing current events-based case study, you must know the scope and depth of material available to the students as they prepare. The material can be gathered ahead of time and provided directly to the students, or the instructor can furnish tools to guide the students in their fact-finding journey. Regardless of how the material is gathered, students will also need guidance on how to use the information they have at their disposal.

Development of the Student Roles

Recent research into engineering education has demonstrated that students learn to address the social factors related to sustainability and ethical issues effectively when engaged in relevant role-playing (McConville et al. 2017; Martin et al. 2019). Role-playing can be incorporated into a case study by forming teams of students and assigning each team member to represent a specific stakeholder involved in the controversy. In the case of the Nā Pua Makani Wind Farm, groups of three students were formed, and each of the three students was instructed to select one of the three roles provided. These roles were developed based on our perception of different groups who may employ an acoustics engineer to represent their concerns:

Role 1: Acoustics engineer for wind farm. You are an acoustical engineer for the company that owns and/or operates the wind farm. You have been involved in the acoustical aspects of this wind farm's development, construction, and operation. You feel strongly that wind energy is a critical component of your state's commitment to increasing its usage of renewable energy and strive to develop efficient and inexpensive noise control measures that will allow the wind farm to continue operating as efficiently as possible without causing harm to the residents.

Role 2: Acoustics engineer for the local municipality. You are an acoustics consultant hired by the local municipality to help them address residents' concerns. Your goal is to evaluate the residents' concerns based on the local regulatory statutes as well as information that has been provided to you by the wind farm developers in the Noise and Sound Level Assessment Report(s), and use this information to help the city council respond to the concerns that have been raised in a way that continues to address the state's required clean energy initiatives, continues to promote the economic well-being of the city, does not put undue cost on the municipality, and can withstand legal challenges by the wind farm.

Role 3: Acoustics engineer for the residents. You are an acoustics consultant hired by the residents to help address their concerns. Your goal is to evaluate the residents' concerns based on the local regulatory statutes as well as information that has been provided to you by the wind farm developers in the Noise and Sound Level Assessment Report(s), and use this information to help the residents have their concerns met in a way that promotes their well-being, is not interpreted as being anti-clean energy, and does not put undue cost on the residents (through energy price hikes or increased local taxes).

Instructions for the Case Study Development and Presentation

On the last day of the course, each group of three students collaboratively presents an acoustical background for their wind farm and then engages in a mock town hall-style meeting with the local concerned residents. During the town hall-style meeting, students conduct an informal discussion to develop a plan for addressing each of the invested parties' concerns and the instructor provides questions when necessary to keep the conversation focused. After completing the discussion, students share their impressions of the experience.

Description of Case Study Requirements

The students were instructed as to what type of information to use in developing their collaborative portion of the presentation as well as their individual, stakeholder specific solutions. Elements include: acoustical background; location; acoustical assessment; regulations; residents' concerns; noise control measures by role; regulations; administrative controls; wind turbine remediation; additional compensation (see Earthrise Commons; earthrisecommons.org).

Outcomes of First Offering

Student Case Study Presentation

Student Reaction

Due to the small class size, the college's student evaluation of instruction was not available. About four months after the course ended, the students were asked to send us comments about what they learned from the course and any perspectives they had gained. One student wrote that the final presentation "helped me a lot to learn about the problems that today's society faces and how it can be improved." Another wrote that, "It's not so simple to just implement clean energy and there's positives and negatives to something that just seems like it should be an obvious solution. The town meeting simulation .... gave me a new perspective on how lawmakers operate and how it might not be so simple to do the right thing..." This student concluded, "Renewable energy actually isn't the perfect solution I had originally thought it was....I still believe it's the correct solution, but it comes with some downsides that professionals will have to continue to iron out and have to explain transparently to the public before simply making the switch to green energy in every circumstance.... I genuinely had a black-and-white perspective before this course." (*All comments used with permission*).

Next Steps/Further Development

The next iteration of this course will be in an asynchronous online modality. Although most of the materials are already available in the college's learning management system, this transition will require revision of some course components. Students will conduct some lab activities individually rather than in a group setting. Other demonstrations have been filmed, and students will be given the data to analyze rather than making the measurements themselves. The case study project will require students to collaborate online to give their final presentations and conduct their discussions, which will then be submitted as a video.

Conclusion

Overall, this course provides the technical background that the students need to analyze a topical Engineering case study and grapple with human well-being and environmental, societal, and ethical issues. Student reactions from this first offering suggest that this scaffolded case study project can help learners meet the two ABET student outcomes. Because the course is open to students in other majors who have completed a physics course, there is the potential to have a mix of Engineering and Environmental Science participants, which would bring multiple perspectives to the discussion to benefit both student groups.

An additional benefit is that students were exposed to noise control engineering. Engineering students typically receive no training in acoustics or noise control engineering. It is not required for licensure as a professional engineer and is not included explicitly in ABET student outcomes (Wang et al. 2021).

Because this is an elective course, it does not obviate the need for ABET outcomes 2 and 4 to be met through other required courses, but it provides an additional opportunity to reinforce these outcomes in a more focused manner. The approach described here provides a general framework and could be adapted to other engineering topics related to a wide range of current societal and/or environmental problems.

References

Accreditation Board for Engineering and Technology. 2023. "Criteria for Accrediting Engineering Programs, 2024–2025". https://www.abet.org/accreditation/accreditation-criteria/criteria-for-accrediting-engineering-programs-2024-2025/

American Chemical Society. 2023. *2023 ACS Guidelines for Undergraduate Chemistry Programs: Working Draft* https://www.acs.org/content/dam/acsorg/education/standards-guidelines/approval-program/guidelines-draft-sept2022.pdf

Association for the Advancement of Sustainability in Higher Education. n.d. https://www.aashe.org/

Bartz, Michael J., and Russell J. Deaton. 1996. "Role Playing in Engineering Education." Paper Presented at *1996 Annual Conference the American Society for Engineering Education*. Washington, District of Columbia. https://doi.org/10.18260/1-2—6275

Macias, Madison, Jorge Morales Guerrero, Edwin A. Schmitt, Anthony M. Levenda, Jean Leon Boucher, and Darshan M. A. Karwat. 2022. The Engineering and Scientific Challenges of Environmental Justice Organizations in the US: A Qualitative Study. *Journal of Cleaner Production* 377. https://doi.org/10.1016/j.jclepro.2022.134463.

Martin, D., E. Conlon, E. Bowe, and E. 2019. The Role of Role-Play in Student Awareness of the Social Dimension of the Engineering Profession. *European Journal of Engineering Education* 44:882–905. https://doi.org/10.1080/03043797.2019.1624691

McConville, Jennifer, Sebastian Rauch, Ida Helgegren, and Jean-Henrick Kain. 2017. Using Role-Playing Games to Broaden Engineering Education. *International Journal of Sustainability in Higher Education* 18:594–607.

Mochizuki, Yoko, and Masaru Yarime. 2016. Education for Sustainable Development and Sustainability Science: Re-purposing Higher Education and Research. In *Routledge Handbook for Higher Education for Sustainable Development*, ed. Matthias Barth, Gerd Michelsen, Marco Rieckmann, and Ian Thomas. London: Routledge-Taylor Francis Group.

United Nations Department of Economic and Social Affairs, n.d. "Sustainable Development." https://sdgs.un.org/goals

Walther, J., S. E. Miller, and N. N. Kellam. 2012. Exploring the Role of Empathy in Engineering Communication Through a Transdisciplinary Dialogue. In *Paper Presented at the 2012 ASEE Annual Conference & Exposition*. San Antonio, Texas. https://doi.org/10.18260/1-2%2D%2D21379.

Wang, Lily M. Beamer, Bryan, Moore, Keegan J., and Katie Krainc. 2021. "Case study-Lesson Plan for Noise Control Engineering Concepts for Use in ABET Accredited Engineering Programs." In *Proceedings of Inter-Noise 2021*.

Wiek, Arnim, Michael Bernstein, Rider Foley, Matthew Cohen, Nigel Forrest, Christopher Kuzdas, Braden Kay, and Lauren Withycombe Keeler. 2016. "Operationalising Competencies in Higher Education for Sustainable Development." In *Routledge Handbook of Higher Education for Sustainable Development,* edited by Matthias Barth, Gerd Michelsen, Marco Rieckmann, and Ian Thomas. London: Routledge–Taylor Francis Group.

CHAPTER 38

STREAM: A Theory of Community-Led Engagement for Health, Wellness, and Healing

Edward Lawson Jr. and Sarah Mead Wyman

Introduction

If you want to go fast, go alone. If you want to go far, go together.

What does it mean for sustainability-minded members of a university to engage intentionally with a regional community? Or for a neighborhood community to interface with representatives from an established academic institution? It's all about the intention, the people involved, and the mode of interactions over time. Historically and conventionally, well-intentioned interventions start by defining existing *problems* in the

E. Lawson Jr.
Department of Black Studies, SUNY New Paltz, New Paltz, NY, USA

Family and Community Engagement Services, Inc. (FACES), Newburgh, NY, USA

E. Lawson Jr.
e-mail: Edlaw@nyfaces.com

S. M. Wyman and J. Korenblat (eds.), *Sustainability Learning for Action and Community Engagement*,
https://doi.org/10.1007/978-3-032-11698-7_38

community. When collaborating teams begin by defining *opportunities* instead, they can leverage a rich survey of community aspirations and facts on the ground through co-design.

This chapter introduces a framework for how to facilitate community-led engagement around health, healing, and environmental/social wellness by recognizing the wisdom within the community. We developed an adaptable STREAM education and engagement modality to guide university students, staff, and faculty in ways to responsibly and ethically bring their resources and knowledge to address complex systemic change toward greater environmental and human thriving. Our community-centered theory acknowledges the diversity of opinions and perspectives within partnering demographics as they build environmental literacy (for all involved) and set an intention for a thriving future. Our holistic, embodied approach integrates informal, arts-based science, technology, engineering, and mathematics (STEM) experiential learning activities according to the vision of the learners themselves.

STREAM Methodology for Sustainability Education and Community Engagement

Our theory-to-practice collaboration is based on a partnership developed between the City of Newburgh, New York Health, Wellness, and Environmental Justice Collective (HWEJC) and the Eddy learning lab at SUNY New Paltz, US. We co-created the STREAM methodology for site-specific learning to spark social transformation through innovative educational opportunities and youth leadership. STREAM positions learners at any stage to move through a relevant curriculum that includes crucial content, critical thinking, problem-solving, and futures-thinking sustainability education competencies (Armon et al. 2020; Brundiers et al. 2021). Our empirical rationale is further lodged in critical sustainability theory with its commitment to social change toward greater equity and attention to how language is used to perpetuate systemic structural inequities and misperceptions about human/nature intersections

S. M. Wyman (✉)
Department of English, SUNY New Paltz, New Paltz, NY, USA
e-mail: wymans@newpaltz.edu

(Cachelin et al. 2019). We also follow transformational learning methodologies with an emphasis on metacognition (thinking about thinking) reflection activities, fostering agency, and an acknowledgment of varied learning styles. Transformational learning approaches are further suited to our work as they can accommodate ethical, values-thinking dimensions and promote dynamic, interactive listening (Mezirow 2009). We add R's to arts-integrated STEAM frameworks to create a STREAM methodology:

- Relationship
- Relevance
- Rigor
- Recreation
- Regeneration

We invite interrogation to expand the STEM acronym by embedding it within

- Systems-thinking & Sustainability
- Transformation
- Environmental Justice & Empowerment
- Mindfulness

The STREAM initiative is designed to rethink teaching and learning beyond institutional boundaries with holistic, art-infused programming in public spaces, by centering underrepresented voices, perspectives, and other forms of knowledge. Together, neighborhood audiences and professional practitioners innovate to identify and define challenges to human thriving and environmental conservation. This improves relationships between and among invested groups who understand the relevance of environmental and social wellness literacy to their lived experience at the local level. STREAM is an instructional methodology: a call to healing and wholeness. It is a means of community engagement that is intentionally community-driven so that knowledge production and learning *belong to those who create it* (Paris and Samy Alim 2017). This mode of applied learning connects earth-care to human care and empowers STEM skill-building (for community members and participating university students) through site-specific storytelling and hands-on activities (Cunnington

Cunnington et al. 2014). Such experiences, held in familiar spaces, can change the narrative of impoverished struggle and open doors to life-long learning, sustainability-informed systems-thinking, change agency, and careers. STREAM positions learners at any stage to move through relevant curricula that involve critical thinking, problem-solving, and futures-thinking, all key competencies for 21st century careers (Brundiers et al., 2021; Wiek et al., 2011). Local change can ripple to larger systems transformation around regional and national commitments to enhanced sustainability-informed community empowerment and educational excellence.

Putting STREAM into Action

Our research-to-practice STREAM pedagogy and partnership inform two learning activities. First, we briefly describe a 3D mapping exercise developed by the Presencing Institute at MIT (Scharmer Scharmer 2018, 92) in which participants visualize the key entities in a system that call for a design response, using tangible, tactile materials to make unfamiliar ideas easier to grasp and discuss (Fig. 38.1). This constitutes our *asset map*. We then describe *hacking a grant* or using the ideas and perspectives gained through 3D mapping to analyze and interpret a Notice of Funding Opportunity (NOFO). Those most directly impacted by the phenomenon the grant addresses lead this process. This ensures funding opportunities center those with the most to gain from successful applications. Second, we present an outdoor creative writing/earth art activity that engages informal STEM learning according to the STREAM methodology. Our goal in each project is to empower individuals by amplifying essential voices and increasing social and environmental literacy around healing and wholeness through collaboratively designed learning experiences. Such community-engaged critical research, propelled by the hopes, dreams, and aspirations of those who live in the area, can help chart a generative path toward solutions (Shorters 2019). This work depends upon deep listening and trust-building over time and coordinating with existing entities already doing the work (Brown 2017; Hooks, bell 1994; Rowe and Hiser 2016; Shorters 2021; Tuck 2009). These networking and collaborative efforts can refocus higher education/workforce priorities and result in shifting resources, including capital and agency (Villanueva 2018).

Fig. 38.1 Building the Habitat for Artists. *Photo credit* Stefanie Singer

On-Site Learning in the City of Newburgh

Our collaborative initiative enables lasting change and innovative social transformation in the City of Newburgh, situated 80 miles/129 km north of New York City on the west bank of the Hudson River. Newburgh's population of nearly 29,000 includes approximately 25% Black (non-Hispanic) and 40% multiracial Hispanic individuals (Datausa n.d.). In North America, these demographics have been historically at the forefront of defining and advocating for environmental justice (Bullard 1990; Montrie 2011; Bullard 1991). Newburgh's cultural richness may be measured further in terms of its Waoranek Lenape heritage, its role in US history as the headquarters of the Continental Army during the Revolutionary War, and a once-active industrial port for textiles, bricks, shirts, ships, leather, mechanical, and many other products.

Newburgh is currently in the throes of rapid gentrification and healing from poly-fluoroalkyl substance (PFAS) and lead contamination with the third highest level of toxicity in the US Our collaborative engagement increases the capability of the multi-site PFAS water contamination study in this historically underserved city, now remediating decades of environmental damage.

The City of Newburgh poverty rate of 26.5 is far above the national average of 12.5% (datausa), a figure which tends to correlate with fewer protections against environmental threats, including disease, toxins, and effects of climate instability. We take Amartya Sen's (1999) definition of *poverty* in terms of relative freedom and access, or "compromised capabilities due to limited access," (3; 20) as an additional feature of structural racism and a significant social determinant of health (Lewis et al. 2022). Our STREAM learning outcomes emphasize engaged, transformative experiences that are joyful and shared so that people living in *frontline* communities (disproportionately exposed to environmental hazards) see their place in the system and appreciate their capacity to be agents of change (Gunn-Wright 2020, Taylor 2014).

Education and Action for Sustainability

Education and action for sustainability depend on building partnerships, listening carefully to each other, and enabling individual citizens to bring their wisdom to the table. The original purpose of the State University of New York (SUNY) was to be a source for community development through education. The power and privilege dynamic between established communities and well-meaning institutions with missions to serve and resources to bestow, including service-learning opportunities for their students, has been well documented and explored (Aslin and Brown 2004; Ersoy 2017; Israel et al. 2013; Rowe and Hiser 2016, Wallerstein et al. 2020). However, most relevant research questions of ethical "co-production" between these entities remain open (Kuttner et al. 2022). The National Institute of Environmental Health Science (2023) has established that "Community-engaged research approaches are essential when addressing environmental justice issues, social determinants of health, and structural/systemic barriers." Our theory of engagement addresses such site-specific markers, including the potential for transformative educational opportunities outside the limits of the traditional school or university system.

Intentional Engagement with Community: Allies for Activation

Traditionally, environmental movements and initiatives have not featured all voices and perspectives. They have typically conceptualized *community* as a homogeneous group and approached individuals as unitary elements of a problem to be solved. The history of extractive researchers who use their subjects for their own advancement has been well documented (Shorters 2021, Tuck 2009). As such, environmental advocacy priorities, policies, and activism have often failed to be inclusive or responsive to the community's needs, interests, concerns, hopes, dreams, and aspirations. Therefore, the initiatives were not as successful as intended. Sometimes, these efforts were ultimately exploitative because the outsiders prioritized providing experiential learning opportunities for their students or funding opportunities through grants rather than authentic social change for the better. Each community holds its distinctive strengths and intelligence and a set of site-specific challenges and resources that must inform and structure experimental innovations (Shorters 2019). By placing the community at the center of socially transformative work at any socioeconomic level, those directly affected can rightfully insist and ensure *Nothing about us without us*. It is imperative that participants hold space for inquiry to work toward understanding the root causes of environmental damage, social inequality, and other intersecting challenges to building

Sustainable Cities and Communities (UN SDG 11) where people can thrive.

Working to understand others' perspectives requires consistent dialogue over time; this process must be iterative and acknowledge the complexities of the systems in which individuals are embedded. Instead of arriving with a set agenda, a supposedly brilliant fix, or an organizational narrative, the allies' priority is to amplify the *intention of the community.* This may require great patience and the forfeit of assumed authority or expertise on the part of the outsiders to avoid a colonizing or ultimately extractive dynamic. Resident community champions, who will not always agree with each other, set the volume level and determine the venue for this communication around prototypes for change. We refer to this initial stage of the process as developing an *opportunity statement* rather than the more common *problem statement*. This involves a deep understanding of the issues as they manifest in a particular community. The main role of the co-conspirators or community allies is to open space for conversation, foster partnerships, build networks, be the connective tissue, enable high-impact practices, and then fall away. Thus, strategic action for social transformation requires forging authentic partnerships through deep listening and co-creating research tools in the locations where those who stand to benefit most directly live.

People in crisis lack access to resources and support to create the communities they want to call home. Individuals whose basic needs are being met are often unaware of others' suffering as it does not impact them. Deep and lasting transformation depends upon the genius of heterogeneous communities and the solutions their members define. All individuals require *environmental and social justice literacy* to understand research findings, and they deserve full access to information generated about their communities. It does not just "take a village" to bring positive change to a community; it takes a WHOLE village, a group effort by those who require positive change and those with the privilege to overlook the root issues. The STREAM model engages communities who have full power and agency in design and decision-making; it provides an opportunity to focus on aspects of sustainability—health, wellness, and healing—in a positive, productive way that is fully transparent and inclusive.

STREAM Theory of Community-Led Engagement

Our goal as academic co-conspirators or allies, in partnership with underserved regions, is to increase engagement in community-led education and action for sustainability. This means collaborating on understanding existing systems, strategizing for lasting change, and working to build strong and healthy cities and neighborhoods. This is a healing-centered approach to social transformation, not a trauma-based one. Trabian Shorters (2019) and Eve Tuck (2009) have developed sound theories of community engagement that inform our research-to-action approach. The components of our STREAM engagement theory include 3 A's:

- Assessment (both before and after the project)
- Analysis
- Action

Strategic Action for Community-Led Social Transformation

1. Build partnership and trust by holding space for discussion driven by community champions.
2. Begin to understand the community by visiting and interacting with residents in a variety of ways.
3. Identify and address site-specific social determinants of health and flourishing through community-led engagement. Craft an *opportunity statement* that reflects existing resources and wisdom. Upon conclusion, report on outcomes using disaggregated data. (**Assessment**)
4. Use critical sustainability methods, including systems thinking and futures thinking to address issues presented and defined by the community. Use a healing-centered theory of engagement focused on existing strengths and aspirations to prototype high-impact activities. (**Analysis**)
5. Set an intention to promote *literacy* in key health, science, and environmental justice concepts through transformative, holistic, and creative learning activities. (**Action**)

Assessment: Assessment involves defining problems and, moreover, recognizing possibilities that may not appear at first glance. Assessment looks to an *opportunity statement* that specifies the challenge the community wishes to address. This process also involves the more conventional aspect of creating a tool to evaluate the project's success. Potential collaborators come to know the target community by listening and interacting so that together, they can identify and define the scope of the problem and the criteria for success. It is essential to meet people on their terms and at the locations they select. Strategies for social transformation depend upon developing a sense of trust and mutual understanding to the greatest degree possible.

Once interpersonal connections are established, assessment begins with such questions as:

- Who are we?
- What assets/resources exist?
- What are the challenges to health, wellness, healing access, and support?
- How can we understand and approach the root cause of the issue?

The first step is to perform a social/environmental inventory: name existing areas of success, barriers to further growth, and other challenges. A SWOT analysis (strengths, weaknesses, opportunities, threats) may be a useful assessment strategy at this stage. It is important to determine the clear boundaries and intention of the project, although this will only become clear through the process itself. An initial rubric that represents project goals (with the understanding that they may shift over time) will guide the process and prepare for evaluation at checkpoints along the way and upon conclusion of the work.

Analysis: Any successful tactic will be tied to specific outcomes that address discrete issues identified by the community. External practitioner allies do not arrive with the intention to organize the community or impose their own agendas, although they are typically full of ideas. Using a critical sustainability approach, participants consider the many intersecting factors that structure or influence current circumstances (Rose & Cachelin 2018). Systems-thinking allows those involved to identify and conceptualize the way overlapping burdens such as poverty, health challenges, learning loss during COVID, and environmental toxins can compound barriers to well-being in certain locations. Participants seek opportunities for transformation, including greater knowledge and capacity, such as leverage points for high-impact intervention and feedback loops (Meadows 2008). Futures-thinking activates creativity to spark emergent solutions in the form of prototypes for change. When a community can envision the stable, secure, and *sustainable* place they would like to call home, they can design a framework to bring that world into being.

Action: Action for community-led engagement begins with human connections established during the assessment stage. Communication must be based on a solid foundation of trust, a desire to understand, and community agreements to hold space for listening and respect for one another's opinions. One may think of creating this partnership in terms of 5 C's:

- Communication
- Consistency
- Collaboration
- Continuity
- Cause (as in determining the root cause of the problem)

Once the hybrid insider/outsider group has established a partnership and opened communication with an appreciation for the varieties of perspectives and local agendas, the real co-design work can begin. Together, the group establishes clear goals and assessment frameworks, even if these will shift throughout the process.

Activity #1: Performing a Needs Assessment and Hacking a Grant

One fifty-seven-year-old community activist, Gabrielle Hill, described the moment she fully conceptualized the process of PFAS and lead particle contamination in her community. She visited her granddaughter's fifth-grade class and watched an activity designed to demonstrate how toxins infiltrate waterways and soil. On a three-dimensional landscape map of Newburgh, the teacher dripped a red dot of food coloring. The children squirted water on the red drop until it thinned and dispersed, running into the creek and earth. This interactive and embodied visualization allowed Gabrielle, alongside the children, to understand better what had happened in their hometown and their own bodies, clearly connecting self and environment.

As a restorative justice leader in her neighborhood, Gabrielle was well-positioned to articulate her community's hopes, dreams, and aspirations. She and Newburgh resident and SUNY educator Edward Lawson co-directed the conversations at Blacc Vanilla Cafe that led to prototyping two arts-based projects and *hacking* a Spencer Vision grant, working through what may have remained opaque and exclusionary language. Through this STREAM-informed process, the NHWEJC and Eddy members set out together to render visible the needs, goals, hopes, and dreams of City of Newburgh representatives. After a year of asset-mapping and trust-building meetings around the city, we were ready. The two groups joined additional residents, including youth, for a 3D mapping activity at the cafe to facilitate conversation and visualize the community as an integrated system. We used everyday objects to identify areas of strength and potential growth and strategized how to complete our grant application.

3D Mapping as Critical Analysis

During our 3D mapping session, we articulated our primary values then sketched out the City of Newburgh as a dynamic system of interrelated stakeholders. Around a large table, we selected objects to represent humans and other entities. Participants created a three-dimensional map where a sprouting potato stood for *youth* and a clutch of keys from a local ex-cop's pocket represented *access* and *gatekeeping* with respect to opportunities for learning and growth. A bunch of clothespins indicated *family* and an inverted jar signified *lawsuits* pending upon the Newburgh Enlarged City School District. The conversation developed around this visualization as a wet paintbrush enacted the sweeping gestures of politicians who often miss the mark. Next, we examined our display from four directions, according to the MIT Presencing Institute's Theory U methodology: East, the place of feeling: What do you love? What frustrates you? South: What hard truths need to be spoken? West, reflection: What assumptions underly the situation? What are systemic barriers that lock us into the current state of operating? North: What is ending, wanting to die? What is wanting to be born? As a team of students, community members, and co-collaborators, we began to *critical path* or identify and define (Shorters 2019) the connections between the motivation to learn and grow and the routes to success that can manifest via youth leadership. Other communities, such as Newark, New Jersey, US, provide wonderful models of similar projects (see This is Newark).

Our conversations and arts-based use of symbolic objects arranged on the cafe table helped generate an intuitive understanding that youth in Newburgh can use their genius to chart paths to success. When we begin by ***living in the questions*** (Eaton et al. 2013; Greene 2013), we start to understand the social system in terms of connections and leverage points for greater impact (Rowe 2019; Meadows 2008). We assess which types of change the culture can handle at present (Shorters 2019). Our mapping exercise worked because we had a history of deep listening and collaboration in that space and other parts of the city—parks, urban farm, library, affordable housing building, local maker space, community center—where community members could build ideas through our STREAM framework.

Next, we worked in mixed groups to identify the grant's priorities, goals, and optimal outcomes. This first step in collectively ***hacking a grant*** involved prototyping a vision for holistic learning and development, both inside and outside the classroom. Our STREAM methodology was still being discovered and developed based on real experiences in the field. Representatives from all constituencies focused on specific aspects of the NOFO, identifying the research question, examining the foundation's mission, outlining the program's goals, and attending to the specific language used in the application. Gabrielle, cafe owner Jerrod Lang, and three young men titled our proposal *Youth Leadership at the Heart of Our Community.*

Activity #2 STREAM: Arts-Based Community Engagement Using Informal STEM Learning and Poetry Writing

Community champions, NHWEJC, and the Eddy at New Paltz developed STREAM storytelling events designed to empower learners and open vistas for future growth through arts-based activities using hands and minds. Participation in creative exploration, whether 3D mapping or engaging in earth art and creative expression, increases the value of partnership in problem-solving and spatial reasoning. These sustainability education skills are crucial for developing twenty first-century career-relevant sustainability competencies (Brundiers et al. 2021). Our initial engagement tools were 6x6' http://www.liptonarts.com/simon-draper-habitat-for-artists-2/ Habitat for Artist sheds prototyped by Simon Draper and built by young Newburgh residents who received stipends for their participation. They also directed the placement of these structures for interdisciplinary education and outreach, to hold space for conversation and learning about food life cycles and soil health, integrating basic Math skills with Biology, Ecology, earth art sculpture, and poetry writing. Newburgh middle and high school students collaborated with Sculpture professors in the construction, emphasizing measurement, process, and technical skills for using equipment safely (saws, drills, tape measure, level, etc.) These experiences would become material for their creative writing and learning activities.

Middle and high school learners co-wrote a collaborative poem at the Armory Unity Center after observing new grass on their circular soil sculpture. Learners also enjoyed embodied informal STEM learning as they calculated the dimensions of the habitat they had constructed several weeks earlier, now decorated with paint and drawings (Wagner & Shahjahan 2015). Youth creatively integrated measurements taken with tape, scales, string, rulers, fingers, and their own eyes. They examined their immediate environment as material for creative expression: an urban vegetable and flower farm with animals. They collected visual and tactile details, considered sustainability themes, and started composing aloud. From lists of words and piles of numbers, they brought a scene to life using both English and Spanish diction. Dimensions of the earth art stacked structure (base, middle, and top circumference, and height), fencing length, and estimated numbers of leaves and cups of mulch rendered their poetry vivid.

Discussion

The process of creating the interpersonal bonds that enable community-led engagement, often around sensitive sites of damage, takes time and clear intention. Regular meetings and shared activities should be documented to establish continuity and build a story. It is essential to keep showing up, to be welcomed into the designated location, to hold space for conversations to evolve and strategies to emerge. In certain circumstances, creating this partnership will prove impossible due to conflicts in values, priorities, time restraints, and allegiances. Allies must be aware of the emotional toll of physical harm, livable-land dispossession, food insecurity, and other failures to care for and protect human populations and their environment.

Conversations will be iterative, progressive, and exploratory. When prototypes for change begin to materialize, the collaborators must remain adaptable, ready to shift, share leadership, engage in non-hierarchical decision-making, and, most importantly, ensure that community members take the lead in any significant planning. Meet people where they are to understand and value *who* they are. Allow space for joy, for failure, and for shifts in goal-setting. This agility comes through building relationships and embarking on a partnership, developed on location, that feels legitimate and productive. Allies committed to *education* as a practice of freedom and a means of social transformation (Hooks, bell 1994) may consider its root terms *educere* meaning "to pull out from within."

Education and action for sustainability that drives community engagement for environmental and social transformation is site-specific, like Gabrielle's transformative understanding of the contamination that had happened to her community's soil, water, and bodies. It's imperative that practitioners identify and address the social determinants of health in a particular location through community-led engagement with the guidance of community champions. These determinants include access to quality education and health care, affordable and secure shelter, nutritious food, workforce training, employment opportunities, and adequate resources and support. Communities must be free of environmental toxins, including violence and racism (Lewis et al. 2022). To reduce health disparities that result from injustice and detrimental determinants of health, communities must be fully engaged in all aspects of intervention implementation.

Populations that have not experienced consistent and authentic delivery of promised resources and support may be unwilling to cooperate with outsiders who arrive offering "solutions" or wish to open doors to conversation. When communities are unwilling to be vulnerable enough to be transparent, allies cannot explore the root cause of the problem with them. Such communities are often the subjects of sociological or clinical studies. Typically, the results of the investigations or the grant money earned by "poverty pimps" does not return to the community. As Erin Bell (2023) explains, and many critics concur, sociologists or public health experts often gather data, analyze it, write the peer-reviewed article, and leave. Community members are typically cut out at that point. In the face of such extractive measures by *professional experts*, apathy and a lack of trust often set in. People do not feel like they have a seat at the table. If they do, they think they are on the menu. Co-conspirators need to be intentional about meeting people where they are: in their schools, living spaces, places of worship, and other community hubs.

A Transparency, Standards, and Accountability (TSA) framework can ensure ethical and effective engagement: Allies need to enable:

- *Transparency* by formulating clear…
- *Standards* in collaboration with all parties and by ensuring…
- *Accountability* for how results will be shared, and inclusive actions will be carried out.

When these practices are followed, participants can then gather the necessary data for evidence-based practice to address environmental and social challenges and co-design frameworks for transformation. Data needs to impact policy to be useful, and for policy to be appropriate and effective, it must involve collaboration with the community. Otherwise, academics participate in merely "performative allyship." All data needs to be disaggregated and analyzed thoroughly, according to various perspectives, categories, and approaches. Furthermore, these partnerships must be contextualized within frameworks of care: participants identify and discuss their values and priorities to avoid the assumptions and presumptions that can misguide activity.

Communities that are not in immediate crisis may still wish to organize around Education and Action for Sustainability (EAfS) to promote social cohesion, mental health, and physical well-being. For example, they might try soil regeneration composting projects, community storytelling engagement efforts, or confront urgent issues such as the environmental impacts of toxic fertilizers or building affordable housing for all. Careful listening and trust-building are also key to these efforts, especially around contentious decisions that can easily become divisive along political party lines. To avoid alienating potential partners, begin with a focus on factual histories of the location and its lived experience rather than adopting a *social justice* mandate for change before the goals of that intended transformation have been defined. This approach can neutralize the process and prove more inclusive in the long term. Providing opportunities for all community members, including youth and elders, to come together delivers additional affordances in knowledge sharing, group identity formation, and the mitigation of isolation and loneliness.

Once the dialogue has been established and enough community champions and other residents have joined the conversation, the theory of community-led engagement framework can be put into practice. The newly established cohort of residents and allies may complete a SWOT analysis of practices, programs, policies, and procedures. The requisite information will not be forthcoming at a town hall where people tend to show up with prepared speeches, ready to argue and defend their positions. Instead, this work should take place in a space where participants have agreed to negotiate, learn, share openly, and compromise. Voluntary participation is ideal; however, some collaborators, such as school board members, will participate in their political capacity as public servants. There will always be detractors and the risk of a *scarcity mentality* due to limited resources and competition for funding. Using a circle method of engagement that emphasizes deep listening and respectful reflection will be the most effective tactic and mitigate "horizontal hostility" or conflicting passions that can manifest among those who care and ostensibly share a common purpose. (Brown 2017).

When co-conspirators or allies who have not experienced economic disadvantage, racism, or other forms of discriminatory treatment enter communities where these rates are high, they need to develop their literacy in the complex underlying social structures that account for today's astounding levels of inequity worldwide. Again, these circumstances will be site-specific and culturally contextualized. To provide a few examples of the US phenomenon, "historically disadvantaged" communities have become additionally disenfranchised as the country became increasingly de-industrialized after WWII and official policies restricted interracial marriage and home ownership. William Julius Wilson has pointed out, in *The Truly Disadvantaged*, that working-class African Americans experienced disproportionate impacts both economically and socially when industrialization became more automated, anti-miscegenation, red-lining, and other racist policies were codified, and manufacturing moved overseas (Wilson 2012; Johnson 2020; Kahloon 2023).

The wisdom, information, and dreams generated through dialogue between community members and "outsiders" provide rich material for change. As critical sustainability practitioners, allies and co-conspirators use critical thinking to consider sustainability challenges in a way that acknowledges and interprets complex, shifting, heterogeneous systems and entrenched social structures based on capitalist expansion at the cost of the planet and its inhabitants. According to Cachelin et al. (2019), critical sustainability recognizes the material ecological necessities that sustain all life on the planet, while also acknowledging and seeking out sociopolitical orientations that support dignity, equity, respect, and rights for all parties. *Environmental Justice* encompasses climate justice, land justice, and food justice, all compelling topics for community conversation. No matter the nomenclature, the reality of economic disenfranchisement and its effects is never news to those experiencing it. A close look at inequality and material suffering in a particular location may be unfamiliar to students and other visiting allies, however, who will need frameworks and guidance for processing and understanding new knowledge gleaned from lived experience across often invisible lines of privilege, opportunity, and need.

Conclusion

When we focus on hopes, dreams, and aspirations in our effort to disrupt normalized inequities in educational access and environmental justice literacy, we uncover new narratives of empowerment and success that can bring communities and institutions together. Intentional engagement between long-standing neighborhood communities and academic institutions depends on deep listening, respect for varying perspectives, and a willingness to prototype, adapt, fail, and try again. Engaging communities within, between, and across established social boundaries, including exclusive academic institutional ones, means jumping into a dynamic flow of learners who will surprise with all they can teach.

Newburgh is poised to tell its own twenty first-century story as a leader in Education and Action for Sustainability. We can move beyond outdated and often harmful community engagement strategies that focus on damage and deprivation to more equitably share knowledge, opportunities, and wealth in all its forms (Tuck 2009; Shorters 2019; Villanueva 2018). Accordingly, allies or co-conspirators can help shift the story via deep listening to and with practitioners as we develop our partnerships, prototypes, and change agent skills through a STREAM methodology. Furthermore, we acknowledge that fostering communication and relationships within the community is as important as establishing the allies' own welcome there (Kuttner et al. 2022).

We met with various barriers and challenges along the way. Our Spencer grant application was not funded, and we could not muster funds, time, or energy to hold a youth leadership summit. One of our Habitat for Artists structures, built on a privately-owned corner with heavy drug dealing, had to be dismantled before we could initiate public programming there because people started living in the small structure. Distance from our campus (twenty-two miles) and pressures on student time made it difficult to bring enough university learners into conversation with Newburgh residents. Yet, we remain as determined as ever that such partnerships are rich with potential, and we continue our exchanges that increase every semester as people in New Paltz and Newburgh recognize the Eddy name and our STREAM work.

A major aspect of our success is bringing representatives, including youth, from the City of Newburgh to our campus, where they visit the Music Department, tour the campus, or attend and present at our Health, Wellness, and Environmental Justice symposia. A great triumph was the unexpected rap performance by City of Newburgh Mayor Torrance Harvey at our fourth *Sound Your Truth* campus event designed to amplify underrepresented voices in celebration of George Floyd's birthday and the Black Lives Matter movement.

Through intentional collaboration and shared vision, Eddy continues to co-create the STREAM methodology that informs our overall approach to (re)generative relationship building on campus and in the regional community. This enables us to use embodied systems-thinking and pro-planet values-thinking in all our capacities. 3D mapping, informal STEM learning, and arts-based opportunities for expression foster skills and empower learners as change makers. Eddy engages STREAM methodologies to build upon relationships between people, connection with their urban environment and natural surroundings, and individual learners' visions for building their own more sustainable cities and communities.

References

Armon, Joan, Stephen Scoffham, and Chara Armon. 2020. *Prioritizing Sustainability Education: A Comprehensive Approach*. Routledge.

Aslin, H.J. & Brown, V.A. 2004. *Towards Whole of Community Engagement: A Practical Toolkit*. Murray-Darling Basin Commission, Australian National University.http://hdl.handle.net/1885/77450.

Bell, Erin. 2023. Conversation.

Brown, Adrienne Maree. 2017. *Emergent strategy: Shaping change, changing worlds*. Chico, CA: AK Press.

Brundiers, K., et al. 2021. Key competencies in sustainability higher education–toward an agreed-upon reference framework. *Sustainability Science* 16:13–29.

Bullard, Robert. [1990] 2019. *Dumping in Dixie: Race, Class and Environmental Quality. 3*$_{rd}$ Ed. Routledge.

Bullard, Robert, et al. 1991. *Principles of Environmental Justice.* [PEJ] EJnet.org: https://www.ejnet.org/ej/principles.pdf.

Cachelin, Adrienne, A. Darling, B. Clark, and K. Hackford-Peer. 2019. Weaving equity and sustainability into the fabric of higher education: The University of Utah experience. *Sustainability* 12 (6): 304–309.

Cunnington, M., A. Kantrowitz, et al. 2014. Cultivating common ground: Integrating standards-based visual arts, math, and literacy in high-poverty urban classrooms. *Journal for Learning Through the Arts* 10 (1):. https://eric.ed.gov/?id=EJ1050589.

Datausa. n.d.. https://datausa.io/profile/geo/newburgh-ny (Accessed February 24, 2025).

Draper, Simon. n.d. *Habitat for Artists*. Lipton Arts http://www.liptonarts.com/simon-draper-habitat-for-artists-2/

Eaton, Marie, Davies, K., Gillespie, M., Harding, K., and Daloz Parks, S. 2013. "Living the Questions: Contemplative and Reflective Practices in Sustainability Education." *Sustainability in Higher Education: Stories and Strategies for Transformation*. MIT Press.

Ersoy, Aksel. 2017. *The impact of co-production: From community engagement to social justice*. Bristol, UK: Policy Press.

Greene, Maxine. 2013, Spring. Afterword: The turning of the leaves: Expanding our vision for the arts in education. *Harvard Educational Review* 83 (1): 251–252.

Gunn-Wright, Rhiana. 2020. A green new Deal for all of us. In *All we can save: Truth, courage and solutions for the climate crisis*, ed. Ayana Johnson and Katharine Wilkinson, 92–102. One World.

Health Science. 2023. National institute of environmental health science. https://www.niehs.nih.gov/about/community/communityforums. (Accessed, 2023).

Hooks, bell. 1994. *Teaching to transgress: Education as the practice of freedom*. Routledge.

Israel, B. A., E. Eng, A. J. Schulz, and E. A. Parker, eds. 2013. *Methods for community-based participatory research for health*. 2nd ed. San Francisco, CA: Jossey-Bass.

Johnson, Walter. 2020 (January 23). "American Bottom." *Boston Review*. https://www.bostonreview.net/articles/american-bottom/

Kahloon, Idrees. 2023. (January 30). https://www.newyorker.com/magazine/2023/01/30/whats-the-matter-with-men *What's the Matter with Men?* The New: Yorker.

Kuttner, Paul, et al (2022). "Establishing principles for community-based research: Story & Power in the community research collaborative." Metropolitan Universities 33 (3), 52–68.https://journals.indianapolis.iu.edu/index.php/muj/article/view/25819 https://journals.indianapolis.iu.edu/index.php/muj/article/view/25819.

Lewis, Crystal, et al. 2022, August. Racism is a public health crisis: How local governments are responding. *Institute for Healing, Justice and Equity.*. https://ihje.org/our-work/reports/racism-is-a-public-health-crisis-report-2/.

Meadows, Donella. 2008. *Thinking in Systems: A Primer.* Chelsea Green.

Mezirow, Jack. 2009. "An overview on transformative learning." *Contemporary Theories of Learning*, 2nd Ed. Ed. Knud Illeris. Routledge.

Montrie, Chad. 2011. *A people's history of environmentalism in the United States.* London: Continuum.

Paris, Django, and H. Samy Alim, eds. 2017. *Culturally sustaining pedagogies: Teaching and learning for justice in a changing world.* Teachers College Press.

Rose, Jeff, and Adrienne Cachelin. 2018. Critical sustainability: Incorporating critical theories into contested Sustainabilities. *Journal of Environmental Studies and Sciences* 8:218–525.

Rowe, Debra and Hiser, Krista. 2016. "Higher education for sustainable development in the community and through partnerships." Routledge handbook of higher education for sustainable development, Eds Mattias Barth, et al. Routledge, 315–330.

Rowe, Debra. 2019. *Education and action for sustainability.* SUNY New Paltz: Keynote address Sustainability Faculty Fellows' January Institute.

Scharmer, Otto. 2018. *The essentials of theory U: Core principles and applications.* Berrett-Koehler.

Rowe, Debra. 2014. (June 4) "Secrets of Newark's water presentation." *This is Newark* (https://thisisnewark.wordpress.com/2014/06/04/secrets-of-newarks-water-presentation/).

Sen, Amartya. 1999. *Development as freedom.* Oxford UP.

Shorters, T. 2021. (June). https://dataqualitycampaign.org/wp-content/uploads/2021/06/DQC-Consumers-Guide-to-Data.pdf *A Consumer's Guide to Data.* Data Quality Campaign.

Shorters, T. 2019. (June 26). https://www.ctphilanthropy.org/sites/default/files/files/events/You%20Can%27t%20Lift%20People%20Up%20by%20Putting%20Them%20Down%20-%20Chronicle%20of%20Philanthropy.pdf 'You Can't Lift People Up by Putting Them Down': How to talk about tough issues of race, poverty, and more. The Chronicle of Philanthropy.

Taylor, Dorceta. 2014. *Toxic Communities: Environmental Racism, Industrial Pollution, and Residential Mobility*. New York University Press.

Tuck, E. 2009. (Fall). Suspending damage: A letter to communities. *Harvard Educational Review* 79 (3): https://www.researchgate.net/publication/268000737_Suspending_Damage_A_Letter_to_Communities.

Villanueva, Edgar. 2018. *Decolonizing wealth: Indigenous wisdom to heal divides and restore balance*. Berrett-Koehler.

Wagner, Anne, and Riyad Shahjahan. 2015. Centering embodied learning in anti-oppressive pedagogy. *Teaching in Higher Education: Critical Perspectives* 20 (3): 244–254. https://doi.org/10.1080/13562517.2014.993963.

Wallerstein, Nina, et al. 2020. Engage for equity: A long-term study of community-based participatory research and community-engaged research practices and outcomes. *Health Education and Behavior* 47 (3): 357.https://doi-org.libdatabase.newpaltz.edu/10.1177/1090198119897075.

Wiek, Arnim, Lauren Withycombe, and Charles L. Redman. 2011. Key competencies in sustainability: a reference framework for academic program development. *Sustainability Science* 6: 203–218. https://doi.org/10.1007/s11625-011-0132-6

Wilson, William Julius. 2012. *The truly disadvantaged: The Inner City, the underclass, and public policy*. U Chicago.

CHAPTER 39

Using Hemp and Public Benefit Contracts to Build a Green Economy by Collaborating with Disproportionately Impacted Communities

Henry Gage Jr. and Sally Warren

Build Green Now (BGN), founded by Henry Gage, Jr. and Sally Warren, PhD, is New York State's leading Industrial Hemp community working group located in Kingston, New York, US, providing strategic planning, outreach and engagement, event planning, market research, business analysis, supply chain design, training, and collaboration engineering to increase opportunities for *Communities Disproportionately Impacted* to participate in and accelerate the multi-billion dollar Industrial Hemp market—Globally.

H. Gage Jr. (✉) · S. Warren
Build Green Now, Kingston, NY, USA
e-mail: henrygagejr@voicesforprogress.net

S. Warren
e-mail: sallywarren@voicesforprogress.net

S. M. Wyman and J. Korenblat (eds.), *Sustainability Learning for Action and Community Engagement*,
https://doi.org/10.1007/978-3-032-11698-7_39

Message in a Bottle

Build Green Now sends out a message to be found and acted upon, just like a message in a bottle, but with much more certainty! Our cry for help is not a single piece of paper to be tossed by waves to be washed up on some random shore; it is a clarion declaration to be acted upon around the world.

> Hemp will be the future of all mankind, or there won't be a future.
> – Jack Herer

Educators play a pivotal role, their choices create ripples, shaping a narrative and positivity that shifts perspectives from despair to hope. This message resonates with anyone who values the legacy we leave for future generations. The work ahead in green building for sustainable cities and communities is considerable, but it becomes more manageable when efforts are shared collectively.[1]

This message provides inspiration and a map for all to collaborate for a better outcome using industrial hemp as a valuable natural resource and a means to develop financial security through green entrepreneurship. A dusting of positivity from small beginnings leads to hopeful and real changes.

View from the Trenches

Chemical spills, land contamination, extreme weather events, unrest, and wars have plunged areas of the world into a desperate fight for survival. Locally, we can point to the preventable 2023 East Palestine, Ohio, accident that led to 38 train cars derailing, carrying substances including vinyl chloride, butyl acrylate, ethylene glycol monobutyl ether, Ethylhexyl acrylate, isobutylen—all released into the atmosphere and all recognized as toxic and dangerous to both humans and the environment. This happened following complaints that could have prevented the accident and toxic contamination.

Additional contributors to this dire ecological situation include overfarming, factory farming, excessive pesticide and chemical fertilizer use,

[1] Build Green Now's commitment to sustainability education and action: https://www.buildgreennow.net/store-sustainability-education-and-action.html

industrial brownfields, and extreme weather events such as California's cycles of droughts and floods, wildfires, hurricanes, and earthquakes.

Build Green Now started Voices for Progress as a Project Management Platform connecting with community members worldwide who serve as grassroots champions for decades. They tell us that political leaders, corporations, and governments have consistently broken promises and eroded trust.

As founders of BGN, we champion legislative changes, promoting shared governance as a tool crafted by people for the benefit of people. We encourage individuals to actively engage in their communities to drive the changes they want to see instead of clinging to the false comfort of the status quo. Achieving community consensus is crucial and opens the door to mutually beneficial collaborations.

Within this document, BGN provides "**Considerations**" and a "**Case Study**" as key resources for community organizers around the world to inspire positive action in a ***post-trust*** era. To transform these resources into effective action plans, it is vital that we embrace a vetted strategic approach. This chapter provides recommendations for such an intervention. This article was started about 1 year ago. It is now January 14, 2026 and the world has changed significantly. We decided to leave a majority of the text as is. However the way forward is through content sharing, collaboration and financial sustainability that supports environmental sustainability. To continue the conversation and to explore ways we can help you support a training vision, let's connect.

Meet Industrial Hemp

Industrial Hemp can be used to make plastics, paper, clothing, homes, and cars, as well as biofuel for renewable energy and running cars. Hemp seeds are a complete protein, containing all nine essential amino acids found in a steak without the need for cattle. Some experts consider hemp seed one of the most nutritionally complete food sources.

Industrial Hemp also rejuvenates the Earth: it was used in Chernobyl to help clean the soil after the nuclear plant accident (1986). It can enrich brownfields while detoxifying them. Furthermore, an acre of hemp can sequester more CO_2 than an acre of trees; while trees take 3–4 years to grow before harvest, hemp can grow to 15 feet in just 3–4 months.

Industrial Hemp is a Public Benefit Everywhere!

According to Adelaida (Addy) Vazquez (Hemp Fusion, Puerto Rico), a contractual partner, growing and processing Industrial Hemp on the island reduces transportation and import costs and thus avoids the negative economic impact of the 1920 Jones Act. The Jones Act, officially known as the Merchant Marine Act of 1920, is a federal law regulating maritime commerce in US waters and between US ports. This truth applies to territories around the world at the mercy of such policies and enforces a need for local farming and manufacturing.

Political corruption has long blocked the use of Industrial Hemp in the US, influenced by corporations that want to protect their synthetic petrochemical markets for nylon and plastics. These companies promote reliance on harmful practices such as deforestation and intensive cotton farming. We can move toward a regenerative future with Industrial Hemp, allowing global communities to heal and thrive, feed, clothe, and house themselves in regenerative ways.

About Build Green Now

At BGN, we prioritize sustainable financial practices, particularly in communities that face the greatest environmental impacts. Our innovative approach centers on ***Public Benefit Contracts*** (PBC), designed to ensure projects deliver significant community benefits and promote social equity. PBCs aim to empower leaders globally, providing essential training and engaging them in negotiations and agreements that ensure public benefit and pre-professional training for youth. We operate according to reciprocal mentoring relationships and use Collaboration Engineering to accelerate the adoption of regenerative best practices informed by the wisdom and desires of those who will co-create and share products and services.

BGN leads the way with templates for Public Benefit Contracts and provides an innovative and practical resource for community leaders. These frameworks eliminate the need for costly corporate certifications or re-incorporations, allowing organizations to improve their communities quickly. Organizations can promptly work toward financial stability and support environmentally regenerative practices by mastering a few essential initial steps, such as contract creation, drafting, and approval.

Public Benefit Contracts Versus Franchises

Public Benefit Contracts (PBCs), present a viable alternative to traditional franchises and corporations, which typically demand substantial capital, extensive legal knowledge, and significant time to establish. These contracts enable more straightforward and less expensive involvement in projects that deliver immediate and enduring community benefits. In contrast to franchises and corporations, which often prioritize wealth accumulation and the safeguarding of legacy knowledge and intellectual property, PBCs are more accessible, especially to those in economically marginalized communities. This accessibility makes them crucial tools for promoting financial stability and environmental sustainability in areas that need it most.

The hemp industry can only take off worldwide if sanctioned by government regulators. In 2022, the BGN team (members of the US Hemp Building Association ICC-11) advocated at the International Code Council (ICC) in Rochester, New York, for the adoption of Industrial Hemp-lime (Hempcrete) into the ***International Building Code*** as residential insulation. The team's efforts were successful, and Hemp-lime was adopted into ***Residential Building Code*** in 2024. At the US National level, along with the work of the Hemp Exemption campaign and the National Hemp Association (NHA), Industrial Hemp is a recognized agricultural crop in the US Farm Bill and other legislation.

Strategic Overview to Rebuild Trust and Promote Green Entrepreneurship

Ensure Community Well-being: Recognize that communities have suffered financial insecurity, health, and safety damages from systemic injustices, harm from power generation, garbage incinerators, petrochemical facilities, gun violence, and more. Start by measuring the impact of these injustices.

Identify Immediate Needs: Avoid entering a community with an unsolicited agenda. Instead, ascertain the immediate needs of communities disproportionately impacted and align with their priorities.

Identify a Primary Demographic: We suggest focusing on progressive individuals, Generation Z, and disproportionately affected communities. While older leaders offer valuable wisdom from experience, they frequently focus on preserving the status quo. In contrast, progressives and Gen Z members understand the immediate need to stop harm and begin repair.

Work through Community Organizations: Engaging with community-based organizations is highly effective for driving social change. Community leaders have established trust and understand their neighborhoods' specific needs and dynamics. By actively participating in their meetings and offering services that directly respond to expressed needs, you can enhance the relevance and impact of your efforts. Working through these organizations ensures that initiatives are community-led and community-approved, increasing the likelihood of successful and sustainable outcomes.

Integrate Sustainability Priorities: It's crucial to challenge the elitist notion that environmental sustainability is a universal priority. Instead, we must recognize that economic struggles and financial constraints often overshadow environmental concerns in specific communities. Integrating sustainability objectives with community priorities is essential because it ensures that sustainability efforts are relevant and directly beneficial to the people involved.

Assemble a Team

> People are more important than ideas!
> - Ed Catmull

Our strategic approach emphasizes the importance of assembling effective teams over merely proposing good ideas. Catmull noted that a well-coordinated team can transform a poor idea into a successful initiative, while a dysfunctional team can undermine a great idea. This principle guides our strategy to ensure community well-being, identify immediate needs, and effectively engage with community organizations.

Leaders in communities disproportionately impacted are more receptive to collaborators who support forming unified, community-based teams. These groups must be aligned in pursuing human rights, equality, and justice.

By demonstrating the public benefits of projects, we strengthen the case for sustainability. As we discuss Industrial Hemp and environmental justice, we focus on regenerative work, acknowledging our debt to future generations and our obligation to repay it fully. This approach aligns with our overarching goal to foster a thriving and just society.

Educational Course Development Opportunities

Build Green Now leverages the regenerative nature of Industrial Hemp to access a multi-billion dollar market, emphasizing professional training and financial sustainability through *Public Benefit Contracts*, ensuring accountability and long-term benefits. This approach makes BGN's initiatives ideal for inclusion in business, environmental design, and strategic research curricula.

As an aspect of Education and Action for Sustainability, according to BGN educators can develop coursework around the economic and environmental impacts of Industrial Hemp. For instance, courses could explore how hemp cultivation stimulates local economies through knowledge-building, apprenticeship training, job creation, and market development centered on regenerative products and services.

Another course could explore hemp-lime's properties as a building material, such as its fire resistance, temperature moderation, and mold resistance due to its high pH (see Wulandana & Giordani). Additionally, the retrofit challenges encountered with the Hemp House #1 prototype in Kingston, New York, serve as a practical case study for courses on project management, financing, workforce development, and business analysis. These courses could explore how BGN tackles social inequality through environmental initiatives and provide a distinctive angle on environmental justice.

Courses focusing on practical applications could be developed to concentrate on implementing new business and training models through deliberate co-creation with end users. This applied learning approach would enable students to directly interact with concepts of sustainable product development and business strategies as demonstrated by regional partners in the field. For example, it is possible to provide an immersive educational experience by obtaining an acre of land, planting Industrial Hemp, harvesting, decorticating, and using the material for market opportunities, including insulation and textiles. All of this can be done by hand without expensive equipment.

By integrating BGN's projects into academic curricula, students gain hands-on insights and actionable knowledge of sustainable practices and strategic business planning, which prepares them to contribute effectively to social, environmental, and economic advancements for their home communities. As mentioned previously... in 2026, Build Green Now launches Voices for Progress at voicesforprogress.net. Visit this site as a next step where you can join or start a topic-focused working group to develop your version of training content. We are creating Project Management 101 (the Voices for Progress way) and need people to customize content for other cultures.

I. Project Management 101

Consideration 1: Establish Priorities, Then Focus

At Build Green Now (BGN), we recognize the critical importance of working directly with people, not just corporate brands, to achieve community-oriented goals that provide public benefit. This approach includes holding corporations accountable for their environmental and social impacts, such as waste, pollution, economic injustices, intellectual property injustices, noise, and other disruptions. We have a project management platform: Voices for Progress.net powered by Maayadata to address this.

In selecting our focus area, we combined the urgent need for affordable housing with the imperative of sustainability. This choice reflects our commitment to simultaneously addressing the housing crisis and environmental concerns. By integrating sustainability into affordable housing initiatives, we aim to create living spaces that are economically accessible and environmentally responsible.

Consideration 2: Apply the BGN 20/80 Rule

While collaborating with communities and discovering what is actionable, we also find it helpful to focus on solutions as much as possible, leading us to the BGN 20/80 Rule.

In community meetings, too often, there is a lot of talking about problem after problem, followed by aspirational statements about what should be done by somebody.

The 20/80 rule increases progress and productivity by shifting awareness from problems to solutions.

Limit descriptions of problems to 20% of the time.
... Spend 80% of the time creating, analyzing, and prototyping solutions!

By deliberately limiting the time allocated to talk about problems, we increase the time spent implementing solutions. This method ensures that our energy and resources are directed more efficiently toward creating tangible improvements, enhancing productivity, and achieving significant outcomes more swiftly.

Consideration 3: From Aspirations to Action Steps

At BGN, we understand that desired goals express dreams that shape our vision. This process begins with a conversation, where we listen to understand core aspirations and apply creative thinking to develop an action plan.

However, past disappointments can leave a residue, causing discussions about a preferred future to remain stuck in "should" or "could" rather than progressing to "will." BGN's approach is crucial: we truly listen to goals and aspirations, free from the acknowledged disappointment of past letdowns. Through dialogue, we craft a roadmap transforming dreams into action. The senior project manager listens to the aspirational goals and then co-designs actionable steps that lead to optimal outcomes.

When establishing priorities, ground them in detailed observations and studies, integrating the values highlighted during the aspiration phase into the strategic planning process. By tailoring decisions to suit the specific context, we ensure that our prototypes effectively drive us toward our desired outcomes.

Consideration 4: Clarify the Meaning of Words: The Need for Facts and Truth

Many have endured decades of loud, glaring, and persistent corruption, political manipulations, media propaganda campaigns, and a collective attack on civil discussion. Restoring objective facts is critical. BGN found success by beginning with the core elements of "Words," "Awareness," and "Meaning" to ensure that all parties operate from a common understanding.

The text arrangement below shows a progression used by BGN that is rooted in beliefs and values, including essential milestones that lead to mutually beneficial positive actions. The following practices strengthen relationships leading to better agreements in the context of both learned and ancestral values and beliefs.

7. Positive Actions with Accountability
6. Negotiated Agreements/Payments
5. The Power of Collaboration
4. Strength in Unity
3. Meaning
2. Awareness
1. Words

Consideration 5: Products and Services

Contractual engagement enables community organizations to supplement their offerings with pretested hemp-based products and services. Ensure that teams offering products and services have a track record of beneficial community results.

For example, BGN focuses on the following foundational regenerative products and services to prepare communities to participate fully in the emerging global Industrial Hemp marketplace:

- Adopt Hemp-Lime into Local Building Code
- Hemp-Lime Training (Building & Fire Officials)
- Apprenticeship Opportunities (multi-generational)
- Eco-Entrepreneurship and Innovation (New Business Incubation)
- Hemp-Lime Insulation Built and Mobile Homes

Consideration 6: Define Equity and Cash Flow with the Help of Trusted Experts

Community leaders trained as Collaboration Engineers can boost efficiency by drafting contracts with a project manager and an attorney. This approach ensures thorough legal review and fosters beneficial partnerships. Often, attorneys will offer services *pro bono* to such community enrichment projects.

Defining equity and cash flow predictions lowers the risk of misplaced trust by documenting expectations in a contract. Reviewing contracts with a trusted advisor and working through important details with an attorney familiar with the body of law that governs the work is an essential step to protecting an initiative's legal standing and reducing liability risks.

Address all anticipated questions of equity and cash flow within the contract. For example, when a training course is created, if someone records it, the contract should define who has the right to monetize the content. Organizations must define these terms at the beginning of the relationship to set appropriate expectations. BGN practices this mentorship model at the onset of all partnerships.

Consideration 7: Proof of Public Benefit is a Powerful Argument

It is essential to define measurable public benefits and include transparent accounting. For example, Greenwashing is a deceptive marketing strategy by which companies falsely portray their products, services, or operations by making unsubstantiated claims to appeal to environmentally conscious buyers and gain a reputation for sustainability.

As the Changing Markets Foundation reported, one of the most striking cases of greenwashing involved H&M. The investigation revealed, in 2021, a whopping 96% of H&M's sustainability claims in its fast fashion lines, such as using organic and recycled materials, were misleading. This extensive greenwashing was part of a broader pattern in the fast fashion industry, which has long been criticized for its environmental impact.

II. Strategic Steps for Community-Led Reform

By outlining the logical steps of community organizing, policy research, strategic planning, drafting legislation, and advocacy, we prepare participants to effectively interact with various governmental branches to achieve community-driven policy changes.

The process begins with **community organizing**, and gathering local support to highlight a specific issue. In this example, we will use affordable housing. Following this, **policy research** helps to identify problems and potential solutions, setting the stage for strategic planning. Detailed action plans are formed here, specifying how identified solutions can be legally supported and implemented.

Next, the focus shifts to **drafting legislation**, which involves translating the strategic plans into proposed laws that can be formally introduced. This phase is crucial and requires precise language to ensure the proposals are clear and enforceable.

Advocacy and **outreach** are essential to persuade legislators to support and pass the drafted legislation. Effective advocacy involves continuous engagement with legislative and executive branches, understanding their distinct roles. The legislative branch is primarily responsible for creating and passing laws, while the executive branch enforces these laws and manages the day-to-day government administration.

III. Leadership—Collaboration Engineering for Community Development

Non-profit executive directors play a crucial role in guiding their organizations toward fulfilling their missions effectively. To enhance their capabilities and boost their productivity, several types of specialized training are beneficial, making these topics excellent for course development.

Consideration 8: Select, Train, and Empower Your Collaboration Advocate

Specialized training in "Collaboration Advocacy" equips leaders with negotiation skills and strategies to forge advantageous agreements with partners, funders, and community stakeholders. This training is essential for crafting Public Benefit Contracts that are tailored to both the organization's objectives and the needs of the community. A critical step is selecting, training, and empowering Collaboration Advocates to effectively implement and manage public benefit contracts that promote environmental sustainability and community development.

Properly training Collaboration Advocates ensures that implementation aligns with local needs, allows for scalability across communities, and fosters innovation—making these contracts effective vehicles for sustainability and community uplift.

IV. Hemp House Project

The Hemp House #1—Kingston, NY Story

Build Green Now (BGN) embarked on a remarkable journey that transformed our organization from advocating for the Jersey City Unity Walk against gun violence to championing sustainable construction practices.

It all started with our search for a home that could serve as an example project for teaching retrofit techniques and demonstrating how to properly restore an older house. This search sparked our shift from affordable housing policy advocacy to hands-on sustainable retrofit building using hemp-lime insulation.

As we dove deeper into researching insulation solutions, we made an alarming discovery—spray foam insulation made with unregulated CFCs could be a significant factor contributing to environmental devastation. This finding comes from David L. Chandler at the MIT News Office in a 2017 publication which stated:

> Compared to carbon dioxide, CFCs can produce more than 10,000 times as much warming, pound for pound, once they are in the air.
>
> (Chandler 2017)

This alarming CFC insulation discovery highlighted the irresponsibility of using traditional methods and prompted our pivot toward advocating for alternative, eco-friendly building materials like hemp-lime. Recognizing the need to put our insights into practice, we began hands-on construction projects that directly benefit communities.

To begin this phase, we raised money and purchased an abandoned house in the sustainability friendly City of Kingston in the Hudson Valley 100 miles/161 km north of New York City. We chose Kingston due to the supportive local building planning department and our focus on hemp-lime as insulation that aligns with New York State's Climate Leadership & Community Protection Act goals (70% renewable energy by 2030).

At the same time, we developed a relationship with Cameron McIntosh (Founder Americhanvre—National Spray Hempcrete Leader). This relationship blossomed into a passion to move the mission forward and build the supply chain we needed.

Cameron, Sally, and Henry negotiated and financed a partial container of hemp hurd from France since it was not available locally. The material arrived ready for mixing with a lime binder. This means it was already ***Decorticated.*** This is a process where the outer shell (fiber) is mechanically separated from the inner core (hurd). The hurd is cleaned, dried, and bagged for delivery. Our plan was to split the cost and work out delivery logistics (port-ship-port-truck-storage-jobsite).

When the project was ready for retrofit, we had five pallets of hemp hurd and four pallets of lime binder.

As logistics planning commenced, Henry and Dir. Knox, the Kingston Building Safety & Zoning Enforcement Director, discussed fire rating requirements and developed training plans for the building safety team and the fire department. With preparations complete, arrangements were made with Executive Director Iversen of Seed Song Farm Center, to host a free community training session on hemp lime at the farmhouse. That afternoon, people were invited to the project site, where they mixed their first batch by hand and installed the material in the walls, one handful at a time.

This marked an exciting milestone, fostering community engagement while gaining hands-on experience with sustainable hemp-lime insulation.

As the senior project manager, Henry arranged logistics to complete the project, leading to a small team of four people who worked for over three weeks to finish the perimeter walls in the two-story house using a paddle mixer (in the backyard) to install material one bucket at a time: a huge art project.

These experiences in sustainable building provided crucial insights and emphasized the need to replace harmful materials. Our practical involvement has helped advance the community's shift toward sustainable living, underlining the importance of eco-friendly materials that sequester carbon and are fully biodegradable.

However, significant work is just beginning. New York state must insulate over 200,000 residential properties/year plus an uncalculated number of commercial buildings if they expect to reach their 2030 energy-saving goals. Insulation is required before conversion to all-electric energy because the typical low-volume eight-nine-degree maximum output from a heat pump requires buildings to be insulated and air-sealed.

Looking ahead, detailed training programs on hemp-lime insulation are necessary to integrate this material into current practices. With climate emergencies impacting the world, everyone can take action from where they are. To prepare New York state, we are collaborating with

Kereem Berry (I.B.E.W. Local Union #86 Journeyman Wireman, Executive Director of the Multi-Craft Apprenticeship Preparation Program, Inc.) to use hemp-lime insulation on commercial buildings in Albany, Kingston, Poughkeepsie, Binghamton, and Rochester.

This is just the beginning of the next phase of our journey, and together, we can pave the way for a greener, more sustainable world. The end (is just the beginning). As a next step, visit: voicesforprogress.net. Join or start a topic focused working group to develop your version of training content. Explore creating your version of Project Management 101 (the Voices for Progress way) customized for your cultural market preferences.

References

Chandler, David. 2017. Cambridge, MA: MIT News Office.

Sherrill, Zia. 2022 (January 20). "Material Collection: HempLime" healthy materialslab.org. New School, Parsons.

Tuck, Eve. 2009 (fall). "Suspending Damage: A Letter to Communities." *Harvard Educational Review* 79(3): https://pages.ucsd.edu/~rfrank/class_web/ES-114A/Week%204/TuckHEdR79-3.pdf

Wulandana, Rachmadian & Glenda Rodrigues Santos Giordani sites.newpaltz.edu/news/2024/03/green-alternatives-building-material/

CHAPTER 40

Formative Participatory Communication and Engagement in Mali and Malawi, West Africa

Carrie Young

When my friend and former *National Geographic* colleague, Josh Korenblat, invited me to contribute a chapter on international sustainability work, I reflected on my formative experience two decades ago in Mali, West Africa, as well as the lessons learned over the years leading to the development of Cornell's first Cross-campus Sustainability Leadership Program. My work and research in Sub-Saharan Africa and my efforts in educating the next generation have been through the lens of respectful communication and dialogue, intentional partnership, and community-led initiatives. In Mali, I learned the value of participatory methods, which place listening, inclusion, and collaboration at their core.

The methodological and philosophical approaches that I use today are influenced by what I experienced living and working in the small rural farming village of Woroni in Mali as an agriculture volunteer for the

C. Young (✉)
Cornell University, Ithaca, NY, USA
e-mail: cey26@cornell.edu

S. M. Wyman and J. Korenblat (eds.), *Sustainability Learning for Action and Community Engagement*,
https://doi.org/10.1007/978-3-032-11698-7_40

United States Peace Corps. During this time, I worked with farmers, women's garden collectives, and school groups. My interest in communication to help strengthen environmental and health outcomes that began in Sub-Saharan Africa also guides the values of the Cornell Atkinson-Environmental Defense Fund Sustainability Leadership Program I helped design and manage.

In Mali, I came to understand many of the daily rhythms, challenges, beliefs, traditions, strengths, and practices of the community. I also came to appreciate the potential for misunderstandings and miscommunication when someone from another cultural background and context attempts to impact the health and livelihoods of others, even with the best intentions.

The village of Woroni sat at the base of the hills, running through southern Mali and into Burkina Faso and the Ivory Coast. The village of mud huts and granaries forms a labyrinth surrounded by fields of wheat, millet, corn, peanuts, cotton, and rice. Beyond these fields are orange groves and waterfalls with root vegetables and ginger planted along the way. The rains in Woroni, as in many places in West Africa during that time, were becoming more unpredictable due to climate change. When the rain did come, it often arrived over the hills in large clouds filled with pink and orange flashes of lightning. First, the wind picked up, blowing dirt from mats and knocking mangos from the tall trees, and then the clouds opened to a downpour that played like fingertips on a drum on my tin roof. When the rains came, it was always a time to celebrate, but now the rains came less often and in more sporadic patterns. One of my dear friends, Owa Traore, said to me one day as we waited patiently for the rain, "We think it's Allah's will that we die now."

Owa was the first of two wives. After about a year of service, I felt comfortable asking her about sensitive topics, such as polygamy, and what the experience was like for her. When I asked her whether it bothered her being one of two wives, she explained to me that while it did bother her, and she wished it didn't have to be that way, there was far too much work to be done on the farm and within the household for one woman. As I witnessed in Mali, women in subsistence farming villages often do disproportionate household work, even when the farming season has ended. They gather wood, fetch water, wash clothes by hand, bathe the children, pound grains, cook food, clean after the meal, and more. These traditions around gender roles have been in place for many generations, as have traditions around eating.

A traditional meal in a farming village in Sub-Saharan Africa often consists of a polenta-like staple food and a green-leaf sauce, usually eaten with the hands while sitting on small wooden stools or a mat on the ground. In Mali, this staple food called *tô* made from corn or millet, was served in a communal bowl with male elders eating first, then other men, then female elders, followed by other women, and finally the children. Because the children eat last, what little protein or vegetables are served with the sauce is often gone, leaving only high-calorie starch for those still growing and needing greater nutrition. This tradition, which has long-standing historical roots in the community, would be difficult to change without the understanding and participation of those who grew up with these practices. These same children have families who often cannot afford to send them to school, and they are needed to help on the farms. Much of their learning takes place through modeling, often starting at a young age. It's not uncommon to see a very small child create a mortar and pestle from old pieces of wood, mimicking the pounding of the grains that their mothers do each morning as the sun rises.

Challenges related to health and nutrition, as well as to water, sanitation, food security, and environmental conservation, were central and constant in the village during my time there. Alongside these profound obstacles, I also witnessed a strength of spirit and will that inspired me and endeared me to these friends who became like family. Their challenges became my challenges, as often happens with those we love. But I quickly came to understand that facing these shortages and threats to well-being was a complicated undertaking, one that required real listening before making assumptions, patience for the slow process of change for communities rooted in tradition and living a subsistence lifestyle, and the will and desire of individuals and the community to determine and make changes so those transformations might be sustained over time.

An example that relates to the importance of listening and understanding before making assumptions happened in the cooking hut of my friend Waraba Watara. I noticed that Waraba had a knife, but the sharp edge was on the top rather than the bottom as it would be in my home country of the US At first glance, this seemed to make no sense. But when I asked Waraba about the knife, she said that because the women don't cut food using a downward motion (as one would on a cutting board) but rather cut directly over the cooking pot, it makes it much easier to hold the vegetable, or whatever you're cutting, in your hand and use an arcing motion that moves toward you. This works best and most safely if the blade is on the top. This technological detail made perfect sense once she explained it. There are no tables in the kitchen huts and no cutting boards, so the Malian knife design worked better for their needs.

Another example is around the issue of sanitation. In the village, there were no toilets; people used the fields and took what was called a *selidaga* with them—a water pot to clean themselves afterward. The left hand, not the right, was always used in this process. While the right hand was considered the clean hand and was used for eating, there was always a great chance for cross-contamination of that hand as people worked throughout the day using both hands. Despite this, many refused to use soap—only water—when preparing to eat. Diarrhea was a common cause of death, especially among children. When I asked a friend why they didn't use soap to avoid eating with a dirty hand and getting sick, they told me that people believed that if they washed their hands with soap, they would never be rich. At first, this didn't make any sense. It sounded like an old wives' tale. But I came to understand that the only soap available in the village was lye soap, which is very corrosive. Because this soap was corrosive to the hands, it made it difficult to work in the fields, making it harder to have a high yield, which, in turn, made it a challenge to provide for the family or profit from any extra food production.

Another experience that has stayed with me and impacted my current work and perspective relates to my efforts to help educate farmers about deforestation. I had two favorite trees that stood side by side just outside of Woroni, and one day, I noticed that one of them had been burned. I went to the farmer who owned the field where the tree had grown and asked him why he had burned the tree. He explained that it was to make more room in his field for crops. I then tried to explain to him the value of the tree canopy and the importance of not killing trees unnecessarily. Without the trees, there would be less and less rain in the area. But what the farmer heard, I learned later, was that by burning this one tree, he had caused less rain in the village—a huge burden to carry and not at all what I meant to say to him.

Language and meaning are complex even for those who share the same culture and context, and effective communication is even more challenging when people come from very different cultures. Years later, as I worked on a large interdisciplinary team to develop a sustainability-focused curriculum and evaluation in Malawi, this lesson remained helpful as communication and language were two of the project's greatest challenges. Finding ways to convey complex concepts such as gender equality and climate change, and to communicate the interrelationship of these issues was difficult, and it raised questions about how creative forms of participatory communication with built-in feedback loops might aid in this process.

These examples have stayed with me as reminders to always listen first and to understand the importance of understanding concepts and expectations from the farmer directly, as well as the essential nature of good communication for accomplishing goals and collaborative work around sustainability issues in the international context.

Sustainability Context for Smallholder Farmers

For smallholder farmers like my friends in Mali, struggling with issues of household food security and health along with the impacts of climate change can be devastating (Wolfe 2013; Yadav et al. 2011). Both climate change and population growth add tremendous pressure to an already challenging situation for those who depend on their environment for their livelihoods (Niles et al. 2015). Food security and biodiversity are closely linked within subsistence farming communities, with households often forced to choose between short-term food sourcing and long-term ecosystem health (Habel et al. 2015; Tscharntke et al. 2012). Extreme weather events, such as flooding and drought, and an increase in unpredictable rainfall patterns, can create an even greater focus on short-term solutions that put sustainable, long-term environmental and human health at risk. Despite—and sometimes because of—the complex challenges facing farmers, many are open to new ideas and innovations, with some now believing there is more risk in not trying new things than in trying them (Young and McComas 2018).

Those working on issues in sustainable development often promote practices such as conservation agriculture and other means of sustainable intensification that reduce chemical inputs and water use while increasing long-term soil health and yield. These methods have been shown to help farmers "produce more food from less land through more efficient use of natural resources and with minimal impact on the environment to meet growing population demands" while simultaneously working to mitigate greenhouse gas emissions (Hobbs et al. 2008, 543).

Participatory communication efforts help create a healthier future by considering communities' many interrelated environmental and cultural factors. These systemic factors include nutrition, feeding practices, water, sanitation, and poverty (Pinstrup-Andersen and Watson 2011). They also involve the socio-ecology of communities: consideration of traditions, values, attitudes, and perceptions around the benefits of local ecosystems (Palomo et al. 2016). These efforts require special attention to the involvement of communities in sustainability activities and are often strengthened by participatory approaches that are holistic and interdisciplinary (Godemann and Michelsen 2011).

The Case of Malawi: Strengthening Health and Environmental Outcomes Through Transdisciplinary Collaboration, Participatory Action Research, Experiential Learning, and Food Sovereignty

Malawi currently faces and will continue facing many challenges in relation to climate change as a multiplier of threats. These include rising overall temperatures and less, more unpredictable, rainfall (Funk et al. 2008). Training within the country has typically been "top-down," but evidence increasingly points to participatory approaches as a more effective means of education on the grassroots level (Masangano et al. 2016). Education efforts around agriculture and interrelated health, climate, and social issues in Malawi are part of a larger *agroecological* framework defined by the FAO:

> Agroecology is based on applying ecological concepts and principles to optimize interactions between plants, animals, humans, and the environment while taking into consideration the social aspects that need to be addressed for a sustainable and fair food system. By building synergies, agroecology can support food production and food security and nutrition while restoring the ecosystem services and biodiversity that are essential for sustainable agriculture. (www.fao.org/agroecology)

In Malawi, as in Mali, at least a third of the population is impacted by food insecurity and malnutrition yearly (Ellis & Manda 2012; Knueppel et al. 2010; National Bureau of Statistics and ICF Macro 2011). Land degradation and deforestation pose serious threats to the well-being of farming communities (Kangalawe et al. 2008; Zulu 2010). Furthermore, there is a lack of educational access to health and nutrition, including that which relates to the crucial years of early childhood and during pregnancy (National Statistical Office and ICF Macro 2011)—issues of social inequality further place the health and livelihoods of those in minority groups at risk. For example, women do much of the labor on the farm and in feeding the household but are often left out of decisions around which foods will be grown and eaten by the family; they often have lower rates of education and literacy and may suffer from the impacts of physical violence (Bezner Kerr 2005; Peterman 2011).

The Participatory Curriculum on Agroecology, Climate Change, Health & Nutrition, and Social Equity was developed in 2015–2016 to address some of the most pressing and complex challenges facing smallholder farmers in Sub-Saharan Africa. These include issues relating to food insecurity, malnutrition, long-term soil health, the impacts of climate change on crop yield and overall household security, social equality and gender equality, violence, alcoholism, and budgeting in the household. This unique curriculum design includes information relating to the core topics—climate change, agroecology and soil health, health and nutrition, and social equality—and information about the connections among these issues.

This innovative, multi-faceted curricular design was one of the project's greatest strengths as well as one of its greatest challenges. Primary questions facing the team early on included how to incorporate these interrelationships into the training and encourage sharing existing innovations and wisdom among farmers. The team had to determine how to design the training so that information and understanding would be passed along farmer-to-farmer after the formal training ended, increasing the chances that the material, ideas, and innovations would be lasting and transformative. The team chose to incorporate storytelling and participatory drama into the training to accomplish these goals. Research by team members and literature highlighted the value of participatory approaches for problem-solving within local contexts to encourage community members to integrate the many complex and interrelated determinants of health outcomes. These methods also increase social interaction, engagement, attitudes, and efficacy around environmental and social issues (Bezner Kerr et al. 2019; Masangano et al. 2016; Godeman and Michelsen 2011; Greenwood 2007; Greenwood and Levin 2007).

The curriculum was developed by smallholder farmers in Malawi and Tanzania who were working alongside faculty members, PhD students, and professionals from these two countries, the US and Canada. The interdisciplinary team came from multiple academic areas, including Communication, Development Sociology, Plant and Soil Sciences, Health and Nutrition, Health Administration, Horticulture, and Theater. Participatory methods were used in both the development and the implementation of the training—encouraging farmers to share knowledge and problem-solve collaboratively as they became teachers themselves. Four concepts—participatory action research, experiential learning, transdisciplinary approaches, and food sovereignty—were central to the framework used to develop and implement the curriculum (Bezner Kerr et al. 2019).

My experiences in Mali and Malawi, as well as additional experience collaborating with Community Markets for Conservation in Zambia, teaching sustainable farming techniques to reduce hunting and poaching around National Parks, all illustrate the central importance of approaches that move away from traditional top-down thinking and assumptions, toward respectful, collaborative approaches (Young and McComas 2018). These methods use a "listen and learn first" model rather than one that assumes a standardized answer for all people and communities. When decisions and pathways forward are co-created by those individuals, households, and communities who will be tasked with living with and implementing sustainable solutions over the long run, they are given their best hope for success.

Nearly two decades after I arrived in the village of Woroni and moved into my little mud house under the big mango tree, I found myself following in the footsteps of my Great-Grandmother Carrie. I went to Cornell University to earn an MPA and PhD in sustainability communication in Sub-Saharan Africa. I was then recruited by the Cornell Atkinson Center for Sustainability to design and manage innovation, cross-campus sustainability leadership, and network programming at the university. As I sat down and put pen to paper to sketch out the outline and foundation of a formalized sustainability leadership program for students and postdocs, communication and collaboration were at the heart of the planning and development. After all these many years and experiences—from small villages to university classrooms—participatory engagement and listening continue to be the foundational building blocks of our activities. I have come to believe that creating a sustainable future will be most possible when respect is given for diverse ways of living and knowing, and this begins with greater harmony within ourselves, one another, and the natural world.

References

"Agroecology Knowledge Hub," Food and Agriculture Organization of the United Nations. https://www.fao.org/agroecology.

Bezner Kerr, R. 2005. "Food Security in Northern Malawi: Historical Context and the Significance of Gender, Kinship Relations and Entitlements." *Journal of Southern African Studies* 31: 53–74.

Bezner Kerr, R., S. Young, C. Young, V. Santoso, E. Lupafya, L. Dakishoni, M. Magalasi, M. Entz, V. Morrone, S. Snapp, B. Nyirenda, E. Mtinda, D. Wolfe, and E. Kalonga. 2019. "Farming for Change: Developing a Participatory Curriculum on Agroecology, Nutrition, Climate Change and Social Equity in Malawi and Tanzania." *Agriculture and Human Values* 36: 3.

Ellis, F., and E. Manda. 2012. "Seasonal Food Crises and Policy Responses: A Narrative Account of Three Food Security Crises in Malawi." *World Development* 40: 1407–1417.

Godemann, J., and G. Michelsen. 2011. *Sustainability Communication—An Introduction*. Springer.

Greenwood, D. J. 2007. "Teaching/Learning Action Research Requires Fundamental Reforms in Public Higher Education." *Action Research* 5 (3): 249–264.

Greenwood, D.J. and Levin, M., 2007. An epistemological foundation for action research. Introduction to action research, pp.55–76.

Habel, J. C., M. Teucher, B. Hornetz, R. Jaetzold, J. N. Kimatu, S. Kasili, Z. Mairura, R. Mulwa, H. Eggermont, W. W. Weisser, and Luc Len. 2015. "Real-world Complexity of Food Security and Biodiversity Conservation." *Biodiversity and Conservation* 24: 1531–1539.

Hobbs, P. R., K. Sayre, and R. Gupta. 2008. "The Role of Conservation Agriculture in Sustainable Agriculture." *Philosophical Transactions of the Royal Society B: Biological Sciences* 363 (1491): 543–555.

Kangalawe, R. Y. 2008. "Land Use and Land Cover Changes and Their Implications on Sustainable Agriculture in the Irangi Hills, Central Tanzania." In *Land and Water Management in Southern Africa: Towards Sustainable Agriculture*. African Books Collective.

Knueppel, D., M. Demment, and L. Kaiser. 2010. "Validation of the Household Food Insecurity Access Scale in Rural Tanzania." *Public Health Nutrition* 13 (3): 360.

Masangano, C. M., H. Chiwasa, D. K. Kambewa, M. Kakwera, M. Chimombo, M. Matita, and W. Gausi. 2016. "Making the Demand Driven Extensions Services Systems Work Through Decentralised Structures: Prospects for the Future Extension Service Delivery in Malawi." *Journal of Agricultural Extension and Rural Development* 8 (12): 240–249.

National Statistical Office (NSO) and ICF Macro. 2011. *Malawi Demographic and Health Survey 2010*. Zomba, Malawi, and Calverton, Maryland, USA: NSO and ICF Macro.

Niles, M.T., Lubell, M. and Brown, M. 2015. How limiting factors drive agricultural adaptation to climate change. *Agriculture, Ecosystems & Environment*, 200, pp.178–185.

Palomo, I., M. R. Felipe-Lucia, E. M. Bennett, B. Martín-López, and U. Pascual. 2016. "Disentangling the Pathways and Effects of Ecosystem Service Co-production." *Advances in Ecological Research* 54: 245–283.

Peterman, A. 2011. "Women's Property Rights and Gendered Policies: Implications for Women's Long-term Welfare in Rural Tanzania." *The Journal of Development Studies* 47 (1): 1–30.

Pinstrup-Andersen, P., and D. D. Watson II. 2011. *Food Policy for Developing Countries: The Role of Government in Global, National, and Local Food Systems*. Cornell University Press.

Tscharntke, T., Y. Clough, T. C. Wanger, L. Jackson, I. Motzke, I. Perfecto, J. Vandermeer, and A. Whitbread. 2012. "Global Food Security, Biodiversity Conservation and the Future of Agricultural Intensification." *Biological Conservation* 151 (1): 53–59.

Wolfe, D. W. 2013. "Contributions to Climate Change Solutions from the Agronomy Perspective." In *Handbook of Climate Change and Agroecosystems: Global and Regional Aspects and Implications* 2, 11.

Yadav, S. S., R. Redden, J. L. Hatfield, H. Lotze-Campen, and A. J. Hall. 2011. *Crop Adaptation to Climate Change*. John Wiley & Sons.

Young, C., and K. McComas. 2018. "Media's Role in Enhancing Sustainable Development in Zambia." In *Climate and Sustainability Communication*, 84–107. Routledge.

Zulu, L. C. 2010. "The Forbidden Fuel: Charcoal, Urban Woodfuel Demand and Supply Dynamics, Community Forest Management and Woodfuel Policy in Malawi." *Energy Policy* 38 (7): 3717–3730.

PART VI

Reimagining and Remixing Materials for Circular Cycles: Sourcing—Design—Disposal

CHAPTER 41

Sculpting Sustainable Futures

Michael Asbill and Emily Puthoff

Introduction

Sculptors have historically had the same life expectancy rates as race-car drivers. This sobering fact becomes apparent when one enters a typical sculpture studio. The toxic aroma of resins, plastics, foam, aerosols, oil, and particulates permeates the hazy atmosphere. The incessant growl of heavy machinery forecloses the possibility of conversation. Waste materials containing poisonous chemicals like formaldehyde are hurled into dumpsters, destined to choke landfills and waterways. Each sculptor competes for space, tools, and fame as an ephemeral 'genius' in the fickle and extractive art market. The field of sculpture is littered with monumental, indestructible expressions of ego that impose themselves on the viewer and landscape.

M. Asbill · E. Puthoff (✉)
Department of Art, Sculpture Program, SUNY New Paltz, New Paltz, NY, USA
e-mail: puthoffe@newpaltz.edu

M. Asbill
e-mail: asbillm@newpaltz.edu

S. M. Wyman and J. Korenblat (eds.), *Sustainability Learning for Action and Community Engagement*,
https://doi.org/10.1007/978-3-032-11698-7_41

The Sculpture Program at SUNY New Paltz actively works to counter this entrenched culture and build a radically regenerative environment where sustainable material use, collaboration, mutual support, and community focus prepare students to sculpt sustainable lives and careers. In this chapter, we detail the methods we employ to shift the culture and curriculum of our program from one that is toxic and damaging to one that is healthy, resilient, and regenerative. Environmental, social, and economic sustainability are embedded in the Sculpture Program in rhizomatic and circular ways: one strategy nourishes another. Our *Sculpture for Non-Humans* project models how sustainability can be integrated into an introductory sculpture class and the program. We attend to materials management, emphasizing *process*, rather than product, and the importance of creating a culture of care and connection in the studio classroom.

Our journey was inspired and informed by numerous catalytic mentors and fellowships. While we would need to write a book to describe their influence fully, we'll briefly highlight them here to inspire further research. Monona Rossol, an industrial hygienist, was one of our first catalysts when she visited the Art Department. She pointed out every material that will "put nodules on your nodules," meaning potential tumors in our bodies. Following her visit, we jettisoned carcinogenic materials such as urethanes and toxic cleaners in favor of less harmful substances like silicone rubber. Linda Weintraub, author of *What's Next, Eco Materialism & Contemporary Art* (2019), inspired us to ask: *How might we grow our art materials? Does art need to live forever? Can art be composed to decompose? What and who is art for?* In addition, we actively leverage community-engaged design and belonging best practices by employing Movement Generation's *Just Transition* principles and social change prototyping ideas from the MIT Presencing Institute's Theory U-School. Finally, at a workshop with Debra Rowe for the Sustainability Faculty Fellows at our institution, we were introduced to the *Change Agent Abilities Required to Help Create a Sustainable Future* (2019). We quickly realized that artists regularly practice change agent skills, such as curiosity, passion, resilience, and adaptability in the studio. The learning outcomes in all our classes now foreground environmental, social, and economic themes and develop skills to empower Art students as change agents.

Learning Experience: Sculpture for Non-Humans

We developed the curriculum for this project at the height of the COVID-19 pandemic. The Sculpture Program's evolving foundational sustainability practices provided the requisite resilience, stability, and adaptability to make curricular and operational changes when the well-oiled systemic gears ground to a halt. Pragmatic decisions about studio and tool use were important when teaching shifted online, but the essential pivot was philosophical in nature. We left the heroic artist image even further behind and elevated other species—flora to fauna—above the human as subjects of our investigations. This was a logical step for us, an act of resistance to the Anthropocene. Emily Puthoff's research in social sculpture focuses on pollinator conservation, and Michael Asbill's continuing work examines the unfolding functional extinction of the North American Ash tree. The name for this project was inspired by an outdoor exhibition called *Owning Earth*, curated by Tal Beery.

Sculpture for Non-Humans is an additive wood fabrication project that requires students to explore local ecosystems by creating artwork for a living (not mythical) resident being.

To create work for a living organism, students use materials and processes to form objects composed to decompose. Instead of plywood, each student works with a locally harvested and milled wood slab to construct their species-specific sculpture. The students research local flora and fauna and site their work geographically to optimize engagement with their selected species. Students are challenged to finish their sculpture projects with natural, non-toxic finishes.

To begin the project, students explore their local ecosystem by taking a sensory walk on campus and looking at their environment carefully. Using intuition and nature-identifying apps, students discover twenty-five non-human species with whom they co-exist. They sketch these beings and list key characteristics such as their color, shape, behavior, forage, and shelter preferences. Looking carefully at a square foot of turf in the campus pollinator meadow reveals a surprisingly abundant and dense concentration of plants and invertebrates. For many, this search changes how they walk through campus, activating curiosity and recalibrating their senses to see things that used to go unnoticed. Students choose three species from their list of twenty-five and prepare a presentation for the class. They consider the following questions: *What do other beings, such as worms, dandelions, or slime mold, look for in a piece of art? How do I create an aesthetic and sensory experience for a non-human? How might I make this wooden sculpture appealing to both humans and non-humans?*

Students continue their research by selecting a single species for which to design their sculptures, and each iterates ideas through a series of sketch reviews. They incorporate the being's aesthetic, sensory, and practical needs into the design. Mentoring the prototype-and-build process requires care and diplomacy. Many students can slip into anthropocentric assumptions about the lives of mice, crayfish, and oyster mushrooms. Our goal of de-centering human experience confounds many. Humans are accustomed to thinking of themselves as exceptional, different, and apart from our world. De-emphasizing human self-expression is a first step in questioning anthropocentrism and naturalized human impacts on the planet. By shifting the focus from artistic individuality in a heroic sense, we can prioritize the aesthetic and practical concerns of other species and decompose the artifice of the human/nature binary.

To model the interspecies research and collaboration that this curriculum requires, the Sculpture faculty co-presents a three-hour *Carving Community Cross-Pollination Workshop* that educates about pollinators and trees through a wood carving exercise. Puthoff begins by sharing her work as a Hudson Valley Bee Habitat co-founder and leads the participants through a series of cross-pollinating conversations and sensory activities. Each participant seated in a circle around a table is invited to draw a pollinator and their own favorite food. As participants share their pollinator and food drawings, they hold a green ball of yarn and toss it to the next participant across the circle while still holding their point of connection on the yarn. As the yarn unfurls across the table, a large green web emerges. While holding the web taut, participants hypothesize about which type of pollination was involved in creating their chosen food (wind, water, or creature). If a creature pollinated their food, the participant drops their piece of yarn. As 75% of food in the North American diet is pollinated by bees, the green web inevitably collapses. This moment illustrates Albert Einstein's warning that if bees disappear from the surface of the earth, we will have only four years of life remaining on the planet. Puthoff then explains the causes of pollinator species decline and ways we can support them. Many factors contribute to the threat to pollinator species, such as toxins, lack of biodiverse agricultural practices, climate change, and our disconnection from the ecosystem, including the pollinators upon which our food, clean air, soil, and water rely.

To build an understanding of our interspecies interdependence and engender community connection, Puthoff then leads a sensory engagement that includes tasting honeycomb. Participants bite into a piece of capped honeycomb to taste the honey's sweetness and seasonal floral notes. As learners feel the texture of the beeswax in their mouths, she demonstrates that honeycomb is one the strongest structures on the planet with its nested, overlapping hexagons. This is also a stunning example of collaboration: honeybees make honeycomb by excreting beeswax from their abdomens while hanging together in a parabolic chain and manipulating the wax with their mandibles and feet. Learners then contribute their chewed beeswax to a collective melting pot to make a beeswax and olive oil finish to seal their wood sculptures and spoons. The embodied web building and honeycomb investigations enliven the senses and build community connection, a core principle of the sustainable sculpture curriculum, melting down the alienating lone-wolf mentality of the traditional sculpture studio to produce a structurally integral, sustaining social network (Fig. 41.1).

Fig. 41.1 *Carving Community Cross-Pollination* workshop detail, honeycomb and hand-carved spoons. *Photo credit* Emily Puthoff

As Asbill introduces the wood carving workshop, he reinforces the theme of interconnection by discussing his research into the functional extinction of ash trees. In our region, these trees have been nearly wiped out by a beetle called the Emerald Ash Borer that was introduced to North America in the 1990s on wooden shipping crates. Asbill explains that he began using fallen ash wood in projects to understand better the broader ecological impact of losing a single species. Through his post-mortem explorations, he realized that every tree embodies a unique archive of its own life, each is an individual, and an ecosystem storyteller. He has also observed that trees are interdependent. In more than one instance, Asbill observed the abrupt collapse of an adjacent species when the resident ash trees died. Suzanne Simmard (2021) and Peter Wohlleben (2016) would reinforce his growing impression that trees may indeed be sentient.

For this phase of the presentation, Asbill displays an ash log that has been milled into cross-sections and thin slabs. Students are encouraged to carefully dismantle the log and examine the annual growth rings to determine the tree's age. By observing the relative width of the rings, students can deduce an abundance or lack of rain in a particular year and distinguish between early and late season growth. While assessing overall growth patterns in that cross-section, Asbill helps students recognize a prevailing lean in the tree and other clues about the surrounding topography. In the first year of growth, a sapling embedded in the wood is identified, along with the knots that indicate the presence of branches of the younger tree. The most devastating observation comes when the bark is pulled away to reveal a beautiful network of paths, a hieroglyphic gallery left by the Emerald Ash Borer. The fact that these beautiful marks were also the cause of the tree's demise manifests as a startling realization that connects one species' death to another's life. Each track is one season deep and the width of the grub's body, which grew as it ate. Returning to the bark, we find small D-shaped exit holes where adult beetles emerged through the bark. Asbill impresses upon the students that the log before them represents the last of this wood, rapidly decomposing in our local forests, and nearing extinction. To engage with this material is to engage with something precious.

The workshop continues with faculty teaching the basics of whittling, shaping what was once the vascular tissue of a living tree into a honey dipper. Each student receives a razor-sharp sloyd knife, shares a hook knife, and is issued a protective Kevlar glove for holding the wood.

Asbill discusses the importance of blade profiles for carving, maintenance, and listening to the wood as you carve. Removing wood slowly and observing grain orientation is a crucial aspect of carving effectively as the knife most easily follows the wood's structural matrices. As participants converse and slowly carve their honey dipper, a more casual discussion about bees and trees continues and slowly transitions to a meditative hush. Students attend to the non-human species before them and allow the wood to express itself, acknowledging the historical path back to its earliest moment of pollination. An additional important outcome of this exercise is an appreciation of the difficulty, complexity, slowness, and sophistication of working with wood in a subtractive mode.

The use of well-maintained carving knives in our workshop reflects a change in our tooling for the wood project. The pragmatic aspects of our initial pandemic curricular shift dovetailed with our evolving sustainability focus. The pandemic precipitated this pivot from power tools to hand tools. At that time, we sent kits of hand tools—Japanese ryoba saws, coping saws, and bench chisels—home with students so that they could work remotely. When we returned to the sculpture studio, we kept the hand tools in the curriculum, as learning to use these tools effectively reduced pressure on our power tools and offered an option for continuing work when they no longer have sculpture studio access. Most importantly, working with hand tools activated a critical slowing down and more embodied connection to the material. Being present in this way also impacts the planning and scale of the finished projects. Students are encouraged to minimize waste and make something modest and refined, experiential, site- and species-specific, and safe for the environment.

Another crucial change in this curriculum and studio practice is harvesting wood from fallen trees. Before the *Sculpture for Non-Humans* project, we used imported plywood for wood fabrication projects in our introductory courses. This is no longer our practice. We now forage and prepare the wood with a portable bandsaw mill as a community-building exercise. Harvesting the wood together brings students into a closer relationship with the forest and each other. Collectively moving a 500-pound log requires careful planning, an understanding of physics, and teamwork. Students are excited and awed when the first cut of the log is made and the inside grain is revealed. Each slab reads like the page of a book. We then carefully stack and dry the wood. Students receive a board of this collectively harvested wood for their project. Initially, we utilized White Ash, but we have expanded into Eastern White Pine, Eastern Hemlock,

Eastern Red Oak, Eastern Red Cedar, Pignut Hickory, Red Maple, Basswood, and Big-Tooth Aspen. While each student works with their unique board, they embrace the living quality of the wood, its memory, and movement.

In designing their sculptures, students are encouraged to think beyond the conventions of carpentry. They embrace the wood's live edge, organic shapes, warpage, and the unique character of their board. We ask students to respond creatively with their concepts, avoiding existing and familiar forms like bird and bat houses, feeders, and baths. Once their design is refined, they build a scale model out of cardboard, which they can disassemble and use as a cut template. This helps students understand how to make the most of their single board.

As students work, we collect every wood scrap as an ongoing studio circularity practice. The wood scraps are not waste but a resource. We convert these bits into charcoal using a retort constructed by Asbill and sculpture student Mariah Day. We use the charcoal in multiple ways. Firstly, students are enthusiastic about drawing with a large chunk of charcoal made in-house and connecting the dots on how this fundamental art material is grown and made. Secondly, we grind the charcoal into biochar, which is inoculated with worm castings (produced in the studio) and used to fertilize our plants in the Eco Art Lab and Natural Dye Garden. Thirdly, rather than using fossilized carbon like coal, we use our cleaner-burning charcoal in lump form as fuel for a forge to heat-bend iron and steel in our welding class.

Students finish their sculptures with materials that will not harm the environment and species that will engage with their work. At first, students protested about not being able to use readily available art materials, such as acrylic paint (plastic), resins, and urethane wood sealers. Their discomfort shifts as they learn to make natural finishes and stains using foraged and food-grade materials. Beet juice is a popular red wood stain. Walnut ink made from boiling down walnut husks offers a rich brown. Rubbing marigold flowers into the grain imbues the wood with an orange hue. Students can seal their projects with natural finishes such as beeswax and olive oil, linseed oil, iron-oxide solution, and a Japanese charring method called Yakisugi.

Students responded to the challenge by creating innovative and unexpected sculptures. Some sculptures communicate with non-human beings, for example, one that can be played like a guiro to call choral frogs. Others are kinetic. For example, a recreational sculpture for a dragonfly is sited in a pond and moves in the wind; a hand-held, butterfly-shaped sculpture containing milkweed seeds stealthy disperses seeds as one walks; and a duck-feeding boat floats in a pond with a sail that informs humans that bread is not good for ducks. Another, shaped like the posterior of a white-tailed deer, issues a warning about oncoming traffic with a tail that flips up when danger is present. Other sculptures mimic natural habitats and provide shelter and safe places for nurseries and nests. A stack of flower blossoms stained red with beet juice and drilled to mimic the black spots on the ladybug beetle offers many crevices for them to shelter and reproduce; stalactite forms hang in rafters to mimic the caves that bats prefer; and small rocky formations charred black provide safe shelter and recreation for millipedes. In each of these sculptures, the students create a work that is in relation to another being.

As evidenced above, students site their projects to optimize the potential for species engagement. The works are placed far afield from the pristine and sterile confines of a white-cube gallery space. Students observe and document the interactions and changes in their work over time. Projects are composed to live with and return to the environment rather than built for an archival art collection. The challenge of a project like *Sculpture for Non-Humans* comes with the myriad ways that it decomposes assumptions about what art is supposed to be, for whom it is made, and why it is made. Most notably, we challenge the persistent stereotype that art is made by a "romantic" solitary creator, absorbed in his genius, producing lasting objects for wealthy patrons. In this outdated scenario, a select few artists are deemed exceptional geniuses while the rest are destined for noble starvation. This exclusive model feeds heavily on extractive and non-sustainable systems. As educators, we choose to invest in a more sustainable and inclusive paradigm that embraces critical and collaborative approaches to making and being so that young artists see alternatives to a consumer and patron-driven art world as they sculpt a more sustainable future for themselves and others.

References

ACPA—"Change Agent Abilities Required to Help Create a Sustainable Future." College Student Educators International (http://www.myacpa.org/) in collaboration with U.S. Partnership for Sustainable Development (http://www.myacpa.org/)

Beery, Tal. 2022. "Owning Earth." Unison Art Center.

Rowe, Debra. 2019. "Change Agent Abilities Required to Create a Sustainable Future," January Institute keynote address. SUNY New Paltz, New York, U.S.

Simmard, Suzanne. 2021. *Finding the Mother Tree: Discovering the Wisdom of the Forest*. Penguin Random House.

Weintraub, Linda. 2019. *What's Next? Eco Materialism & Contemporary Art*. The University of Chicago Press.

Wohlleben, Peter. 2016. *The Hidden Life of Trees: What They Feel, How They Communicate*. Greystone Books.

CHAPTER 42

Finding and Working with Wild Clay

Anat Shiftan

Introduction

Sourcing clay locally constitutes a radical shift in contemporary ceramic studio practice and teaching, benefiting planet and individuals, from early learners to professional artists. Locally sourced clay can reshape studio practice with lessons in resisting extractive mining practices that harm the planet. In this chapter, we discuss working with locally available clay deposits as an act of resistance to climate instability and large-scale mining and commerce. While naturalists often advocate the *take nothing, leave nothing* "campsite" approach to nature, this exercise recognizes that humans can use natural resources ethically in ways that minimize environmental harm and acknowledge the limits of our shared commons (Hardin 1968). Students are introduced to historical ceramics practices that inform their work in our university art studios. They learn about ancient craft-based civilizations and more recent leaders of the Arts & Crafts movement who were alarmed by the pollution caused by industrialization and burgeoning capitalism's impact on social structures and daily life. Exploring the connections between place, material, and making

A. Shiftan (✉)
Department of Art, Ceramics, SUNY New Paltz, New Paltz, NY, USA
e-mail: shiftana@newpaltz.edu

S. M. Wyman and J. Korenblat (eds.), *Sustainability Learning for Action and Community Engagement*,
https://doi.org/10.1007/978-3-032-11698-7_42

leads learners to understand the political and ethical implications of their own acts of making and being in the world.

Materials are typically sourced commercially and transported from all over the world into the studio lab. To dig and use local clay is to oppose unsustainable practices that contribute to pollution and excessive energy usage. In a world of rapid global warming, increasing pollution and waste, action must be taken to preserve our environment. By skipping the transport and the industrial processing of packaged clays from distant locations, we demonstrate our ability to change our practices to benefit a more sustainable world. There are many valuable aspects to foraging and processing local wild clay in an Art class: the students become aware of their place in the natural world and renew connections between us and our ecosystem. Furthermore, students learn that they can make a difference. They recognize their potential influence, dispel concerns about impurities of locally sourced clay, and become leaders and caregivers of the planet.

As the raw stuff of ceramics transforms by hand and fire, we witness a natural, scientific, and magical process: alchemy in clay. The possible aesthetic outcomes of material mixes fired in the kiln are endless. Learning the chemistry of materials when heated to high temperatures over time informs the student about ancient geological and ceramic processes and bridges interdisciplinary learning between Art, Geology, and Earth Science. Students see what happens to clay matter and can apply this experience to the phenomena of erosion and transformation of materials in heat-work. Cross-disciplinary, informal STEM learning has been shown to be enriching, and this is a perfect example of an artistic application of scientific and pedagogical research in natural minerals.

While in the West and global North, the mainstream Enlightenment view sees humanity and nature as opposed, we need to demonstrate the interconnectivity between natural resources and human beings (Clarke 2017). Design was often defined in terms of controlling or mastering nature rather than working with its qualities while being mindful of minimizing impacts. When students perceive culture and nature as mutually reliant, they gain a sense of connection to and respect for the Earth and nature, leading to ethical materials use. While our university is located in the relatively rural Mid-Hudson Valley of New York, US, many of our students' families are suburban and urban. For some, this interaction with nature prompts a great shift in perspective. Learners can enjoy a basic introduction to their region's local landscape and history and see the evidence of geological processes. Learning to "read" the landscape from a geological perspective is another way to connect with nature and by extension, care for it (Clarke 2017). Beyond this aspect of empowerment, students' experiences in nature also have mental and physical health benefits. They will often continue this digging process, expand their visits beyond urban settings, and make the natural environment their "second home" after the course ends.

Learning Experience

We dig wild clay at a riverbed and bring it into the classroom or studio, remove impurities such as large stones or sticks, sulk it and sieve it, test it for workability and durability at various kiln firing temperatures, and form it into a pinch pot and a coil pot. These are historically, the basic forming-with-clay techniques. We also discuss modifying and testing our found clay to improve its properties, such as workability and firing temperature. Finally, for the advanced students, we explore altering the clay's constitution and add minerals in proportions to make a usable glaze.

This activity is formative for students who might be otherwise focused on the product rather than the process. Making a ceramic object is about working with the materiality of the clay and letting it spin and unfold possibilities in the hand. Learners confront the issues of sourcing materials, sustainable practices, and connection with nature. At first, students become fully engaged in their natural environment through embodied experiences of digging and processing the clay. These technical activities involve informal STEM learning as students measure and observe the qualities of wild clay and the effects of heat exposure with attention to environmental impacts (see Earthrise Commons; earthrisecommon s.org for details). STEM expands into STEAM (arts-informed science, technology, engineering and mathematics) learning by focusing on relations between the human mind, body, and nature by harnessing the knowledge and energies of physics, geology, and chemistry. Learning history, craft, and interdisciplinary connections demonstrates how the power of creativity supports the appreciation of revisiting nature, human collaboration, and care for the land (Fig. 42.1).

Students who work closely with clay experience the most basic form of human expression: making an object with the hand. Using one's hands to create a physical object is a fundamental need and an expression of our humanity. According to histories of early cultures, basketry and pottery

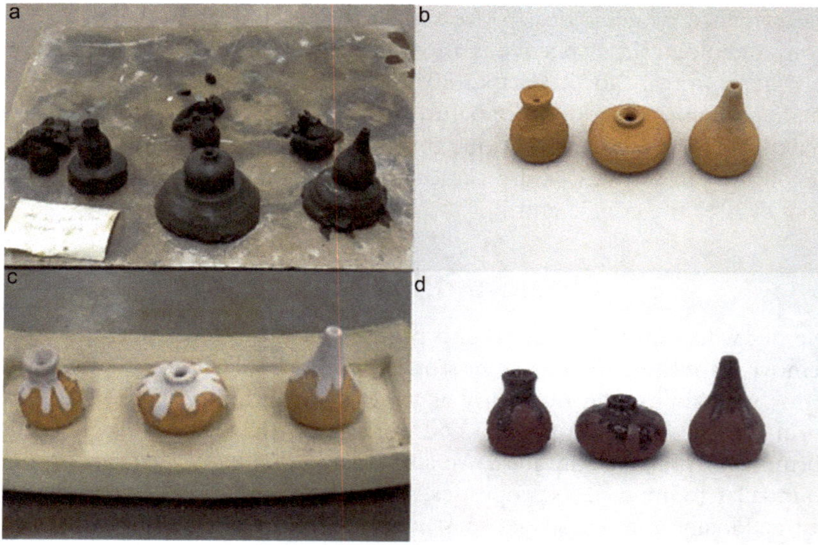

Fig. 42.1 Process *(Photo credit* Anat Shiftan): **a** The raw clay is formed, **b** The clay is fired to 1665 Fahrenheit, **c** The bisque clay is raw glazed, **d** The bisque is fired to 2200 Fahrenheit, and the glaze is molten into glass

encouraged gathering and storing nutrition supplies so that cooking eventually became possible and allowed humans to settle and evolve into complex communities.

The Assignment

In preparation for this project, students complete readings and discuss the impacts of industrialization and capitalism on the natural environment and social justice issues. This activity is appropriate for children and adults, high school, and university students. The project would be fully developed in a ceramics studio with a lab and kilns, yet parts can be achieved in any classroom. People of any age will feel empowered by digging clay and forming it with their hands into an object.

We specify learning outcomes that emphasize values thinking, systems thinking, and interpersonal sustainability competencies (Brundiers and Barth 2021). Students who participate in this ceramics experience will:

1. Develop a strong sense of personal identity and form mature, respectful, and collaborative relationships with others (intrapersonal and interpersonal development).
2. Demonstrate an understanding of and commitment to justice, equitable treatment of others, civic responsibility, and personal responsibility (social responsibility).
3. Integrate and apply knowledge and skills gained across curricular, co-curricular, and other applied learning experiences to advance personal and professional growth (life competencies).
4. Engage others in informed debate and principled dissent, accept, and appreciate other views, think critically, and view leadership as a process, not a position (cognitive skills).
5. Participate thoughtfully and respectfully in cultural exploration, engagement, acceptance, and collaboration as members of an inclusive community (diversity and inclusion).

Literature Review: Building Context

When we research and discuss indigenous perspectives and the Arts & Crafts movement's emphasis on simplicity and natural motifs in relation to traditional handmade ceramics, students gain a fresh perspective on industrially manufactured objects. By embracing these art historical ideas with a critical eye cast on the present, students enhance their respect for the earth and slower processes, as they become more environmentally responsible individuals. They engage in systems thinking and values thinking as they understand complex social systems involved in material practices and, by extension, impacts on human creativity and productivity.

We discuss the history of ceramic pottery production which dates to around 12,000 years ago and continues today, starting with indigenous people using materials from their immediate environment and forming pots by hand using coiling techniques. The "beginnings of ceramics production" is a common story around the world where indigenous people found the materials for their ceramic pots in their local environment and used their hands to create beautiful utilitarian objects. In fact, these were the first containers that allowed cooking food on the hearth and precipitated the development of civilization. Circling back to connect the discussion to mining our own clay, students review the work of Pueblo Potters of the Southwest, who source local clay and use low-tech methods of production and firing. Pueblo potters do not use a wheel but construct pots using the traditional horizontal coil method or freely form the shape. After the pot is formed, the artist completes the piece with a natural polishing stone, such as a river stone, and then paints it with slips or engobe.

Readings from the theory of decorative arts and the Arts & Crafts movement introduce additional historical topics. These include writings by nineteenth and twentieth-century critics, environmentalists, and social justice activists Mahatma Gandhi, John Ruskin, and William Morris. Students learn how cultural and political leaders, designers, and artists have used ecological subjects (cotton production, flower, leaf, tree motifs, etc.) to emphasize the natural world's importance for humanity and resist separation from it. In Gandhi's case, students learn about the deep connection of crafts and material production to individual identity and nationhood. Ruskin and Morris decried the pollution from factories that spoiled their air and waterways, including London's Thames River. Students discuss and respond to the problems that come with excessive extraction of materials from our planet and the cost of industrialism, capitalism, and colonialism.

For example, British colonial activities in the areas of cotton and textile commerce in Bengal, India, demonstrate how sourcing materials, craft, industry, ecology, capitalism, and colonialism interact. Here, the British "stole" the cotton production system from the small craft cottage industries in Bengal by flooding the Indian markets with cheap textiles made in Britain, using Indian-sourced cotton. The colonial power forced the sale of raw cotton to European markets. British factories produced cheap textiles in England by exploiting child and women's labor and polluting the environment, only to ship the cheap cotton products back to India. This extractive, imperialist economic system destroyed the local Indian industry and created dependence upon the capitalist machine. Gandhi argued that the loss of the Indian cotton industry was not just an economic disaster, but the loss of the craft also constituted a cultural destruction of the Indian people (Adamson, 2010: 192–198). Today, the craft-based cotton spinner is on the national flag of India.

A careful look at the history of the European and US Arts & Crafts movement and its response to labor exploitation through industrialism and capitalism reveals that the seemingly neutral disciplines of the decorative arts and the craft field carry, in fact, a political message. In the nineteenth-century, critic John Ruskin blamed industrialization for the loss of connection with nature and for the loss of individual expression through craftmaking. He represents nature as a spiritual space juxtaposed with the current affairs of industrial England (Sawyer 2012). In his book *The Stones of Venice*, Ruskin demonstrates that in Venice's medieval architecture, one can see how society was already suffering the loss of its connection with nature. Decorative ecological motifs such as leafy or animal ornamentation are incorporated into the architecture and bring elements of the natural world back into the urban space. As Ruskin discusses Gothic architecture, he laments the loss of such expression of the individual craftsman. Some argue that he was too romantic about the stone carving craftsmen of the time. But this culturally specific valuation of beings on a personal level demonstrates an ethic of care easily lost in modern industrialized society. When Ruskin reviews the variation and lack of regularity in patterns of ornament in Gothic architecture, he assumes an ideal collaborative project where artisans work together to create something monumental while enjoying the freedom to express themselves individually. Students tend to resonate with his critique of the loss of individual expression on the level of ornamentation in the new industrial order of capitalism, where the machine dehumanizes the laborer while both the human expression of species being and the connection to nature are lost.

Speaking from a Marxist perspective, William Morris railed against the Industrial Revolution and its destructive environmental byproducts by creating an alternative to large industry. He advocated for the return to small cottage industry operations that utilized the various craft disciplines to produce work with value both in quality and content. Morris's hand-printed garden-like wallpaper patterns may seem decorative, yet introducing these motifs into the home actively critiques the loss of human connection with the ecological environment and celebrates natural beauty. In that way, the decorative artifacts he produced convey political content. Students encounter this lesson by way of an online lecture by Sam Jacob (2019), where he demonstrates, step by step, how the patterns of Morris' wallpaper that seem merely decorative are used to bring the lost natural world into the domestic space and reconstruct, in an Arts & Crafts home, a sense of warmth that was often lost in the Victorian age of cold, formal interiors.

When students learn about these movements and understand that artistic expression is political and can deliver social commentary, they see that their work can make a strong statement and even incite social change.

Furthermore, they understand that every human activity (and non-activity) has consequences for humanity, the environment, and society. At this moment, the links between *me* (the learner) and *what I do* and *what I use* take on ethical dimensions. The outcome of these choices and actions is an object or image made by *me* that transforms from a private experience (*I am doing this for myself*) to an ethically consequential expression of the self within the world we inhabit (a communal *we*). As David Gall explains, "Art education's response to capitalism's indifference to global and local well-being requires the incorporation of a dialectic of disinterested and immersive aesthetics that complement each other" (Gall 2015, 142). In other words, students discuss objective criteria for beauty while immersing themselves in working with materials; this coincides with critical sustainability priorities.

In our experience, students respond strongly when they realize that their current feelings about the way the world is evolving are not new. They feel a loss of hope relative to the state of the ecosystem and social systems that are impacted by it. Global warming and the human strife that accompanies it cause great anxiety. Students are not sure how their future will be. However, they feel empowered when they realize they can express thoughts and feelings artistically; they can advocate through their work. Since meaning is created through connections, learning about nineteenth and twentieth-century thinkers and applying these ideas to our times allows students to recognize their ability to have a voice that has meaning and that matters. Sustainability solutions depend not on new technological solutions but on what's better and different—and this often involves a return to past values and practices.

The Process of Finding, Processing, Testing, and Making an Object with Wild Clay

The introduction to the ceramic lab involves a visit to the clay storage area and a discussion about available materials and clay bodies. Students list all the materials in the lab and learn where they come from, how much they cost, and the production methods used to extract and process these materials. This throws light on mining and commerce issues related to environmental justice and the preservation of natural resources. A review of recent cost hikes for mined materials such as Lithium and Cobalt illuminates contemporary issues of supply and demand, as seen with the influence of the development of the electric vehicle (EV) industry.

Lithium, a main flux in many ceramic glazes, is almost impossible to obtain. If a source is found, it may cost 500% more than it used to due to its role in producing car batteries. A discussion of working conditions on geographical sites where those materials are often mined can help develop an awareness of issues related to child labor, consumerism, and social justice on the global level.

Sourcing Wild Clay

Gather necessary tools: buckets, shovels, and other equipment. The next step is to go for a walk and find local clay. Clay is easy to find on riverbeds but can also be found elsewhere. Depending on the location, some geological hikes will afford additional materials. In the New Paltz area, for example, we find clay deposits, quartz and shale. Remember to record findings, attend to informal STEM learning opportunities, and reflect on the process in a journal.

1. **Clay Sourcing:**
    - Locate clay in riverbeds or other locations.
    - Use Geology Department resources to find available deposits.
    - Collect materials and bring them to the studio.
2. **Clay Processing:**
    - Soak clay in water until it reaches a workable, yogurt-like consistency.
    - Sieve clay to remove impurities.
    - Dry clay evenly to ensure plasticity. This can be done by spreading the clay on canvas or plaster slabs.
3. **Clay Testing:**
    - Test clay for plasticity by making pinch and coil pots.
    - Create shrinkage clay bars for mass evaluation.
4. **Evaluation of Clay:**
    - Measure clay mass loss and water absorption.
    - Relate activities to STEM learning and art connections.
5. **Experimentation with Clay:**
    - Prepare shrinkage clay bars and test water absorption.
    - Explore clay as a glaze component and test glazes.

Materials Management: Making Art

Students tend to make artwork that focuses thematically on the need to connect, conserve, and care for the land we live on. For example, MFA student Rebbeca Verrill created unfired clay forms of fruit and vegetables that contained seeds of various useful plants. The forms were meant to be used as *seed bombs*, objects that could be hurled into open spaces where they would generate plants. Rebecca wanted her thesis to bear fruit as an act of resistance to waste. BFA student Tom Lent used locally dug clay for his pots and local shale as his glaze material. Ari Pirwitz made unfired hollow, large-scale, corpse-like porcelain bird forms, and filled each bird's body with locally found plastic bags and containers. Ari placed these beautiful birds in front of our library and allowed the forms to break down as it rained. As the clay forms deteriorated and dissolved, the plastic debris was exposed. Ari thus spoke to the need to protect our creatures and the harmful role of pollution in wildlife survival. These art projects are visual expressions of ethical action in material use and care for the earth (*examples used with permission*).

Materials Management: Disposal

At the end of this process, students may keep the clay for further exploration or return it to the riverbed. As they decide how to conclude this experience, students address their responsibility toward our material world. They ask, *Should we dispose of the clay in the garbage? Should we utilize all that we have taken? Should we return the unused material to the same spot as a gesture of respect? How might we reconcile our need for self-expression with the disruption of the natural order caused by sourcing art materials from nature?*

Conclusion

In his book *Collapse, How Societies Choose to Fail or Succeed*, Jared Diamond (2011) provides an example of how excessive sourcing of a society's natural resources threatens its survival. He recounts the societal collapse of the Easter Islands. The work done to erect the famous monolithic human figures on the islands required massive deforestation: "The overall picture for Easter is the most extreme example of forest destruction in the Pacific, and among the most extreme in the world:

the whole forest gone, and all of its tree species extinct. Immediate consequences for the islanders were loss of raw materials, loss of wild-caught foods, and decreased crop yields" (Diamond 2011, 107). When students encounter this information, they construct new conceptualizations for parallel threats to their ecosystem. They use their artistic practice to explore more sustainable ways of building and maintaining community with each other and their environment.

Students who source their materials to make art feel empowered. They experience a shift from the traditional human-versus-nature dichotomy to a unified view of themselves and nature. Their appreciation deepens for a shared and limited commons of space and resources (Hardin 1968): our earth. This new understanding of the sustainability impact of local materials management gives their concerns a voice. With this activity, the learner can appreciate their agency to make change.

References

Adamson, Glenn. 2010. *The Craft Reader*. Berg.

Brundiers, Katja and Matthias Barth, et al. 2021. Key Competencies in Sustainability in Higher Education: Towards an Agreed-Upon Reference Framework. *Sustainability Science* 16: 13–29. https://link.springer.com/content/pdf/10.1007/s11625-020-00838-2.

Clarke, David A. G. 2017. Educating Beyond the Cultural and the Natural: (Re)Framing the Limits of the Possible in Environmental Education. In *Reimagining Sustainability in Precarious Times*, Ed. Malone, K., Truong, S. and Gray, T. London: Springer.

Diamond, Jared. 2011. *Collapse: How Societies Choose to Fail or Succeed*, rev. ed. Penguin.

Gall, David A. 2015. A Dialectic of Disinterested and Immersive Aesthetics: Santiniketan Art Education and Labor Translated. *International Journal of Art and Design* (15 February), Wiley. https://doi.org/10.1111/jade.01769.

Hardin, Garrett. 1968. The Tragedy of the Commons. *Science* (13 December). https://www.investopedia.com/terms/t/tragedy-of-the-commons.asp.

Jacob, Sam. 2019. *What is Ornament*? https://vimeo.com/399852147.

Morris, William. 1888. *The Revival of Handicraft*. Fortnightly.

Ruskin, John. 1900. "The Nature of the Gothic" in *The Stones of Venice*. London: George Allen

Sawyer, Paul. 2012. Ruskin the First Ecologist and the Future of Landscape. *YouTube*. https://www.youtube.com/watch?v=gGMBLlo8USc.

CHAPTER 43

Sustainability-Focused Engineered Product Design

Jared Nelson and Lisa C. Hix

To move us toward a more sustainable future, our world needs creative change-makers who are confident, competent in their thinking practices, and grounded in a framework of sustainability principles to guide their decisions. Future engineers and designers must lead toward sustainability action and industry influence by developing common competencies transferable to many applications. Student engineers can be prepared and inspired to pursue careers in design and engineering that will positively impact society and the environment. Educators in higher education and workforce developers can instill curiosity and foster creative, problem-solving, and future-thinking skills, as well as opportunistic abilities to continue to learn in every situation. Rooting the study of product design

J. Nelson (✉) · L. C. Hix
Department of Sustainable Product Design & Architecture, Keene State College, Keene, NH, USA
e-mail: jared.nelson@keene.edu

L. C. Hix
e-mail: lhix@keene.edu

S. M. Wyman and J. Korenblat (eds.), *Sustainability Learning for Action and Community Engagement*,
https://doi.org/10.1007/978-3-032-11698-7_43

and the practice of Engineering in values found in typical general education learning outcomes allows educators and trainers to progress toward this sustainable future.

This case study presents a curriculum developed in the Sustainable Product Design and Innovation (SPDI) Bachelor of Science program at Keene State College, New Hampshire, US. Sustainability principles, research methods, guidelines, and applications are woven through all the courses in the program. Students are encouraged to build on their interests and stay curious while they engage in open-ended design projects, learn new skills, build their knowledge, and develop their values. With course content in digital technology, classical and digital manufacturing methods, materials, product design, and engineering processes, students find their calling through three pathways—product design and prototyping, manufacturing engineering, or general engineering. The product design process is a vehicle for students' minds-on/hands-on experiential learning in extensive maker space/manufacturing labs. A cross-disciplinary approach includes coursework in business management and occupational health and safety so students can see connections and build empathy with other professions. Sustainability issues reach across many areas; understanding business operations and methods used in other professions increases a person's ability to be a more effective changemaker. A strong liberal arts educational core at our institution strongly complements this integration with learners' SPDI education.

The SPDI program has integrated sustainability principles alongside materials and production design decisions and into a single, entry-level course that provides an important basis for understanding sustainability that can impact students' broader course of study. The case study that follows integrates Keene State College's College-Wide Learning Outcomes and a study of material life-cycle assessment (LCA) into an introductory Product Design class. Students must address each outcome and relate them to their designs throughout the semester while becoming comfortable building LCA into their design process. By having students see the inherent benefit in such learning outcomes (critical thinking, creative inquiry, civic engagement, intercultural competence, well-being, and sustainability), they become more involved in the entire design process, develop a wider understanding of sustainability and an appreciation for their own role in the learning experience and their future profession. Further, students see the impacts of integrating LCA as a design tool rather than considering it as a separate afterthought. This course content starts the process of normalizing a sustainability-focus in the product design and engineering professions. Effective steps toward this goal are shared below.

Integration of CWLOs—Developing Sustainability

To begin, it is important to establish the long-term educational goal which is *to prepare and inspire students to pursue societally positive careers in design and engineering while instilling curiosity and fostering creative, inclusive, sustainable, and opportunistic abilities to learn in every situation.* When students engage in formal and informal educational experiences, they build the skills to become life-long learners through self-directed means. We use as many tools as possible for students to see how their academic work has direct and useful application to their forthcoming professional careers. This approach draws not only from other disciplines (e.g., sciences, art, communication, business, and philosophy), but also from each instructor's own professional experiences. Through emulation, students develop their abilities to listen and think to understand. When they understand the topics at hand, students come away confident that they have the skills and resources to be successful, paradigm-shifting designers and engineers. Connections such as these must be at the core of engineering-based sustainability education and are central to the case discussed herein. Students are provided with the engineering tools necessary to develop and succeed in sustainable economies as engineers and researchers while fostering interaction with industry.

An introductory course in product design, such as *IISPDI 151: Product Design Principles*, offers opportunities to introduce, discuss, and evaluate the six KSC College-Wide Learning Outcomes (CWLOs) and how to interpret them:

- **Critical Thinking**: Keene State College students will evaluate evidence, consider multiple perspectives, choose, and defend a position from several alternatives, and analyze complex problems.
- **Creative Inquiry**: KSC students will engage in research, scholarly activity, or creative work to contribute new knowledge, art or expression and reflect on the process of inquiry or self-expression.
- **Intercultural Competence**: KSC students will reflect critically on their own culture and on the intersectionality of culture and social

location, demonstrate knowledge of a diversity of cultures, and communicate effectively with people from a variety of backgrounds.
- **Civic Engagement**: KSC students will demonstrate knowledge of one or more social or environmental issues including relevant cultural, political and policy contexts; take action individually or collectively to address issues; and reflect on the ethical dimensions of civic engagement.
- **Commitment to Well-Being**: KSC students will reflect critically on their own well-being and that of the larger world, demonstrate knowledge of issues that impact health and wellness, advocate for themselves, and commit to one or more practice that promotes well-being.
- **Sustainability**: KSC students will explore their place in interconnected natural and human systems; evaluate the personal, social, and environmental impacts of their choices; and apply their knowledge and skills for building a just, resilient, and thriving world.

While sustainability is one of the CWLOs, integrating all six would allow for discussion and a more comprehensive understanding of the three pillars of sustainability: people, profit, and planet. Thus, as we transitioned to our second of three projects in the semester, students addressed each CWLO with a reflective assignment followed by a 20–30-minute discussion. Sustainability, critical thinking, and creative inquiry were generally easier for SPDI students to identify with and relate to, allowing the group to get used to the format of reflection and sharing. Students were able to consider how they integrate these skills into product design easily and comfortably. Anticipating that the concepts of civic engagement, intercultural competence, and well-being would not be as familiar, the reflection portion was shifted to address each a bit differently.

Each assignment follows as presented to students with brief comments included:

CWLO Assignment #1: Critical Thinking

Review the Critical ThinkingCWLO and consider these questions:

- In your life, where have you learned this skill?
- Where have you used this skill?

- How can you further develop this skill?
- How is this skill useful in this project?

Comments: Even though this was thought to be a familiar area for most students, the majority of reflections were rather superficial. Students were able to generally identify where they used critical thinking and that it was crucial to the design process, but it became clear in our discussion that they were unsure how to define the concept. Much of the discussion was spent drawing out this definition and the following assignments were adjusted to allow students to first identify their definition of each CWLO.

CWLO Assignment #2: Creative Inquiry

- How do you define Creative Inquiry? Does this align with the Creative Inquiry CWLO?
- In your life, where have you learned and used this skill?
- How can you further develop this skill?
- How is this skill useful in this project?

Comments: This assignment elicited fewer superficial responses, and some students showed significantly more depth in their reflections. Generally, students commented that while critical thinking was a science-based, analytic activity, creative inquiry was more artistic or generative. The entire discussion was spent deconstructing this notion and reconstructing the independence of critical thinking and creative inquiry from academic categories. This discussion was a turning point with the class as they bought into the sustainability-focused learning framework being co-created.

CWLO Assignment #3: Sustainability

- How do you define Sustainability? Does this align with the Sustainability CWLO?
- Do you live your life sustainably? How so?
- How can you further develop this skill?
- How is this skill useful in this project?

Comments: After the discussion of creative inquiry, the buy-in to this framework seemed even more evident. Many students chose this course, or SPDI, because of the identifiable sustainability component. Each student answered the questions more deeply and the degree of self-reflection was much more apparent. Generally, all students easily grasped the environmental aspects of sustainability, while only a few noted the economic and social aspects. Thus, most of the discussion investigated these less visible aspects and what they mean. This was amplified by a planned discussion of sustainability that launched the life-cycle assessment portion of the research/ideation project. Students felt heavily invested in this area and even foreshadowed how the remaining CWLOs might relate to our approach to sustainability as designers.

CWLO Assignment #4: Civic Engagement

- How do you civically engage in your world? Does this activity align with the Civic Engagement CWLO?
- How is this civic engagement useful in this project?

Comments: The reflections continued to show more thought and depth. Generally, students focused on engaging in their community in political ways (for example, voting, lobbying for change, protesting, etc.), and discussed other social avenues for engagement. This assignment evidenced more reflection by each student into how civic engagement factored into their research/ideation project. For example, a student whose group was researching re-envisioning food storage containers (for example, Tupperware®) commented:

> I would say that just through the goal of researching a problem, listening to user feedback, and drawing conclusions on how you can make improvements it makes every one of us a better critical thinker. This is also a CWLO [civic engagement], but it directly aligns with being aware of what's going on in the world around you. Some of the projects may also have a direct connection to creating a product that is environmentally friendly or which helps a specific subset of people who might be otherwise marginalized in product design. As far as Tupperware goes, you can make an argument that motivating people to cook at home more is a great thing for a community as a whole. Healthier food choices lead to lower rates of obesity and dietary related illnesses, which leads to lower healthcare costs, etc.

This example (*used with permission*) clearly demonstrates the interrelationships of the CWLOs and their importance in design.

CWLO Assignment #5: Intercultural Competence

Review the Intercultural Competence CWLO. If we use the definitions suggested by Wittkower (2018), where technology is "a way to get something done" and discrimination is "when a morally irrelevant characteristic is allowed undue influence in a determination of individual or distributive justice." Then, a discriminatory technology is one that promotes or deepens social inequity.
With these concepts in mind take a few minutes and address these items:

- Identify and share one engineered discriminatory technology. (Propaganda items are not considered technologies and should not be considered. If you are unsure, ask!)
- How do we as designers address discrimination in technology?
- How does intercultural competence make us better designers?

Comments: Engineering and design have tremendous social and environmental responsibilities that are often ignored, as evidenced in Ruha Benjamin's *Race After Technology* (2019), and this assignment challenged students to identify examples. Some students were apprehensive about addressing this topic, while others immediately appreciated it. During the following week, several students asked for guidance and discussion, which led to reflections that showed creativity and diverse lenses. While most students only reflected on one discriminatory technology, several identified more. Further, students were able to identify the importance of how intercultural competence, or lack thereof, can influence the research portion of the design process, as noted in this example (used with permission):

> Knowing, researching, and understanding a wide variety of cultures and backgrounds allows designers to dig deeper into what society really needs and can benefit from. This prompts us to not just design from a superficial standpoint but opens up the possibility to create something that will be

truly meaningful and impactful for people who may have often been overlooked prior. Seeking to listen to the concerns of those individuals who are not directly in front of us brings a clearer and well-rounded perspective to the design process itself.

CWLO Assignment #6: Well-being.

> Reflect on how you are committed to the well-being of yourself and those around you. How does this commitment make us better designers?

Comments: This assignment completed the CWLO exploration for most students by clearly showing their ability to see how each of these outcomes positively impacts their own lives, as noted in this example (*used with permission*):

> [A] person will make decisions based on trying to benefit their well-being entirely, even if that decision has known downfalls (i.e. instant gratification). It's impossible to insert yourself into another person, therefore we are all committed to either promoting or hurting the well-being of oneself directly. Designing to accommodate the well-being of others also falls into this category. If I designed a feature that I wouldn't necessarily need but others do, I made that decision because seeing others being cared for [because of] what I did is a good deed to do.

The discussion followed suit, allowing the class to revisit the emphasis on well-being over the semester.

User Research/Ideation Project: As a culmination of these assignments, each student group was asked to collectively review and discuss how they applied and related to each CWLO while working through a user research/ideation project. This project collates significant user experience data to generate product functional requirements. During their final project presentation, groups included a review of this CWLO work. In most cases, the comments were general, though they demonstrated a clearer understanding of each CWLO. Groups related well to critical thinking and creative inquiry as they noted such questions as "*Are we arriving at the right conclusions?*" and "*Are we asking the right questions?*" respectively. Pertinent questions and comments also followed from the sustainability, civic engagement, and well-being CWLOs. Several groups addressed intercultural competence thoughtfully, and one group even noted that "*we are now aware that we DIDN'T address the challenges that certain individuals may face using our product*." Overall, addressing these topics in the second project allowed students to practice and implement these considerations in their third project, life-cycle assessment, while redesigning a lamp.

Lamp Redesign Project & Life-Cycle Assessment

After the work outlined above, students begin the lamp redesign project. Groups dissect a found lamp, assess materials and impacts, and each student designs and builds a sustainability-informed light fixture. This project builds students' awareness of design for disassembly to recapture materials; they calculate energy use comparisons, work with a new light source (LEDs), identify materials and manufacturing methods, design and prototype using reclaimed materials, and are introduced to life-cycle assessment (LCA) parameters and methods. Over the past 18 years, Dr. Hix has experimented with various simplified LCA methods to find ones that would work well in undergraduate introductory product design courses. The *Okala Practioner: Integrating Ecological Design* has been helpful in the classroom for introducing concepts and comparing existing products' environmental and human health impacts. The creators of the Okala methodology contributed to the Sustainable Minds LCA software, which provides a cloud database and a user interface that students find engaging.

To provide a baseline for comparison to the student's redesign, small groups of students started by disassembling a used incandescent or fluorescent light fixture collected by faculty from a local transfer station, thrift store, or direct donation. Photos of each step in the process are taken throughout the disassembly, and students consider ways to make the product easier to disassemble. Each light fixture piece is weighed and analyzed to identify materials and manufacturing processes to build an accurate bill-of-materials (BOM) to enter into the software. This analysis of each part is an opportunity to learn material identification through qualitative and quantitative comparisons of material finishes and properties such as density and hardness. As students complete their disassembly and BOM, they use this information to establish a life-cycle assessment baseline score with the Sustainable Minds software (referenced below). Students also enjoy taking things apart and often gain new experience with hand tools, building an understanding of how things are made and how to repair a product.

Sustainable Minds is a software tool that enables a standardized and consistent way for rapid iteration and evaluation of product concepts in the earliest design stages. It utilizes databased material and processing options to align with the *Okala Practioner: Integrating Ecological Design* methodologies. The guide is "organized for working designers, educators, and students in all design disciplines with practical methods for designing products, services, and systems with low impacts to ecological health and human health." Integration of the comprehensive impacts of products and services allows designers to achieve higher levels of sustainability. While the guide provides methods for quick assessment and calculation, using Sustainable Minds will enable students to compare designs quantitatively for ecological impact, leading to more responsible decisions.

So, as students transition to entering their BOM into the Sustainable Minds software, they are also required to read the complementary first chapters of the Okala guide. Further, they are quizzed on these readings to ensure they absorb this important background on the LCA framework they use. The project's initial phase is completed with each group presenting the work of their disassembly, their BOM, and their resulting impact score, along with their suggestions for making disassembly easier. The energy use is calculated with a worksheet that includes the lamp wattage, number of lamps, and estimates of annual usage in hours to calculate the kWH of electricity used per year. The country of origin is found on the product label, and the number of miles and mode of transportation to a central US distribution hub are calculated with a map phone app and entered into the software. These data quantify a baseline impact score per year of use for the product disassembled. After each group has shared their LCA inputs and results, the discussion focuses on the comparison of the impact scores and the effects of various factors. This allows each group to investigate and discuss the logic behind some of these inputs, which grants students insight into their design choices.

The second phase of this project involves each student designing and prototyping a light fixture using found materials. Focus is placed on the overall structure, creating the form, reducing glare, and exploring the reflection, diffusion, and translucency of the light off or through the material. LED strips with connectors to USBs are required for their new design to create a built-in kWH reduction in their LCAs with the same hours used and more flexibility with the design. Students can choose the light strip length they need and learn simple low-voltage wiring.

Project guidelines (outlined below) intentionally give students wide latitude in their design:

a. Come up with a design intent for your lamp using mostly reclaimed, reused, or recycled materials.
b. Design for the same function of either ambient, task, or accent lighting, and from the ceiling, table or wall, based on the function of the light fixture that was disassembled as a team.
c. Please use this opportunity to explore creating exciting visual interest with materials, translucency, and light!
d. Paper or fabric can be used if they are well away from the electrical components.
e. Plan for use of LED light strips which will be supplied in class. Do not plan to use an incandescent or fluorescent lamp (bulb)!
f. Work to make something that people will love and make them smile when they see it and want to keep it around and take care of it.
g. Use the life-cycle assessment (LCA) modeling in Sustainable Minds as a design tool to see any change in impacts (recalling it is only a guide). You may find that some of these bio-based, found or re-purposed materials are not characterized in SM so you will have to make a note and use the best substitution. Note: reused materials are noted in Sustainable Minds as “secondary” not primary.
h. Please enjoy exploring exciting visual interest with materials, translucency, and light as you create your light fixtures.

After each student completes their light fixtures, they are required to perform a similar LCA and compare it to their baseline. This usually enables students to see lower impact scores with their designs and identify what causes these scores to be lower. Students have incorporated DIY (do-it-yourself) and ready-to-assemble kit strategies and local manufacturing to reduce transportation impacts. Finally, students present the story arc of their project work with a poster that highlights the baseline, ideation, and redesign phases, as well as a comparison of impact scores. Emphasis is placed on showing at least three eco-design strategies from *Okala* and how they were used in their redesign.

This project tends to be a highlight of the semester for students who commonly express that they enjoyed the practical use of life-cycle assessments. Students often also note that they enjoy making something that is useful; they regularly use their unique lamp or give it as a gift to others.

Final Thoughts

By focusing on instilling curiosity and fostering creative, problem-solving, and future-thinking skills, as well as opportunistic abilities to continue to learn in every situation, educators can work to develop sustainability-minded designers and engineers. These designers and engineers have an inherent understanding of the three key areas of sustainability, environmental, social, and economic, and this allows them to positively impact future developments in climate action, creating a circular economy of materials and improving human health and well-being and as they recognize responsible production and consumption (UN SDG 12) within reasonable profitability parameters. Educators must also remain opportunistic and identify where these skills may be integrated into the classroom and research labs. The cases herein exemplify such opportunities and demonstrate that the process of students becoming sustainability-focused designers and engineers is under way.

References

Benjamin, Ruha. 2019. *Race After Technology: Abolitionist Tools for the New Jim Code*. Cambridge: Polity.

Sustainable Minds. "Design Greener Products Right, from the Start." Sustainable Minds. http://www.sustainableminds.com/. Accessed 8 March 2024.

White, Philip, Louise Pierre, and Steve Belletire. 2013. *Okala Practitioner: Integrating Ecological Design*. S.l.: Okala Team.

Wittkower, D.E. 2018. Technology and Discrimination. In *Spaces for the Future: A Companion to Philosophy and Technology*, ed. J. C. Pitt and A. Shew, pp. 37–64. New York: Routledge. (PDF here)

CHAPTER 44

De-Fetishizing Commodities for Sustainable Futures: How Commodity Chain Analyses Expose the Hidden Social and Environmental Costs of Consumption

Melissa Yang Rock

Introduction

In an age of relative prosperity and hyper-consumption, disposability is a convenience, whereas waste, generally, is an afterthought. If we are to get serious about social and environmental sustainability, then economic growth reliant on the ever-expanding consumption of short-lived or single-use, low-quality products must be critically interrogated. Fast fashion, plastic beverage containers and cutlery, shopping bags, and individually wrapped items all support our "grab and go" cultural habits

M. Y. Rock (✉)
Department of Geography and Environmental Studies, SUNY New Paltz, New Paltz, NY, USA
e-mail: rockm@newpaltz.edu

S. M. Wyman and J. Korenblat (eds.), *Sustainability Learning for Action and Community Engagement*,
https://doi.org/10.1007/978-3-032-11698-7_44

and our everyday choices. But we can make different choices individually, and we can demand more sustainable and accessible options through government and corporate investments for the collective good.

Traditional commodity chain analyses trace objects from production to consumption, and provide the impression that once commodities are purchased, consumed, and tossed out, they cease to exist (Denchak 2018). Yet these items (and their packaging) do not disappear once they are disposed of; rather they often go on to pollute communities and environments for generations to come. In the following learning experience, students are challenged to develop a commodity chain analysis from production to post-consumption to thoroughly investigate the true costs of our current consumer practices—costs not paid by the consumer or factory owner but borne by the people and places where labor and materials are extracted. These locations and communities double as potential pollution sites during production and post-consumption. The final Human Geography project in this chapter connects economic globalization, environmental sustainability, and social equity. This assignment helps learners understand that commodities are crystallizations of social and environmental relationships between people and places throughout the globe, including ourselves as consumers.

The relationships between people and consumer products are opaque, hidden in the physical manifestation of the item, which seemingly takes on its own agency. The social relationship we cultivate with *things for sale* is also known as *commodity fetishism*. In my classes, learners are encouraged to think critically and problem-solve. At the same time, they trace the commodity chain of one mainstream personal hygiene item and one sustainable alternative to explore both individual and societal costs and benefits of the product, including packaging, transportation, and post-consumer life. Through the exploration of alternatives, students can learn how socially and environmentally sustainable goods may be developed and manufactured to account for their post-production lifecycle rather than being designed for the dump. Lastly, we discuss how the uneven landscape of costs and benefits is structurally organized and subsidized through state, national, and global policy and practice—and how the development of widespread sustainable commodities cannot be prompted by consumer demand alone but by policy and regulation that properly incentivizes sustainable product development while disincentivizing planned obsolescence and a consumer culture of disposability.

The Assignment: Project Framing

Commodities are the product of "social relations between individuals at work" (Marx). That is, they derive their value from human labor and material resources, which are invested in their very form. Once commodities hit the market, however, they become objects to be bought, sold, and consumed—seemingly divorced from the labor power, social relations, material resources and extraction processes utilized in their production. As a result, physical commodities become objects that appear to have mysterious origins.

When we purchase these items, we act as if we are merely buying something of use or that will provide a moment of (fleeting) contentment. In doing so, we invest in contrived social relationships with *things*. We assign economic value to the product rather than the people and places that made it. Once the price is set, the time, effort, and resources put into the item are obscured from the social and environmental relational meaning.

This project attempts to de-fetishize commodities by unpacking the hidden relationships embedded in their *everyday low, low prices!* Indeed, these items or units of labor that we make meaningful by including them in our daily lives' material and symbolic practices *embody* relatively complex and far-reaching social, economic, political, and environmental *relationships*. Put more simply, commodities—the stuff we buy—connect us in very real and tangible ways to people and places around the globe. Whether we are aware of it or not, our lives (and the future of our planet) are intimately intertwined through our active and passive participation in global capitalism.

Learning Activity: Commodity Chains and De-Fetishizing Commodities

Students are encouraged to pick simple everyday commodities rather than big-ticket electronics or other highly complex items. This ensures the research project is achievable in the time allowed (one semester), but it also makes sense because these ordinary commodities are purchased more regularly. We premise our research on a consumptive, personal hygiene need, so the alternative is not necessarily to go without (i.e., deodorant, toilet paper), but to consider the longevity of use so that one might reduce the amount of waste over the long term. Some products cannot be reused once consumed, but we might consider whether the container that holds or wraps it is even necessary. Where containers are imperative to the item's function, we ask whether that part may be reused or repurposed. When that is not an option, students explore recyclability and compostability.

The assignment prompts learners to delve into the intricate web of processes, people, and places involved in bringing a product to market, including extraction, manufacturing, distribution, and disposal. By examining each stage, students gain insight into social and environmental impacts, such as labor conditions, resource depletion, pollution, transportation methods and the post-consumer "life" of a product. By tracing the materials management journey from production to post-consumption, we de-fetishize the commodity and expose the social and environmental costs embedded within it.

Comprehending the commodity chain fosters critical thinking skills. It empowers students to make informed choices as they question the ethical implications of our consumption patterns and they explore alternative, more responsible approaches to meet our needs. Further, this assignment highlights corporate responsibility, revealing tactics such as planned obsolescence that work against sustainable practices and environmentally responsible consumer choice.

Critically investigating a product's commodity chain is not a straightforward process: most complex social and environmental relationships are stretched across time and space and thus often hidden from view. Thus, several key considerations of the project research foci are worth highlighting.

Guiding Concepts

Global Supply Chain:
Using Immanuel Wallerstein's *World Systems Analysis* (2004), we highlight the inter-regional and transnational global division of labor such that *core* (developed) countries often capitalize on *semi-peripheral* and *peripheral* countries for their low-wage labor and cheap raw materials. Then, utilizing research and development housed in these relatively wealthy and developed regions, *quinary* sectors develop, negotiate, and regulate the sale of high-profit, value-added products back to the semi-periphery and periphery (Fig. 44.1).

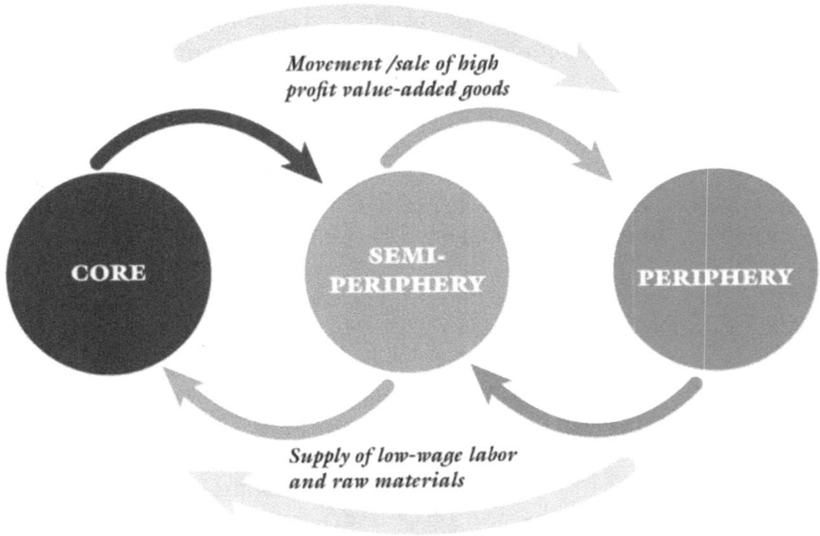

Fig. 44.1 World systems analysis: Economic flows

The United States, a *core country*, has a very diversified economy, including the primary (resource extractive) sector that involves industries such as mining, forestry, and agriculture; the secondary (manufacturing) sector that includes equipment, food, electronic, or chemical manufacturing; the tertiary (service) sector comprised of retail, restaurant, hotel, and sales; the quaternary (information and knowledge) sector featuring health care, education, theater or financial industries; and the quinary sector that focuses on the high-level decision making within the quaternary sector.

In contrast, countries in the *periphery* tend to rely primarily on extractive industries as their main source of income and export. They will likely have some basic secondary (manufacturing) industries to build from their primary industrial base. *Semi-peripheral* countries tend to have slightly more diversified economies, including primary and secondary industries, expanding into the tertiary service sector as they are able.

As such, the global supply chain is not just a collection of lines on a map but a journey within geopolitical, and economic relationships that have important implications for transnational social, environmental, and economic well-being. Understanding the contemporary global economic landscape helps place commodity chain analyses in a geopolitical context.

Labor:

> Hi, I'm Jenny. I'm the Sales Manager. At Lifeng we are happy to make jeans for the whole world. We never miss a deadline, even if that means our workers keep working all night. (China Blue 2005)

Early in the semester as we discuss economic globalization, students watch a PBS Documentary titled *China Blue* (Peled 2007) that follows a seventeen-year-old girl, Jasmine, as she migrates from her home village to find a factory job in a southern Chinese city. The harsh working and living conditions in the blue jeans factory expose some of the unaccounted-for production costs. Gone are the days when the clothes we buy are hand-sewn or tailor-made. These types of clothes are reserved for those with deeper pockets. Thus, few of us understand the mental and physical demands made by compartmentalized, assembly line mass-produced or fast fashion clothing production—where workers labor at stations tediously sewing together fronts with backs, affixing zippers, hemming seams, or cutting loose threads for long, relentless hours.

The daily rigors (and compensation) of a workplace depend greatly on the value added to the commodity by the labor involved. Such surplus value is not quantified by how hard one works but rather by the value of the skills one brings to the production. In a post-Fordist global economy, corporations tend to site manufacturing operations in countries of the semi-periphery or periphery where labor costs are relatively low and union representation is suppressed or non-existent. The de-skilling associated with modern manufacturing has created bountiful low-skill, low-wage, dead-end jobs (as there are few opportunities to rise through the ranks for most employees). As such, embellished pockets or functional zippers may seem an important addition to the final product (for the consumer), however, the additional value in monetary terms—reflected in the employee's pay—is minuscule. The fragmentation of the work into segments that require fewer skills and less complexity likewise leads to a disconnect between the person and the product—both in the processes of making and consuming it.

Resource Extraction:
The process of sourcing materials is also fundamentally important to analyzing a commodity's social and environmental sustainability. These prompts ignite the students' critical analysis of their selected product:

- Who is involved in the materials extraction process, and what is the cost to their health and welfare?
- How is this resource cultivated or created? Has the land been converted into monoculture systems? Are pesticides used?
- Is the resource easily renewable? If not, do extraction practices allow for a balance of regrowth and regeneration? Or do the extraction practices focus on maximizing profit?
- Where are such material resources located—in urban areas or rural areas, in developed *core* countries or in developing ***peripheral*** countries? What do these landscapes look like once the resource is removed?
- Who lives in or near these sites of extraction or production, and what is the impact on their health and wellbeing?
- What new issues and complications arise from resource extraction or materials processing processes—and for whom?

Often, resource extraction, especially when executed to maximize profits, involves patterns of environmental destruction, labor exploitation, and community fragmentation. This creates new problems for others' shoulders to bear (Baichwal 2006).

Transportation:
Contemporary globalization is further marked by modern technologies and innovations that have succeeded in shrinking time and space. One of the most influential factors in bringing commodities from sites of production to sites of consumption is the advent of ***containerization***: the use of large-scale container ships that bring vast quantities of varied commodities across oceans to and from ports across the globe. This bulk conveyance significantly reduces transportation costs, which can lead to lower prices for the buyer and increased profits for the seller. These lowered costs do not factor in the environmental impact contained within a single commodity's journey across the globe. Considering transportation methods and geographical routes in a commodity chain is not

so much an exercise in the detailed particulars of a product's journey but rather a mental exercise in thinking through the *what* and *how* of commodity production and transportation in our global economy while also considering whether other models exist (or could) and to what and whose benefit.

Reduce, Reuse, Recycle + :

The three Rs (3Rs) of waste management for environmental sustainability are well known: reduce, reuse, and recycle. If these practices were comprehensively adopted in a shift toward a circular economy, this could address up to 70% of global greenhouse gas emissions (EPA 2024b; King 2022; McGinty 2021; UNDP 2023; Ellen MacArthur Foundation 2023). The 3Rs are not an equally weighted triad but part of a sustainable materials management hierarchy that should prioritize waste reduction and reuse of materials as the necessary *first steps* toward sustainability.

The 3Rs provide a basic roadmap for what we do once we have purchased a product. But let's throw in three other actions to consider before any purchase: (1) call out corporate practices of *planned obsolescence,* which purposefully develop and sell products that become obsolete within a short period, as well as corporate rules and regulations that prevent or create hurdles to consumers' ability and right to repair their devices; (2) interrogate fads and trends that tend to promote *perceived obsolescence* which works by trying to entice us to abandon our belongings—tossing them when they are still perfectly useful and functional; and (3) consciously cut consumption by resisting or refraining from making unnecessary purchases.

Waste Management?

Post-consumption considerations of a product's lifecycle challenge students to consider what happens to the remains of a commodity when all parts are not fully consumable, compostable, or recyclable. Most trash will end up in landfills, but a significant portion will spill or seep out to pollute our land, air, and water in other ways. Although the earth may seem vast, it is still finite.

Engaging Concepts, Going Deeper in the Classroom

Global Supply Chain: Manufacturing

As the global supply chain stretches the factory floor to become a complex web of transnational interactions, the local impacts of manufacturing are engulfed in the swift pace of *just-in-time* production. Nonetheless, manufacturing processes sit at the intersection of local environmental and labor laws, which themselves are subject to international geopolitical influences and supranational economic institutional agreements. The academic term for negative local impacts is *negative externalities*, when the cost of an action is borne by a third party. In this case, the true costs of manufacturing cannot be accounted for by merely adding up the capital investments of resource extraction and processing, labor, production space, and machinery. Additionally, environmental pollution as well as short- and long-term healthcare costs are subsidized by the suffering of local(e)s via their once healthier bodies, land, air, and water.

Labor

The film *China Blue* (Peled 2007) ends with Jasmine reflecting on the people who wear the clothes she has made. She shares her thoughts and journal entries with the camera: “I’ve been very curious about who wears the ‘big and fat’ jeans… are they girls and boys my age? How fortunate they must be that they don’t have to work like us.” I highlight this yearning for connection as Jasmine is aware of these invisible ties that bind but about which the consumer is often oblivious. She and her friends go so far as to wonder about the possibility and repercussions of slipping a note into one of the finished products.

Jasmine: Jade, I want to slip a letter into one of the pockets. It’s for the person who buys the jeans.
Jade: But what if the buyer discovers and tells the boss?
Jasmine: It is not likely. I’ll put it in the pocket of the jeans ready for shipping. No way anyone will catch it.

Finally, the words of the hypothetical note from Jasmine narrate the closing scenes of the documentary.

> Hi friend. Hello from China. My friends say I’m cute. They call me Little Jasmine. I left home four months ago. Now I live in a factory. I miss my

> parents, my sister, my grandma, and my kitty. I'd like to show you my beautiful Sichuan Province. Do you like Sichuan food? My friends and I made the jeans for you. I hope you like the pants. Liping made the back pockets. Orchid installed the zipper. And I cut the loose thread. They say we're lucky to have this job. I'm glad I can help my family. You can write to me at Lifeng Factory, dorm room #402. PS. My friends are wondering why you guys are so big and tall?

Taking inspiration from *China Blue,* students research the working conditions of the people who make their selected product for the research project. They may not discover the exact working conditions within the branded company they chose but often extrapolate from the information they find about general location patterns, regulations, and processes for specific commodity production.

Transportation

Taking transportation into account quickly intertwines with many other facets of manufacturing, such as the material composition of products (for example, plastic vs. glass vs. aluminum bottles) and its impact on fuel consumption or the likelihood of reuse or recycling. For example, glass is more likely to be recyclable and recycled than plastic but weighs more, and thus its transportation will contribute to more greenhouse gases if that delivery uses fossil fuels. Transnational shipping becomes necessary when companies relocate factory production to newly industrialized or developing countries with lower labor costs and lax environmental regulation or enforcement (Zhou 2017; Lartey et al. 2020). Spreading the supply chain out also increases the potential for local and global events to impact the timely movement of goods—whether component parts or finished products. For instance, the Covid-19 pandemic exposed the fragility of global supply chains (McGillivray 2021). Geopolitical tensions can also restrict access to shipping routes (Bednarski et al. 2023). Or when infrastructure is damaged or impaired, as in the case of the March 2024 Baltimore bridge collapse (Grabenstein 2024), supply chain resilience is cast into doubt (Liu 2024). Thus, understanding transportation methods and routes challenges students to use systems thinking to analyze the stability or vulnerability of a specific supply chain.

3Rs– > 5 + Rs

Often, the 3Rs of environmental sustainability are discussed in the context of waste management. Instead, students can enlarge the framework to question the prevailing capitalist model of growth-oriented economics (Marquis 2024) by discussing degrowth principles and strategies (Hickel et al. 2022; Hickel 2020). In concert with prioritizing such principles, we can reset product design and integration toward a circular economy to keep technical materials in use as long as possible and ensure that nutrients from biodegradable materials are returned to the Earth to regenerate nature (Ellen MacArthur Foundation 2023). By engaging with more Rs, such as respect, rethink, refrain or refuse, repurpose, regenerate, restore, replant, right to repair, and rot, we can consider how to live well with our commodities, such that each meets us in its various forms and state of assembly, disassembly and reintegration into our lives, communities, and ecosystems. With this new framework in mind, consider the following 5 Rs: refrain, reduce, reuse, recycle, and right to repair.

Refrain

Refraining from or resisting unnecessary consumption is a self-reflective practice that demands that we differentiate between desire vs. need while weighing a more complete understanding of the impacts of consumption against personal gain or satisfaction. This does not need to be a guilt-inducing exercise but one where we reflect on what we value and assess the true cost of this purchase. Perhaps continuing to use your 10-year-old to-go hot beverage container *is* preferable to buying a brand-new Stanley Cup (or whatever is fashionably *on trend* this month). Conscious or mindful consumption reminds us to take a moment to comprehensively consider and weigh the hidden costs embedded in our purchases (Becker 2020).

Reduce & Reuse

When we purchase goods, we can opt for quality, reusable products over single-use or flimsy products. In purchasing durable goods, we can develop a healthier relationship with these commodities and by extension the human and environmental relationships represented by them. Our attitude toward (disposable) products mimics our ephemeral valuation and earthly (dis)connection with each other and the planet.

Thus, we must be mindful that the durable choice usually isn't usually the most affordable choice—and that can be a significant barrier for people who do care and want to make socially and environmentally responsible choices but do not have the financial resources to do so. This challenge in our contemporary capitalist economy currently manifests significant and growing income inequality (Horowitz et al. 2020).

Recycling

Glass and aluminum are relatively easy, efficient, and cost-effective to recycle (Wakefield 2022). However, plastic, despite being ubiquitous, has a much lower recycling success rate (Sullivan 2022; Tabuchi 2024; EPA 2024a). In a May 2022 report by Beyond Plastics, plastic recycling has never been a promising prospect. The US achieved a paltry 5-6% recycling rate in 2021. The remaining waste ends up in landfills (80%) or incinerated (14%), which contributes to plastics being littered throughout our land and ocean scapes and coursing through our bodies (Beyond Plastics 2022; Ritchie et al. 2023; Yee, et al. 2021). Plastics are a particularly pernicious and persistent material—lasting anywhere from 30 years (plastic bags) to 500 years (coffee pods, disposable diapers, and plastic toothbrushes). So, when it comes to the 3Rs, reducing or replacing the consumption of plastic-containing products is a key solution.

Right to Repair

"'Planned obsolescence' is another word for *designed for the dump*. It means they actually make stuff that is designed to be useless as quickly as possible so we will chuck it and go buy a new one" (Leonard 2020). Products can be designed for short lifespans with constructions that aren't modular or don't have parts that are easily interchangeable and thus not easily repairable. Numerous states have successfully passed "right-to-repair" laws, which "require manufacturers to make the parts, tools, and documentation needed to diagnose, maintain, and repair consumer electronic devices and appliances available to independent repair shops and consumers at fair and reasonable prices" (White House 2023).

Waste Management

The establishment of Municipal Solid Waste Landfills (MSWLF) helped to eradicate many of the public health problems that emerged in US cities around the time of the nineteenth-century Industrial Revolution. Cities lacked a public works system for street cleaning, trash collection, water treatment, and waste removal. By the late 1800s, the easy spread of rodent-borne and mosquito-borne infectious diseases, as well as other recurrent epidemics, gave rise to the establishment of waste management systems, including the construction of potable water and wastewater treatment systems, as well as a local system of solid waste management via municipal waste dumps (Ross and Law 2023).

Concern over landfills' negative local environmental impacts was slow to develop and translate into regulatory frameworks, improved management systems, and enforcement. At the same time, global shipping became a much more cost-effective method for transporting all sorts of commodities across the globe—including trash. Although the number of landfills dramatically increased in the US over the decades, new locales for waste disposal were simultaneously sought overseas. With the many concerns associated with waste management systems, including ground and surface water pollution, e-waste, the dangerous nature of sanitation work, and the exportation of waste, the post-consumer lifecycle of a commodity becomes much more complex.

Countries like China and the Philippines have moved away from accepting the globe's trash, recycling, and e-waste. Instead, "countries of trash origin" are challenged to find more palatable ways to manage their abundant waste (Tabuchi and Corkery 2021; Mosbergen 2019; Campbell and Christiensen 2016).

The Folly of "Consumer Citizenship"/"Democracy of Consumption"

On its surface, this assignment appears to be designed to prompt students' critical thinking and analysis of their own consumer choices and "vote with their dollar" to let corporations know what they care about and thereby influence their production decisions. And while making informed purchases is valuable, there is a limit to the potential impact. Further, this kind of "democracy of consumption" is not only a diversion but also deceptive. David Harvey expands upon this concept in *A Brief History of Neoliberalism* (2007), where he charges China's former premier leader Deng Xiaoping with cultivating a "democracy of consumption" in China to obscure social injustices, following the lead of Margaret Thatcher (United Kingdom) and Ronald Reagan (United States) by doggedly pursuing strategic neoliberal policy as part of their economic reform agenda.

The idea that we express our democratic values through consumption is absurd and deceptive. To be clear, the perception of a democracy of consumption—or democracy *through* consumption highlights the *illusion* of democracy through consumer choice. This harkens back to Soviet Era communism when citizens could only purchase goods at state-sponsored shops and thus had very few (if any) options. When the Union of Soviet Socialist Republics (USSR) dissolved in 1991, citizens had more political freedoms alongside greater consumer choice—and while they can be linked, they are neither interchangeable nor mutually exclusive. Capitalism does quite well under authoritarian rule (and even under socialist market economies like China). Thus, students should not confuse consumer choice with democratic values. Further, we must be careful not to conflate consumer spending power with political power—though we often do so.

Moreover, the advice to "vote with your dollar" as an equitable, democratic option is misleading: people living in democracies are unlikely to have an equal number of "votes" or dollars to spend in the marketplace. In a 2023 survey, 78% of people in the US reported living paycheck to paycheck (Batdorf 2024). People who must debt-finance their purchases typically pay a higher price because of interest rates, payday loan fees, adverse credit score impacts, and psychological stressors associated with such precarity. As Barbara Ehrenreich (2014) succinctly explains in *The Atlantic:* it's expensive to be poor. With such financial constraints, individuals who opt to buy more expensive durable goods will pay more for items up front, but lower-income folks will pay more for multiple cheaper, poor-quality items over the long term (Deer 2022). As such, tactics such as "shopping your values" has the potential for only limited impact.

Conclusion: The Takeaways

To move toward a more socially equitable and environmentally sustainable society, we must examine the hyper-consumptive practices we've grown accustomed to because they are habits we've developed in response to a system designed with economic growth as its main priority. Considering the hidden costs embedded in products for sale and revealed through a commodity chain analysis, we can make individual choices that reflect our priorities where possible, while pushing for structural reforms that benefit the collective good. And though the *Human Geography* assignment detailed above prompts students to choose between a *mainstream* vs. an *alternative* item—the choice is not binary but multifaceted.

References

Batdorf, E. 2024. Majority of Americans Live Paycheck-to-Paycheck Statistics 2024. *Forbes Advisor.* April 2. https://www.forbes.com/advisor/banking/living-paycheck-to-paycheck-statistics-2024/.Accessed 17 June 2024.

Becker, Joshua. 2020. Reasons to escape excess consumerism—and how to do it. Forbes. April 30, 2020. https://www.forbes.com/sites/joshuabecker/2020/04/30/escape-excess-consumerism/?sh=47fca55e2fbb

Bednarski, L., Roscoe, S., Blome, C., and M.C. Schleper. 2023. Geopolitical Disruptions in Global Supply Chains: A State-of-the-Art Literature Review. *Production Planning & Control*, 1–27. https://doi.org/10.1080/09537287.2023.2286283

Beyond Plastics. 2022. The Real Truth About U.S. Plastics Recycling Rate. *The Last Beach Cleanup..* Bennington College. May.https://www.lastbeachcleanup.org/_files/ugd/dba7d7_9450ed6b848d4db098de1090df1f9e99.pdf.Accessed 17 June 2024.

Campbell, K., and K. Christensen. 2016. Where does America's e-waste end up? GPS tracker tells All. *PBS NewsHour.* May 10, 2016. https://www.pbs.org/newshour/science/america-e-waste-gps-tracker-tells-all-earthfix

Baichwal, Jennifer, Director. *Manufactured Landscapes.* Edited by Roland Schlimme. Distributed by Zeitgeist Films.

Deer, R. 2022. The Price of Poverty. *Tribune Magazine.* UK (June 23). https://tribunemag.co.uk/2022/06/inflation-prices-cost-of-living-crisis-energy-food-housing.Accessed 17 June 2024.

Denchak, M. 2018. Reduce, Reuse, Recycle. Most of All, Reduce. *Natural Resources Defense Council,* July 31. https://www.nrdc.org/stories/reduce-reuse-recycle-most-all-reduce. Accessed 17 June 2024.

Ellen MacArthur Foundation. 2023. What Is a Circular Economy? *Ellen MacArthur Foundation.* https://www.ellenmacarthurfoundation.org/topics/circular-economy-introduction/overview. Accessed 17 June 2024.

EPA. 2024a. Plastics: Material-Specific Data. Environmental Protection Agency: Facts and Figures About Materials, Waste and Recycling. April 2. https://www.epa.gov/facts-and-figures-about-materials-waste-and-recycling/plastics-material-specific-data.Accessed 17 June 2024.

EPA. 2024b. What is a Circular Economy? Environmental Protection Agency: Circular Economy Basics. June 12. https://www.epa.gov/circulareconomy/what-circular-economy.Accessed 17 June 2024.

Ehrenreich, B. 2014. It is Expensive To Be Poor. *The Atlantic, Business.* January 13. https://www.theatlantic.com/business/archive/2014/01/it-is-expensive-to-be-poor/282979/.Accessed 17 June 2024.

Grabenstein, H. 2024. How baltimore's key bridge collapse will affect supply chains and the economy. *PBS NewsHour*. March 28, 2024. https://www.pbs.org/newshour/nation/how-baltimores-key-bridge-collapse-will-affect-supply-chains-and-the-economy

Harvey, D. 2007. *A Brief History of Neoliberalism.* Oxford University Press.

Hickel, J. 2020. What Does Degrowth Mean? A Few Points of Clarification. *Globalizations* 18(7): 1105–1111. https://doi.org/10.1080/14747731.2020.1812222Accessed 17 June 2024.

Hickel, J., Kallis, G., Hickel, J., O'Neill, D. W., Jackson, T., Victor, P. A., Raworth, K., and D. Ürge-Vorsatz. 2022. Degrowth can work—Here's how science can help. *Nature* 612 (7940): 400–403.

Horowitz, J., Igielnik, R., and R. Kochhar. 2020. Trends in Income and Wealth Inequality. *Pew Research Center*. January 9. https://www.pewresearch.org/social-trends/2020/01/09/trends-in-income-and-wealth-inequality/.Accessed 17 June 2024.

King, S. 2022. Reduce, Reuse, Recycle: Why All 3 R's Are Critical to a Circular Economy. *Scientific American.* December 13. https://www.scientificamerican.com/article/reduce-reuse-recycle-why-all-3-rs-are-critical-to-a-circular-economy/.Accessed 17 June 2024.

Lartey, T. A., Amankwah-Amoah, J., Danso, A., Adomako, S., Khan, Z., & S.Y. Tarba. 2020. Environmental sustainability practices and offshoring activities of multinational corporations across emerging and developed markets. *International Business Review*. https://doi.org/10.1016/j.ibusrev.2020.101789

Leonard, A. 2020. The Story of Stuff, Referenced and Annotated Script. The Story of Stuff. https://www.storyofstuff.org/wp-content/uploads/2020/01/StoryofStuff_AnnotatedScript.pdf.Accessed 17 June 2024.

Liu, Z.Z. 2024. Baltimore bridge collapse tests U.S. supply chains. *Council on Foreign Relations*. April 18, 2024. https://www.cfr.org/expert-brief/baltimore-bridge-collapse-tests-us-supply-chains

Marquis, C. 2024. In defense of degrowth. *Harvard Business Review.* June 11, 2024. https://hbr.org/2024/06/in-defense-of-degrowth

McGinty, D. 2021. 5 Reasons to Shift from a 'Throw-It-Away' Consumption Model to a 'Circular Economy'. *World Economic Forum: Nature and Biodiversity.* February 22. https://www.weforum.org/agenda/2021/02/change-five-key-areas-circular-economy-sustainability/.Accessed 17 June 2024.

McGillivray, G. 2021. An expert explains: What COVID did to global supply chains. *World Economic Forum*. July 30, 2021. https://www.weforum.org/stories/2021/07/covid-19-pandemic-global-supply-chains/

Mosbergen, Dominique. 2019. Here's why America is dumping its trash in poorer countries. Mother Jones. March 9, 2019. https://www.motherjones.com/environment/2019/03/heres-why-america-is-dumping-its-trash-in-poorer-countries/

Peled, Micha X. 2007. Teddy bear films, independent television service, and bullfrog films. China Blue. Oley, PA: Bullfrog FIlms.

Ritchie, H., Samborska, V., and M. Roser. 2023. Plastic Pollution. *Our World in Data*. https://ourworldindata.org/plastic-pollution.Accessed 17 June 2024.

Ross, D. E., and H. J. Law. 2023. Evolution of Municipal Solid Waste Management in North America. *Waste Management & Research*. 41 (7): 1177–1187. https://doi.org/10.1177/0734242X231173308

Sullivan, L. 2022. Recycling Plastic is Practically Impossible—and the Problem Is Getting Worse. *National Public Radio: All Things Considered*. https://www.npr.org/2022/10/24/1131131088/recycling-plastic-is-practically-impossible-and-the-problem-is-getting-worse.Accessed 17 June 2024.

Tabuchi, Hiroko. 2024. There's an explosion of plastic waste. Big companies say 'We've Got This.' *The New York Times*. April 5, 2024, sec. Climate. https://www.nytimes.com/2024/04/05/climate/plastic-recycling.html

Tabuchi, Hiroko, and Michael Corkery. 2021. Countries tried to curb trade in plastic waste. The U.S. is shipping more. The New York Times. March 12, 2021, sec. Climate. https://www.nytimes.com/2021/03/12/climate/plastics-waste-export-ban.html

Tabuchi, H., and M. Corkery. 2024. Countries Tried to Curb Trade in Plastic Waste. The US is Shipping More. *The New York Times*, March 12. https://www.nytimes.com/2021/03/12/climate/plastics-waste-export-ban.html.Accessed 17 June 2024.

United Nations Development Program. 2023. What is Circular Economy and Why Does It Matter? *UNDP Global Climate Promise*. April 24. https://climatepromise.undp.org/news-and-stories/what-is-circular-economy-and-how-it-helps-fight-climate-change.Accessed 17 June 2024.

Wakefield, F. 2022. Top 25 Recycling Facts and Statistics 2022. *World Economic Forum*. June 22. https://www.weforum.org/agenda/2022/06/recycling-global-statistics-facts-plastic-paper/.Accessed June 17, 2024.

Wallerstein, Immanuel. 2004. *World Systems Analysis*. Duke UP.

White House. 2023. Readout of the White House Convening on the Right to Repair. *The White House Breifing Room*. October 25. https://www.whitehouse.gov/briefing-room/statements-releases/2023/10/25/readout-of-the-white-house-convening-on-right-to-repair/.Accessed 17 June 2024.

Yee, M.S., Hii, L.W., Looi, C.K., Lim, W.M., Wong, S.F., Kok, Y.Y, Tan BK, Wong CY., and C.O. Leong. 2021. Impact of Microplastics and Nanoplastics on Human Health. *Nanomaterials* (Basel). Feb 16, 11 (2): 496. https://www.ncbi.nlm.nih.gov/pmc/articles/PMC7920297/

Zhou, Yue Maggie. 2017. When some US firms move production overseas, they also offshore their pollution.*The Conversation.* May 19, 2017. https://theconversation.com/when-some-us-firms-move-production-overseas-they-also-offshore-their-pollution-75371

CHAPTER 45

The DIRT Project: Empathy, Experimentation, and Belonging

Emilie Houssart

> ...let us make no mistake: the climate crisis is also a crisis of culture, and thus of the imagination.

Amitav Ghosh, *The Great Derangement: Climate Change and the Unthinkable.*

Soil is alive. The interdependent relationships of soil life forms are more vast, complex, and varied than we will ever understand. If considering the stars gives us a sense of perspective, learning about life underfoot can transform the way we move through the world. As an aspect of education and action for sustainability, interactive art projects based on soil education and natural cycles can build connection and belonging within a campus biome while providing critical distance from inherited attitudes toward land. This chapter presents soil-based and soil-related creative activities that foster an embodied understanding of complex,

E. Houssart (✉)
Art Department, SUNY New Paltz, New Paltz, NY, USA
e-mail: houssare1@newpaltz.edu

S. M. Wyman and J. Korenblat (eds.), *Sustainability Learning for Action and Community Engagement*,
https://doi.org/10.1007/978-3-032-11698-7_45

Fig. 45.1 *Convergences #2,* (2015) Andrea Frank

interdependent systems. Working with local, responsibly foraged organic matter to create biodegradable constructions encourages learners at any stage to think in alignment with natural cycles: digestion, breathing, life and death, decay and regeneration. As my artistic practice expanded into interactive public works, I began to design open-ended, play-based, experiential projects for community spaces and classrooms that promote critical thinking in tandem with bodily intelligence, *change agent* skills, interpersonal solidarity, and individual well-being (Fig. 45.1).

Origins of The DIRT Project

The DIRT Project is an experimental journey of collaborative learning based in rebuilding lost relationships with invisible organisms, those underfoot as well as inside the body. It began in response to a growing consciousness of my lifelong immersion in extractive Eurocentric systems that deplete natural resources and promote destructive, anthropocentric behaviors. I embarked on a process of awakening to what it means, in local land history terms, to be a Dutch Brit with a Huguenot last name who immigrated to violently appropriated land renamed the Hudson Valley of New York, US. My research into commercial food systems and interactions with *DIRT Project* participants of all ages led me to conclude that many people have little relationship with the plants and other organisms in their immediate vicinity, including in their bodies, and little connection with the organisms they eat. Soil, the home of plants and bodies, is often viewed as a necessary inconvenience in people's daily lives.

The DIRT Project originated with a linguistic surprise. I remember visiting the US from Europe in 2007 and being shocked at hearing the word *dirt* used for the living earth—a word typically associated with the unwelcome, the unclean. The word *dirt* provides a jumping-off point to examine colonizer language and its impact on our ways of living, including how we coexist with soil. Deeper research into the term *dirt* confirmed additional linguistic and cultural associations: stemming from Old Norse, via Old Dutch, the term denotes filth and excrement and often connotes disgust in mainstream US culture. The word *soil*, in contrast, was normalized into invisibility for me in the UK; its roots are in Anglo-Norman French for *seat* or *ground*, though it is also used to connote filth in certain contexts. My artistic practice is rooted in absurdist, anti-fascist works by writers such as Eugène Ionesco, Dario Fo, and Günter Grass, who engage cultural symptoms of emergent fascist ideology. I understood the normalized use of the word *dirt* to be an example of the passive ways in which mainstream culture perpetuates damage begun generations ago, often upholding racist and classist connotations (such as the turn of the 20th century "war on dirt" connecting whiteness to cleanliness (Zimring 2016, 109), that become harder to identify as malignant as time and distance from their origin increases. Attention to the language around soil terminology, including the commercial forces shaping language, seemed to be one key to addressing contemporary relationships with land, which are essential when shifting to more sustainable ways of being.

My research into industrial farming, local land history, and Environmental Justice led to an interactive museum project on the theme of *DIRT*. I was curious about how experiencing landscape art in museums is disconnected—and contributes to disconnection—from the land itself. Even as the ground pulses with life beneath our feet, we are distanced from it by assumed attitudes of institutional reverence and artistic elevation that often preclude viewer engagement on equal footing, reinforced by multiple factors including museum architecture, framing, and money—as well as by cultural, often Eurocentric and Romantic attitudes embedded in the works themselves. This museum culture is shifting to some extent. I intended to complicate the neatly separated categories of landscape, public, and artwork in the museum and see what might happen. I also intended to rebuild reverence for land—based not on what it looks like but on what it *is* and our living relationship to it, past, present and future.

As a 2021 Artist in Residence at the Samuel Dorsky Museum of Art, I took over a small room with an exhibition called *DIRT: Inside Landscapes*. A definition and etymology of *dirt* accompanied the introductory educational text about the history of Lenape and European settler ways of treating the land under the reader's feet. A chronological display of local landscape artworks from the museum's collection lined two walls. These ranged from Hudson River School founder Thomas Cole's journal (1828) to Jean-Marc Superville Sovak's *a-Historical Landscape no. VII* (2019), one of a series of altered engravings of Romantic Hudson Valley vistas, into which the artist inserted prints of enslaved people whose stolen labor built great wealth for landowners and who were excluded from these picturesque scenes. Artwork labels prompted visitors to contemplate each artist's relationship to the land as evidenced in their work—a gentle exercise in critical distance and a warm-up for an interactive prompt. On the wall across from these works, a living map of visitors' thoughts, poems, drawings, and memories associated with the word *dirt* emerged over six months. Following the shape of each contribution, I cut the paper responses into unique organic forms. On the wall, these were pieced together to create a giant subterranean organism under a horizon of handwritten questions ranging from *Do you like mud?* to *Is domination a sign of strength?*, in the form of the iconic Shawangunks ridgeline visible from the museum entrance. On the gallery wall, people's responses to *dirt* ranged widely, bringing up memories of joyful childhood play and deep family connections as well as traumatic, shame-based experiences and a sense of revulsion.

A key connection between extractive colonizer mentality and the individual museum-goer's connection to the earth was made in Peter Iannarelli's work, installed in the center of the room. *This Land Is My Land and This Land Is Your Land* (2007) comprises two large horizontal plexiglass boxes, perhaps grave-sized, placed next to each other with a small gap, each with a thin layer of dirt spread evenly on top. The artist collected this material during walks around the town of Beacon, NY, and included a cigarette butt and small particles of trash. The dirt had been sterilized before entering the museum's collection and was visibly dry and lifeless. Yet, despite the museum's standard artwork protocols, visitors felt powerfully drawn to it. Museum stewards reported that people's demeanor often changed upon spotting the dirt; that it drew people into the room and made them feel at home in a heavy, institutional space. At the end of each day, imprints of fingertips were discernible around the edges of the dirt before it was reset.

Close by, between the Nyquist-Harcourt Wildlife Sanctuary and New Paltz Gardens for Nutrition, my installation *DIRTdoors* invited people to interrupt a scenic walk by bending low to grasp a brass doorknob, opening a door into the turf to reveal the earth underneath, and to spend time with billions of their (mostly invisible) subterranean neighbors. Outdoor interactions at *DIRTdoors* brought related findings: some people were unwilling to stop and engage in an unfamiliar activity; many opened the doors with curiosity to varied delight or disappointment (there was no grand surprise!—just dirt), and young children were unafraid to look closely, to smell, to touch, and to imagine the unseen.

Based on these early findings, *The DIRT Project* evolved to continue the exploration beyond the museum. I began to develop an interactive curriculum based on the principle that everyone has a right to a healthy relationship with their neighbor organisms. This right has been steadily eroded by colonizer culture. Extractive (and highly destructive) practices that treat living entities as resources for profit, including land and labor, continue in the present moment quite intentionally, woven into the fabric of our daily lives. Commercial forces create dependency upon these systems and normalize them into invisibility. Through experiential projects, I intend to gently build public awareness of inherited attitudes and systems, foster open-ended dialogue around our relationships with the land, and open channels to make active choices and explore alternatives through the framework of the Principles of Environmental Justice (1991). Alongside public workshops for all ages, I started to design a set of *DIRT* activities that could work in an institutional classroom.

Learning Experience: The Dirt Project Enters the Classroom

Setting an Intention
In order to bring the ethos of *The DIRT Project* into the college art classroom, I thought about its central dynamic in terms of a psychological shift away from Eurocentric principles. How could we as a class move beyond inherited concepts of human identity separate from earth, and other species as simple, inferior beings, at best useful as resources for our use or aesthetic appreciation? We would work to build an understanding of self as a multitude, as part of an interdependent ecological system, in a way that would build simultaneous connections with other humans, coexisting species, and the local environment. By gently rupturing ordinary campus experiences, we would open spaces for learning and informed choices so that students could come to independent conclusions about how to move toward reciprocity. The studio classroom would become a fertile site for sharing, imagining, and exploratory action. To this end, I developed several *DIRT* activities. Described here is a "forked" set of projects: *Habitats for Hybrids* and a *Sound Your Truth* collaboration, as well as an online workshop called *Dirt Dinner Party*.

Course Context

Design: Form is an Art Foundations 101 class intended to introduce Art majors to a variety of techniques, processes, and studios in 3D and 4D (time-based) work. In 2023, I based two overlapping sustainability assignments in this class around foraging and freeform weaving to engage with intersectional sustainability principles, biocycle design, and the campus ecosystem. Either could be adapted to the circumstances of any college or community program with access to green space. Students were allowed to work on a collaborative public engagement sculpture or an individual project. The assignments began together with an introduction to campus as an interwoven living network, ancient and constantly evolving, but most importantly, vibrating with life *now* and *all the time* all around us, through us, with us, and including us. Our process began with outdoor foraging activities and included invisible life forms underfoot and all living campus organisms, including students.

In the classroom, we discussed responsible foraging practices and plant identification as well as safety protocols (for example, ticks, poisonous plants and fungi, tools, not getting lost, and letting someone know where you were going if foraging alone). We walked to the far end of campus, on and off paths, wandering with curiosity, noticing as a group, and sharing associated thoughts or memories. We saw many living members of our interspecies community, observed others by smell, tracks, and sound, and made space for yet more through heightened awareness, combining collective knowledge with imagination. We gathered samples of weaving materials to try out: outer iris leaves, cattails, common reeds, grapevines, and willow, plus some long grasses, English ivy and garlic mustard stalks. Many students are nervous about the uncontrolled outdoors, and some of the work involved gently addressing these fears, which were heightened by COVID-era sanitized living.

Back at the studio, we observed the different textures and forms and experimented with testing flexibility and strength under wet and dry conditions. We identified "native", "non-native" and "invasive" plants and began a collective map of relevant campus organisms. The assignments were based on the principle that some local creatures had millions of years of habitat-building experimentation and knowledge to draw upon—an ideal place for us to start learning about biodegradable constructions—and that we humans have been weaving for at least 12,000 years. Students were encouraged to use their bodily intelligence and multiple learned techniques to create original, responsive, freeform structures.

In our research, we looked at examples of creatures that make highly customized habitats: the Australian mallee fowl's 300-pound compost heap, which both protects and incubates eggs until they hatch; penduline tits, whose soft hanging nests are carefully sited over water and often have false entrances; various weaver birds and their intricate group-living habitats; tiny, camouflaged hummingbird constructions with spiderweb elasticity to accommodate growing young, and mud dauber wasp nests with individual larvae chambers, each containing a paralyzed spider for a future meal.

We discussed related human constructions, including Romanian architect Ileana Mavrodin's cob houses; Indigenous North American basketry; traditional Colombian *bahareque* homes made from sticks and mud;

contemporary artists Doug and Mike Starn's *Big Bambú* installations; *Mycoknit*, mycelium-infused knitted yarn forms, by Felecia Davies and Benay Gürsoy (SOFTLAB and ForMat); *Myconoe*, a mycelium canoe designed and grown by student Katy Ayers (Central Community College, Columbus, Nebraska); and the aggregate architecture experiments of Karola Dierichs and Achim Menges at the Institute for Computational Design and Construction (Stuttgart) that use star-like, recycled plastic "granules" inspired by sand and snowflakes. We reviewed video tutorials for simple techniques covering basic coiling, twining, random weaving and tension weaving. All our materials would be responsibly foraged and eventually returned to their original sites to biodegrade.

At this point, the two earth-based assignments diverged.

I. **Habitatsfor Hybrids**

Students who chose the *Habitats for Hybrids* assignment were prompted to make experimental, hanging habitats for imaginary hybrid creatures, following a logic based on selected attributes and tendencies of two well-researched creatures of interest whose characteristics would be imaginatively combined. This premise required learners to create a new form based on hypothesizing and empathizing with their fictional hybrid's unique needs, values, and behaviors. Students foraged for organisms suitable for weaving an environment for living—including clay, feathers, and petals—for binding, comfort, camouflage, and perceived visual appeal. Process "failures" were discussed publicly as valuable class learning experiences; material successes and breakthroughs were generously shared within the group. The playful nature of the assignment allowed for considerable freedom of experimentation, extending the boundaries of students' assumptions and expectations.

II. ***Sound Your Truth* Collaboration**

Sound Your Truth is a student-led gathering designed to strengthen the community by centering the experiences of underrepresented voices on campus to promote empathy and solidarity. This annual event, organized by Eddy at New Paltz, is held outdoors around George Floyd's birthday in October. Floyd was an African American person whose egregious public suffocation by police in Minneapolis, Minnesota, US, in

May 2020 ignited anti-racist protests worldwide. The event involves an open mic and active listening practices, supported by creative projects that help define and hold the space. In my class, the collaborative project for this event would be a site-specific sculptural installation made from campus-foraged materials.

Collectively, students in this group envisioned a large-scale structure that could contain one or two seated people on woven floor mats to create a focused listening space in the crowd and a private space to reflect and consider what to share. The exterior of the woven matrix would serve as a place for people to hang written "truths," as an alternative for those uncomfortable with public sharing at a microphone, or for attendees to preserve what they heard in written form. The structure was designed with three large coiled horizontal rings, the widest at the base, and connecting ribs. The students collected grapevine for the coils and random-weave connections, which was abundant in October and still carried some late fruits. I supplied ribs cut from invasive honeysuckle bushes in my yard. More project specifics emerged from contact and experimentation with the materials, and the design evolved accordingly. Groups of students collected cattails (*Typha*) to make dipping pens and common reeds (invasive *Phragmites australis*) to weave a soft floor mat. Students wove a tension tray from grapevine and sticks, lining it with soft reed seed heads to hold slips of paper (torn from upcycled paper bags) and reed pens. Spiraling grapevine tendrils, which reminded one student of her Afro-textured hair, were used to attach written papers to the structure, and the same student spent hours collecting and juicing grapes to make ink. Finally, the group named the project *Nested Truths*. In place of a critique, the students covered two shifts hosting the public engagement at the event, presented documentation to the class, and wrote a feedback statement about their experience, with suggestions for future iterations.

III. **DirtDinner Party.**

A playful, hour-long Zoom workshop called *DIRT Dinner Party* built awareness and connection around soil microorganisms while students were deprived of regular social engagement during the COVID pandemic. We assembled one evening at 8 p.m. and introduced ourselves; then, we began by brainstorming personal associations with the word *dirt* and sharing our findings. We discussed soil and the unique, slow-forming, complex network of interdependent relationships particular to each site. Everyone was asked to gather their favorite bowl and spoon and share a little about those objects. Then, using our bowl and spoon, we each collected a spoonful of "dirt" from directly outside our residences and brought it back to the group to report observations about the site, smell, texture, etc. Next, we spent ten minutes on a scavenger hunt around our living environments, collecting anything that could be offered to the earth as a feast—from food morsels and flower petals to fingernail clippings and hair from the bathroom drain. We reassembled, compared our findings, planned the following steps, and went offline.

Returning outside, each participant laid a place setting, as for an honored dinner guest, on the ground at the collection site. We moistened the spoonful of soil with water, mixed our collected offerings into the bowl, and placed it on our place settings. We spent some time with our dinner arrangements and photographed them in situ. When we felt ready, we spooned the mixture into the hole where the soil was collected and covered it with any surface matter that had been removed (moss, in my case).

Finally, we returned to the video call to share our reflections. I observed that the group had become quite peaceful, and some bonding had occurred. The anxiety usually lurking behind video meetings seemed to have dissipated. At this stage, participants did not have many words for what had occurred; yet they did state that they had enjoyed an act of generosity and quiet attention and were keen to share their photos with the group. This is three years ago; I still think about my spoonful of soil when I step outside and see the dinner party spot.

Discussion

I wanted to base our class explorations on assumptions about life and complexity, even where neither was immediately discernible. All projects began with the revelatory fact that there are more microorganisms in one's body than human cells and that therefore, a sense of isolation, or what Robin Wall Kimmerer (Potawatomi Nation) refers to as *species loneliness* or "a deep, unnamed sadness stemming from estrangement from the rest of Creation, from the loss of relationship" (2013, 208), is based on an inherited myth. We carry our inborn interspecies community around with us day and night; from most perspectives, we *are* an interspecies community. Every time we eat, we feed a whole community. Walking around campus, we considered how many living forms coexist under our feet at any moment and in our reciprocal exchange of breath with the trees we pass. Throughout the project, we valued collaborative interdependence as a way of life, while students were invited to imagine new definitions of "sustainability" and "campus."

Our students hail from New York State's urban, suburban, and rural areas and bring a wide range of experience in terms of land contact and comfort with the less-manicured outdoors. Our exploratory group dynamic created a space where students could name and examine discomfort, for example, around contact with soil and insects. Wandering across the grounds and into the woods was empowering for some who had been mainly conditioned to follow paths, a behavior often reinforced by campus design, and unnerving for others, for whom the woods had negative associations. Students challenged perceived social pressures about seeming "weird" for wandering slowly without a firm destination, stopping to observe their surroundings, and collecting weaving materials. In my opinion, social media video culture has reinforced the common early-adulthood sense of being under judgmental surveillance whenever in public. To some extent, these projects challenged that burden by making locating and touching campus organisms an essential component of the assignments.

Perhaps these projects' most apparent short-term learning outcome was the rapid collaborative, often joyful dynamic in the groups, as the assignments nurtured both individual and collective exploration and identity. From my observations, several factors contributed to this general sense of generosity and openness. Overcoming social stigmas, openly addressing feelings of isolation, and strengthening a sense of place on campus helped to create bonding and trust in and out of the classroom. It was noticeable that time outdoors and multisensory contact with natural materials had a direct, positive effect on morale and focus levels. Students gained familiarity with local flora and distant green spaces visible from paved areas and formed memories there, increasing a sense of home on campus. They spent time thinking about their status as multispecies organisms with reciprocal relationships across a campus biome constantly in flux. This contributed to a radical new sense of belonging and identity for many. One student wrote (*used with permission*):

> The way you have taught [this class] has consistently prompted us to question our position/identity within the place we are inhabiting and creating.

> I am internalizing this way of thinking, and it's made me look into the idea of ecology and interdependence in many places. I'm thinking more about contributing to people and environments … I'm grateful for what I have begun to learn, both about the physical/material creation of art and about being a part of the world.

The intentionally open-ended, playful nature of the prompts helped empower students to explore actively and adapt their designs to the characteristics of found, living materials, engaging humor, and imagination with either a fiction-based prompt (in the case of *Habitats*) or an opportunity for hands-on community care (*Sound Your Truth*). Students reported feeling less pressure to produce a polished object for my assessment, and more able to explore seriously with an eventual outcome based on process and adaptable design. Several students also noted their satisfaction at not producing waste during art making. Their low-impact works were made with local materials in alignment with natural cycles, using timeless techniques, and could be composted at any time back at the site of origin. We considered the soil organisms that decomposing projects might feed, nourishing the ground and the next generation of plants. All three projects involved a unique act of generosity, constituting a significant factor in building class morale.

For the *Sound Your Truth* project, I noted that students understood the potency of supporting the health of their campus community by working together to create an interactive public offering/experience using living materials from that same ecosystem: their campus biome. Seeing their work become a successful platform for meaningful social equity engagement was truly rewarding.

Sound Your Truth (SYT) students noted in their feedback (*used with permission*):

> Working SYT is definitely one of the highlights of my freshman year here at New Paltz. Through the project, I was able to bond with my classmates and become closer with them. Listening to the opinions of everyone in the group really inspired me to work my hardest.
>
> My favorite part was seeing how excited people were about our project. Whether it was to sit inside, write their truths, or learn about the materials, it warmed my heart that we were able to cross paths in such an impactful way.

For *Dirt Dinner Party* participants in the online workshop, the students' homes were activated as sites of generous intervention. This had a potency I had not fully anticipated and supported the formation of an impromptu community ready to embark on a joint experiment. In addition, it allowed for a highly personal and customized soil interaction, forging connections that exceeded my expectations, especially through contact with the participants' personal dinnerware and leaving a memory—as well as physical traces of the offering and even students' DNA—at a particular site that they would pass each day.

Conclusion

Learning modes that engage the body and psyche alongside the intellect are powerful tools in trying to move beyond a culture of domination into more reciprocal and fulfilling ways of being. In the short term, these projects aim to create a positive college learning experience of generative and challenging art-making and community building, with longer-term progress towards interspecies empathy and related behavioral shifts. My goal is to create opportunities for open-ended, personal connections to grow in an academic setting, while fostering critical thinking around inherited attitudes toward other species—making space for students to form environmental relationships and practices independently. From my observations, the parallel *Habitats for Hybrids* and *Sound Your Truth* projects created a class experience that worked toward a sense of reciprocity and belonging on a shared, living campus and promoted confidence, interpersonal relationships, and a general sense of well-being while building a varied set of multidimensional creative skills. An emphasis on creative risk-taking and intentional dual intelligence ("bodily" and "rational") combined effectively with a class culture of sharing resources and process discoveries to create a productive, collaborative group dynamic. Making collective decisions at the start of each project—such as organizing foraging expeditions or pivoting activities in response to weather conditions and group needs and desires—contributed to groundedness in time, community, and place. This, in turn, facilitated classes conducive to the robust exploration of concept and design with neighbor organisms. Some students began to notice their surroundings in new ways, alongside composting facilities around campus and areas of green space.

Some challenges that arose related to the perceived value of compostable work, particularly concerning dirt and signs of decay. Some students were deeply attached to their woven structures and were excited to keep them, place them outside for other species to discover, or observe them composting on-site; others felt that the pieces were not worth saving and did not care what happened to them, or rejected them once they had touched soil or started to dry out and therefore become undesirable. It seems likely that they would have treated the work differently if the same intricate structures had been woven from plastic or wire. I intend to explore this theme openly during future project iterations. Another aspect that needed closer attention was an initial frenzy of harvesting observed in a few students once they overcame their hesitations about touching plants. More care and time are needed for students to internalize respectful attitudes when working with other life forms; a more experiential approach to delivering this content would better support the theory. I would build time for this into future assignments and employ a combination of macro imagery of the symbiotic workings inside plant roots with an initial walk during which we sense and draw but collect nothing. We will practice slow, deep awareness of our shared biome before thinking about harvesting "materials" for weaving—perhaps this word is best avoided altogether. It is an ongoing challenge to balance complex considerations in introductory projects, from everyday products that may contaminate offerings of hair, to industrially poisoned soils, from both personal safety and wider systemic Environmental Justice standpoints. A collaborative, joyful, and non-judgmental approach in regenerative making and discussion of earth and climate issues seems key in transcending the guilt, hopelessness, and anger that can accompany learning in this area, where a sense of powerlessness and locked-in hypocrisy can negate positive intentions and forestall action.

I propose that students have a *right* to a reciprocal relationship with the land they live with, a right *not* to produce unnecessary waste with assignments, and that we can, and often do, unintentionally further alienate art students from their environments by carrying on as usual with unsustainable material practices and habits that reinforce species isolation. What if art education included responsible production and sourcing and the cyclical repurposing of materials as a baseline? How can our classes empower students to give more than they take? I'm fortunate to work with two progressive departments where this is already happening: locally fallen trees are milled on-site with the students for interspecies projects in Sculpture, and a pollinator dye garden is cultivated for shared use in the Natural Ink and Dye course in Printmaking. My projects will continue to evolve in dialogue with students, educators and *DIRT Project* participants, in and beyond the university setting.

References

Kimmerer, Robin Wall. 2013. *Braiding Sweetgrass: Indigenous Wisdom, Scientific Knowledge, and the Teaching of Plants*. Milkweed Editions.

Zimring, Carl A. 2016. *Clean and White: A History of Environmental Racism in the United States*. New York University Press.

Correction to: From (Teaching About) Existential Threats to (Teaching) Existential Resilience and Sustainability

Christine Wamsler

Correction to:
Chapter 9 in: S. M. Wyman and J. Korenblat (eds.), *Sustainability Learning for Action and Community Engagement*, https://doi.org/10.1007/978-3-032-11698-7_9

In the original version of Chapter 9, on page 84, there should be four bullet points instead of five. Secondly, the reference on page 90 begins with author's first name rather than last name. The chapter has been updated with the changes.

The updated version of this chapter can be found at
https://doi.org/10.1007/978-3-032-11698-7_9

S. M. Wyman and J. Korenblat (eds.), *Sustainability Learning for Action and Community Engagement*,
https://doi.org/10.1007/978-3-032-11698-7_46

Afterword

Imagine it's late summer 1940, and you find yourself trekking the hilly countryside of Southwestern France. You and your trusty German Shepherd, named Robot, have reached a hill near the quaint town of Montignac—smooth terrain compared to the rugged landscapes further upstream in the valley. The world is at war, yet surrounded by the serene green hills and birdsong, you can't help but feel tranquil. Suddenly, Robot yelps in pain. He has slipped on loose dirt and fallen into a hidden hole, disappearing. You call for him. No response. Heart thumping, you rush to rescue him. And there, catching your breath, patting your wise and grateful companion, something draws your eye: an entrance to a cave.

Intrigued, you return with three friends and homemade lanterns. You venture inside the cave. Flickering light reveals paintings and engravings contoured on the walls. You can see animals, dazzling, fleet, and mighty—horses, deer, extinct cattle known as aurochs, bison, lions, and even a woolly rhinoceros. In this prehistoric space, you sense the presence of those who once visited here with hands stained in yellow and red ochre. You try to imagine what life may have been like for the people who inhabited the verdant hills of this region thousands of years ago when it was the Ice Age. According to Marcel Ravidat, an apprentice mechanic and minor fabulist, this is the true story of how the legendary prehistoric art of Lascaux Cave was discovered: by pure chance, by Robot the dog and some curious teenagers. At first, they kept their find an entrepreneurial secret, charging local children a pittance to view the cave paintings. Yet,

S. M. Wyman and J. Korenblat (eds.), *Sustainability Learning for Action and Community Engagement*,
https://doi.org/10.1007/978-3-032-11698-7

Fig. 1 Lascaux IV, International Center for Cave Art (Centre international de l'art pariétal) presents a complete facsimile of all the decorated parts of the Lascaux cave (bull room, axial diverticulum, passage, well, apse and nave). *Photo credit* Hugo Martin/Alamy

over time, the magnitude of their find dawned on them. They could no longer keep this dark cathedral an enchanting secret. They told their local schoolmaster about it, who contacted a well-known archaeologist. Soon, a research team would descend into the cave and account for roughly 600 paintings and 1,500 engravings of animals and abstract marks. Their illumination devices helped reanimate the animals from their firelight-painted origins thousands of years ago. Some researchers have suggested that, possibly, the abstract marks around the animals were maps of the starry night. These researchers began translating a birth record of humankind's creative and symbolic thinking, accumulated by prehistoric artists over a lively tradition of countless generations (Ministère de la Culture 2024) (Fig. 1).

Few of us have the sheer luck to stumble upon a great archaeological find and touch the source of the human spirit with our hands. Yet, as educators, we can facilitate this opportunity. We can co-create a similar experience with our students by asking them to reflect upon how their inner world connects with the world outside. John Cimino, a New York-based thought leader has done precisely this in his visits to classrooms and think tanks around the globe prompting students and business leaders alike to explore their inner Lascaux:

> In the stillness of the semi-dark, I turned on a projector, and we looked at an image of one of the Lascaux Cave drawings from 17,300 years ago. I asked them to tell me what they saw, and then how and why these pictures might have come to be in such a deep, dark place within the earth. They began to speculate. They said that the pictures are probably "life lessons about how to survive," perhaps a way of paying homage to the animals their lives depended on. Then I asked them to imagine a place deep inside themselves where they keep their own cave drawings, their own life lessons and precious images. We talked about what they chose to let into that place, and how, over a lifetime, that space might enlarge to contain a vast collection of amazing things.

When I interviewed Cimino about this project in September 2023, he told me that this cave art made in Paleolithic France had been studied from multiple angles to decode its purpose, its significance to the people who created it, and how these depictions of animals became vehicles to transmit cherished lessons and vivid stories between generations. We can create an inner Lascaux with our students. According to Cimino, we curate our experiences like these cave artists: We choose what to let in and keep out to maintain our mental and philosophical well-being. Our internal Lascaux becomes a library of ideas for creative work, reminiscent of Walt Whitman wandering about and weaving people and his travels into his poetry. He'd write his impressions in his notebook and press leaves between its pages.

The project-based learning for sustainability education, action, and community engagement offered in this book can begin with a simple yet profound question that connects us with our core humanity and unbridled possibility: *What can you find in your Lascaux?* This inner place is made of people, places, and animals we've met in our lives, and it's made vibrant and alive by our thoughts, feelings, and imagination. Cimino notes that it's not just that we are interdependent: we are *inter-accommodating*, as inventor Buckminster Fuller put it in his 1982 book *Synergetics: Explorations in the Geometry of Thinking*. Our imagination and creativity can shape our worldview like hands upon spinning clay. Even if we're not shaping clay or spelunking in metaphorical caves, we can hold space for discussions that encourage attentive observation and imaginative thinking in all educational environments. We can nurture these skills to address the societal, environmental, economic, and spiritual divides that make our human-caused crises seem insurmountable. Physical presence isn't needed for connection; even physicists theorize this. In quantum mechanics, particles can be entangled at a distance.

This idea that we are all deeply connected lies at the heart of sustainability education. Sustainable development relies on our ability to understand and articulate our common project's meaning, and for this to happen, we need to work toward a shared understanding of the concepts involved (Landorf et al. 2008). Critical Sustainability experts have noted that sustainability is about protecting the environment and creating a more just and equitable society (Rose and Cachelin 2018; Gunn-Wright 2024). Education can also help learners develop a sense of interconnectedness with the natural world and each other (Clarke and Mcphie 2014). Further, education can help learners develop the knowledge, creative and critical thinking skills, values, and habits of hope and resilience that grow from this practice, which they need to understand and respond to ongoing sustainability challenges (Landorf et al. 2008).

Care ethics philosopher Nel Noddings suggests that an education that promotes caring relationships with ourselves, others, and the rest of nature also permeates our treatment of crafted things: "We know that well-tended engines purr, polished instruments gleam, and fine glassware glistens. The care we exert induces something like a response from fields of ideas and from inanimate objects." She asks, "Do our students hear enough—or anything at all—about these wondrous events?" (Noddings 2005, chap. 2, para. 21). For educators, our learning material is the foundation of our craft. The care and effort we put into selecting and presenting this content can help bring it to life, and this endeavor asks for problem-solving and meaning-making. Sustainability education goes deeper than solving current and future problems; it involves caring for our past and reconnecting with our ancestors, who could see life in all things, even a seemingly lifeless stone or cave wall. They recognized a connection to a living, co-creative world and how to make meaning from their experiences.

Indeed, Robin Wall Kimmerer says that Indigenous peoples understand that the land "calls forth our deepest moral responsibility: It is home to plants and trees—our non-human relatives. It is our home, our ancestral connection and identity, a healer, a sustainer, a source of knowledge. Land is enspirited and sacred. In the Western worldview, however, land is property. It is capital, and raw materials to be taken and used."

Kimmerer warns, "We are losing two hundred species each day across the globe. What comes next will be defined by our relationship with our land." (Kimmerer 2022).

The severity of this wildlife loss cannot be overstated. Scientists believe we are experiencing a Sixth Great Mass Extinction event on Earth. To provide some context, the most recent mass extinction occurred around 66 million years ago when a giant asteroid hit the Yucatan peninsula, leading to the demise of the dinosaurs. But in this time, which some call the *Anthropocene*, humans are causing a mass extinction by burning fossil fuels— among many activities that fail to see and honor the consequential inter-accommodating nature of life.

Elizabeth Kolbert (2014) shares another alarming example of this ongoing extinction crisis from the lush rainforests of Panama, once home to the iconic Panamanian Golden Frog. Suddenly, these creatures began to vanish. The loss was caused by a fungus introduced through human activities. This led scientists David Wake and Vance Vredenburg to speculate the demise of the golden frog was one signal that we are already experiencing a global mass extinction event caused by humans. Amphibians, known as the *great survivors* for their ability to endure on Earth for over 300 million years, are now under severe threat, with an unprecedented two out of five amphibian species facing extinction. Kolbert writes, "Wake and Vredenburg argued that, based on extinction rates among amphibians, an event of a similarly catastrophic nature was underway. Their article was illustrated with just one photograph of about a dozen mountain yellow-legged frogs—all dead—lying bloated and belly-up on some rocks" (Kolbert 2014, 6).

The biodiversity crisis expands far beyond amphibians. It's accelerating due to human neglect, disregard, and purposeful destruction. One-fifth of reptile species, one-third of mammal species, and one-sixth of bird species are also at risk of extinction. Coral reefs are vital ecosystems for tropical, colorful marine life and support a quarter of all ocean species; currently, they're experiencing dramatic declines in size and have already suffered from mass bleaching incidents (Kolbert 2014). The world's most abundant creatures, insects, also face mass extinction of two out of every five species. For insects, "The main drivers of species declines appear to be in order of importance: 1. Habitat loss and conversion to intensive agriculture and urbanization 2. Pollution by synthetic pesticides and fertilizers 3. Biological factors, including pathogens and introduced species, and 4. Climate change" (Bayo and Wyckhus 2019).

Humans have created many of these crises by trying to solve a particular problem rather than *solving for a pattern*—a term minted by the Kentucky naturalist Wendell Berry in 1981. In contrast to problem-solving, pattern-solving reminds us to consider what we wish to study in context and protect the integrity of its environment and ecology. For example, pesticides may solve a problem for humans, but these chemicals harm the pattern, the integral ecosystem. To solve for a pattern, we can adopt a Deep Ecology mindset. We let go of our tendency to see the world as a collection of separate and disconnected entities. Instead, we view the outside world as reflecting our interiority—our minds, a constantly flowing, entangled, and radiant place. This mindset of gentle observation allows us to see how ecosystem entities accommodate each other within the patterns and rhythms that animate our planet. By embracing this orientation, we become closer to one with the phenomena we study. We see how each part contributes to the ensouled whole. "The pattern that connects" is a phrase by anthropologist Gregory Bateson, introduced in his classic books *Steps To An Ecology of Mind* (2000) and *Mind and Nature: A Necessary Unity* (2002).

At the same time, Social Ecology views nature through the prism of human utility and social systems. Ecocriticism is a field of study that maintains equilibrium between these opposing viewpoints, recognizing the complex interplay between human society, cultural stories and words, and the rest of the natural world that, in the age of climate change, is becoming a tumultuous actor unleashed from human designs. All these entities work together in intricate ways to shape our world (Heise 2006, Latour 2018).

Sustainability topics like climate change and mass extinction can prove daunting for many learners. Sometimes, it helps to zoom in on individual beings and their stories. Consider Sudan, a 45-year-old Northern White Rhino who was the last male of his kind. In 2018, photojournalist Ami Vitale photographed a touching moment between Sudan and his caretaker, Joseph Wachira, in Kenya. The photograph shows Joseph kneeling before an ailing Sudan, placing a comforting hand on a wrinkled, armored head. This storytelling image captures the vulnerability of an endangered being facing the tragic fate of his distant family lineage: the woolly rhinoceros of Lascaux Cave. But in this same image, we glimpse humankind's vulnerability, too—our capacity to become tender, to be shaped by a compassionate relationship with the world. "Without rhinos

Fig. 2 Laikipia, Kenya. 31st Mar, 2018. A caretaker touches Najin, one of the world's last two remaining female Northern White Rhinos, in Ol Pejeta Conservancy in Laikipia county, north Kenya. Caption by Alamy contributor. *Credit* Lyu Shuai/Xinhua/Alamy Live News

and elephants and other wildlife," Vitale says, "we suffer a loss of imagination, a loss of wonder, a loss of beautiful possibilities. When we see ourselves as part of nature, we understand that saving nature is really about saving ourselves" (Vitale). Only two female Northern White Rhinos are left today, Najin and her daughter, Fatu. "Both are guarded in a semi-wild enclosure," and poignantly, like Sudan, they "have had their horns sawed off to deter poachers" (Ritchie) (Fig. 2).

While Northern White Rhinos are disappearing, Southern White Rhinos are rebounding from what could have been the same fate. "Numbers grew 1000-fold within a century" (Ritchie) due to conservation efforts. We can find stories of hope where people invest in the knowledge, skills, values, and habits of mind that collectively create a cultural narrative that all life is worth nurturing. Where can we begin? First, as educators, we can design project-based learning where learners can seek knowledge and discover their inner Lascaux. Then, no matter the subject, poetry or science, we can guide them to begin with an open heart, mind, and wonder, possibly following Walt Whitman's first steps:

Beginning My Studies

Beginning my studies the first step pleas'd me so much,
The mere fact consciousness, these forms, the power of motion,
The least insect or animal, the senses, eyesight, love,
The first step I say awed me and pleas'd me so much,
I have hardly gone and hardly wish'd to go any further,
But stop and loiter all the time to sing it in ecstatic songs.
—Walt Whitman, *Leaves of Grass*, 1865, 1871.

References

Bateson, Gregory. 2002. *Mind and Nature: A Necessary Unity (Advances in Systems Theory, Complexity, and the Human Sciences)*. New edition. Hampton Press. ISBN: 978–1572734340.

Bateson, Gregory. 2000. *Steps to an Ecology of Mind: Collected Essays in Anthropology, Psychiatry, Evolution, and Epistemology*. 1st ed. Chicago: University of Chicago Press. ISBN: 978-0226039053.

Bayo, Francisco S., and Kris A. Wyckhus. 2019. "Worldwide Decline of the Entomofauna: A Review of Its Drivers." *Biological Conservation* 232 (April): 8—27. https://www.sciencedirect.com/science/article/abs/pii/S00063207 18313636.

Berry, Wendell. 2009. *The Gift of Good Land: Further Essays Cultural and Agricultural*. Berkeley: Counterpoint.

Cimino, John. 2018. "Leadership and the Inner Work of Art." Leader to Leader 88 (Spring), https://www.renaissancecenter.org/post/big-data-101-an-introduction-to-data-query-engines. Accessed June 24, 2024.

Cimino, John. 2023. Founder and president of Creative Leaps International and Renaissance Center for Interdisciplinary Thinking, Knowledge Integration and Advanced Applications of Imagination. Interview with the author. September 29.

Clarke, David A. G., and Jamie Mcphie. 2014. Becoming Animate in Education: Immanent Materiality and Outdoor Learning for Sustainability. *Journal of Adventure Education and Outdoor Learning* 14 (3): 198–216. https://doi.org/10.1080/14729679.2014.919866.

Fuller, Richard Buckminster. 1982. *Synergetics: Explorations in the Geometry of Thinking*. Macmillan Publishing Company. ISBN: 978–0020653202.

S. M. Wyman and J. Korenblat (eds.), *Sustainability Learning for Action and Community Engagement*,
https://doi.org/10.1007/978-3-032-11698-7

Gunn-Wright, Rhiana. 2024. "Why the Climate Movement Must be a Movement for Justice." Talk presented at Vassar College, September 12. Poughkeepsie, NY.

Heise, Ursula. 2006. The Hitchhiker's Guide to Ecocriticism. *PMLA-Publications of the Modern Language Association of America* 121 (2): 503–516. https://doi.org/10.1632/003081206X129684.

Kolbert, Elizabeth. 2014. *The Sixth Extinction: An Unnatural History*. N.p.: Henry Holt and Co., Kindle.

Landorf, H., S. Doscher, and T. Rocco. 2008. Education for Sustainable Human Development: Towards a Definition. *Theory and Research in Education* 6 (2): 221–236. https://doi.org/10.1177/1477878508091114.

Latour, Bruno. 2018. *Down to Earth: Politics in the New Climatic Regime*. 1st ed. Polity.

Ministère de la Culture. "The discovery | Lascaux cave." https://archeologie.culture.gouv.fr/lascaux/en, Accessed September 4, 2024.

Noddings, Nel. 2005. *The Challenge to Care in Schools: An Alternative Approach to Education*, 2nd ed. New York: Teachers College Press. Kindle.

Ritchie, Hannah. 2022. "The State of the World's Rhino Populations." Our World in Data, October 14. https://ourworldindata.org/rhino-populations. Accessed September 5, 2024.

Rose, Jeff, and Adrienne Cachelin. 2018. Critical Sustainability: Incorporating Critical Theories into Contested Sustainabilities. *Journal of Environmental Studies and Sciences* 8 (4): 518–525. https://doi.org/10.1007/s13412-018-0502-9.

Taft School. 2022. "Morning Meeting: Robin Wall Kimmerer on Restoring Our Relationship with the Land." *Taft School*. https://www.taftschool.org/news/featured/~board/news/post/morning-meeting-robin-wall-kimmerer-on-restoring-our-relationship-with-the-land.

Vitale, Ami. 2019. "Life-Changing Lessons of the Last Male Northern White Rhino." *National Geographic*, September 17. Accessed September 5, 2024).

Whitman, Walt. 1998. "Beginning My Studies," *Leaves of Grass*. Project Gutenberg, May. https://www.gutenberg.org/files/1322/1322-h/1322-h.htm

Index

S. M. Wyman and J. Korenblat (eds.), *Sustainability Learning for Action and Community Engagement*,
https://doi.org/10.1007/978-3-032-11698-7

D

T

U

V

W

Kolibri 360 GmbH
Ferdinand-Jühlke-Straße 7
99095 Erfurt, Deutschland
produktsicherheit@kolibri360.de